Therapeutic Uses of Trace Elements

Therapeutic Uses of Trace Elements

Edited by

Jean Nève
Free University of Brussels
Brussels, Belgium

Philippe Chappuis
Lariboisière Hospital
Paris, France

and

Michel Lamand
National Institute for Agronomic Research
Theix, France

Plenum Press • New York and London

Library of Congress Cataloging-in-Publication Data

Therapeutic uses of trace elements / edited by Jean Nève, Philippe
Chappuis, and Michel Lamand.
       p.   cm.
    "Proceedings of the Fifth International Congress on Trace Elements
in Medicine and Biology, held February 4-7, 1996, in Meribel,
France"--T.p. verso.
    Includes bibliographical references and index.
    ISBN 0-306-45485-8
    1. Trace element deficiency diseases--Congresses.. 2. Trace
elements--Therapeutic use--Congresses.   I. Nève, Jean, 1951-   .
II. Chappuis, Philippe. III. Lamand, Michel. IV. International
Congress on Trace Elements in Medicine and Biology (5th : 1996 :
Méribel, France)
    [DNLM: 1. Trace Elements--therapeutic use--congresses. 2. Trace
Elements--pharmacology--congresses.   QU 130.5 T398 1996]
RC627.T7T44 1996
616.3'96--dc20
DNLM/DLC
for Library of Congress   96-41996
                                                               CIP

Proceedings of the Fifth International Congress on Trace Elements in Medicine and Biology,
held February 4 – 7, 1996, in Méribel, France

ISBN 0-306-45485-8

© 1996 Plenum Press, New York
A Division of Plenum Publishing Corporation
233 Spring Street, New York, N. Y. 10013

Printed in the United States of America

PREFACE

Organized by the French Speaking Society for Study and Research on Essential Trace Elements (SFERETE), the Fifth International Congress on Trace Elements in Medicine and Biology "Therapeutic Uses of Trace Elements" was held February 4–7, 1996, in Meribel (Savoy, France). This resort is situated in the heart of the Three Valleys domain, at the gateway of the beautiful Vanoise National Park. More than 250 participants covering six continents attended the meeting. This volume contains the text of plenary lectures and of several oral and poster communications.

Trace element deficiencies are not only encountered in developing countries or during malnutrition. Subclinical features are also observed in developed societies where they constitute a background for an impressive number of pathological states. Preventive and curative treatments with commercial products are often prescribed without reliable studies about their clinical interest or potential efficiency. By contrast empirical approaches such as the catalytic therapy, nutritional and pharmacological aspects of trace elements were emphasized on a scientific basis to favor their rational therapeutic use.

Discussions focused not only on elements which are essential when specific intracellular homeostatic regulations are available, but also on their ligands, through which the same metal ion (i.e., copper) may display either beneficial or detrimental effects. A rationale for their pharmacological use, certainly too much ignored, was also reported. The book goes on by questioning trace element requirements throughout different periods of life. About RDA specifications, it was suggested to distinguish between individual and population requirements. In elderly people, the importance of trace elements and their role in connection with the free radical theory of aging have to be confirmed and should result in adequate recommendations for this population. Pathological aspects of zinc, copper, and selenium in relation to inflammatory conditions and infections were quite well documented (sepsis, digestive diseases, cirrhosis). In diabetic states, a more unified theory of trace element interactions with intracellular insulin signal transduction pathway is now emerging for zinc, selenium, chromium and vanadium. Some comprehensive reviews were also presented on the role of trace elements in various pathologies: pharmacological effects of zinc, copper, and selenium in dermatology, anticarcinogenic effects of selenium and effects of fluoride and strontium on some bone diseases such as osteoporosis and Paget's disease. Numerous epidemiological studies also indicated that specific trace elements-related diseases are suspected to affect large groups of populations: iodine deficiency is still a concern in Europe and zinc, selenium, or other micronutrients deficiencies may induce a risk of cerebrovascular diseases or cancer for other groups. Finally, recent advances in the physiopathology of copper metabolism were presented in the satellite workshop "Molecular Basis of Copper Metabolic Disorders." Researchers who identified and cloned the gene for Wilson disease, and others who looked after different intracellular copper transporters, developed new metal transport concepts in the cell. They put forward a genetic response on a link between different copper transport genes and between different trace metal metabolisms such as iron and copper.

The editors thank all contributors who allowed a rapid publication of this book. They are also grateful to the referees who reviewed the papers. Finally, the meeting was made possible thanks to the members of Scientific and Organizing Committees and with the financial help of public and private sponsors. Their names can be found hereafter.

J. Nève, P. Chappuis, and M. Lamand

Fifth International Congress on Trace Elements in Medicine and Biology
Meribel, France, 4–7th February 1996

Scientific Committee

G. Berthon (F), G. Boivin (F), P. Chappuis (F), A. Favier (F), M. Hagueneau (F), S. Hercberg (F), M. Lamand (F), P. Marie (F), R. Milanino (I), J. Nève, President (B), J.R.J. Sorenson (USA), G.C. Sturniolo (I), P. Walravens (USA).

Organizing Committee

M. Accominotti (F), A. Alcaraz, General Secretary (F), J. Arnaud (F), F. Baruthio (F), M. Bost (F), P. Chappuis, President of the SFERETE (F), V. Ducros (F), A. Favier, President (F), O. Guillard (F), M. Lamand (F), J. Nève (B), F. Nabet (F), J. Poupon (F), A.M. Roussel (F), D. Vitoux (F), R. Zawislak (F).

Public and Private Sponsors

The French Ministery of Foreign Affairs and the Joseph Fourier University, Grenoble (F). The Volvic Centre for Research on Trace Elements (F) and the Labcatal Laboratory, Montrouge (F), and the following firms: Aguettant, Analab, Becton-Dickinson, Behring, BIO2, Boehringer, Boiron, Centre National des Biologistes, Crédit Lyonnais, Fisons, Fumouze, Jobin-Yvon, Johnson and Johnson, Kontron, Lavoisier Tec & Doc, Lero, Les Granions, Nestlé, Olympus, Perkin Elmer, Randox, Richelet, Roche Posay, Roucous, Sanofi, Servier, Spin, Varian.

CONTENTS

IV. TRACE ELEMENTS IN ENDOCRINOLOGY AND NUTRITION

V. PHARMACOLOGICAL APPLICATIONS OF TRACE ELEMENTS

VII. STATUS, EPIDEMIOLOGY OF TRACE ELEMENTS, AND INTERVENTION STUDIES

VIII. GENETIC DISORDERS OF COPPER METABOLISM

RISK ASSESSMENT FOR ESSENTIAL TRACE ELEMENTS IN HUMANS

Walter Mertz

12401 St. James Road
Rockville, Maryland 20850

1. INTRODUCTION

Almost a century ago, the great French scientist Gabriel Bertrand established what is known as Bertrand's Law, on the basis of his many observations on the effects of trace elements on growth and metabolism of plants (1), providing the scientific basis for the total dose-response curve of all living matter to all essential nutrients. In its most basic terms the law states that for each biological system there is a range of exposure, compatible with and essential for optimal function, and that below and above that range function deteriorates, resulting in disease and, ultimately, death. Thus, essential trace elements, like all essential nutrients, present two risks: One of deficiency and another of toxicity. Because both must be considered before any nutritional intervention can be safely implemented, the definition of the range of safe and adequate intakes for all essential elements for all systems of interest is the supreme challenge in our field.

In the past this challenge was met by two groups of scientists who worked independently and had little or no communication with each other. Nutritionists sought to establish intakes adequate to prevent specific deficiency diseases, and toxicologists determined exposure levels that would produce signs of toxicity. Because of the relatively crude criteria used, severe disease or death, as indicators for deficient or excessive exposure, toxic and deficient levels often differed by whole orders of magnitude, and there was little need for communication between the two approaches. That situation has changed drastically by the introduction and acceptance of much more sophisticated criteria of adequacy which now include functions such as immunocompetence, intellectual and emotional development and even risk reduction for certain metabolic and neoplastic diseases. The application of such criteria on both sides of the dose-response curve has raised the estimates of adequate intakes and lowered those of excessive exposure, resulting in substantial narrowing of the range of safe and adequate intakes for many elements. In one extreme case, zinc intake of infants and children, the definitions of safe intakes by nutritionists and toxicologists even overlapped, and no agreement could be reached for this element that is most important for growth and development (2). A conference of nutritionists and toxicologists in 1992 called for a reevalu-

Therapeutic Uses of Trace Elements, edited by Nève et al.
Plenum Press, New York, 1996

ation of the criteria used in risk assessment for essential trace elements and improved communication between toxicologists and nutritionists in setting their standards and recommendations (3).

The following discussion of risk for deficiencies and toxicities of essential trace elements is based on the acceptance of these postulates:

- The intact organism, unlike isolated organ systems, cell cultures or subcellular preparations, possesses efficient homeostatic regulatory mechanisms to maintain the "milieu interne". Therefore, results of deficiency or toxicity studies cannot be extrapolated from one system to another.
- Biological effects of different valence states and species of an element can be markedly different from each other; results obtained from one or not necessarily valid for others.
- Biological effects depend on the route of exposure, e.g., inhalation, oral intake as part of the diet, vs. intake in drinking water.
- Biological effects are strongly influenced by dietary interactions with synergists and antagonists. Depending on those, one level of exposure to an element can be deficient, adequate or toxic.

2. RISK OF DEFICIENCIES

2.1. From New Definitions of Requirement

According to the latest WHO/FAO/IAEA Expert Consultation (4) trace element deficiencies are defined as resulting in consistent impairment of physiologically important functions. Such impairment is not identical with the specific, classical deficiency diseases, but is considered a valid risk indicator for their development. The functions impaired as a result of a deficiency may be specific for a particular element, as is the case for thyroid function and iodine, or they may be non-specific, for example, reduced growth and development or poor reproductive performance. Immune competence, antioxidative functions, glucose tolerance, even cognitive and emotional functions are under discussion as valid criteria. Any of those can be accepted if they are consistently produced by induction of a deficiency and consistently prevented or cured by supplementation with the element under study(5).

Previous experience in humans and animals has clearly demonstrated multiple levels of requirement for any one trace element, depending on the criteria of adequacy applied. Recent studies have quantified these levels of copper requirement by showing that the laboratory rat needs three times more copper in order to maintain optimal immune functions than it needs to meet the classical criteria of adequacy, such as growth or organ concentrations of the element (6). Although this factor of three may not be applicable to other species, there is convincing evidence for a higher requirement in man as well, when immunologic criteria are applied. There is also strong evidence from many human studies for a decline of immune and metabolic functions with advancing age. Some of those, for example, the progressive impairment of glucose tolerance, have been considered by some as "normal, age related", but the beneficial effects of nutritional interventions have clearly proven important nutritional components. Applying the new criteria of adequacy (maintaining physiologically and clinically important functions) to this situation would lead to the conclusion that risks for deficiency of essential elements and other nutrients are greater than previously accepted, especially in the aging population.

We must immediately emphasize that this statement does not call for immediate intervention with any one trace element or other nutrient. We know that most physiological functions depend on more than one element; this is especially proven for immune functions which require a spectrum of elements, including Fe, Cu, Zn, Se (7). As will be discussed later, any excess of one can interfere with the utilization of another, equally important element.

2.2. From Environmental Causes

Environmental causes that raise the risk of deficiencies relate either to inadequate concentrations of elements in the soil or to problems of bioavailability of otherwise adequate levels. Large areas in all continents are deficient in iodine and fluorine, so that the feedstuffs and foods cannot furnish the requirement for animals and humans. Selenium also belongs to this category, but it presents problems of bioavailability as well. Bioavailability depends on the soil pH and on interactions with competing elements (8). In general the cationic soil elements, such as Fe, Cu, Co, Cr and Zn require a more acidic pH; Se is better available from a more alkaline milieu. The decline of Se status in foods, animals and people in several areas worldwide has been attributed to the gradual acidification of the soil by acid rain (9). The Cu-Mo antagonism is known to account for Cu deficiency in animals, and interactions with sulfur and heavy metals can depress the availability of Se in heavily fertilized or contaminated soils.

The impact of environmental deficiencies on human health is most severe in isolated areas depending entirely on locally produced foods, as has been the case in the Keshan disease areas of China. Free circulation of foods imported from adequate regions is an effective prevention, as has been shown in the South Island of New Zealand and in Finland. Fortification of staple foods or drinking water with the missing elements (e.g. F and I in the USA), fertilization of agricultural soils (e.g. Se in Finland) or supplementation of meat and dairy animals and poultry (e.g. I in the former GDR) are effective Public Health measures to reduce the risk of deficiencies to the population, but are subject to considerable political controversies. This leaves the alternative of supplementing individuals at risk. Selenium supplementation of children has eradicated the Keshan disease in all but the most inaccessible parts of China. On the other hand, the value of individual supplements to prevent the risk of *marginal* deficiencies as they may occur in developed countries (e.g. Se in parts of Europe) is still under much discussion and research; much will depend on the ultimate acceptance of the new, stricter criteria of requirement, as discussed above.

3. RISK OF TOXICITIES

3.1. From New Definitions of Toxicity

Just as the criteria for nutritional adequacy have become more stringent during the past few decades, so have the criteria for safety. Avoidance of clinical disease from toxicity has been complemented as a goal by much more sophisticated objectives, such as protecting developmental, metabolic and intellectual functions from excessive exposure. These new criteria have lowered the recommended safe levels of exposure for essential trace elements into ranges of intakes that were considered the exclusive domain of the nutritionists in the past (3). Nutritionists now accept the interference of one element with the absorption or utilization of another, essential element as the first manifestation of toxicity, even when no immediate clinical disturbances are detectable. These changes are illustrated by comparing official recommendations for zinc intake. A WHO Expert Committee stated in 1973: "Clinical observations in patients given therapeutic zinc for impaired wound healing indicate that approximately 200 mg of elemental zinc can be taken by man for prolonged periods of time in divided daily doses without apparent toxic effects." (10). The zinc chapter of the 1989 Recommended Dietary Allowances (11) concluded as follows: "For these reasons, chronic ingestion of zinc supplements exceeding 15 mg/day is not recommended without adequate medical supervision." The modern recognition of multiple micronutrients being essential for most health-related functions is at the root of the concern with interactions among elements. If only one of the micronutrients involved in maintaining a particular function is marginal or deficient, raising its intake can be expected to produce beneficial effects. If, on the other hand, there is a marginal or deficient intake of two or more, supplementation with just one will be ineffective or, through interactions may further depress the impaired function. This may be the interpretation of the generally negative effects of zinc supplementation alone (12) on immune functions reported from North America, where the copper status is generally marginal (13).

Any nutritional intervention, therefore, must be concerned with restoring and maintaining the normal balance among micronutrients, especially among trace elements. In practical terms this postulate suggests that in most cases supplementation with a balanced mixture of micronutrients in moderate amounts is preferable to high amounts of just one.

3.2. From Environmental Causes

The following discussion of trace element toxicity will be restricted to exposure via the food chain, from the soil to plants and animals into the human diet. It excludes airborne or effluent pollution from industrial sources and internal combustion engines, occupational exposure, as well as excessive intakes from drinking water. These problems are the domain of toxicology and are being intensively studied by toxicologists.

As to the food-borne excesses, research in animal nutrition and production has identified many areas worldwide with excessive concentrations of trace elements in the soil and agricultural crops. Although animals living in such environments can be severely affected, overt toxicity in the human populations of these areas is relatively rare. Selenium toxicity in animals occurs in many areas worldwide, but only isolated cases of adverse effects in humans have been described in areas of China and Venezuela (14). Excessive molybdenum in the soil has created severe problems of animal nutrition through interaction with copper, but there is only one known area, the Ankavan province of Russia, where this condition has a negative impact on substantial parts of the population: Elevated xanthine oxidase activity and uric acid concentrations in blood and urine, resulting in a gout-like syndrome in 31% of the population studied (15). Fluorine excess affects both animals and humans, albeit to a different degree. Fluorosis in cattle and sheep is a life threatening disease. In contrast, the most common sign in human populations is mottling of the teeth, of cosmetic, but not of health concern. Skeletal fluorosis with bone deformities is known to occur in areas of India and China, but is believed to depend on additional nutritional factors, especially calcium intake (16). The effects of soil pollution by mining and industrial byproducts during past centuries in Europe have created much concern about potential risks to the exposed populations. Extensive studies in several such areas of Central Europe have found minimal, if any effects on trace element concentrations in foods and no effects on human health (17–20).

Finally, the effects of acid rain on the bioavailability of soil elements must be taken into account. Changes of the soil pH will make most of the cationic elements , including the heavy metals, more available to the plants, while reducing the bioavailability of the anionic forms, such as selenium (8). The long range effects for human health are being studied.

4. CONCLUSION

If the new, more stringent criteria for adequacy are accepted, estimated requirements for some trace elements will increase, raising the risk of marginal or inadequate intakes. Even by the traditional criteria, the supply of selenium, fluorine and iodine is marginal in parts of Europe. Any intervention to reduce the risk of marginal deficiency must be designed so as not to increase the risk of marginal toxicity by creating imbalances among trace elements. The definition of the levels of the individual trace elements and their daily intake that constitute balance should be an important goal of trace element research.

5. REFERENCES

1. G. Bertrand, in *8th International Congress of Applied Chemistry*, vol. 28, New York, pp. 30–49 (1912).
2. B.A. Bowman and J.F. Risher, in *Ris Assessment of Essential Elements*, W. Mertz, C.O. Abernathy and S.S. Olin, eds., ILSI Press, Washington, D.C. pp. 63–73 (1994).
3. W. Mertz, C.O. Abernathy and S.S. Olin, eds., *Risk Assessment of Essential Elements*, ILSI Press, Washington, D.C. 300 pp. (1994).
4. Joint FAO/WHO/IAEA Consultation, *Trace Elements in Human Nutrition and Health*, World Health Organization, Geneva, in press.
5. W. Mertz, *Nutrition Reviews* **51**, 287–295 (1993).
6. R. G. Hopkins and M. L. Failla, *J. Nutr.* **125**, 2658–2668 (1995).
7. R.K. Chandra, in *Trace Elements in Nutrition of Children-II*, R.K. Chandra, ed., Raven Press, New York, pp. 201–213 (1991).
8. W.H. Allaway, in *Trace Elements in Human and Animal Nutrition, 5th Edition*, Vol. 2, W. Mertz, ed., Acad. Press, San Diego, CA, pp.465–488 (1986).
9. D.V. Frost, in *Selenium in Biology and Medicine, Part A*, G.F. Combs, Jr., J.E. Spallholz, O.A. Levander and J.E. Oldfield, eds., AVI, New York, pp. 534–547 (1987).
10. WHO Expert Committee, *Trace Elements in Human Nutrition*, WHO Technical Report Series No. 532, Geneva, Switzerland, 65 pp (1973).
11. Subcommittee on the 10th Edition of the RDAs, *Recommended Dietary Allowances, 10th Edition*, National Academy Press, Washington, D.C., 285 pp. (1989).
12. R.K. Chandra, L. Hambreaus, S. Puri, B. Au and K.M. Kutti, *FASEB J.* **7**, A723 (1993).
13. L.M. Klevay, S.J. Reck and D.F. Barcome, *J. Am. Med. Assoc.* **241**, 1916–1918 (1979).
14. O.A. Levander, *Ann. Rev. Nutr.* **7**, 227–250 (1987).
15. V.V. Kovalskij and G.A. Yarovaya, *Agrokhimiya* **8**, 68–91 (1966).
16. K.A.V.R. Krishnamachari, in *Trace Elements in Human and Animal Nutrition, 5th Edition, Vol. 1*, W. Mertz, ed., Acad. Press, San Diego, CA, pp. 365–415 (1987).
17. M. Anke, L. Angelow, M. Müller and M. Glei, in *Trace Elements in Man and Animals- TEMA-8*, M. Anke, D. Meissner and C.F. Mills, eds. Verlag Media Touristik, Gersdorf, Germany, pp.180–188, (1993).
18. M. Müller, C. Thiel, M. Anke, E. Hartmann and W. Arnhold, *IBID*, pp.211–214.
19. M. Simonoff, L. Razafindrabe, G. Simonoff, P. Moretto and Y. Llabador, *IBID*, pp. 216- 218.
20. K. Karlowski, M. Wojciechowska-Mazurek, K. Starska and E. Brulinska-Ostrowska, *IBID*, 259–260.

2

DIETARY REQUIREMENTS OF TRACE ELEMENTS

A Brief Overview of Population and Individual Requirements as Specified in Some Recent International and National Recommendations

R. M. Parr

International Atomic Energy Agency
P.O. Box 100, A-1400 Vienna, Austria

1. INTRODUCTION

During the past decade, new estimates of desirable levels of dietary intake of essential trace elements have been developed by several different national and international groups of experts. Their work has been stimulated partly by scientific advances in our knowledge of the role of trace elements in human nutrition and health, and partly also by the need for more accurate and informative food labelling. Although these experts have all had access to more or less the same, or very similar, sets of scientific data on which to base their conclusions, their recommendations show large numerical differences arising from different philosophical approaches and assumptions. Within a short publication such as this it is not feasible to present a comprehensive review of this subject. Instead, this report places emphasis on the recently developed recommendations of WHO, FAO and IAEA (1) and draws some comparisons with similar recommendations from the USA (2), ILSI Europe (3), and the UK (4).

2. OVERVIEW OF RECENT INTERNATIONAL RECOMMENDATIONS BY WHO, FAO, AND IAEA

For many years, the main recommendations of WHO and FAO on the subject of essential trace elements in human nutrition were contained in a 1970 report on various micronutrients, including iron (5), and in a 1973 report on 17 other trace elements (6). The first more modern review of this subject was made in a 1988 report (7), which provided much more detailed guidance on iron requirements (Table 1). This is now followed by a new publication

Therapeutic Uses of Trace Elements, edited by Nève et al.
Plenum Press, New York, 1996

Table 1. Estimated basal[a] dietary requirements for iron (mg/day) for diets of differing levels of bioavailability (adapted from ref. (7))

| Group | | | Low bio-availability 5% | | | | Intermediate bio-availability 10% | | | | High bio-availability 15% | | | |
| | | | Requirement to prevent anaemia | | Basal requirement | | Requirement to prevent anaemia | | Basal requirement | | Requirement to prevent anaemia | | Basal requirement | |
Sex	Age (yr)	Wt(kg)	median	incl. variability	median	incl. variability	median	incl. variability	median	incl. variability	median	incl. variability	median	incl. variability
M&F	0.25-1	8	11	14	17	21	5.5	7	8.5	11	3.8	5	5.5	7
M&F	1-2	11	6.5	8	10	12	3.5	4	5	6	2.2	3	3.3	4
M&F	2-6	16	7.5	9	11	14	3.5	5	5.5	7	2.5	3	3.7	5
M&F	6-12	29	12.5	16	19	23	6	8	9.5	12	4.2	5	6.3	8
F	12-16	51	22	27	32	40	11	13	16	20	7.3	9	10.8	13
F	Menstruating	55	17	(29)[b]	(25)[b]	(48)[b]	8	14	12.5	(24)[b]	5.6	10	8.3	16
F	Post-menopausal	55	10	13	15	19	6.5	6	9.5	9	3.4	4	5.1	6
F	Lactating	55	14	17	21	26	7	9	10.5	13	4.7	6	7	9
M	12-16	53	19	24	29	36	9.5	12	15	18	6.5	8	9.7	12
M	16+	65	12	15	18	23	6	8	9	11	4.1	5	6.1	8

[a]Normative storage requirements were not derived for iron. They might be some 50% higher than the basal requirements suggested here.
[b]Values in parentheses represent levels of intake that are deemed to be very unlikely on usual dietary patterns.

Table 2. Estimated basal and normative population requirements for copper (mg/day), iodine (μg/day), selenium (μg/day) and zinc (mg/day) (adapted from ref. (1))

Sex	Age (yr)	Wt(kg)	Cu		I[a]	Se		Zn-low[b]		Zn-moderate[c]		Zn-high[d]	
			basal	norm.	Rec.	basal	norm.	basal	norm.	basal	norm.	basal	norm.
F	0-0.25	5		0.33-0.55[e]	50	3	6	7.1[e]		3.1[e]		1.2[e]	
M	0-0.25	5		0.33-0.55[e]	50	3	6	8.0[e]		3.4[e]		1.3[e]	
M&F	0.25-0.5	7		0.37-0.62[e]	50	5	9	4.7[e]		1.9[e]		0.7[e]	
M&F	0.5-1	9		0.60	70	6	12	8.0	11.1	3.4	5.6	2.2	3.3
M&F	1-3	12	0.50	0.56	90	10	20	7.9	11.0	3.4	5.5	2.1	3.3
M&F	3-6	17	0.51	0.57	90	12	24	9.2	12.9	3.9	6.5	2.5	3.9
M&F	6-10	25	0.67	0.75	120	14	25	10.7	15.0	4.6	7.5	2.9	4.5
F	10-12	37	0.68	0.77	120	16	30	12.0	16.8	5.1	8.4	3.3	5.0
F	12-15	48	0.88	1.00	150	16	30	14.7	20.6	6.3	10.3	4.0	6.1
F	15-18	55	1.01	1.15	150	16	30	14.6	20.6	6.3	10.2	4.0	6.2
F	18-60	55	1.01	1.15	150	16	30	9.4	13.1	4.0	6.5	2.5	4.0
M	10-12	35	0.64	0.73	150	16	30	13.4	18.7	5.7	9.3	3.6	5.6
M	12-15	48	0.88	1.00	150	19	36	17.4	24.3	7.4	12.1	4.7	7.3
M	15-18	64	1.17	1.33	150	21	40	18.7	26.2	8.1	13.1	5.1	7.8
M	18-60	65	1.19	1.35	150	21	40	13.4	18.7	5.7	9.4	3.6	5.6
Pregnancy													
1st trimester			1.01	1.15	200	18	39	10.7	14.7	4.6	7.3	2.9	4.4
2nd trimester			1.01	1.15	200	18	39	13.3	18.7	5.7	9.3	3.6	5.6
3rd trimester			1.01	1.15	200	18	39	18.7	26.7	8.0	13.3	5.1	8.0
Lactation													
0-3 months			1.11	1.25	200	21	42	21.3	25.3	9.1	12.7	5.8	7.6
3-6 months			1.11	1.25	200	25	46	19.6	23.3	8.4	11.7	5.3	7.0
>6 months			1.11	1.25	200	26	52	15.5	19.2	6.6	9.6	4.2	5.8

a. Recommended intakes of iodine in mg/day (for practical purposes, serving the same objectives as the normative population requirements).

b. For a diet of low zinc bioavailability (15%).

c. For a diet of moderate zinc bioavailability (30-35%).

d. For a diet of high zinc bioavailability (50-55%).

e. For formula-fed infants only.

Table 3. Estimated basal and normative population requirements (mg/day or μg/day) of adults proposed by WHO/FAO/IAEA for *other* essential trace elements (adapted from ref. (1))

Element	Basal	Normative	Comment
Boron	~0.75 mg	~1.0 mg	Tentative
Chromium	~25 mg	~33 mg	Tentative
Manganese	—	—	No values proposed
Molybdenum	50 mg	—	Tentative, assuming 25 % CV of population intakes
Nickel	100 mg	—	Tentative (if animal data can be extrapolated to humans)
Vanadium	~10 mg	—	Tentative

(1), which provides detailed guidance particularly on copper, iodine, selenium and zinc (Table 2), and less detailed guidance on some other essential trace elements (Table 3).

The 1988 publication introduced two important concepts into the terminology of trace element requirements, namely the *basal* and *normative* requirements. These concepts have been further developed in the new 1996 report to have the following meanings. *Basal* requirements refer to the intakes needed to prevent pathologically relevant and clinically detectable signs of impaired function attributable to inadequacy of a nutrient. *Normative* requirements are the intakes needed to maintain a level of tissue storage or other reserve that is judged to be desirable.

A further distinction should be drawn between the average requirements of *individuals*, and the lower levels of the safe ranges of *population* mean intakes needed to meet these requirements (referred to here respectively as the individual and population requirements). In the approach adopted in the new WHO/FAO/IAEA report, the population requirement is calculated as the average individual requirement plus two standard deviations of the variation of individual *intakes* in a population (generally a 20 % or 25 % CV). Conceptually, this is very different from the approach followed in the USA and UK recommendations, which is to calculate the population requirement (respectively the RDA and RNI) by adding two standard deviations of the variation of individual *requirements* (usually not stated explicitly, but typically around 15 % CV).

The WHO/FAO/IAEA approach also differs significantly from most national recommendations in recognizing that there can be large differences in *bioavailability* of trace elements in different kinds of diets, particularly for the essential trace elements iron and zinc. Three different levels of bioavailability are considered in Tables 1 and 2 – low, moderate and high. In the case of low bioavailability diets, it is apparent that some of the required levels of dietary intake are very *unlikely* to be achievable on normal dietary patterns. All the values in Tables 1–3 refer to populations. However, the WHO/FAO/IAEA report also provides information on estimated average requirements of *individuals* (both basal and normative). The *population* requirements are generally higher than average *individual* requirements by 47 % for selenium, 67 % for copper and 100 % for zinc.

These different approaches to the definition of dietary requirement lead to a high level of complexity in deciding which value to use in any specific situation. Some of these complexities are illustrated in Figure 1 for the case of zinc. The values illustrated (which refer only to adult males, typically in the age range 20–50 years) show that, depending on the type of requirement chosen, e.g. (i) basal or normative, or (ii) whether for an individual or a population, or (iii) whether the diet is of high or low bioavailability, the estimated requirement may range between 1.8 and 18.7 mg/day. A further complication in applying these require-

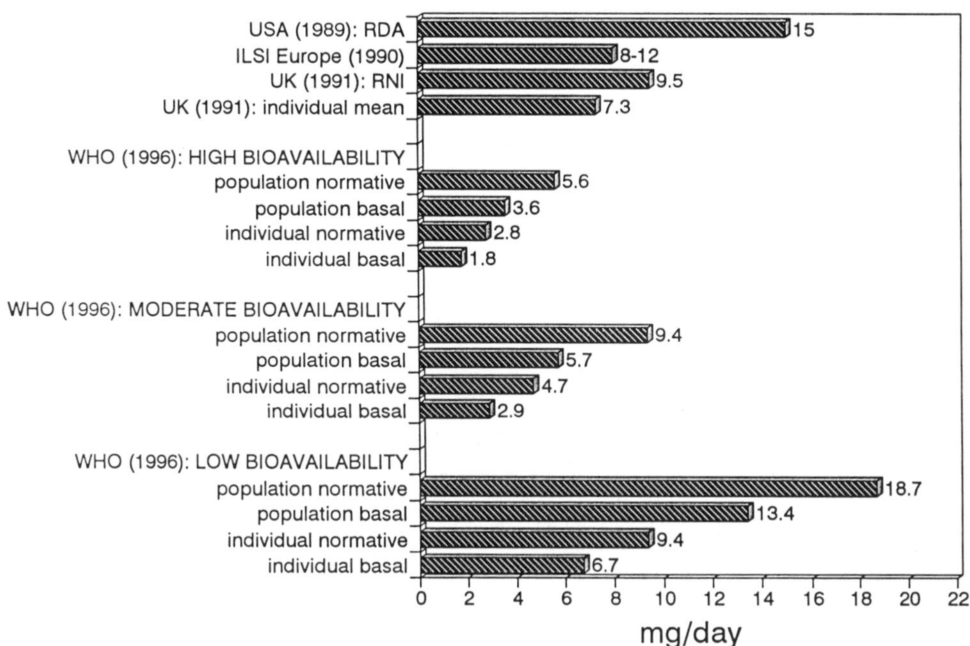

Figure 1. Estimated dietary requirements for zinc in adult males: a comparison of different kinds of estimate prepared from values quoted in references (1–4).

ments is that account should also be taken of the different body weights that have been assumed , e.g. 79 kg (2), ~74 kg (3) and 65 kg (1) for adult males.

During the last 6 years the author and his colleagues have had the opportunity of examining more than 400 publications on dietary intakes in different countries and creating a database of the values reported (8). Some of the data are illustrated in Figure 2 for the trace elements copper, selenium and zinc. In this figure the dietary intakes have all been normalized to the respective WHO/FAO/IAEA *normative* population requirements (in the case of zinc, account is also taken of likely differences in bioavailability). From these data it would appear that, for all three trace elements, around 20–30 % of the reported dietary intakes were *below* the proposed normative requirements. However, as discussed below, the practical significance of these findings cannot be established with confidence unless supported by additional data.

3. DISCUSSION

The interpretation of dietary intake data is complicated by the many different kinds of reference values that can be used as the basis for comparison. In the approximately 400 publications reviewed for inclusion in the IAEA database (8), a majority of the authors used the US RDAs (2) as the comparator. However, many of these investigators appeared to fall into the trap of interpreting intakes below the RDAs as indicating that *most* members of the population were probably suffering from a trace element deficiency. This is a *wrong* interpretation. Intakes below the RDAs are simply an indication of an increased *risk* of deficiency by possi-

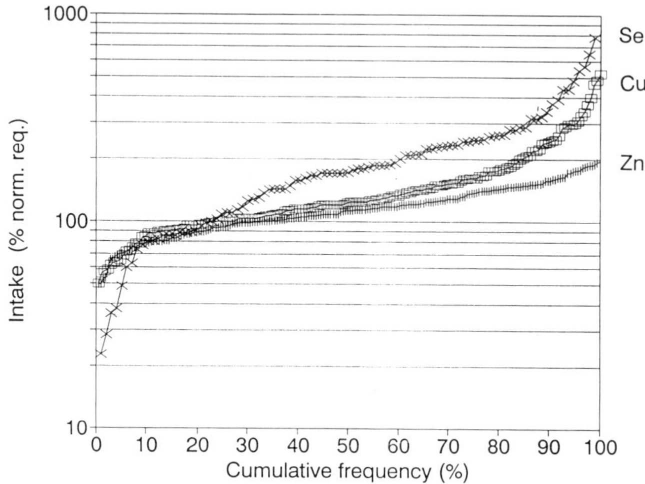

Figure 2. Cumulative frequency distributions of dietary intakes of copper, selenium and zinc compared with the relevant WHO/FAO/IAEA normative population requirements (adapted from a global database reported in ref. (8)).

bly only a *small* fraction of the population. The same interpretation applies to the UK RNIs and the WHO/FAO/IAEA basal and normative population requirements (Tables 1 and 2).

4. CONCLUSIONS

Measurements of dietary intake – provided they are conducted and interpreted correctly – can certainly be very useful in helping to identify segments of a population, or geographical regions, in which trace element intakes are excessively high or low. However, such studies are generally not sufficient *in themselves* to establish a reliable diagnosis of nutritional adequacy or inadequacy. In the words of the WHO/FAO/IAEA report (1) ".. analyses of diets ... are inappropriate indices of possible trace element deprivation. An unequivocal diagnosis is rarely achieved without careful monitoring of metabolic, functional or clinical responses to supplementation."

5. REFERENCES

1. Trace Elements in Human Nutrition and Health (A report of the World Health Organization prepared in collaboration with the Food and Agriculture Organization of the United Nations and the International Atomic Energy Agency), World Health Organization, Geneva, 1996.
2. Recommended Dietary Allowances (10th Edition), National Academy Press, Washington (1989).
3. Recommended Daily Amounts of Vitamins & Minerals in Europe (Report of a Workshop Organised by ILSI Europe) Nutr. Abstracts & Reviews (A) **60**, 827–842 (1990).
4. Dietary Reference Values for Food Energy and Nutrients for the United Kingdom, Report on Health and Social Subjects 41, HMSO, London (1991).
5. WHO Technical Report Series 452, World Health Organization, Geneva (1970).
6. WHO Technical Report Series 532, World Health Organization, Geneva (1973).
7. FAO Food and Nutrition Series 23, Food and Agriculture Organization, Rome (1988).
8. IAEA Report NAHRES-12, International Atomic Energy Agency, Vienna (1992).

3

PHARMACEUTICAL FORMS CONTAINING TRACE ELEMENTS FOR HUMANS

Jean Nève

Free University of Brussels
Belgium

Nutritional or therapeutical doses of trace elements are administered to humans though different forms including modified and fortified diets or medicinal-type preparations. The latter actually are the most common. However, the diversity of their presentations, characteristics and even legal status involves many questions. As trace element bioavailability in the preparations may considerably differ, it is of interest not only to assess them comparatively but also to optimise the forms proposed. Some factors affecting bioavailability will be reviewed showing examples of procedures adopted to document and/or solve practical problems (1).

1. THE CHEMICAL FORM

Preparations contain different chemical forms and oxidation states of trace elements. Active inorganic forms are either cations (e.g.: Zn^{2+}, Cr^{3+}, Mn^{3+} or Fe^{3+} chlorides; Fe^{2+}, Fe^{3+}, Zn^{2+}, Mn^{3+} or Cu^{2+} sulfates; Cu^{2+}, Zn^{2+} or Mn^{3+} oxides), anions (potassium I^- or F^-; calcium F^-; sodium selenite, selenate or molybdate) or even elemental forms, i.e. at zero oxidation state (Fe, Se). These entities are cheap and usually have well established properties such as purity, stability, incompatibilities, solubility or toxicity. Most inorganic forms are soluble in water which is favourable for optimal absorption (2). Some are only soluble in acid pH (oxides, carbonates), but other precipitate in alkaline pH, which reduces absorption. The oxidation state plays a role in the absorption of some elements, for example Fe which is only effectively absorbed as Fe^{2+}. In contrast to Fe^{3+}, the species does not easily precipitate at alkaline pH values resulting from duodenal secretions. Success of the administration of Fe^{3+} therefore depends on its reduction by substances present in the digestive tract (ascorbic acid, cysteine, fructose, glutathione, etc), a phenomenon that is facilitated by stomach acidity (3). Selenite or selenate, two oxidation forms of Se, have different pharmacokinetic properties. Selenate is absorbed from the ileum faster and to a larger extent, apparently by a carrier-mediated mechanism, whereas the absorption of selenite may be by diffusion. However, urinary excretion is three times higher for selenate and peak urinary excretion occurs earlier with the first derivative. This is linked to a similarity in handling of selenate and sulfate (4). Preparations containing elements at the elemental state may be poorly absorbed. Ducros et al. (5) however showed that a commercial complex of elemental Se and amylose sold in France for years is as efficiently absorbed as sodium selenite. This directly poses the problem of overexposure as Se content of the form is rather high (960 μg Se/vial).

Simple organic forms are also well-defined compounds. They generally are salts with an organic acid (ascorbate, aspartate, citrate, fumarate, gluconate, glycerophosphate, lactate, orotate, picolinate, pidolate or pyroglutamate, succinate, tartratre, etc). Complexes do not exist as such for trace element supplementation, but are present in food matrices. Organic salts are sometimes preferred to inorganic molecules but the interest of the "vector" molecule is diversely documented. Considerable differences are frequently evidenced when comparing chemical compounds of an element. For example, when using "tolerance tests" (measurement of the variation in serum concentration after a challenge dose of an element) for comparing the absorption of oral Fe-containing liquid forms, the best results are obtained with ferrous succinate. This salt is slightly better than ferrous sulfate whose performance is comparable to ferrous lactate, fumarate, glutamate or gluconate. In contrast, ferrous citrate or tartrate and ferric citrate, sulfate or carbonate show bad performances (3). Using improved "tolerance tests" based on modern pharmacokinetic concepts, Nève et al. (6,7) demonstrated a clear superiority in terms of comparative bioavailability of Zn gluconate over Zn sulfate. This favourable effect was also evidenced for other salts such as dipicolinate, citrate, pantothenate or orotate (6). In another study based on an obsolete methodology, Zn sulfate was found comparable to organic salts like acetate, aminoacetate and Zn-methionine (8).

More complex forms for supplementation are the aminoacids selenocystine and selenomethionine, which are Se-analogs of the sulfur aminoacids, and Se-enriched yeast, obtained by growing brewer's yeast in a Se-rich medium, followed by lyophilization of the cells that captured and metabolised the element. This process gives rise to a powder containing both organic and inorganic combinations, including Se-aminoacids, selenosulfides, selenite, selenate, etc (9). As no standardisation exists for manufacturing these forms (N.B.: Cr-enriched yeast is also available), their characteristics may greatly vary. A comprehensive review was recently devoted to the differences in the effects on current indicators of Se status (plasma and erythrocyte Se and activities of the enzyme glutathione peroxidase, GSH-Px, in plasma, erythrocytes and platelets) of different chemical forms proposed for Se supplementation, i.e.: selenite, selenate, Se-methionine, Se-enriched yeast and Se-rich wheat or meat (9). Briefly, organic Se forms (Se-yeast, selenomethionine and food-Se) increase blood Se more rapidly and to a greater extent than inorganic forms. However, no difference in the response of both plasma and erythrocyte GSH-Px activity (a functional indicator of Se status) is observed. In contrast, platelet GSH-Px is more sensitive to the chemical form of Se. Indeed, saturation of platelet GSH-Px activity (a good indicator of Se bioavailability) occurs at lower plasma Se levels when selenite or selenate is used than with the organic forms. Inorganic forms therefore appear more available for platelet GSH-Px than organic ones, although these last and more particularly selenomethionine, increase Se stores in proteins more than inorganic forms. Another example concerns the treatment and prophylaxis of iodine deficiency, which in developed countries generally consists in administration of iodised salt (with KI or KIO_3), bread or water. In developing countries, supplementation is efficiently achieved by slowly absorbable iodised oil delivered by intramuscular injections or orally. The form is obtained by covalently linking iodine (I_2) to fatty acids esters of vegetable oil (poppyseed or lipiodol, walnut, soybean). Injected oil is stored at the site of injection, progressively deiodinated in muscle or in the bloodstream and presented to the thyroid as iodide. With oral administration, deionination occurs in the digestive tract and in blood and a part of iodinated fatty acids are stored in adipose tissues. Single high oral doses (1 to 2 ml) provide adequate iodine for 2 to 3 years, but may cause toxicity due to the rapid deiodination in the digestive tract (10). However, lower doses (0.1 to 0.25 ml) are capable of correcting iodine deficiency for about one year without significant side effects (11). A better effectiveness of iodised oil consisting in triacylglycerol esters of fatty acids than of ethyl esters was recently demonstrated (12).

Previous examples demonstrate that the chemical form modulates trace element bioavailability to a variable extent. This effect is dramatic for Co which is the sole element that has to be administered directly under its biologically active form, the cobalamines or vitamin B_{12}. Even if Co salts were used in the past at relatively high doses (20 to 30 mg/d) as pharmacological agents for treating some forms of anaemia, it is difficult to justify the presence of Co salts at lower doses in multi-element preparations (13).

2. THE ADMINISTERED DOSE

The reference ranges for trace element intakes supposed to be optimal from a nutritional point of view are undoubtedly those of the well-known "Recommended Dietary Allowances (RDA)" tables that are far from being perfect but for which some agreement exists. Commercial products delivering doses within RDA are therefore typical "nutritional supplements", i.e. preparations devoted to the prevention and correction of trace element deficiencies regardless their origin is nutritional (insufficient intake, increased dietary losses, etc), physiological (pregnancy, growth, lactation) or pathological (presence of diseases, etc). It is well known that RDA limits are fairly broad and that the allowances take into account the variable bioavailability of elements in food matrices. It is however important to remind that usual dietary intake is not a negligible source of trace elements and that meeting 100 % (sometimes more) of the RDA with a trace element preparation is very often exaggerated. This is the case for Mn preparations that are offered to the consumer although Mn deficiency is very rare in humans. Exaggeration of doses not only exposes to toxicity problems but also to pharmacological interactions. The opposite, i.e. administration of doses far lower than the RDA (e.g., homeopathic doses), is very often insignificant from a nutritional point of view. Concerning a therapeutic concept developed in the thirties in France (the "catalytic" therapy), which involves the treatment of functional disturbances with relatively low doses preparations of selected minerals, it can easily be calculated that both nutritionally significant and totally insignificant doses are dispensed through this approach (14).

Several studies demonstrate a dependence on dose of trace element absorption. Administration of aqueous solutions of Zn consistently results in an absorption above 50 % and fractional absorption does not change much with increasing Zn doses, resulting in a linear increase in the amount of Zn absorbed. In contrast, the absorption of Zn from meals or total diets shows a different picture as percentage absorption gradually decreases with increasing Zn content and the relationship between Zn content and absorbed Zn indicates a saturation of absorption for relatively high supplies (15). Concerning Mn, for which the percentage absorption is low, most often below 10 %, neither the mode of administration nor the quantity supplied seem to have any significant impact on the percentage absorption (15,16). Concerning Fe, the percentage absorption seems to be higher for low intake levels than for relatively high Fe supplies. Moreover, reductions in Fe stores are correlated with increases in Fe absorption and therefore body Fe stores affect Fe absorption (17). The problem of the dose has to be considered taking into account the biological efficacy: indeed, considering that three oral Fe-based formula with different concentrations (5, 50 and 100 mg Fe) were found as efficient for correcting hemoglobin levels in Fe-deficient anaemic subjects, it seems preferable to choose the low dose formula. The importance of the dose was underlined by Mertz (18) who examined the results of 15 controlled studies supplementing defined Cr^{3+} compounds to subjects with impaired glucose tolerance. Even if 12 of them produced beneficial effects, he noted that the quantity of Cr used ranged from 50 to 2000 µg/d and further indicated that no exact dose-response studies to Cr have been performed in humans, and no efforts have been made to determine the lowest effective dose.

Besides nutritional-type preparations, there are forms that are presented as drugs, i.e. with properties or therapeutic indications different from the nutritional properties. The frontier between the two kind of preparations is sometimes not easy to establish. Essential trace elements are generally administered at doses higher than in nutritional complements, for example Zn doses of 50 mg three times per day to induce intestinal cell metallothionein and block intestinal absorption of Cu in Wilson's disease (19) or Se doses around 200 µg/d or more for immunosuppression or adjuvant cancer treatment. Trace elements that are not considered as essential or for which either proofs of essentiality are low or that usually are not subject to deficiency problems are also used as therapeutic agents: F for the prevention of dental caries or to treat osteoporosis, Au as an immunosuppressive agent in the treatment of rheumatoid arthritis, Li in the prophylaxis of bipolar affective disorders, etc. These indications have to be carefully documented and evaluated by competent commissions on all aspects including adequation of the dose to the alleged therapeutic effects. The case of Li is particularly illustrative as its therapeutic effect depends on the maintenance of a steady-state serum Li concentration of 0.5 to 0.8 mmol/L. This recently established range is based on randomised prospective studies showing that it is as effective as, or even more effective than, serum concentrations in the previously recommended range of 0.7 to 1.2 mmol/L (20). The long-lasting controversy concerning the efficacy of F as a pharmacological agent for stimulating bone formation, through its peculiar mitogenic dose-dependent action on the osteoblast cell line, is also well known. Doses of 50 mg NaF/d are now recommended instead of higher doses that were prescribed in the past and induced morphologically abnormal bone areas (21). Moreover, as it is also important to provide Ca supplements together with F therapy to avoid an increase in eroded surfaces, the combination of Ca carbonate with Na monofluorophosphate in effervescent tablets is a way of enhancing the compliance to Ca supplementation (21). Another element of interest is V: even if its essentiality for humans remains uncertain, V salts like meta or orthovanadate (V^{5+}) and vanadyl sulphate (V^{4+}) have demonstrated interesting antidiabetic effects in animals, but the rather high oral doses used posed toxicity problems rendering V unsuitable for human use. More recently however, two groups administered smaller doses of vanadyl sulphate (100 mg/d) or sodium vanadate (125 mg/d) to diabetic patients for a period of 2–3 weeks and reported encouraging results (22).

3. THE INTERACTIONS WITH OTHER CONSTITUENTS

Nutritional and pharmacological interactions considerably influence the success of the intervention. Food-type components have complex effects that were largely described. Carbohydrates such as lactose stimulate the absorption of Zn and Fe, while fructose has a negative effect on Cu absorption (23). Aminoacids such as cysteine, histidine and methionine form mixed chelates with elements like Zn or Fe, which constitute metabolically different forms than inorganic sources, and modify absorption and bioavailability. An organic acid like ascorbic acid is beneficial to Fe absorption because not only it facilitates the reduction of ferric ions, but it also forms complexes with ferric ions which retain the soluble character of

Fe at less acid duodenal pH values (3). In contrast, ascorbic acid decrease Se absorption when co-administered as selenite, and Cu bioavailability is unfavourably influenced by high ascorbic acid supplies (24,25). Picolinic acid (a tryptophane metabolite) enhances Zn and Cu absorption (24). Also, citric, lactic, malic, pyruvic, succinic or tartaric acids increase Fe absorption (3).

Other active constituents of the preparation (e.g., vitamins and other minerals) may as well interact with the trace element. The reciprocal inhibition by folates of Zn absorption by formation of an insoluble chelate of molar ratio 2:1 has been the subject of many studies with controversial results (26). Among other explanations, it has been suggested that the interaction has no practical consequence as the chelate, which is insoluble at acid pH, redissolves at intestinal pH. The influence of the consumption of other minerals is also largely demonstrated: for example, regular consumption of Zn or Ca supplements can cause Cu deficiency, regular administration of Fe can deplete Zn stores, Zn supplementation can inhibit Ca absorption, and Ca supplementation can affect Fe status. Such possibilities actually depend on many factors and generalisation of observed effects is difficult. The chemical similarity has suggested that Mn shares or competes for absorptive mechanisms with Fe. Addition of Mn to a Fe solution or a meal depresses Fe absorption in a way that gives the impression that the body cannot distinguish between Fe and Mn (15). Davis et al. (27) supplemented women for 120 days with Fe (60 mg), Mn (15 mg) or a Fe-Mn combination and demonstrated that Fe alone improves serum ferritin but decreases serum Mn and Mn-dependent superoxide dismutase activity in lymphocytes, that Mn alone improved these last two parameters, and that the combination improved both Fe- and Mn-dependant parameters. Such results stress on the interest of balanced multi-element supplements instead of mono-element preparations. This was supported by a recent investigation in pregnant women where isolated supply of folates or of Fe caused Zn deficiency, whereas a preparation combining Zn with these nutrients improved Fe, folates and also Zn status (1).

4. THE GALENIC FORM AND THE ADMINISTRATION CONDITIONS

Preparations for oral use are most common for trace element supplementation, but their pharmaceutical characteristics were diversely studied. Sustained effect formulations for Fe developed to improve gastrointestinal tolerance may have a lower bioavailability as assessed by tolerance tests which demonstrate a relative bioavailability from 46 to 100 % for ferrous ions contained in rapid release forms whereas it ranged from 31 to 47 % for slow or controlled release preparations (3). A gastric delivery system for Fe has recently been described, which retains ferrous sulfate in the stomach while releasing it slowly over several hours: not only it decreases gastrointestinal side effects, but, if taken with food which strongly inhibits Fe absorption, the Fe is retained in the stomach while the bulk of the meal passes on to the small intestine (29). Zn in a gastro-resistant tablet was found comparatively less available than in a soft capsule containing the same salt at similar concentration (6). Enterocoated tablets are also useful in F therapy for osteoporosis as the element at relatively high doses can irritate the gastric mucosa causing severe bleeding or peptic ulcer. It has been observed that 40 % of patients complain about gastrointestinal side effects when using non enterocoated tablets versus about only 10 % with enterocoated preparations (21). Such tablets do not release all the F they contain.

Impaired bioavailability due to the presence of inert excipients (lactose, starch, silica, Mg stearate, etc) in oral forms for Zn supplementation was also demonstrated (6). A classical example of the dependence on the galenic form of a therapeutic effect is the case of Zn lozenges. A study first showed that lozenges containing 23 mg Zn as gluconate were effective in reducing the length of the common cold. Subsequent trials using other lozenges formulations were however unable to confirm the results. It was later demonstrated that the activity of Zn, which depends on the presence of free Zn ions in the mouth, is completely abolished by the presence of additives used to mask the unpleasant taste of Zn in these preparations. Indeed, derivatives such as citric and tartaric acid, mannitol and sorbitol excessively chelated Zn rendering it poorly available (28).

About optimal conditions of administration, few work has generally been done, except on the interest of ingesting during meals formulations with a bad gastro-intestinal acceptability (e.g., causing nausea, pain, diarrhoea or constipation). This time, slow-release preparations may offer some advantages as they guarantee continuous and long lasting release in the intestine. By using isotopic labelling to compare the absorption of two preparations containing 50 mg Fe as ferrous sulfate, one in the form of an elixir, the other in the form of a controlled release preparation, it was demonstrated that they were equivalent (absorption yield around 5 %) when subjects ingested them in fasting state, but that the rate of absorption decreased less for the controlled release form than for the elixir when the form was taken during a meal (2). A more recent refinement to Fe supplementation programs that has been largely discussed is the administration of the element less frequently than once daily. The concept was based on studies indicating that the administration of oral Fe impairs the absorption of a subsequent Fe dose. Cook and Reddy (29) measured Fe absorption from 50 mg radiolabeled ferrous sulfate in groups of subjects taking supplements either daily or weekly and were unable to demonstrate any significant absorptive advantage in giving Fe less often than once daily. They also discussed the fact that less frequent administration of Fe supplements (e.g., once a week) will not satisfy Fe requirement in situations in which Fe supplementation is required, particularly in preschool children or during pregnancy. Using improved tolerance tests to compare absorption of Zn in a commercial form when taken during a meal, it was demonstrated that the effect of the meal was minor in terms of bioavailability (28 % decrease in the area under curve as compared to the form taken at fasting state) in spite of the significant increase in the lag time (6). Such results undoubtedly depend on the kind of food consumed with the supplement. For example, authors showed that taking Zn sulfate during a meal did not produce any modification in plasma Zn although the subjects very well responded to the treatment when ingesting the same form without a meal (30).

5. THE INDIVIDUAL PARAMETERS

They have to be considered not only by manufacturers who want to propose administration conditions adapted to specific groups (e.g., pregnant women, children, elderly, patients with pathologies, etc), but also by prescribers of trace element preparations. Subjects with pronounced deficiencies will generally absorb to a greater extent trace elements such as Fe, Se or Zn while decreasing their elimination rate (23,31,32). Age can also exert some influence on absorption. Time needed for digestion increases with age whereas acid production decreases, particularly in cases of atrophic gastritis that is frequent in elderly. The decrease in stomach acidity has been related to impaired Fe absorption (33). The atrophy of intestinal mucous membrane in the course of ageing also influences Zn absorption although it seems that the organism is able to maintain Zn balance by adapting Zn excretion (34). Nevertheless, Zn supplementation is less efficient in older subjects as compared to younger ones. Pregnancy and lactation are also physiological states that have been the scope of many studies: Fe absorption seems to be increased, particularly during the 3rd trimester of pregnancy while Zn absorption seems to be unaffected. Several pathological states have also been diversely implicated in modifications of trace element requirements or metabolism: Chron's disease, sprue, enteropathic acrodermatitis, Menkes' disease, phenylketonuria, inflammatory rheumatic diseases, etc (35). The recent suggestion of Guigoz et al. (36) illustrates the complexity of the problem: they indeed indicated that Zn supplementation in the elderly should only be decided after considering three different situations: (i) a good health status, for which Zn requirements do not differ from young adults, (ii) the presence of social stress or acute diseases, which momentarily increase Zn requirement, and (iii) the presence of chronic diseases for whom prolonged supplementation may be necessary. Finally, it is also well known that drug therapy with agents such as corticosteroids, nonsteroidal antiinflammatory drugs, inhibitors of the angiotensine converting enzyme, tetracyclines, levodopa and analogs, penicillamine or quinolones can sometimes considerably affect trace element status causing biochemical and clinical manifestations of trace element deficiencies (1). For example, ethambutol can cause optical neuropathies related to induction of a Zn deficiency state, and captopril or penicillamine can induce ageusia or alopecia by depleting Zn stores. The opposite is also true as trace element administration concurrently with drug therapy can cause interactions, sometimes with clinical significance. This has been well documented for Fe supplements (37). As both treatments may be affected, it has to be recommended to separate by 2 or 3 h the administrations of elements and drugs able to interfere.

6. CONCLUSIONS

Trace element supplementation can no longer be considered as an empirical or esoterical intervention. Although some latitude is acceptable for nutritional complements provided the chemical form, dose, presentation and route of administration do not differ too much from accepted standards, a great attention should be devoted to the style of documentation each time specific indications are alleged. Numerous methods nowadays exist that give clear indications about the characteristics of products proposed to the consumer and their scientific value can be assessed with greater accuracy than in the past.

7. REFERENCES

1. J. Nève, J. Poupon, V. Ducros, C. Charlot and A. Favier, in *Les oligoéléments en nutrition et en thérapeutique*, P. Chappuis and A. Favier, eds., Lavoisier, Paris 189 (1995).
2. J. Cook and M. Reusser, *Am. J. Clin. Nutr.* **38**, 648 (1983).
3. E. Harju, *Clin. Pharmacokin.* **17**, 69 (1989).
4. J. Nève, in *Proceedings of the STDA'S fifth International Symposium,* Brussels 123 (1994).
5. V. Ducros, A. Favier and M. Guiges, *J. Trace Elem. Electrolytes Health Dis.* **5**, 145 (1991).
6. J. Nève, M. Hanocq, A. Peretz, F. Abi Khalil and F. Pelen, *J. Pharm. Belg.* **48**, 5 (1993).
7. J. Nève, M. Hanocq, A. Peretz, F. Abi Khalil, F. Pelen, J.P. Famaey and J. Fontaine, *Eur. J. Drug Metab. Pharmacokin.* **16**, 315 (1991).
8. A. Prasad, F. Beck and J. Nowak, *J. Trace Elem. Expl. Med.* **6**, 109 (1993).
9. J. Nève, *J. Trace Elements Med. Biol.* **9**, 65 (1995).
10. B. Hetzel and J. Dunn, *Annu. Rev. Nutr.* **9**, 21 (1989).
11. R. Tonglet, P. Bourdoux, T. Minga and A. Ermans, *N. Engl. J. Med.* **4**, 236 (1992).
12. C. Furnée, G. Pfann, C. West, F. van der Haar, D. van der Heide and J. Hautvast, *Am. J. Clin. Nutr.* **61**, 1257 (1995).
13. J. Nève, *J. Pharm. Belg.* **46**, 271 (1991).
14. J. Nève, *Porphyre* **256**, 38 (1990).

15. B. Sandstrom, *Proc. Nutr. Soc.* **51**, 211 (1992).

16. L. Hurley and C. Keen, in *Trace Elements in Human and Animal Nutrition*, vol. 5, W. Mertz, ed., Academic Press, San Diego 185 (1987).

17. M. Gavin, D. McCarthy and P. Garry, *Am. J. Clin. Nutr.* **59**, 1376 (1994).

18. W. Mertz, *J. Nutr.* **123**, 626 (1993).

19. G. Brewer, *Drugs* **50**, 240 (1995).

20. M. Peet and J. Pratt, *Drugs* **46**, 7 (1993).

21. J.P. Devogelaer and C. Nagant de Deuxchaisnes, *Clin. Rheumatol.* **14**, 26 (1995).

22. S. Brichard and J.C. Henquin, *Trends Pharmacol. Sci.* **16**, 265 (1995).

23. E. Morris, in *Trace Elements in Human and Animal Nutrition*, vol. 5, W. Mertz, ed., Academic Press, San Diego 79 (1987).

24. H. Sandstead, *Am. J. Clin. Nutr.* **35**, 809 (1982).

25. E. Finley and F. Cerklewski, *Am. J. Clin. Nutr.* **37**, 553 (1983).

26. J. Arnaud, A. Favier, M. Herrmann and J. Pilorget, *Ann. Nutr. Metab.* 36, 157 (1992).

27. C. Davis and J.L. Greger, *Am. J. Clin. Nutr.* **55**, 747 (1992).

28. J. Zarembo, J. Godfrey and N. Godfrey, *J. Pharm. Sci.* **81**, 128 (1992).

29. J. Cook and M. Reddy, *Am. J. Clin. Nutr.* **62**, 117 (1995).

30. J. Keyzer, E. Oosting, B. Wolters and F. Muskiet, *Pharm. Weekblad* **5**, 252 (1983).

31. J. King, *J. Nutr.* **120**, 1474 (1990).

32. R. Wapnir, *Protein Nutrition and Mineral Absorption*, CRC Press, Boca Raton (1990).

33. R. Russel, *Am. J. Clin. Nutr.* **55**, 1203 (1992).

34. J. Turnlund, N. Durkin, F. Costa and S. Margen, *J. Nutr.* **116**,1239 (1986).

35. A. Peretz, B. Cantinieaux, J. Nève, V. Siderova and P. Fondu, *J. Trace Elem. Electrolytes Health Dis.* **8**, 189 (1994).

36. Y. Guigoz, *Facts and Research in Gerontology* 265(1992).

37. N. Campbell and B. Hasinoff, *Brit. J. Clin. Pharmacol.* **31**, 251 (1991).

4

METAL–LIGAND INTERACTIONS AND TRACE METAL BIOAVAILABILITY

Guy Berthon

INSERM U305
38 rue des Trente-six Ponts
31400 Toulouse, France

1. INTRODUCTION

The use of metalloelements in human medicine is common practice. As a function of their physico-chemical properties, these elements play distinct roles with respect to life and health. They may therefore be prescribed for a large range of applications, at different doses, and in different chemical forms. Metals considered as essential are frequently administered at dietary levels to compensate for deficiencies. They may also be used at higher doses, to take advantage of their intrinsic pharmacotoxicological properties. In contrast, non-essential metals are used exclusively for their pharmacotoxicological capacities in therapy, or for their special physico-chemical properties in diagnosis.

The chemical forms under which metalloelements are administered—from totally dissociable salts to very stable and/or inert complexes—should logically be adapted to the specific objectives to be met. While this rule is generally observed in the design of pharmacological and diagnostic metal-containing agents, there is often a less well-established basis in the production of dietary supplements. First, such preparations usually contain different total amounts and chemical forms of each element, and their formulations and recommended conditions of administration may vary considerably (1). Also, as most of the time no information is provided regarding the bioavailability of their contents, these preparations implicitly—and wrongly— associate "total amounts" and adequacy of the essential elements supplied. The problem is still complicated by the lack of unified legislation among different countries. Thus, a number of dietary supplements are available over the counter whose regular intake may induce metabolic disorders in the long run. Even for medical practitioners, the distinction between essentiality and toxicity as well as between nutritional and pharmacological relevance of metalloelements is often unclear.

In view of the potential hazards linked to the administration of metal-containing preparations in terms of public health, an effort of clarification is necessary regarding: (i) the classification of elements and its implications for the use of metalloelements in human medicine, including corresponding doses; (ii) the chemical forms under which metal ions may be administered as a function of the expected effects, i.e. the influence of the ligands with which they are associated on their bioavailability. Bioavailability has been defined as the extent to which an element is absorbed and utilised by an organism (2), or may interact with it to produce a concomitant response (3). A short account is therefore given in this mini-review on the fundamental processes of metal metabolism as well as on the possible techniques to assess bioavailability in vivo and in vitro. Regarding the latter point, emphasis has been put on the potential applications of computer-aided speciation.

2. CLASSIFICATION OF TRACE METALS AND IMPLICATIONS FOR THEIR THERAPEUTIC USE

2.1. Classification of Metalloelements

Most of the classical works in inorganic biochemistry range elements in two categories, opposing essentiality to toxicity (see e.g. ref. 4). Others consider the distribution of elements into essential, contaminants and toxic (5), or into essential and not required (6). As noted by Underwood (5), the classification of an element as toxic has limited value because all elements are toxic if ingested or inhaled at sufficiently high levels or for long enough periods. The distinction of essential, beneficial, contaminating and polluting elements (7)—recently reworded as essential, beneficial, neutral and detrimental (8)—appears more realistic in this respect, and has the advantage to put the classification of elements in its evolutionary perspective: a living species coming into contact with an unknown element first sees it as exclusively toxic, i.e. *detrimental*. If the environmental level of the element remains constant, surviving generations of this species will progressively adapt to its toxicity until they render it harmless, i.e. *neutral*. Once this stage reached, sooner or later the species will take advantage of the element and make it *beneficial*; before—eventually—this element becomes necessary to life itself, i.e. *essential*.

This evolutionary process, from which specific homeostatic regulations progressively emerge with time, has important implications for the biochemistry of metalloelements. In particular, the behaviour of a given element with respect to health at its usual (constant) environmental level —i.e. its status in the essential>beneficial>neutral>detrimental classification—depends on its ecological age. Consequently, as was recently pointed out by R.J.P. Williams and da Silva (6), not only "completely foreign compounds will always be a source of risk", but "even very ordinary elements taken in wrong amounts must stress (..) homeostatic balances". In other words, outside usual environmental levels for essential and beneficial elements and virtual zero concentrations for detrimental elements, health effects due to a given element will exclusively depend on its concentration.

2.2. Implications for the Therapeutic Administration of Trace Metals

The graphical representation of the above principle—which implicitly refers to the former Bertrand's Law (5)—is known by specialists as Venchikov's curve (9). Originally relative to useful trace elements, this curve has since been generalised (7, 10) (see Figure 1).

2.2.1. Essential and Beneficial Elements. According to Venchikov, three zones of concentration may be distinguished for each essential or beneficial trace element: a zone of *biotic effect* in which health is improved with increasing concentrations until a plateau is reached, which represents optimal supplementation and normal function, and whose

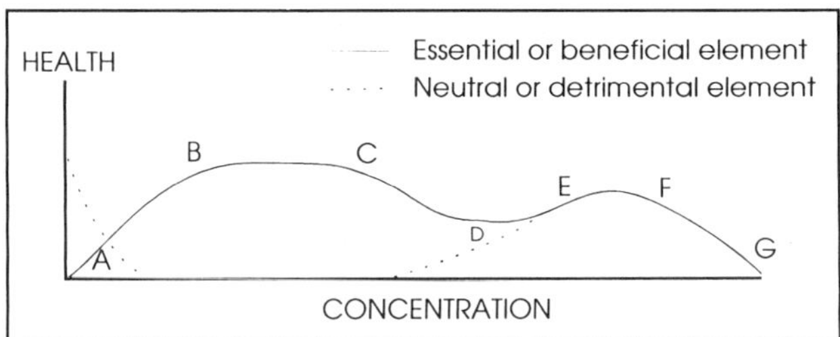

Figure 1. Generalised Venchikov's curve (from ref. 10).

width is determined by the homeostatic capacity of the organism; a zone of *inaction* where health declines with further increasing doses because of body overload of the element; and finally, at much higher concentrations corresponding to drug macrodoses *independent of a deficiency state*, a zone of *pharmacotoxicological effect* where the toxicity of the element serves to stimulate or help in some way the defense mechanisms of the host. Eventually, still higher doses of the element cause an irreversible reaction leading to death.

Such curves differ from element to element and may vary depending on interactions among elements. In particular, in accordance with the above-mentioned role of evolution towards specificity, essential elements are expected to benefit from better homeostatic controls and safety margins between optimum and toxic concentrations than beneficial elements (7, 11). (For example, 0.1 ppm of Se is beneficial whereas 10 ppm is carcinogenic (4, 11).)

2.2.2. Neutral and Detrimental Elements. For not required elements, health is optimal in the total absence of these and starts deteriorating as soon as their concentrations are raised, but as with useful elements, there is a dose interval within which these can be used as drugs (e.g. Pt, Au, Bi,...) or diagnostic agents (Ga, In, Tc, Gd,...). Indeed, "very few chemicals are without any potential value if they are used in controlled amounts and are administered (...) locally" (6). An extreme example of this is the use of radioactive elements to screen diseases and to kill tumour cells locally. Emphasis, however, must always be put on risk/benefit analysis (6).

2.3. Administered Doses versus Available Amounts: The Notion of Bioavailability

As a function of the therapeutic use of a metalloelement (i.e. nutritional, pharmacological or diagnostic), specific concentrations of this element may be needed in particular compartments of the body. Whereas systemic retention is required for a nutrient, localised accumulation may on the contrary be desirable for a drug or a contrast agent, mainly for specificity reasons but also to maintain toxicity within acceptable limits (see above).

For nutritional applications, guidelines for dietary requirements of essential elements are provided in the form of practical allowances by the World Health Organisation (WHO), or country by country as, for example, US RDA's, Canadian RNI, etc. (12). All these data, however, are only indicative—especially in the therapeutic context—as they highly depend on the bioavailability of elements from the average diets taken for the corresponding evaluations (13). The published figures should therefore be put in perspective with the specific bioavailability of each element from the compound actually administered. In connection with this, the important role of interactions between elements and other constituents (see e.g. ref.

14) and among elements themselves (15) makes multi-element supplementation preferable to mono-supplementation from a metabolic standpoint (except for the case of deficiency in a particular element (1)). In practice, however, the separate administration of a single element in an empty stomach represents the ideal case, based on the principle that it will always be simpler to control the bioavailability of a given metal ion in the presence of one ligand than in a complex mixture involving many other metal ions and ligands.

It seems from the above that using a single metal as a drug should a priori be easier than metal supplementation. However, in view of the fact that the high concentrations generally required for drug efficiency are closer to toxic levels, bioavailability is still a more important parameter in that case, whatever the mode of administration of the element. This holds in particular for metal ions used as contrast agents for which minimum ligand exchange and rapid clearance are imperative. Reference doses are established on the basis of classical CI or LD_{50}/CD_{50} ratios (11). These aspects will not be developed here, however, as emphasis is expected to be put mainly on nutritional considerations.

3. METAL-LIGAND INTERACTIONS AND METAL METABOLISM

3.1. Gastrointestinal (GI) Absorption

3.1.1. Modes of Transport of Metal Ions. During the absorption process through which substances are transported from the intestinal lumen to body fluids and tissues, metal ions undergo physical and chemical changes which depend on the composition (pH, ligands, etc.) of the physiological compartments that they successively penetrate. Two major steps are involved: (i) luminal events, i.e. transformations of the substance from its dietary form to its absorptive form, and (ii) mucosal events, i.e. passage of the substance through the intestinal mucosa (16). The transport routes for this passage may be transcellular or extracellular. Transcellular routes include lipid and aqueous routes for non-ionic and water-soluble solutes respectively, and a carrier route for hydrophilic substances too large for the aqueous route. The whole phenomenon implies the passage of the substance across both the mucosal and the serosal membranes of the cell, together with translocation within the cell interior. The main extracellular route is the paracellular route, via the junctions between adjacent cells (17).

Metal ion absorption across the intestinal cell can include both diffusion and active transport processes: (i) Passive diffusion is an energy-independent process. It depends on the metal concentration on both sides of the enterocyte membrane and on the relative solubility of the metal ion in the lipid bilayer. (Solvent drag via the paracellular pathway, which depends on water flow, is also considered a passive transport (16, 17).) (ii) In facilitated diffusion, membrane carriers transfer the metal ion across the membrane. This process also is energy independent and activated by the ion concentration gradient between the two sides of the membrane, but it is more rapid than simple diffusion. Saturable carrier proteins embedded in the cell membrane assume specific conformational states on each face of the membrane, which make metal ion binding sites first available at one and then at the other side of the membrane (16). (iii) Active transport proceeds against a concentration gradient, is saturable and energy dependent, and the carrier protein involved in it shows specificity for the substrate. The energy is provided by hydrolysis of ATP into ADP via an ATPase (16).

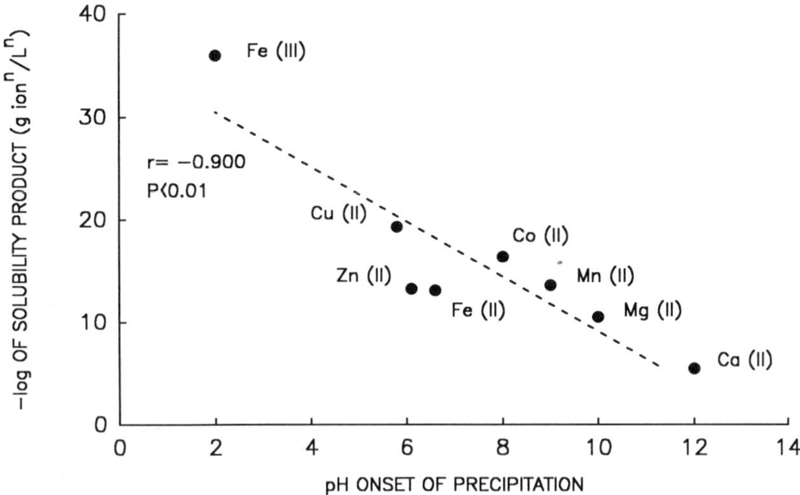

Figure 2. Relationship between the approximate pH of precipitation and cologarithm of solubility products of various metal hydroxides. (from ref. 18)

3.1.2. Metal-Ligand Interactions.

3.1.2.1. Solubility. The main criterion for a metal ion to be taken up by mucosal cells is the solubility of its forms in contact with the cell surface. In the absence of potential ligands, water molecules which coordinate metal ions at low pH progressively dissociate as the pH increases to give rise to hydroxides of sparing solubility (Figure 2). Even fairly soluble hydroxides of calcium, magnesium, manganese and iron(II) can coprecipitate on insoluble matrices in the alimentary canal and, under these conditions, iron is rapidly oxidized to iron(III) at pH > 4 (18). In normal nutrition, the digestion process—through intestinal secretions and enzymatic action—"frees" the metals from macronutrients (e.g. protein) (16, 19). Some of the ligands released (amino acids, sugars, etc.) form low-molar-mass (l.m.m.) chelates which maintain metal ions soluble in the small intestine, while other substances (phytate, polyphenols, etc.), on the contrary, tend to limit metal absorption by precipitation (16). Metal ion supplementation mixtures administered with meals are subject to the same competition from dietary ligands. By definition, metal salts are therefore best absorbed when given in the fasting state (see, e.g., iron (20)—see also 2.3).

Modifying the oxidation state of metals can also have an effect on metal absorbability via solubility. For example, ascorbate increases non-heme iron absorption by reducing ferric iron to the more soluble ferrous iron. However, ascorbate reduction of the cupric ion to the less soluble cuprous form may decrease copper absorption (16). (Administration of ascorbate during the postabsorptive period, in contrast, greatly enhances copper tissue utilisation (21).

3.1.2.2. Metal Complex Formation. Metal complexes formed in the g.i. lumen can directly influence metal ion mucosal uptake. Such complexes are normally due to food constituents, but may also be formed with drugs or synthetic ligands. Resulting effects can range from strongly inhibitory for some dietary factors (e.g. phytate, tannins, oxalate) to positive for others (citrate, malate, lactate, etc.) or strongly positive for ligands specifically used to stimulate metal ion absorption (e.g. EDTA) (19).

Table 1. Preference of metal ions for certain donor atoms in ligands (taken from ref. 24–adapted from Fraústo da Silva and Williams, 1993)

Metal ion	Donor atoms
Mg^{2+}	**O** (carboxylate, polyphosphate; negatively charged)
Ca^{2+}	**O** (carboxylate, carbonyl, phosphate)
Mn^{2+}, Fe^{2+}	**O** (carboxylate, phosphate); **N** (His); **S** (Cys, Met, sulphide)
Fe^{2+} (special)	**N** (polypyrroles)
Cr^{3+}, Mn^{3+}	**O** (phenolate, hydroxamate, hydroxide)
Fe^{3+}, Co^{3+}	**O** (carboxylate); **N** (polypyrroles); **S** (Cys, sulphide)
Ni^{2+}	**S** (Cys); **N** (polypyrroles)
Cu^{+}, Cu^{2+}	**N** (His, amines, ionized peptide bond); **S** (Cys)
Zn^{2+}	**N** (His, amines); **S** (Cys); **O** (carboxylate)

Likewise, metal-ligand interactions can influence metal ion absorption through mucosal events. Some ligands reside in the mucosal membrane to enhance metal ion uptake by the cell. Others occur in the cytoplasm or organelles to translocate the metal to the serosal membrane, or in the serosal membrane to enhance its expulsion from the cell towards mesenteric capillaries (17). Once in the cell, the metal ion may also be trapped and then later sloughed off into the intestinal lumen. For example, ferritin and metallothionein (MT) proteins are synthesized intracellularly in response to iron and zinc status, respectively. Ferritin serves as a storage protein to protect cells from oxidative damage from free ionic iron. Metallothionein can bind zinc and copper, high intestinal MT levels being associated with decreased zinc absorption (16).

Metal-ligand interactions involved in the absorption of all essential and beneficial metal ions have been reviewed recently (22).

<u>3.1.2.3. Ligand Selection for a Better Metal Absorption.</u> Whereas intracellular processes are difficult to influence exogenously, metal absorption is a priori easier to favour via the administration of specific metal-ligand mixtures in an empty stomach. For most trace metals, the objective is to neutralise the ionic charge of the metal ion by associating it with an anionic ligand, so as to induce passive diffusion of the resulting complex through the mucosal membrane (11). (For example, administration of histidine with zinc results in a higher zinc plasma level in humans (23). A similar effect has been noted for amino acids and iron, as well as for citrate and histidine in manganese intestinal absorption: all these ligands increase the initial rate of intestinal absorption for these elements (18).) General rules exist to help select appropriate ligands for a given metal ion: broadly the affinity between metal cations and specific ligand donor atoms follows the trends "non-polarisable, small cations coordinate to non-polarisable donor atoms", and "polarisable, large cations coordinate to polarisable, large donor atoms" (24) (see Table 1). Even quantitative estimations may be drawn from a stability ruler defined recently (see ref. 25—in particular Table 2 in that ref.). However, as complex formation is usually very pH dependent (25), ligand selection is not straightforward. Finally, it must be noted that electrically neutral complexes that are expected to promote metal absorption by passive transport may be so lipophilic that they precipitate in the g.i. fluid and instead reduce absorption.

3.2. Retention and Excretion Processes

Coadministration of a metal ion with a ligand capable of enhancing its g.i. absorption does not automatically result in a better bioavailability for this metal ion. Care must be taken that the ligand does not induce a parallel increase in its excretion. For example, zinc absorption is primarily a passive process (17) that is favourably influenced by the formation of neutral complexes with dietary amino acids (26), particularly histidine—see above (23). However, high oral doses of histidine can induce excessive zinc urinary excretion so that plasma zinc is actually decreased (23, 27, 28). High urinary zinc losses are also induced by cysteine (29, 30), but not by glycine (29). Copper urinary excretion is less sensitive than that of zinc, even though copper plasma level has been reported to fall rapidly and consistently following infusion of amino-acid based parenteral mixtures (31, 32).

In blood plasma, the major fraction of trace metals is firmly incorporated in metalloproteins (e.g. α_2-macroglobulin, ceruloplasmin). The rest, in the form of labile complexes with other types of proteins (e.g. albumin, transferrin) and to a much lesser extent (~1%) l.m.m. ligands, is at equilibrium with traces of hydrated ions. Any increase in the level of strong l.m.m. ligands shifts this equilibrium in favour of the ultrafiltrable pool, at the expense of labile protein complexes—and, in the longer run, metalloproteins. This shift induces important changes in metal metabolism: schematically, electrically charged species are excreted via the kidneys, whereas neutral species tend to diffuse into tissues for retention or biliary excretion—with possible intestinal reabsorption (enterohepatic circulation). Corresponding mechanisms have been reviewed recently (33).

4. METHODS FOR EVALUATING TRACE METAL BIOAVAILABILITY

Different methods are available to assess the extent to which a given element is retained by tissues to fulfil its assigned physiological or pharmacological roles. The information obtained is never absolute, however, as all techniques are based on distinct criteria.

4.1. *In Vivo* Investigation Techniques

4.1.1. Balance Studies. This method is used to determine the needs and bioavailability of elements in dietary supplies. Comparison is made of the contents of food in a certain element with those of feces and urine. Balances, however, largely depend on usual intakes of a given element, and biased results may derive from individual capacities of adaptation to different regimens. Also, other excretion routes (e.g. zinc in sweat) are neglected. This method is tedious and lacks precision (1).

4.1.2. Tolerance Tests. Short-term tolerance tests consist in pharmacokinetic studies following oral administration of an element. Classically, resorption is derived from the area under the curve of plasma levels vs time plots. This method is useful for comparative studies, but provides no absolute value of resorption yields. It also generally requires much higher doses than usually ingested. Tests may be done over longer periods (up to months) to monitor the contents of various biological media (plasma, urine, red cells, etc.) in an element being administered at repeated and regular doses. Such tests are of special interest when combined with functional monitoring (1).

4.1.3. Repletion of a Biological Function. In this method, bioavailability is assessed from the recovery of an element-dependent biological function after supplementation of subjects initially deficient in the element in question. The necessity of a prior deficiency state makes this method difficult to use in practice (1).

4.1.4. Isotopic Labeling Techniques. Isotopes allow to trace metabolic routes of elements in vivo. However, the validity of the measurements depends on there being exchange between the extrinsic tracer added to food and the native element (1, 34). The isotope may be either radioactive (β or γ emitter) or a stable enriched isotope. In spite of their high cost, stable isotopes have the essential advantage of avoiding risks for humans. (For details, see ref. 1).

4.2. *Ex Vivo* Methods

Cell cultures or fragments of organs can also be used for mechanistic studies on metal ion metabolism (1)—e.g., intestinal fragments for the study of g.i. absorption processes (17). Corresponding results are sometimes at variance with in vivo tests (1).

4.3. *In Vitro* Methods

Methods for measuring bioavailability in vitro have been widely sought for numerous reasons. First, in vitro studies are faster and much less expensive than in vivo investigations. Second, they are devoid of risks such as those associated with radioisotope use in humans (34). Above all, they can a priori provide information on the molecular processes that condition bioavailability. Such techniques, however, are still in their infancy. For example, attempts at finding correlations between metal bioavailability and formation constants of metal complexes with food components have been unsuccessful (18, 34, 35). A large number of other important reactions occur in the g.i. fluid which determine the overall degree of complexation (36). More generally, speciation of metal ions in the successive biofluids that they penetrate would be necessary to understand the processes through which they do so. However, experimental speciation of metal l.m.m. (ultrafiltrable) fractions is beyond the limits of analytical techniques. Computer-aided speciation is therefore the only method to obtain the required data.

4.3.1. Computer-Aided Speciation. To be applicable in vivo, speciation calculations must meet two criteria: relevance and reliability. This implies the combination of an appropriate simulation program with a correct model. The choice of a simulation program is a function of the nature of the biofluid investigated and of the type of information required. Building a simulation model requires (i) a realistic selection of the most important ligands in the biofluid in relation to the problem to be solved, as well as a correct estimation of their concentrations; (ii) a rigourous selection—or determination whenever necessary—of stability and solubility product constants for the complexes formed by the metal ion wth the above selected ligands. Most of the applications of this technique have been relative to the gastrointestinal fluid and blood plasma.

4.3.1.1. Gastrointestinal Applications. Usually, g.i. calculations are used to plot complex percentage profiles relative to a metal ion with a single ligand (i.e. both being considered to be taken in an empty stomach—see above) as a function of the pH within the 2–8 range. A classical example of application of this technique is the investigation of the influence of various ligands on iron absorption (11, 37), from which ascorbate was expected to be the most efficient; a result that has largely been confirmed experimentally. Tridimensional adaptations of these graphs may be used to analyse the influence of a third reactant on the interactions between metal and ligand (38).

4.3.1.2. Blood Plasma Applications. Blood plasma calculations are different in nature. First, the pH is fixed at 7.4. Second, most of plasma innumerable components occur at fairly constant concentrations which can a priori all be taken into account. The main difficulty in this case is to select the most critical reactants for the question under consideration. Another important question is how to account for metal-protein equilibria. Fortunately this issue has been solved in the ECCLES program (39) by using free reactant concentrations as input data (40). Another advantage of ECCLES is the great number of species that it can accommodate. This program has been used in most of the investigations relative to metal-ligand interactions in medicine (41). In particular, the simulation of the influence of amino acids in a nutritive mixture on the mobilisation of plasma zinc and copper has been of great significance for the interpretation of total-parenteral-nutrition induced deficiencies in these two metals (42).

5. CONCLUSION

Too frequently, the use of metalloelements in medicine is still lacking rigour. An effort of clarification was therefore necessary regarding the fundamental principles on which this use should be based. In particular, in view of the recommendations recently made against unscrupulous companies in this market (1), it must be emphasized that: (i) pharmacological uses of essential and beneficial metalloelements should be distinguished from nutritional supplementation that merely aims at satisfying dietary requirements, and (ii) not required elements should only be used at drug—or contrast agent—appropriate macrodoses, for well-defined applications.

Bioavailability is the key parameter in the use of metalloelements for therapeutic or diagnostic purposes. This parameter largely depends on the nature of the ligand with which the metal ion is administered. Prevailing metal-ligand interactions in the biofluids that the element penetrates do indeed condition the fate of this element in vivo.

For nutritional applications, a number of in vivo methods are available to assess bioavailability so that global intakes can be adapted to effective needs. Unfortunately, all of these methods are based on different criteria, which sometimes leads to conflicting results. The use of metalloelements at drug doses regardless of a deficiency state does not fall into this category and must systematically be submitted to classical risk/benefit analyses. For all applications, in vitro methods have a promising future in ligand selection provided related calculations fulfil the two imperative conditions of relevance and reliability. A still important shortcoming of these techniques is their present lack of consideration of kinetic factors. This problem is being currently addressed in specialised laboratories (43).

REFERENCES

1. A. Favier and J. Nève, in *Handbook of Metal-Ligand Interactions in Biological Fluids; Bioinorganic Medicine*, vol. 1, G. Berthon, ed., Marcel Dekker, New York, pp. 549–563 (1995).
2. B. O'Dell, *Nutr. Rev.* **42**, 301–308 (1984).

3. C. Exley and J.D. Birchall, *J. Theor. Biol.* **159**, 83–98 (1992).

4. E. Frieden, *Biochemistry of the Essential Ultratrace Elements*, Plenum Press, New York (1984).

5. E.J. Underwood, *Trace Elements in Human and Animal Nutrition*, Academic Press, New York (1977).

6. R.J.P. Williams and J.J.R. Fraústo da Silva, *The Biological Chemistry of the Elements*, Clarendon Press, Oxford (1991).

7. D.R. Williams, in *An Introduction to Bio-Inorganic Chemistry*, D.R. Williams, ed., C.C. Thomas, Springfield, IL, pp. 5–12 (1976).

8. R.B. Martin, in *Handbook of Metal-Ligand Interactions in Biological Fluids; Bioinorganic Chemistry*, vol. 2, G. Berthon, ed., Marcel Dekker, New York, pp. 827–833 (1995).

9. A.I. Venchikov, *Voprosy Pitaniya* **19**, 3–11 (1960).

10. G.L. Christie and D.R. Williams, in *Handbook of Metal-Ligand Interactions in Biological Fluids; Bioinorganic Medicine*, vol. 1, G. Berthon, ed., Marcel Dekker, New York, pp. 29–38 (1995).

11. A.M. Fiabane and D.R. Williams, *The Principles of Bioinorganic Chemistry*, The Chemical Society, London (1977).

12. L.M. Klevay, in *Handbook of Metal-Ligand Interactions in Biological Fluids; Bioinorganic Medicine*, vol. 1, G. Berthon, ed., Marcel Dekker, New York, pp. 287–291 (1995).

13. A.E. Harper, in *Trace Elements in Human Health and Disease*, vol. 2, A.S. Prasad and D. Oberleas, eds., Academic Press, New York, pp. 371–378 (1976).

14. K. Yokoi and H.H. Sandstead, in *Handbook of Metal-Ligand Interactions in Biological Fluids; Bioinorganic Medicine*, vol. 1, G. Berthon, ed., Marcel Dekker, New York, pp. 437–444 (1995).

15. N.W. Solomons and M. Ruz, in *Handbook of Metal-Ligand Interactions in Biological Fluids; Bioinorganic Medicine*, vol. 1, G. Berthon, ed., Marcel Dekker, New York, pp. 428–436 (1995).

16. C. Serfaty-Lacrosnière, I.H. Rosenberg and R. Wood, in *Handbook of Metal-Ligand Interactions in Biological Fluids; Bioinorganic Medicine*, vol. 1, G. Berthon, ed., Marcel Dekker, New York, pp. 322–330 (1995).

17. N.J. Birch, in *Handbook of Metal-Ligand Interactions in Biological Fluids; Bioinorganic Chemistry*, vol. 2, G. Berthon, ed., Marcel Dekker, New York, pp. 773–779 (1995).

18. R.A. Wapnir, in *Handbook of Metal-Ligand Interactions in Biological Fluids; Bioinorganic Medicine*, vol. 1, G. Berthon, ed., Marcel Dekker, New York, pp. 338–345 (1995).

19. B. Lönnerdal and B. Sandström, in *Handbook of Metal-Ligand Interactions in Biological Fluids; Bioinorganic Medicine*, vol. 1, G. Berthon, ed., Marcel Dekker, New York, pp. 331–337 (1995).

20. S.R. Lynch, in *Handbook of Metal-Ligand Interactions in Biological Fluids; Bioinorganic Medicine*, vol. 1, G. Berthon, ed., Marcel Dekker, New York, pp. 392–398 (1995).

21. R.A. Wapnir, in *Handbook of Metal-Ligand Interactions in Biological Fluids; Bioinorganic Medicine*, vol. 1, G. Berthon, ed., Marcel Dekker, New York, pp. 399–406 (1995).

22. G. Berthon, ed., *Handbook of Metal-Ligand Interactions in Biological Fluids; Bioinorganic Medicine*, vol. 1, Part Three, Chapter 2, Marcel Dekker, New York, pp. 322–444 (1995).

23. J. Schölmerich, A. Freudemann, E. Köttgen, H. Wietholtz, B. Steiert, E. Löhle, D. Häussinger and W. Gerok, *Amer. J. Clin. Nutr.* **45**, 1480–1486 (1987).

24. P. de Oliveira, H.A.O. Hill and L.-L. Wong, in *Handbook of Metal-Ligand Interactions in Biological Fluids; Bioinorganic Chemistry*, vol. 1, G. Berthon, ed., Marcel Dekker, New York, pp. 42–62 (1995).

25. R.B. Martin, in *Handbook of Metal-Ligand Interactions in Biological Fluids; Bioinorganic Chemistry*, vol. 1, G. Berthon, ed., Marcel Dekker, New York, pp. 33–41 (1995).

26. E. Giroux and N.J. Prakash, *J. Pharm. Sci.* **66**, 391–395 (1977).

27. R.I. Henkin, H.R. Keiser and D. Bronzert, *J. Clin. Invest.* **51**, 44a (1972).

28. R.M. Freeman and P.R. Taylor, *Amer. J. Clin. Nutr.* **30**, 523–527 (1977).

29. A.A. Yunice, R.W. King, Jr., S. Kraikitpanitch, C.C. Haygood and R.D. Lindeman, *Amer. J. Physiol.* **235**, 40–45 (1978).

30. S.H. Zlotkin, *J. Pediat.* **114**, 859–864 (1989).

31. N.W. Solomons, T.J. Layden, I.H. Rosenberg, K. Vo-Khactu and H.H. Sandstead, *Gastroenterology* **70**, 1022–1025 (1976).

32. A. Askari, C.L. Long and W.S. Blakemore, *J. Parent. Ent. Nutr.* **3**, 151–156 (1979).

33. Z. Gregus and C.D. Klaassen, in *Handbook of Metal-Ligand Interactions in Biological Fluids; Bioinorganic Medicine*, vol. 1, G. Berthon, ed., Marcel Dekker, New York, pp. 445–460 (1995).

34. P.E. Johnson, in *Handbook of Metal-Ligand Interactions in Biological Fluids; Bioinorganic Medicine*, vol. 1, G. Berthon, ed., Marcel Dekker, New York, pp. 346–350 (1995).

35. E. Giroux and N.J. Prakash, *J. Pharm. Sci.* **66**, 391–395 (1977).

36. G. Berthon, C. Matuchansky and P.M. May, *J. Inorg. Biochem.* **13**, 63–73 (1980).

37. J.N. Cape, D.H. Cook and D.R. Williams, *J. Chem. Soc. Dalton Trans.* 1849 (1974).

38. C. Blaquière and G. Berthon, *Inorg. Chim. Acta* **135**, 179 (1987).

39. P.M. May, P.W. Linder and D.R. Williams, *J. Chem. Soc. Dalton Trans.* 588 (1977).
40. P.M. May, P.W. Linder and D.R. Williams, *Experientia* **32**, 1492 (1976).
41. G. Berthon, ed., *Handbook of Metal-Ligand Interactions in Biological Fluids; Bioinorganic Chemistry*, vol. 2, Part Five, Chapter 3, Marcel Dekker, New York, pp. 1184–1298 (1995).
42. G. Berthon, *Inorg. Chim. Acta* **79**, 46–48 (1983). (See also ref. 41, pp. 1240–1251.)
43. P.M. May, in *Handbook of Metal-Ligand Interactions in Biological Fluids; Bioinorganic Chemistry*, vol. 2, G. Berthon, ed., Marcel Dekker, New York, pp. 1291–1297 (1995).

5

INFLUENCE OF NUTRITIONAL STATUS ON SELENIUM PHARMACOKINETICS

V. Ducros,[1] P. Faure,[1] M. Ferry,[2] F. Couzy,[3] I. Biajoux,[2] and A. Favier[1]

[1] Laboratoire de Biochimie C
Hôpital Michallon
BP 217; 38043 Grenoble cedex 9
France
[2] Service de Gériatrie
Centre Hospitalier
26953 Valence cedex 9
France
[3] Nestlé Research Centre
1000 Lausanne 26
Switzerland

1. INTRODUCTION

Several studies of selenium (Se) retention in rats showed that whole body retention of ^{75}Se tracer given as sodium selenite was considerably increased in Se-depleted rats (1–2). We were interested in studying the Se kinetics of a Se deficient human subject by using a safe stable isotope labeling. The deficiency was documented by plasma Se and plasma glutathione peroxidase (GSHPx) measurements. Results were compared with those obtained in subjects with normal Se status, of the same age, and sex, and living in the same area.

2. MATERIALS AND METHODS

2.1. Subjects

One young woman aged 29 y showing low Se status (plasma Se = 0.56 µmol/L, plasma GSHPx = 155 U/L) was enrolled and nine healthy young women aged 31–40 y served as controls (plasma Se = 1.07 ± 0.06 µmol/L, plasma GSHPx = 391 ± 21 U/L; values expressed as mean ± S.E.M.). All subjects were free of inflammatory disease, cancer, digestive malabsorption, liver or nephrotic disease. None has had a surgery during the last three months, or a treatment with minerals or vitamins.

Therapeutic Uses of Trace Elements, edited by Nève et al.
Plenum Press, New York, 1996

2.2. Se Status Determination

Total plasma selenium was measured by electrothermal atomic absorption spectrophotometry as previously described (3) and plasma GSHPx was assayed using the modified method of Günzler with tert-butylhydroperoxide (Sigma chimie, Saint-Quentin-Fallavier, France) as substrate.

2.3. $Na_2{}^{74}SeO_3$ Preparation

A ^{74}Se stable-isotope tracer solution was prepared from metal selenium (98.2 % ^{74}Se, Medgenix Diagnostics, Ratingen, Germany). Metal Se was transformed into sodium selenite as previously described (4). An accurate dose of 100 µg of ^{74}Se in 10 milliliters of saline was administered intravenously to each fasting subject. A catheter was placed for the first day to collect blood samples. Five milliliters of blood samples were drawn in trace element-free heparinized plastic tubes at 1, 2, 4, 5, 6 h post-injection, then at day +1, +2, +3, +8, +14, +30 and finally one time per month for the next six months. Plasma was separated after centrifugation at 1500 g for 10 min and stored at -20°C until analysis.

2.4. Sample Preparation and Analysis

A half ml of plasma was spiked with 0.5 ml of enriched ^{82}Se (100 ng/ml). Enriched ^{82}Se (atomic abundance 96.7%) in elemental form was purchased from the Oak Ridge Laboratory (Oak Ridge, TN) and transformed into selenite before to be used as internal standard. A microwave digestion (5) was applied to these samples. Selenium isotopic ratios were determined by gas chromatography-mass spectrometry (6) on a Nermag R 10–10C (Quad service, Argenteuil, France) equipped with a DN 200 gas chromatograph (Perkin Elmer, Norwalk, CT, USA).

2.5. Data Analysis

From measured isotopic ratios ^{74}Se/^{82}Se and ^{80}Se/^{82}Se, the quantities of natural selenium and enriched ^{74}Se were calculated in each sample. ^{74}Se isotope enrichments were expressed by the ^{74}Se/^{N}Se mass ratio, where ^{N}Se was the natural Se in the sample. This definition of enrichment is identical to the tracer-to-tracee molar ratio which is widely used to express the stable isotope enrichment, particularly in the application of tracer kinetic and compartmental modeling methodologies (7).

The semi-log plasma kinetic curves of the ^{74}Se/^{N}Se mass ratio over 4 months were plotted for the Se deficient woman and the control group (Figure 1). A mathematical approach was used to convert these data into a pharmacokinetic model (PHARM IV software from Roberto Gomeni). The experimental curves were found to fit to a double-exponential equation describing a two-compartment kinetic. The size of the two exchangeable pools, their half life and the exchange fluxes between these pools were calculated from these equations.

2.6. Statistics

Means, standard error of means (S.E.M.) of the control group were carried out using PCSM (Programme Conversationnel des statistiques pour les Sciences et le Marketing) software (Deltasoft, Meylan, France).

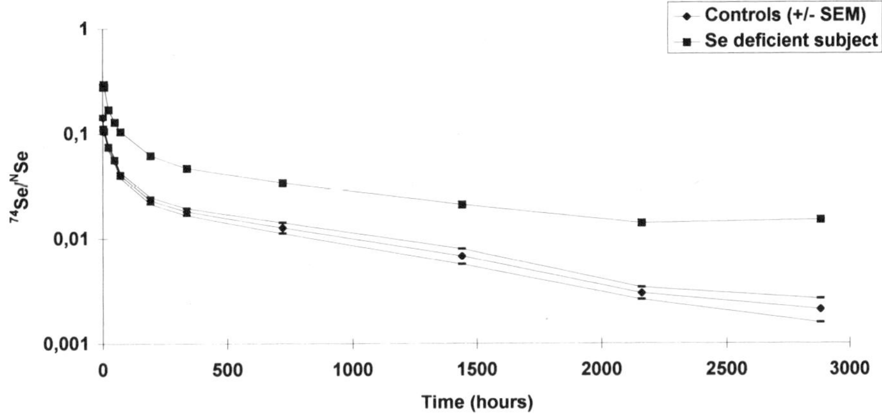

Figure 1. Plasma kinetic curve of the ^{74}Se enrichment over time.

3. RESULTS

In the Se deficient subject, the sizes of the two pools were lower than in controls. The first pool size was reduced by 43 % and the second one by 30 % when the pool size was expressed by kilogramm of body weight (Qa/kg and Qb/kg) (Table 1). The half life of the two pools (t1/2A and t1/2B) was increased, in particular t1/2B was found almost twofold greater than in the control group (Table 2). The transfer rate constants between the two pools showed a slower ^{74}Se redistribution from the tissular pool to the plasma one (Table 2). The elimination rate constant (Kel) was lower in this Se deficient subject. Other kinetic parameters such as the area under curve (AUC) and mean residence time (MRT) equally demonstrated a greater retention of ^{74}Se in the Se deficient woman than in normal subjects (AUC= 111.4 and MRT= 1589 h for the Se deficient subject; AUC= 31.9 ± 2.6 and MRT= 915 ± 54 h for the control group).

4. DISCUSSION

This study showed that Se metabolism in humans is clearly depending on the Se status. The reduction of the size of the first pool (Qa) in the Se deficient case was of the same extent (60 %) than the reduction of the total Se plasma content, so the Se plasma pool seems to be a good indicator of the first pool size. Experimental short-term Se restriction in humans also demonstrated a decrease of the pool size in Se restricted group although the pool definition

Table 1. Size of the Se exchangeable pools in a deficient subject and in controls. Results are expressed as mean ± SEM for the control group

Se pool	Se deficient subject	Se control subjects
Qa (µg)	334	835 ± 39
Qb (µg)	1205	2473 ± 233
Qa/kg (µg/kg)	8.5	14.9 ± 0.8
Qb/kg (µg/kg)	31	44.1 ± 4.4

Table 2. Kinetic parameters such as half life, elimination rate, rate-constants of transfer between the pools obtained from the model. Results are expressed as mean ± SEM for the control group

	Se deficient subject	Se control subjects
t1/2A (h)	31.3	27.6 ± 1.5
t1/2B (h)	1217	720 ± 39
Kel (h^{-1})	0.0027	0.0040 ± 0.0002
Kab (h^{-1})	0.0153	0.0161 ± 0.0011
Kba (h^{-1})	0.0047	0.0066 ± 0.0006

was different (9). Consequently the pool size has to be considered as a marker of Se status as well as the other Se status indicators. The decrease of the Se elimination rate constant, the increase of the area under the curve in the Se deficient case confirmed that homeostatic mechanisms take place to preserve Se when the Se supply is low. Another important result is that the retention of the [74]Se tracer was not the same between the two pools during Se deficiency. In particular the exchange rates between the two pools showed that the [74]Se tracer was mainly retained in the second pool. This confirmed that regulation mechanisms exist, which provide Se to certain tissues with priority in order to get biologically active Se from selenite.

5. REFERENCES

1. R.F. Burk, D.G. Brown, R.J. Seely, C.C. Scaief III, *J. Nutr.* **102**, 1049–1056 (1972).
2. D. Behne, T. Höfer-Bosse, *J. Nutr.* **114**, 1289–1296 (1984).
3. J. Arnaud, A. Prual, P. Preziosi, A. Favier, S. Hercberg, *J. Trace Elem. Electrolytes Health Dis.* **7**, 199–204 (1993).
4. V. Ducros, M.J. Richard, A. Favier, *J. Inorg. Biochem.* **55**, 157–163 (1994).
5. V. Ducros, D. Ruffieux, N. Belin, A. Favier, *Analyst* **119**, 1715–1717 (1994).
6. V. Ducros, A. Favier, *J. Chromatogr.* **583**, 35–44 (1992).
7. C. Cobelli, G. Toffolo, D.M. Bier, R. Nosadini, *Am. J. Physiol.* **253**, E551-E564 (1987).
8. V. Ducros, P. Faure, M. Ferry, F. Couzy, I. Biajoux, A. Favier, *Br. J. Nutr.* (submitted).
9. R.F. Martin, M. Janghorbani, V.R. Young, *Am. J. Clin. Nutr.* **49**, 854–861 (1989).

CONTRIBUTION OF THE AVERAGE MEAT CONSUMPTION IN SWITZERLAND TOWARDS FULFILLING THE REQUIREMENTS FOR IRON AND ZINC

M. Leonhardt and C. Wenk

Institute of Animal Sciences
Group Nutrition Biology
Swiss Federal Institute of Technology Zurich
8092 Zurich, Switzerland

1. INTRODUCTION

Iron deficiency anaemia is the most prevalent nutritional deficiency. In developing countries the prevalence is especially high. However, even in industrial countries it remains high in children below two years of age, adolescent girls, menstruating women and pregnant women. One reason is that the major part of the world's population consumes a diet with low quantities of meat. Their major sources of iron such as cereals and vegetables, contain a poor bioavailable iron (1,2). Meat, fish, poultry and offal are the only foods that contain the higher bioavailable heme iron besides the inorganic iron (nonheme iron) (1). Meat, especially red meat, is not only an important source of bioavailable iron but also zinc (3,4). Yokoi et al. (4) showed that frequent red meat intake by premenopausal women was associated with higher serum ferritin concentration and a "normal" plasma zinc disappearance. They suggested that avoidance of red meat increases the risk of iron and zinc deficiencies.

The objective of our study was to predict the contribution of the average meat consumption (beef, pork and chicken) in Switzerland towards fulfilling the requirements for absorbed iron and zinc.

2. MATERIALS AND METHODS

From July 1994 until March 1995 the following meat cuts were purchased in Zurich (Switzerland): pork (chop and shoulder), beef (fore-rib and shoulder) and chicken (breast and thigh). The sample size of each meat cut was 25 pieces. From beef fore-rib and pork chop, the musculus longissimus dorsi was separated and the other meat pieces were trimmed of visible fat and connective tissue. Iron (= total iron), heme iron and zinc concentrations were analyzed in all samples. For iron and zinc determination, the lyophilized samples were ashed and cooked subsequently with 25% hydrochloric acid. Iron and zinc were measured with an atomic absorption spectrometer (Perkin Elmer 5100 PC) in the Swiss Federal Research Station for Animal Production (RAP, Posieux, Switzerland). Heme iron was determined by the alkaline haematin method from Karlsson and Lundström (5).

Based on average meat consumption (9) the intake of bioavailable iron was estimated according to the "modified Monsen-model" (1,6). For calculating the intake of bioavailable zinc, the bioavailability data obtained from studies, carried out by Gallaher et al. (7) and Sandström and Cederblad (8) were considered.

3. RESULTS AND DISCUSSION

In Figure 1, the iron, heme iron and zinc content as analyzed is shown. Beef was the meat cut with the highest amount of iron and zinc. Iron and zinc content of beef shoulder and fore-rib were nearly similar, whereas the amount of heme iron differed. Pork muscles were very inhomogeneous in the examined trace element concentrations. While pork shoulder muscles contained relatively large quantities of iron, heme iron and zinc, the m. longissimus dorsi was poor in these trace elements. Chicken had the lowest iron, heme iron and zinc content. Furthermore, chicken thigh had higher amounts than chicken breast. Altogether, meat cuts that are rich in iron and heme iron, are also good sources of zinc.

Table 1 shows the average meat consumption in Switzerland in 1994 (9). Pork was the most consumed meat, followed by beef and chicken. For calculating the average trace ele-

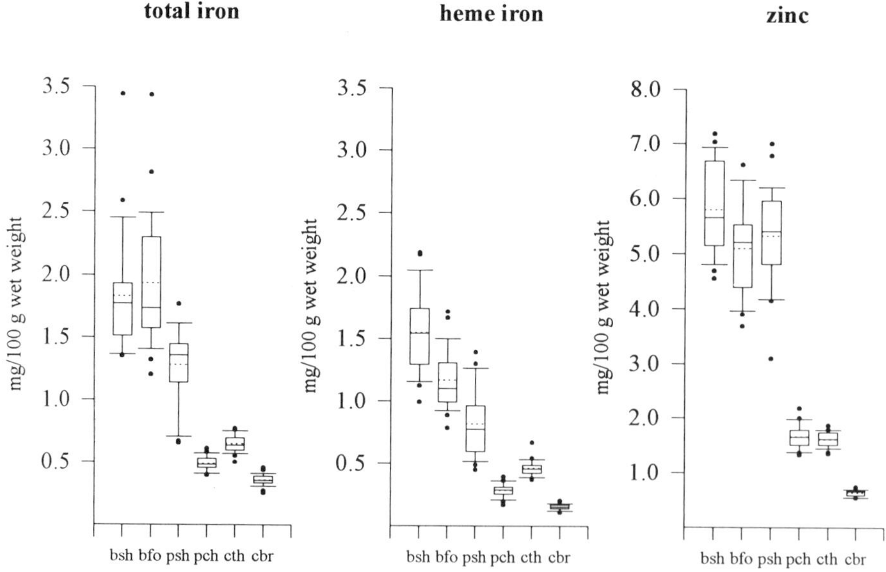

Figure 1. Total iron, heme iron and zinc content of different meat cuts: bsh = beef shoulder, bfo = beef fore-rib, psh = pork shoulder, pch = pork chop, cth = chicken thigh, cbr = chicken breast. The lower boundary of the box indicates the 25th percentile, a line within the box marks the median, and the upper boundary of the box indicates the 75th percentile. The whiskers above and below the box extend to the 90th and 10th percentiles. Points out of this range are graphed separately. The dotted line indicates the mean.

ment intake, the mean of the two median values for each animal species was used. With a daily meat consumption of 96 g, the iron intake was about 1.1 mg/d and the zinc intake was 3.7 mg/d. Recommendations (10) given by the German Nutrition Society (DGE) for daily iron intake were met to 11% (men) and 7% (women) and for zinc to 24% (men) and 30% (women).

The contribution of the average meat consumption towards fulfilling the iron and zinc requirements was much greater, if the bioavailability was considered. Taking into account a zinc bioavailability from meat of about 20–36% (7,8), the daily requirement for absorbed zinc (2.5 mg/d) was covered to 30–50% by meat consumption. This calculation does not consider that the zinc absorption is correlated negatively with the zinc content of a meal and that there is a great individual variation in zinc absorption. In addition dietary inhibitors might also reduce zinc absorption.

The intake of bioavailable iron provided by the average meat consumption, was calculated with the modified Monsen model (1,6). This model takes into account the physiologic iron stores, form of iron and presence of dietary absorption enhancers: the nonheme iron absorption varied among 2–20% of iron intake, depending on iron status and intake of enhancing factors. The heme iron absorption varied among 15–35%. It is a simple model to estimate dietary iron supply, but the effect of certain inhibiting substances and the variability in adaptation among individuals or populations are not considered (11). The daily requirement for bioavailable iron is about 1.0 mg (men) and 1.5 mg (women). The average meat consumption met the requirement for absorbed iron to 10–30% and 7–20% for men and women, respectively.

Meat (heme iron) is important in decreasing the prevalence of nutritional iron deficiency, but also may contribute to an undesirable increase in body stores in iron-replete individuals (11). Bezwoda et al. (12) have shown that heme iron absorption was about 20%, independent from the heme iron content of the meal. Cook (11) concluded that the adaptive response to variations in heme iron intake is minor than to variation in nonheme iron intake. Different studies indicated that there might be an association between high body iron stores and risk of myocardial infarction and cancer in men (13, 14).

There is no exact information towards sex related differences in meat consumption in Switzerland in 1994. Jacob-Sempach (15) investigated the nutrient intake of 213 women aged 25–35 years: 20% of these women did not eat meat at all or only once a month. This agrees with the results of a study from the Swiss Meat Board (GSF) (16), which showed that women are more sceptical about meat consumption than men. The results of these studies suggest that men consume more meat than women. From the view of trace element supply quite the opposite is desirable.

Considering the bioavailability of the examined trace elements, meat is an important iron and zinc source in Switzerland. People who are at risk of iron-deficiency anaemia and

Table 1. Iron, heme iron, nonheme iron and zinc intake as calculated from the average daily meat consumption in Switzerland in the year 1994 (9)

Meat cuts	Meat consumption g/d	Iron intake mg/d	Heme iron intake mg/d	Nonheme iron intake	Zinc intake mg/d
pork	50	0.5	0.3	0.2	1.8
beef	31	0.5	0.4	0.1	1.7
chicken	15	0.1	<0.1	<0.1	0.2
sum	96	1.1	0.7	0.3	3.7

perhaps also zinc deficiency (children below two years of age, adolescent girls, menstruating women and pregnant women), should prefer meat rich in iron (beef and pork shoulder). People who have high iron stores (mostly men) should perhaps avoid a very high consumption of these meat cuts that are high in iron and especially high in heme iron.

4. REFERENCES

1. C.E. Carpenter and A.W. Mahoney, *Crit. Rev. Fd Sci. Nutr.* **31**, 333–367 (1992)
2. A. Lerner and T.C. Iancu, *Front. Gastrointest. Res.* **14**, 117–134 (1988)
3. T. Hazell, *J. Sci. Food Agric.* **33**, 1049–1056 (1982)
4. K. Yokoi, N.W. Alcock and H.H. Sandstead, *J. Lab. Clin. Med.* **124**, 852–861 (1994)
5. A. Karlsson and K. Lundström, *Meat Sci.* **29**, 17–24 (1991)
6. E.R. Monsen, L. Hallberg, M. Layrisse, D.M. Hegsted, J.D. Cook, W. Mertz and C.A. Finch, *Am. J. Clin. Nutr.* **31**,134–141 (1978)
7. D.D. Gallaher, P.E. Johnson, J.R. Hunt, G.I. Lykken and M.J. Marchello, *Am. J. Clin. Nutr.* **48**, 350–354 (1988)
8. B. Sandström and A. Cederblad, *Am. J. Clin. Nutr.* **33**, 1778–1783 (1980)
9. C. Wenk, in *Wieviel Landwirtschaft braucht der Mensch*, C. Wenk, ed., Forum Davos, vdf Hochschulverlag AG, pp. 139–158 (1995)
10. German Nutrition Society (DGE), *Empfehlungen für die Nährstoffzufuhr*, Umschau-Verlag, Frankfurt (1995)
11. J.D. Cook, *Am. J. Clin. Nutr.* **51**, 301–308 (1990)
12. W.R. Bezwoda, T.H. Bothwell, R.W. Charlton, J.D. Torrance, A.P. Macphail, D.P. Derman and F. Mayet, *S. Afr. Med. J.* **64**, 552–556 (1983)
13. J.T. Salonen, K. Nyyssönen, H. Korpela, J. Tuomilehto, R. Seppänen and R. Salonen *Circulation* **86**, 803–811 (1992)
14. R.G. Stevens, D.Y. Jones, M.S. Micozzi and P.R. Taylor, *N. Engl. J. Med.* **319**, 1047- 1052 (1988)
15. S. Jacob-Sempach, *Dissertation*, Swiss Federal Institute of Technology No.11228 (1995)
16. Swiss Meat Board (GSF/IHA), *Image Fleisch*, No. 614.456 (1987)

7

IN VITRO AND *IN VIVO* BIOAVAILABILITY IN RAT OF FOUR DIFFERENT IRON SOURCES USED TO FORTIFY DRY INFANT CEREAL

F. Ros, M. J. Periago, G. Ros, and J. Rodrigo

Department of Food Science and Nutrition
Faculty of Veterinary Science
Murcia University
30071 Murcia, Spain

1. INTRODUCTION

Iron deficiency in infants is recognised as the most common nutritional deficiency of the world, in developing and industrialised countries (1, 2). The fortification of infant cereals has been used for many years to prevent iron deficiency in babies aged from six to twelve months, and many different iron sources have been tested, mainly based on their proportion of available iron and their technological characteristic during food processing (3). However, the bioavailability of iron varies considerably due to the enhancing and inhibitory effects of other food components. In addition, its bioavailability is determined by the kind of iron source and is strongly influenced by the physical characteristics of the iron compounds. Hydrogen-reduced iron has been used in the infant cereal industry since 1972, but its availability is considered poor and depends on particle size. For this reason, new compounds such as ferrous fumarate (4), $NaFe^{3+}EDTA$ (5, 6) and dry haemoglobin (7–9) with higher availability of iron are under consideration as new sources of iron in the infant cereal industry. The aims of the present study were to evaluate the *in vitro* and *in vivo* availability of iron in an infant cereal flour (commercially called "Eight Cereals") fortified with four different iron sources (hydrogen-reduced iron, ferrous fumarate, $NaFe^{3+}EDTA$ and dry haemoglobin).

2. MATERIAL AND METHODS

2.1. Material

The samples used in the present study were five formulations made of a combination of eight cereal flours, an infant follow-up formula, casein and different iron sources as shown in Table 1. The four iron sources selected were: hydrogen-reduced iron (with a particle size

Therapeutic Uses of Trace Elements, edited by Nève et al.
Plenum Press, New York, 1996

Table 1. Formulation and iron *in vitro* availability of the five diets tested

Diet	Infant cereal[1] g/Kg	Formula[2] g/Kg	Casein g/Kg	Iron source3 g/Kg	*In vitro* availability (%)
A	345.9	592.0	63.0	0.082	2.66 ± 0.16^c
B	349.2	590.0	60.5	0.26	2.93 ± 0.37^{bc}
C	349.9	589.9	59.6	0.62	19.2 ± 1.1^a
D	332.0	591.0	24.0	53.0	1.67 ± 0.39^c
E	347.0	591.0	51.0	11.0	4.35 ± 0.76^b

[1] Infant cereal: eight cereals.

[2] Follow-up formula for infant feeding from 4 months.

[3] Different diets according to the iron sources using as fortificant: A, hydrogen-reduced iron; B, ferrous fumarate; C, $NaFe^{3+}EDTA$; D and E, dry haemoglobin (Aprored®) with different concentrations.

[a-c] Different letters within the same column mean significant differences at *P<0.05*.

of 5–10 μm), ferrous fumarate, $NaFe^{3+}EDTA$ and dry porcine haemoglobin (Aprored®). Each of these sources was added to the infant cereal to provide 20 mg of iron per 100 g of flour, giving the diets A, B, C and D, respectively. Diet E was formulated using dry haemoglobin to give a final concentration of 5 mg of iron per 100 g of infant cereal.

2.2. Methods

In vitro iron availability was determined in the five diets using the technique described by Miller et al. (10). The diets were mixed with 200 ml of distilled water, digested with pepsin and then hydrolysed for a second time with bile salts and pancreatin. During the second digestion the samples were dialysed to ascertain the proportion of available mineral. Iron concentrations were determined using a Perkin Elmer flame atomic absorption spectophotometer. *In vivo* iron bioavailability was calculated following a balance experiment with weaning Sprague-Dawley rats. The rats were conditioned for 3 days and the experiment lasted 14 days using metabolic cages, which were placed in a room with controlled temperature (20 ± 2 °C) and moisture (75 ± 5 %). The diets and water were provided *ad libitum* to the animals, and the food intakes, the faeces and urine were taken every day. On the last day of the experiment, an anaesthetic mixture was admistered to the animals, which were then sacrificed. The blood of the animals was analysed for the following haematological parameters: total red cells (RGR), haemoglobin (HbB), haematocrit value (HCT), total serum iron and transferrin iron binding capacity (TIBC). Liver, spleen, heart, right kidney, the first 20 cm of the small intestine and two bones (right tibia and right femur) were excised to examine the iron content.

3. RESULT AND DISCUSSION

The *in vitro* availability of iron in the five tested diets is shown in Table 1. Diet C, formulated with $NaFe^{3+}EDTA$ gave the highest value of *in vitro* iron availability with 19.2 % of dialysed iron. $NaFe^{3+}EDTA$ has been considered as a good source of iron in diets with factors which inhibit iron absorption, such as cereals (6, 11). However, to provide a good proportion of iron in the diet, the amount of this compound needed would lead to overconsumption of EDTA since its concentration in infant cereals would be above the admissible daily intake allowed for the ethylenediamino tretracetic acid (5).

Table 2. Iron balance carried out during 14 days with weaning rats (Sprague-Dawley)

	Diet A[1]	Diet B	Diet C	Diet D	Diet E
Intake (mg)	25.4 ± 2.0^b	27.2 ± 1.1^b	24.1 ± 1.9^b	35.7 ± 1.6^a	15.2 ± 1.4^c
Faecal excretion (mg)	17.1 ± 0.7^c	21.2 ± 1.3^b	15.6 ± 1.8^c	25.2 ± 1.7^a	8.8 ± 0.7^d
Urinary excretion (µg)	70 ± 33^b	125 ± 31^b	711 ± 100^a	128 ± 21^b	60 ± 14^b
Apparent absorption (% intake)	32.5 ± 5.5^b	22.3 ± 2.9^c	35.3 ± 5.0^{ab}	29.5 ± 3.1^{bc}	41.9 ± 3.2^a
Retention (% intake)	32.2 ± 5.3^b	21.8 ± 3.0^c	32.3 ± 4.9^b	22.9 ± 3.1^c	41.5 ± 3.2^a
Urinary (% intake)	0.3 ± 0.1^b	0.5 ± 0.1^b	3.0 ± 0.5^a	0.4 ± 0.0^b	0.4 ± 0.1^b

[1] Different diets according to the iron sources using as fortificant: A hydrogen-reduced iron; B ferrous fumarate; C NaFe^{3+}EDTA; D and E dry haemoglobin (Aprored®) with different concentrations.

[a-d] Different letters within the same row mean significant differences at $P<0.05$.

Generally speaking, the availability of haeme iron is higher than that of inorganic iron, and ranges between 10 - 25 % (12). The *in vitro* availability of iron from dry haemoglobin did not produce the results to be expected from the haeme iron, perhaps due to the methodology which did not involve a specific membrane receptor to improve the availability of haeme iron, as described in *in vivo* studies. Diet E showed higher *in vitro* iron availablet han in diet D, both fortified with dry haemoglobin. However, in diet E the dry haemoglobin was added to give a final concentration of 5 mg of iron per 100 g of infant cereal, whereas in diet D the iron concentration was four times greater.

Table 2 shows the values of the mineral balance after 14 days. Diet E with the lowest concentration of iron gave the highest value of apparent absorption (41.9 %), whereas the ferrous fumarate gave the lowest (22.3 %). We agree with Buchowoski et al. (13), who observed that the availability of iron is not related to the content of iron in the diet. In the present study the highest value of apparent absorption was observed in the diet with the lowest iron content. Urinary excretion during the mineral balance did not differ between samples, with the exception of the rats fed with NaFe^{3+}EDTA, which gave an iron urinary excretion of 711 µg. This may be due to the binding effect of EDTA with the different mineral elements, particularly iron (14), and also because 5% of EDTA is excreted with the urine (15). Thus the iron that is bound to the EDTA and is absorbed can also be eliminated in the urine.

Table 3 shows the iron content in the different organs and bones of the rats. In general, the iron content varied according to the iron used to fortify the diet. Dry haemoglobin increased the content of iron in the liver and femur, whereas diets fortifed with the other three iron sources (hydrogen-reduced iron, ferrous fumarate and NaFe^{3+}EDTA) gave higher levels of iron in the spleen. On the other hand, the iron content in kidneys was higher in animals fed

Table 3. Fe content in different organs (µg/g of wet weight) and bones (µg/g of bone) of the rats

	Diet A[1]	Diet B	Diet C	Diet D	Diet E
Liver	125 ± 13^b	126 ± 13^b	94.7 ± 7.4^b	147 ± 54^{ab}	195 ± 33^a
Heart	54.3 ± 6.2^a	69.4 ± 6.0^a	62.7 ± 3.7^a	58.2 ± 6.0^a	57.6 ± 4.6^a
Kidney	75.3 ± 11.6^b	91.2 ± 8.5^b	117.3 ± 1.4^a	87.8 ± 7.9^b	77.9 ± 4.9^b
Spleen	168 ± 15^a	172 ± 26^a	169 ± 25^a	143 ± 22^{ab}	129 ± 22^b
Small intestine	70 ± 14^a	63 ± 27^a	61 ± 17^a	57 ± 8.4^a	86 ± 18^a
Tibia	104 ± 20^a	84 ± 16^a	72 ± 16^a	91 ± 6.6^a	87.0 ± 9.6^a
Femur	75.0 ± 17^b	115 ± 12^a	86 ± 12^b	117 ± 17^a	115 ± 12^a

[1] Different diets according to the Fe sources using as fortificant: A hydrogen-reduced iron; B ferrous fumarate; C NaFe^{3+}EDTA; D and E dry haemoglobin (Aprored®) with different concentrations.

[a-b] Different letters within the same row mean significant differences at $P<0.05$.

Table 4. Haematological parameters in rats

	Diet A[1]	Diet B	Diet C	Diet D	Diet E
HbB (g/dl)	14.6 ± 0.7^a	14.1 ± 1.3^a	14.6 ± 0.8^a	13.5 ± 0.8^a	14.0 ± 0.6^a
RGR ($\times 10^6$/mm^3)	6.5 ± 0.1^a	6.5 ± 0.5^a	6.5 ± 0.4^a	7.2 ± 1.1^a	7.2 ± 1.1^a
HCT (%)	44.0 ± 1.9^a	43.0 ± 3.7^a	42.3 ± 2.8^a	48.5 ± 7.2^a	48.5 ± 7.2^a
TIBC (mg/dl)	593 ± 128^{ab}	454 ± 55^{ab}	429 ± 39^b	557 ± 48^{ab}	614 ± 153^a
Fe (mg/dl)	47 ± 14^a	76 ± 32^a	58 ± 13^a	69 ± 22^a	68 ± 11^a

[1] Different diets according to the Fe sources using as fortificant: A hydrogen-reduced iron; B ferrous fumarate; C NaFe^{3+}EDTA; D and E dry haemoglobin (Aprored$^{®}$) with different concentrations.

[a-b] Different letters within the same row mean significant differences at $P<0.05$.

with diet C (NaFe^{3+}EDTA), which could be due to the high percentage of iron in the urine of this group of rats. As regards the iron content of the heart, small intestine and tibia, no significant differences were found between the five diets tested. However, the content of iron in the small intestine could be considered high since the mucosal intestinal cells are involved in iron absorption.

Table 4 shows the haematological parameters of the rats determined at the end of the *in vivo* bioavailability study. The haematological parameters did not show significant differences between groups. All the animals used in the experiment showed normal values of haemoglobin (HbB), red cells (RGR), haematocrit value (HCT), transferrin iron binding capacity (TIBC) and serum iron. For this reason, the haematological values indicate the good nutritional status of the animals after the *in vivo* experiment.

4. CONCLUSIONS

We may conclude that dry haemoglobin, used in a concentration of 5 mg/100 g of dry cereal, can be considered as good iron source in the infant cereal industry. However, more studies should be carried out including technological tests and *in vivo* studies with human, to assess the organoleptic and technological properties of the infant cereal and any toxicological risk for babies. Those animal fed with NaFe^{3+}EDTA showed also good iron absorption although the EDTA exceeded the recommended limits, so it cannot be considered for iron fortification.

5. ACKNOWLEDGMENT

Hero Spain S.A. supported this study and helped to prepare the infant cereal. Nestle$^{®}$ Spain provided the samples of the follow-up formula and Aprocat$^{®}$ the dry haemoglobin compound.

6. REFERENCES

1. WHO, in *Health promotion Glossary A*, Discussion Document 79.995, Copenhaguen, (1985).
2. P. Musgrove, in *9th World Congress of Food Science and Technology*, Budapest, (1995)
3. R.F. Hurrell, in *Iron nutrition in infant and childhood*, A. Sytekel, ed., Raven Press, New York, pp. 147–178 (1984)
4. R.F. Hurrell, E.F. Diane, J. Burri, P. Whittaker, S.R. Lynch and J.D. Cook, *Am. J. Clin. Nutr.* **49**, 1274–1282 (1989).

5. R.F. Hurrell, in *Nutritional Anemias*, S. Fomon and S. Zlotkin, eds., Raven Press, New York, pp. 193–208 (1992).
6. R.F. Hurrell, S. Ribas and L. Davidsson, *Br. J. Nutr.* **71**, 85–93 (1994).
7. J.A. Asenjo, M. Amar, N. Cartagena, J. King, E. Hiche and A. Stekel, *J. Food Sci.* **50**, 795–799 (1985).
8. E. Hertrampf, M. Olivares, F. Pizarro, T. Cayazzo, T. Walter and G. Heresi, in *Recent knowledge on iron and folate deficiencies in the world*, S. Hercberg, P. Galan and H. Dupin, eds, Colloque Inserm, París, pp. 647–652 (1990).
9. T. Walter, E. Hertrampf, F. Pizarro, M. Olivares, S. Laguno, A. Letelier, V. Vega and A. Stekel, *Am. J. Clin. Nutr.* **57**, 190–194 (1993).
10. D.D. Miller, B.R. Schricker, R.R. Rasmussen, D. Van Campen, *Am. J. Clin. Nutr.* **34**, 2248–2256. 1981
11. C. Martínez-Torres, E.L. Romano, M. Renzi and M. Larysse, *Am. J. Clin. Nutr.* **32**, 809–816 (1979).
12. S.J. Fairweather-Tait, *Food Chem.* **43**, 213–217 (1992).
13. M.S. Buchowoski, A.W. Mahoney and M.P.V. Kalpalathika, *Nutr. Res.* **9**, 773–783 (1989).
14. P. McPhail, R.W. Charlton, T.H. Bothwell and W.R. Bezwoda, in *Iron fortification of foods*, F.M. Clydesdale and K.L. Wiemer, Academic Press Inc., Orlando, pp. 55–71 (1985).
15. E. Candela, M.V. Camacho, C. Martínez-Torres, J. Perdomo, G. Mazzari, G. Acurero and M. Layrisse, *J. Nutr.* **114**, 2204–2211 (1984).

8

IMPROVEMENT OF Ca AND P CONTENTS AND *IN VITRO* AVAILABILITY IN SOLE FISH-BASED INFANT BEIKOSTS

I. Martínez, M. Santaella, G. Ros, and M. J. Periago

Department of Food Science Nutrition
Faculty of Veterinary Science
University of Murcia
30071 Murcia, Spain

1.INTRODUCTION

Infancy is the period of the life in which nutritional demands are the highest, since weight triples and length doubles during the first year of life (1). Due to the rapid growth rate, high amounts of Ca and P are needed to maintain optimal bone development (2). The mineral matrix of fish bone has higher concentrations of both elements than the edible part and can be used to improve the content of these mineral in some infants foods. The aim of the present study was to ascertain the addition of fish bone in the Ca and P contents and in their *in vitro* availability in three fish-based homogenised infant beikosts. The main difference among these samples was the inclusion or not of fish bones. The beikosts investigated were: sole beikost without bone (SB), sole beikost with bone (SBB) and sole and hake beikost with bone (SHB). We also evaluated the content and *in vitro* availability of other minerals (Fe, Zn, and Mg) to have a better knowledge of the nutritional value of these beikosts

2. MATERIALS AND METHODS

2.1. Samples and Samples Preparation

Figure 1 shows the flow diagram of the different steps in the manufacture process of the sole fish-based infant beikost. Two fish species, sole (*Solea vulgaris vulgaris*) and hake (*Merluccius merluccius*), were used to prepare three samples of minced fish, which were mixed with water and cooked to give the three fish purées (sole purée without bone, SP; sole purée with bone, SBP; and sole and hake purée with bone, SHP). The last purée was elaborated with a mixture (1:1, w:w) of sole muscle with bone and hake muscle without bone. Fish purées were homogenised with a blender and included in the beikosts formulation in a 24%

Therapeutic Uses of Trace Elements, edited by Nève et al.
Plenum Press, New York, 1996

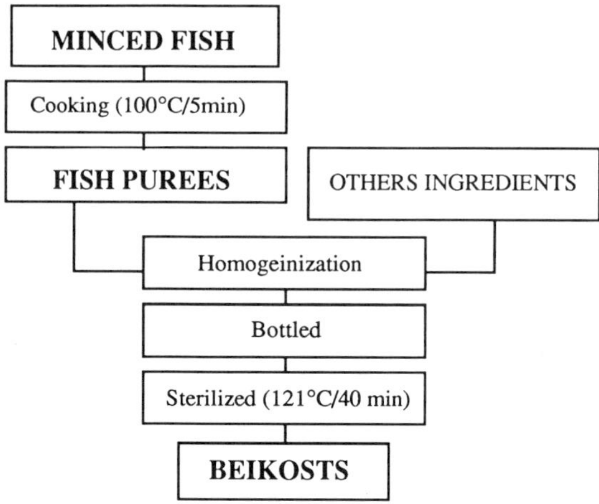

Figure 1. Flow diagram used in the beikosts manufacture process.

wet weight basis. The other ingredients were included in the proportion showed in Table 1, obtaining the following sole fish-based beikosts: sole without bone (SB), sole with bone (SBB) and sole and hake with bone (SHB). Beikosts were experimentally manufactured by Hero España, S.A. (Alcantarilla, Spain).

2.2. Methods

The three fish purées and the three beikosts were analysed for their mineral content and their *in vitro* availability. Fe, Zn, Mg and Ca contents were determined using flame atomic absorption spectrophotometry (3), whereas P content was analysed by Chapman and Pratt's method (4). Fe, Zn, Mg, Ca and P *in vitro* availability was determined following the method described by Miller et al.(5).

2.3. Statistical Analysis

Statistical analysis was performed using Systat software, version 5.0 (6). Tukey's test with a significance level of 5% was used to compare individual pairs of means.

Table 1. Raw materials used for beikosts manufacture and percentage of them at the finished product

Raw materials	Percentage in Beikosts
Fish purée[1]	24
Potato *(Solanum tuberosum)*	8.15
Butter	2.86
Powdered milk	1
Tomato concentrated	1
Onion *(Allium cepa)*	0.53
Celery *(Apium graveolens)*	0.53
Salt	0.25
Water	61.68

[1]The fish pureé was different for each type of beikosts

3. RESULTS AND DISCUSSION

Table 2 shows Fe, Zn, Mg, Ca and P contents, dialyzed (percentage and amount) and Ca/P ratio in fish purées. Ca and P content increased markedly after bone addition. Ca content was 67.7 mg/100 g, when only sole without bone (SP) was used to formulate fish purée, increasing significantly when sole bone was included into formulation (231 mg/100 g in SHP and 330 mg/100 g in SBP). The percentage of Ca and P dialyzed was low in all the samples assayed, but the final dialyzed amounts were higher in the purées with bone added. According to the National Research Council of the American Association of Sciences (7) the Ca/P molar ratio in a food should be 1. In the purées studied the Ca/P molar ratios were in all cases lower than 0.5, going the best ratios the fish purées with bone (0.43 to 0.47 in SBP and 0.34 to 0.38 in SHP) than the purée without it (0.15 to 0.19 in SP). The improvement of Ca/P ratio in the purées can be attributed to the bone addition because of the Ca and P content of the bone matrix. Related to Fe, Zn or Mg no important changes were observed after bone addition.

Fe, Zn, Mg, Ca and P contents, percentage, and dialyzed amount in fish-based infant beikosts are shown in Table 3. Fe and Zn contents were low in the three beikosts assayed, because fish is a poor source of both elements (8) and there is no ingredient in the beikosts (Table 1) that provides a high amount of both elements. Due to the low Fe and Zn contents (0.26

Table 2. Fe, Zn, Mg, Ca and P contents (X ± SD) in fish purées and their *in vitro* availability (percentage and dialyzed amount), and Ca/P molar ratio

Minerals	Without bone SP[1]	With bone SBP	SHP
Fe			
(mg/100g)	0.37 ± 0.01^{b}	0.18 ± 0.05^{c}	0.72 ± 0.13^{a}
% of DM[2]	4.30 ± 0.35^{a}	1.93 ± 0.75^{b}	0.68 ± 0.19^{c}
mg DM/100g	t	t	t
Zn			
mg/100g	0.36 ± 0.00^{a}	0.48 ± 0.12^{a}	0.44 ± 0.02^{a}
% of DM	4.96 ± 0.54^{a}	0.47 ± 0.10^{c}	1.52 ± 0.11^{b}
mg DM/100g	t	t	t
Mg			
mg/100g	21.8 ± 2.2^{b}	32.5 ± 1.3^{a}	32.9 ± 1.1^{a}
% of DM	22.5 ± 3.9^{a}	18.13 ± 0.45^{b}	16.6 ± 0.15^{b}
mg Dm/100g	4.91 ± 0.62^{a}	5.88 ± 0.21^{a}	5.46 ± 0.14^{a}
Ca			
mg/100g	67.7 ± 9.8^{c}	330 ± 48^{a}	231 ± 18^{b}
% of DM	8.3 ± 2.2^{a}	4.54 ± 0.21^{b}	5.08 ± 0.56^{b}
mg DM/100g	5.6 ± 1.0^{b}	14.9 ± 1.5^{a}	11.7 ± 2.2^{a}
P			
mg/100g	401.3 ± 5.9^{c}	673.47 ± 0.66^{a}	629 ± 18^{b}
% of DM	3.31 ± 0.64^{b}	6.21 ± 0.07^{a}	2.94 ± 0.07^{b}
mg DM/100g	13.3 ± 2.4^{c}	41.83 ± 0.49^{a}	18.53 ± 0.71^{b}
Ca/P ratio range	0.15–0.19	0.43–0.47	0.34-0.38

[1] Purées abbreviation: SP= sole purée without bone; SBP= sole purée with bone; SHP= sole and hake purée with bone.
[a-e] Different characters in the same row are significantly different for ($p < 0.05$).
[2] DM: Dialyzed mineral.

Table 3. Fe, Zn, Mg, Ca and P contents (X ± SD) in fish-based beikosts, and their in vitro availability (percentage and dialyzed amount), and Ca/P molar ratio

Minerals	Without bone SB[1]	With bone	
		SBB	SHB
Fe			
mg/100g	0.34 ± 0.01^a	0.39 ± 0.08^a	0.26 ± 0.01^b
% of DM[2]	10.05 ± 0.07^a	9.45 ± 0.02^a	9.85 ± 0.98^a
mg DM/100g	0.03 ± 0.00^a	0.03 ± 0.00^a	0.02 ± 0.00^b
Zn			
mg/100g	0.28 ± 0.01^a	0.26 ± 0.00^a	0.22 ± 0.01^b
% of DM	19.22 ± 0.16^a	18.56 ± 0.19^a	22.01 ± 0.59^a
mg DM/100g	0.05 ± 0.01^a	0.04 ± 0.01^a	0.04 ± 0.01^a
Mg			
mg/100g	15.66 ± 0.57^a	16.22 ± 0.49^a	15.78 ± 0.02^a
% of DM	33.61 ± 0.25^a	20.01 ± 0.30^c	24.2 ± 1.4^b
mg DM/100g	5.26 ± 0.15^a	3.24 ± 0.29^b	3.83 ± 0.29^b
Ca			
mg/100g	39.0 ± 4.3^b	82.7 ± 4.2^a	78.9 ± 1.1^a
% of DM	16.3 ± 1.3^b	20.25 ± 1.9^a	21.4 ± 1.9^a
mg DM/100g	$6.36 \pm 0.34b$	16.7 ± 2.9^a	16.9 ± 1.5^a
P			
mg/100g	223 ± 29^{ab}	243.9 ± 5.4^a	223 ± 10^b
% of DM	2.15 ± 0.06^a	2.19 ± 0.06^a	2.41 ± 0.04^a
mg DM/100g	4.81 ± 0.54^a	5.36 ± 0.07^a	5.40 ± 0.21^a
Ca/P ratio range	0.13-0.21	0.30-0.35	0.33-0.37

[1] Beikosts abreviation: SB= sole beikost without bone; SBB= sole beikost with bone: SHB= sole and hake beikost with bone. [a-e] Diferents characters in the same row are significantly different (p < 0.05)
[2] DM: Dialyzed mineral.

to 0.39 mg/100 g for Fe, and 0.22 to 0.28 mg/100 g for Zn) and the low dialyzed percentage (9 - 10 % for Fe and 18 - 20 % for Zn), very low dialyzed amounts of these trace elements were estimated in these beikosts. The percentage of Mg dialyzed decreased after bone addition, perhaps due to the high levels of Ca and P, that could affect to the Mg availability (10). This behaviour is mainly observed in SBB beikost, which is the one with the highest levels of Ca and P, and the lowest dialyzed amount of Mg. Ca content was significantly higher (p < 0.05) in the beikosts with sole bone (78.9 mg/100 g in SHB and 82.7 mg/100 g in SBB) than in the beikost without bone. The percentages of dialyzed Ca were higher in all the beikosts than in their purées (Table 2). This increase might be due to the powdered milk , used for beikosts formulation (Table 1), which is rich in Ca and lactose. Lactose is known to increase Ca availability (10). Ca dialyzed amounts were higher in the beikosts SHB and SBB than in SB, as result of the higher content of this mineral in those beikosts. P levels were fairly constant among beikosts, increasing after bone inclusion. P dialyzed percentages and P dialyzed amounts were also slightly higher in the samples with bone than in those without it. However, no significant differences (p < 0.05) were observed among all the samples assayed. Ca/P ratio were low in all the beikosts, being higher in the beikosts SHB and SBB, both with bone. This is important because of the role of a good ratio in optimal bone mineralization (9).

The estimation of the amount of Mg, Ca and P dialyzed from fish purées compared with the dialyzed amount of these minerals from the rest of ingredients of the beikosts are represented in Figure 2 (a, b and c, respectively). Mg dialyzed amounts (Figure 2a) from fish

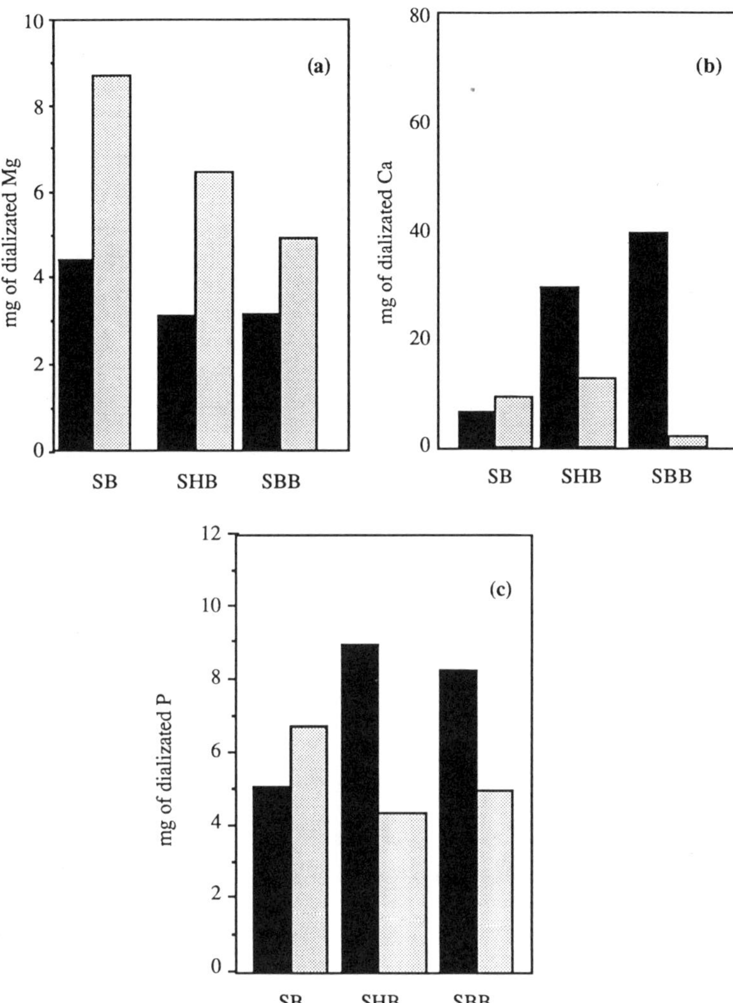

Figure 2. Representation of the dialyzed amount of Mg (a), Ca (b) and P (c) from fish pureé (black square) and the dialyzed amount from the other ingredients (gray square) in the three beikosts: sole without bone beikost (SB), sole and hake beikost with bone (SHB) and sole with bone beikost (SBB).

purées were lower than the amounts from the other ingredients. Moreover, this parameter decreased after bone addition. Bone addition in fish purées led to an increase in the content of dialyzed Ca (Figure 2b), which came mainly from fish purées in the beikosts rather than from the other ingredients. The same behaviour was observed for P (Figure 2c). Figure 3 shows the Fe, Zn, Mg, Ca and P provided by the consumption of a 250 g jar of each beikost in relation to the recommended dietary allowances (RDA) (8). The beikosts investigated showed very small contribution to the Fe and Zn requirements. Bone addition not only increased Ca and P contents but also improved the daily Ca and P contributions.

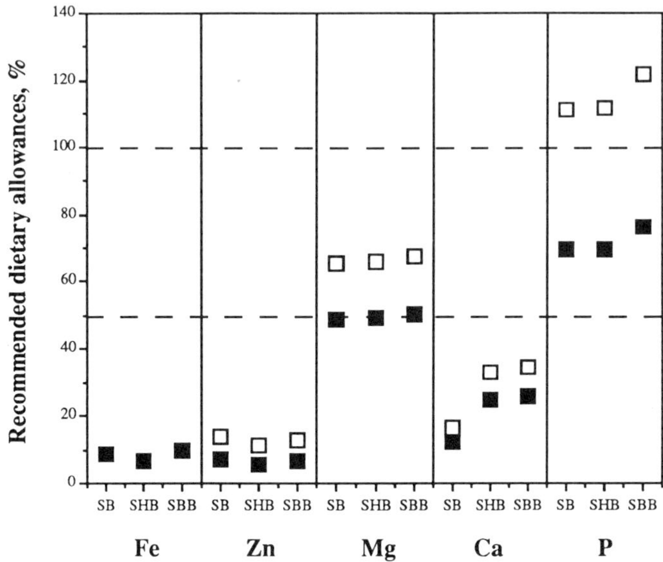

Figure 3. Percentage of recommended dietary allowance (RDA) according to Fe, Zn, Mg, Ca and P content of a 250 g jar bottle of each beikost (SB, sole without bone; SHB, sole and hake; SBB, sole with bone), for two groups of children (white square, 6–12 months; black square, 1–3 years).

4. CONCLUSION

Therefore, we inferred that bone inclusion caused not only an increase of the Ca and P content in beikosts, but an increase of *in vitro* bioavailability as well. In addition, it gave a better molar ratio Ca /P. The other elements analysed (Fe, Zn and Mg), were not affected by bone addition. Beikosts with fish bone could improve the daily Ca and P intakes, according to RDA for infants aged between 6 months to 3 years.

5. REFERENCES

1. J.A. Milner, *J.Pediatr.* **117**, 1447–155 (1990).
2. M.L.A. Cruz and R.C. Tsang, in: *Calcium Nutriture for mothers and children*. R.C. Tsang and F. Mimorini, eds., Raven Press Lid, New York, pp 1–11 (1992).
3. Anonymous, *Operator's mannual*, Bodenseewerk Perkim-Elmer & Co. GmbH, Uberlingen, Germany (1978).
4. H.H. Chapman and D.C. Pratt, Division of Agricultural Science, University of California, Davis C. A. ed., (1961).
5. D.D. Miller , B.R. Schricker, R.R. Rasmussen and D. Van Capen *Am. J. Clin. Nutr.* **34**, 2248–2256 (1981).
6. H. Wilkinson and P. Howe. Systat for Windows 5.0. Evanston, I. L. (1992)
7. National Research Council. *Recommended Dietary Allowances*. 1ª Edición española de la 10ª edición original. National Research Council. Ediciones Consulta, S.A. Barcelona. Spain (1991).
8. M.P. Navarro, *Rev. Agroquim. Tecnol. Aliment* . **31/3**, 330–342 (1991).
9. L.L. Hardwick, M.R. Jones N. Brautbar and D.B.N. Lee, *J. Nutr.* **121**, 13–23 (1990).
10. L.Gueguen, *Cahiers de Nutrition et de Dietetique* **25(4)**, 233–236 (1990).

SELENIUM AVAILABILITY AND PROTEIN DIGESTIBILITY IN HOMOGENISED INFANT FOODS

Josefina Ortuño, Gaspar Ros, María Jesús Periago, Carmen Martínez, and Ginés López

Departamento Bromatología e Inspección de Alimentos
University of Murcia
30071 Murcia, Spain

1. INTRODUCTION

The understanding of the role of selenium (Se) in health has grown during the latest years, and as any essential trace element has multiple metabolic functions determined mainly on dosage. Thus, low intakes can result in clinical signs of deficiency, whereas excessive intakes may produce signs of toxicity (1). It has become also obvious that the greatest health risks of Se deprivation are mainly found among infants and children (2). Although breast milk or infant formula continue to provide the main source of energy and nutrient intakes of infants (3), at the age of 6 months, usually, begin to eat semi-solid foods as a complementary alimentation called beikosts (4). Because of the variability of ingredients in this kind of food the information on Se content is not consistent or frequently studied, in contrast to accurate data on the Se content of adult foods. Only few studies have been published on Se content in beikosts foods and on the effects of this food source on bioavailability (5–7). Therefore, some of these types could be important as a source of biologically available Se (2). Since an alternative to the *in vivo* measurement of the availability of trace elements is the use of *in vitro* techniques (8), we conducted the present *in vitro* study (9) to estimate the Se availability in infant foods and their relationship with protein digestibility.

2. MATERIALS AND METHODS

2.1. Samples

The infant foods selected for this study were of three types: **vegetables** (mixed vegetables, MV), **meats** (chicken, veal and vegetables, CVV) and **fish** (hake with rice, HR). Sam-

Therapeutic Uses of Trace Elements, edited by Nève et al.
Plenum Press, New York, 1996

ples were manufactured from January to July 1995 by Hero España, S.A. (Alcantarilla, Murcia, Spain). For each sampling, three samples were assayed.

2.2. Methods

Se Content was determined by hydride generation atomic absorption spectrometry (HG-AAS) after dry ashing of 25–35g the sample in a muffle furnace. First 10mL of selenium mineralization coadjuvant and volatilization inhibitor ($Mg(NO_3)_2$+MgO, 10+1/100, w/v) were added to the samples and dried overnight in a forced-draft oven at 100 °C. Then, ashing was carried out in a muffle furnace (Nabertherm, model L3/S; Bermen, Germany) with the temperature/time program: 140 °C, 1/2 h; 150 °C, 1 h; 180 °C, 1/2 h; 200°C, 2 h; 225 °C, 1/2 h; 250 °C, 2 h; 300 °C, 3 h; 350 °C, 1 h; 400 °C, 1/2 h and 450 °C, 20 h. Analysis of wholemeal flour (CRM 189) = 134 ± 7 ng/g (certified value = 132 ± 10 ng/g) showed that the method is reliable.

Simulated Gastrointestinal Digestion Procedure. The portion of available Se was estimated following the method of Miller *et al.* (9). One hundred fifty grams of homogenised infant foods were adjusted to pH 2.0 with 6 M HCl. Then, 5 mL of pepsin suspension were added to the baby food, which was then incubated in a shaking waterbath at 37 °C for 2 h. After the pepsin digestion, triplicate 40 g samples of the digest were transferred to 200 mL plastic bottles. Dialysis tubing (Visking 9–36/32", Medicell, London, U.K.) containing 50 mL water and an amount of $NaHCO_3$ equivalent to the measured titratable acidity were placed in the bottles. The bottles were consequently incubated in a shaking water-bath at 37 °C for 30 min. 5 mL pancreatine-bile extract mixture were added to each bottle and then were incubated for three hours at 37 °C. After the incubation the contents of the dialysis tubes (dialysates) were weighted and analysed for the Se content by hydride generation atomic absorption spectroscopy. The portion of available Se, estimated as the percentage of dialysable Se, was calculated as follows:

$$\text{Se available (\%)} = \frac{\text{Se in dialysate}}{\text{Se in infant food}} \times 100$$

For evaluating this *in vitro* method, the total dialysable Se (TDSe) and protein digestibility (PD) during the *in vitro* digestion were estimated as described by Shen *et al.* (10). Digestion fractions were subjected to dialysis for 7 days at 4 °C against Mili Q water. The water was constantly mixed and replace twice a day. After dialysis the samples were freeze-dried and weight. The *protein contents* of the respective freeze-dried samples were calculated by converting the nitrogen content determined by Kjeldahl digestion (Method 955.04)(11) and multiplying it by a factor of 6.25. These values were considered as non-digested protein. Se remaining in these samples after dialysis was considered as non-dialysable Se. PD was calculated using the following equation:

$$\text{PD(\%)} = \frac{\text{protein in infant food} - \text{protein in digested fraction}}{\text{protein in infant food}} \times 100$$

The percentage of TDSe in digested fractions was calculated as follows:

$$TDSe(\%) = \frac{\text{total infant food Se} - \text{non dialysable Se}}{\text{total infant food Se}} \times 100$$

3. RESULTS AND DISCUSSION

Table 1 presents the mean Se content and the Se availability of the beikosts assayed. The Se contents of infant foods showed be ingredient dependent and tend to show the same general pattern as those described in adult foods, where fish, meat and cereal products contain higher levels than fruit and vegetables (12). The lowest Se content was observed in MV (20–83 ng/g d.w.) and the highest in HR (> 250 ng/g d.w.). In addition MV samples gave the widest variability which might be related to the biologically available Se in the soil (13), and also to the harvest period. However, MV showed higher Se availability (> 100 %) than the other two beikosts (< 40 %), in agreement with previous data from adult foods (12, 14).

The second part of the study concentrated on the methology of the Miller's *in vitro* method for mineral availability, focused mainly on the digestion process and the dialysis steps.

Among the beikosts studied, the TDSe of MV is not presented at the gastric and intestinal digeston, because of the wide variability and no consistent data. In fact, the variability depends mainly on the Se concentration. Since no clear data was obtained, the experiment was run again and the results were confusing. The TDSe of the gastric digest of the two experiment for MV were 23.03 % and 40.91 % and, of the intestinal digest were 59.64 % and 71.30 %, respectively. Once again raw ingredients (vegetables) lead as the main factor that determine Se availability. Therefore, for this type of beikosts, a determination of Se content and Se availability in raw materials should be made for each poduction in order to obtain specific data of the lot (final product). Only PD of MV showed the same trend than that of HR and CVV that will be described ahead. Figure 1 and 2 show PD and TDSe in gastric and intestinal digest after the simulated digestion procedure in HR (Fig. 1) and CVV (Fig. 2) beikosts. Beikosts proteins are hydrolysed to peptides along the gastrointestinal digestion with pepsin and pancreatin-bile extract (10), so PD increases with the digestion process in the two beikosts, with HR showing the highest values (67 % and 76 % in gastric and intestinal fractions, respectively). The TDSe also increases with the digestion process in the two beikosts. The high correlation (r = 0.99 for p < 0.001) between both parameters can be explained by the fact that nearly all the Se in animal tissues is associated with protein (15). In gastric digest the TDSe were very similar (20 %), but intestinal digest showed a wide variability, ranging from < 40 % for CVV to > 60 % for HR. It has to be noted that after the entire gastrointestinal digestion the TDSe in CVV is similar to the Se bioavailability (Table 1). However, that value is twice higher in HR. According to Shen *et al.* (10) this difference can

Table 1. Selenium (Se) content and *in vitro* availability
of three beikosts

	Beikosts type[1]		
	MV	CVV	HR
Se content (ng/g d.w.)	20.3-82.6	114.7	258.9
Se availability (%)	111	29.5	35.6

[1]Infant Beikosts abbreviation = MV: mixed vegetables; CVV: chicken-veal-vegetables and HR: hake with rice.

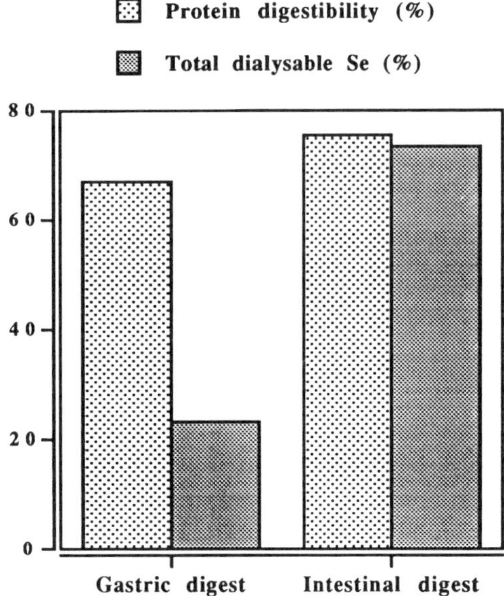

Figure 1. Protein digestibility (%) and total dialysable Se (%) in gastric and intestinal digest after the simulated digestion procedure in hake with rice beikost (HR).

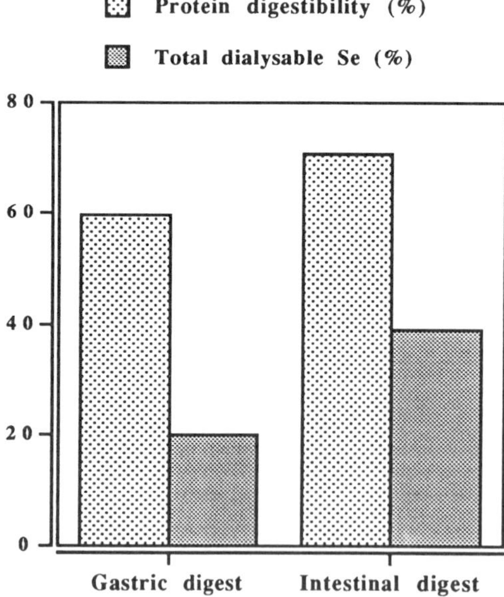

Figure 2. Protein digestibility (%) and total dialysable Se (%) in gastric and intestinal digest after the simulated digestion procedure in hake with chicken veal and vegetables beikost (CVV).

most probably be attributed to the fact that the *in vitro* procedures, as well as Miller's (9), does not comprise end-point equilibrium dialysis against deionized water (in our study only a 3 h dialysis step against the gastrointestinal digestion mixture) but are based on the equilibrium dialysis of minerals and trace elements across a semipermeable membrane. In the continous *in vitro* methods there is a removal of dialysable components, so in general, the dialysability of some trace elements (Ca, Mg, Fe and Cu) determined with the continuos *in vitro* method is higher than determined with the equilibrium *in vitro* method (up to a factor 3 to 4) (16). It is concluded that the removal of dialysable components in the continuos *in vitro* method has a marked influence on the dialysability of minerals and trace elements (16).

In conclusion the Miller's *in vitro* method used to determine the Se availability in the present study must be optimized, probably in a continuous system or for a longer time. It should be pointed out as well that Se availability depends on concentration and chemical form of Se in raw materials, therefore the knowledge of both parameters would be interesting in the study of Se availability in the final product.

4. REFERENCES

1. G.F. Combs, *Scand. J. Work Environ. Health* **19**, 119–121 (1993).
2. R.E. Litov and G.F. Combs, *Pediatrics* **87**, 339–351 (1991).
3. Committee on Medical Aspects of Food Policy (COMA), in *45 Weaning and The Weaning Diet*, London, pp.64–69 (1995).
4. F. Rincón-León, P. Abellán-Ballesta and G. Zurera-Cosano, *Journal of Micronutrient Analysis* **8**, 43–53 (1990).
5. I. Lombeck, K. Kasperek, HD. Harbisch *et al.*, *Eur. J. Pediatr.* **125**, 81–88 (1977).
6. J.A.T. Pennington and B. Young, *J. Food Compos. Anal.* **3**, 166–184 (1990).
7. A. Alegría, R. Barberá, R. Farré *et al.*, *Die Nahrung* **39**, 237–240 (1995).
8. J.B. Luten, W. Bouquet, M.M. Burggraaf and J. Rus, in *Trace element: Analytical Chemistry in Medicine and Biology*, P. Brätter and P. Schramel, eds., Walter de Gruyter *et al.*, Berlín/New York, pp. 509–519 (1987).
9. D.D. Miller, B.R. Schricker, R.R. Rasmussen and D. Van Campen, *Am. J. Clin. Nutr.* **34**, 2248–2256 (1981).
10. L. Shen, P. Van Dael and H. Deelstra, *Z. Lebensm. Unters Forsch.* **197**, 342–345 (1993).
11. AOAC, *Official Methods of Analysis*, 14th edn. S. Williams, ed., Arlington, VA (1990).
12. L.H. Foster and S. Sumar, *Nutrition and Food Science* **5**, 17–23 (1995).
13. O.A. Levander, in *Selenium-tellurium in the environment*, Industrial Health Foundation, Pittsburgh, pp. 26–53 (1976).
14. G.F. Combs and S.B. Combs, *The role of selenium in nutrition*, Academic Press, New York (1986).
15. R.F. Burk and K.E. Hill, *Annu. Rev. Nutr.* **13**, 65–81 (1993).
16. M.G.E. Diepenmaat-Wolters and H.A.W. Schreuder, in *Bioavailability'93*, part. 2, U. Schlemmer, ed., Ettlingen, Germany, pp. 43–47 (1993).

10

DIETARY INTAKE OF TOXIC TRACE ELEMENTS IN INFANT FEEDING

I. Navarro Blasco,[1] I. Villa Elízaga,[2] and A. Martín Pérez[1]

[1] Pediatric Research Unit
Department of Chemistry
University of Navarra
Pamplona, Spain
[2] Department of Pediatrics
Gregorio Marañón Hospital
Madrid, Spain

1. INTRODUCTION

Given the evident toxicological impact of certain elements on lactants, and considering that the absorption of these is significantly higher than in adults, it is desirable that infant formulae are proportionally similar or inferior in levels of concentration to those supplying human milk (1–7).

2. MATERIAL AND METHODS

This investigation included 82 infant formulae of 10 differents households. The infant formulae were milk based (Adapted-starting-formula; follow up formula, specialized-infants intolerant to milk proteins or lactose, infant with fat absorption problems and infant with inborn errors of metabolisms-formula; premature or low birth weight infants formula) and soya formulae (both of them powder and liquid) (8). The recipients were opened in the clean room using vinyl talc-free gloves and plastic material to make the sampling. All of the material which came into contact with the sample such as the containers used, were made of low density polyethylene and were previously cleaned in nitric acid (5%) during six days and later wiped three times with ultrapure water before utilization.

The digestion of infant formulae samples were done with subboiling nitric acid in a closed acid-decomposition system and high-pressure Teflon digestion bomb and microwave energy (Milestone mls 1200). Aluminum concentration was determined by Electrotermal Atomization Atomic Absorption Spectrophotometry (Graphite furnace-AAS), and Anodic

Therapeutic Uses of Trace Elements, edited by Nève et al.
Plenum Press, New York, 1996

Stripping Voltammetry (ASV) was used for Lead and Cadmium. The IAEA milk powder A11 was used as a Standard Reference Material for quality control.

3. RESULTS AND DISCUSSION

The possibility that certain trace elements can create health problems in infants justify the comparison of dietary intake provided by the infant formulae investigated and the *Provisional Tolerable Weekly Intake (PTWI)* of various metallic contaminants established by the Joint FAO/WHO Expert Committee on Food Additives. Considering that the newborn in each age period observes the same feeding regimen, in comparison with the PTWI, the weekly intake of toxic elements investigated was calculated with the distinct types of infant formulae, observing separately the special case of preterm newborns. It is necessary to consider that the newborns which consume the follow up formula accompany their feeding with beikost which can increase the intake of toxic trace elements.

Table 1 contains the weekly intake and the calculated percentage of the PTWI of aluminum with distinct types of infant formulae. The adapted -starting- formula contain small amounts of aluminum, about 4% of the PTWI. The special formulae are in an intermediate level (11–12% of the PTWI), and soya contributes with a large intake (15% of the PTWI). No commercial infant formula was higher than the PTWI and only a soya formula came close to the toxic concentration.

The infant formulae investigated were responsible of lead intakes under the PTWI, as seen in Table 2. The follow up and starting adapted formulae brought less lead, about 24% of the PTWI. The newborns fed with soya formulae were the major consumers of lead (66% of the PTWI). Within the infant formulae studied, 5 surpass such concentration (a starting adapted formula, a follow up formula, a specialized formula , two soya formula).

No type of infant formula contributed to a cadmium intake higher than the PTWI. The major intakes were in the starting adapted formulae and the lowest in the special formulae without lactose (Table 3). Only one starting adapted formula was above this concentration.

As indicated in Table 4, the formulae for preterms also brought a low intake of aluminum (8% of the PTWI); furthermore, they contributed to a low lead content in the diet of these newborn in all infant formulae investigated (21% of the PTWI).

In conclusion, none of the investigated formulae brought an intake of the studied toxic elements (Al, Pb and Cd) that surpassed the PTWI, even though five samples with different infant formulae surpassed the concentration limit in lead and one in cadmium.

Table 1. Percentages of the PTWI of aluminum proportioned to the distinct types of infant formulae

Age	St	A	F	L S	HA S	Diet S	So
0-2 week	4	4		10	13	10	14
3-4 week	4	5		12	14	11	16
2 month	4	5		11	14	11	15
3 month	4	4		11	13	11	15
4-5 month	4	4		10	13	11	14
6 month	3	4	4	10	11	11	14
> 7 month			3				

St, Starting formula; A, Adapted formula; F, Follow up formula; L S, infants intolerant to milk proteins or lactose Specialized formula; HA S, infant with fat absorption problems Specialized formula; infant with inborn errors of metabolisms Specialized formula; So, Soya formula

Table 2. Percentages of the PTWI of lead proportioned to the distinct types of infant formulae

Age	St	A	F	L S	HA S	Diet S	So
0-2 week	38	22		32	28	33	62
3-4 week	40	26		37	32	39	71
2 month	43	25		37	30	35	73
3 month	37	26		34	34	37	67
4-5 month	38	24		33	29	35	67
6 month	32	22	26	31	28	37	58
> 7 month			22				

St, Starting formula; A, Adapted formula; F, Follow up formula; L S, infants intolerant to milk proteins or lactose Specialized formula; HA S, infant with fat absorption problems Specialized formula; infant with inborn errors of metabolisms Specialized formula; So, Soya formula.

Table 3. Percentages of the PTWI of cadmium proportioned to the distinct types of infant formulae

Age	St	A	F	L S	HA S	Diet S	So
0-2 week	5	20		4	6	13	7
3-4 week	6	23		4	6	15	8
2 month	6	23		4	7	14	8
3 month	6	22		4	6	15	8
4-5 month	5	21		4	6	14	8
6 month	4	20	10	4	5	14	7
> 7 month			8				

St, Starting formula; A, Adapted formula; F, Follow up formula; L S, infants intolerant to milk proteins or lactose Specialized formula; HA S, infant with fat absorption problems Specialized formula; infant with inborn errors of metabolisms Specialized formula; So, Soya formula

Table 4. Percentage of the PTWI of distinct toxic trace elements proportional to the preterm formula

Weight	% ISTP		
	Aluminum	Lead	Cadmium
2,0–2,5	9	23	13
2,5–3,0	8	19	11
3,0–3,5	8	21	12
3,5–4,0	8	20	11
4,0–5,0	8	20	11

4. REFERENCES

1. WHO (World Health Organization), *Toxicological evaluation of certain food additives and contaminants: 30ª report of the Joint FAO/WHO Expert Committee on Food Additives*, WHO Technical Report Series nº 751 (1987).
2. WHO (World Health Organization), *Toxicological evaluation of certain food additives and contaminants: 30ª report of the Joint FAO/WHO Expert Committee on Food Additives*, WHO Technical Report Series nº 776 (1989).
3. F.H. Nielsen, *Comprehensive Therapy* **17**, 20–26 (1991).

4. A.B. Gruskin, *Adv. Pediatr.* **35**, 281–330 (1988).
5. WHO (World Health Organization), *Environmental Health Criteria 3. Lead*, World Health Organization, Geneva (1977).
6. K.R. Mahaffey, *Dietary and Environmental Lead: Human Health Effects*, Elsevier Scientific, New York (1985).
7. L. Friberg, T. Kjellstrom and G. Nordberg G, in *Handbook on the toxicology of metals*, L. Friberg, G. Nordberg and V. Vouk, eds., Elsevier Sciences Publisher, Amsterdam, New York (1986).
8. I. Navarro, *Doctoral Thesis*, 170–171 (1995).

11

STUDY OF MAGNESIUM ABSORPTION USING ^{25}Mg STABLE ISOTOPE AND INDUCTIVELY COUPLED PLASMA/MASS SPECTROMETRY TECHNIQUE IN RAT

Charles Coudray,[1] Jean Claude Tressol,[1] Elyette Gueux,[1] Enny Sominar,[1] Jacques Bellanger,[1] Denise Pepin,[2] and Yves Rayssiguier[1]

[1] Laboratoire des Maladies Métaboliques et Micronutriments
INRA de Theix-Clermont-Ferrand
63122 Saint Genès Champanelle, France
[2] Laboratoire d'Hydrologie
Institut Louis Blanquet, Faculté de Pharmacie
Université Blaise Pascal
63000 Clermont-Ferrand, France

1. INTRODUCTION

Magnesium metabolism is regulated at the intestine and kidneys by controlling the fraction of Mg absorbed from the total dietary intake and by renal homeostasis (1,2). As interest in Mg dietary requirements and metabolism has grown, the need for safe and convenient techniques for measurement of Mg absorption and bioavailability has increased. Balance studies are imprecise, labor intensive, give little information on Mg metabolism and do not consider the endogenous fecal excretion (3). Although kinetic analysis has been performed with the short-lived ^{28}Mg radioisotope, the use of radio isotopes in humans is hazardous and restricts the experiment to a few days duration and is being supplanted by stable isotope methods (4). The use of extrinsic labeling presumes that the administered isotope behaves in the same way and that its absorption is the same as that of endogenous forms of Mg. The validity of the extrinsic labeling approach is now well established (5,6,7). Stable isotopes have been analyzed by two different analytical techniques; neutron activation and mass spectrometry. Although thermal ionization mass spectrometry (TIMS) is the reference technique, inductively coupled argon plasma mass spectrometry (ICP/MS) is also being widely developed (8). ICP/MS has many advantages in stable isotope measurement and has been applied to metabolic studies of many different minerals. In the present work, the feasibility of using a Mg stable isotope and ICP/MS technique to study Mg absorption and metabolism was explored in adult rats and the optimum dosage of the isotope was investigated.

Therapeutic Uses of Trace Elements, edited by Nève et al.
Plenum Press, New York, 1996

2. MATERIALS AND METHODS

2.1. Animals

Male Wistar rats, aged 7 weeks were used. Animals were fed a semi-synthetic diet containing 20% casein and 65% Starch. The Mg concentration in the diet was 1073 mg/Kg. After an adaptation period of 6 days, the animals were housed in metabolic cages and the first six day period of conventional balance study was begun by collecting the faces and urine of each rat quantitatively. The rats had free access to feed and demineralized water.

2.1. Stable Isotope Preparation and Administration

^{25}MgO was obtained from Chemgas (Boulogne, France), it contained 98% of ^{25}Mg. One hundred mg of the oxide was dissolved in one ml of HNO_3 50% and the solution was made up to 20 ml with demineralized water. Rats were divided into two groups and two doses of enriched ^{25}Mg (6 either 12 mg) were administered orally. Faces and urine were collected daily before and on isotope test days and for the next 5 days. Blood was also sampled at H0, H4, H8, H12 and D1, D3, and D6, on the eyes under light diethyl ether anesthesia. A second conventional metabolic balance study of 6 days duration was also conducted immediately after the first period.

2.3. Stable Isotope Analysis

Fecal and urine Mg concentrations were determined by both atomic absorption spectrometry (Perkin Elmer) and ICP/MS (Qlasmaquad II, Fisons, France). Plasma and red blood cell Mg concentration and isotope ratios were determined by ICP/MS because the sample volume was limited. Plasma and urine were diluted and analyzed, whereas red blood cell and faces were first ashed at 500 °C for 10 hours, dissolved in 1% HNO_3 and analyzed.

3. RESULTS

3.1. Metabolic Balance Study

Two successive balance periods were investigated before and after stable isotope administration. The results shown in Table 1 indicate good repeatability and that animal handling during blood sampling was without effect. The apparent absorption of Mg of about

Table 1. Magnesium intake, faecal and urinary excretion in two successive metabolic balance periods (Data are expressed as mean ± SD)

	Mg intake (mg/day)	Faecal Mg excretion (mg/day)	Fractional absorption (%)	Urinary Mg excretion (mg/day)	Mg balance (mg/day)
First balance period	22.0 ± 2.3	10.7 ± 2.9	51 ± 11	4.1 ± 1.5	+ 7.1 ± 1.8
Second balance period	22.6 ± 1.7	11.5 ± 1.9	49.3 ± 5.9	4.27 ± 0.72	+ 6.8 ± 1.1

Table 2. Comparison between apparent and true absorption of magnesium in metabolic balance and stable isotope studies in rat (Data are expressed as mean ± SD)

	Apparent absorption	True absorption
Rats receiving 6 mg [25]Mg	53 ± 11%	63 ± 14%
Rats receiving 12 mg [25]Mg	46 ± 12%	54 ± 12%

50% obtained in the present study is close to values previously reported. The Mg balance was positive by 7 mg in both metabolic studies.

3.2. Stable Isotope Study

As indicated in Table 2, fractional absorption, using the stable isotope, was 62.5% and 54.1% in rats receiving 6 and 12 mg [25]Mg respectively. In contrast, fractional absorption was 49.4% and 48.2% for the two periods of classic balance studies. This difference represents, in large part, the fecal excretion of endogenous Mg.

3.3. Stable Isotope Enrichment

Figure 1 shows the [25]Mg enrichment percent in urine and faces over six days. The maximum of enrichment in urine was at 12 hours after stable isotope administration. The [25]Mg excreted in the faces collected on the first day was more than 80% of administered [25]Mg in rats receiving the dosage of 6 mg, and about 65% in rats receiving the dosage of 12 mg. In both cases, a plateau was attained on day three after [25]Mg administration. Stable isotope administration also resulted in measurable isotopic enrichment of plasma and erythrocytes (Table 3), with an optimum enrichment between H4 and H8 and between H12 and H24 for plasma and erythrocytes respectively.

4. DISCUSSION

Interest has been generated in the relation of Mg status to optimal health, because inadequate Mg intake is thought to contribute to the pathogenesis of various chronic diseases particularly in the older population (9). Balance studies are inadequate tools to determine human requirements for Mg (3). While the short-lived [28]Mg radioisotope has long been used for Mg kinetic analysis, especially in animals, its use in human subjects is hazardous and un-

Table 3. Time course evolution of isotopic enrichment percent of [25]Mg/[26]Mg (Mean ± SD, %) in plasma and red blood cells in rats

	Time (hours) after [25]Mg administration					
	H4	H8	H12	H24	H72	H144
Plasma						
6 mg [25]Mg/rat	84 ± 14	69.1 ± 8.6	55.9 ± 7.5	34.9 ± 3.6	14.6 ± 6.2	7.8 ± 1.2
12 mg [25]Mg/rat	144 ± 28	118 ± 21	97 ± 19	66 ± 12	27.0 ± 4.1	14.4 ± 2.1
Red blood cells						
6 mg [25]Mg/rat	13.3 ± 4.2	20.4 ± 5.0	23.3 ± 2.6	25.7 ± 8.0	13.1 ± 2.2	7.3 ± 1.6
12 mg [25]Mg/rat	21.40 ± 0.90	31.2 ± 2.6	48.4 ± 2.7	51 ± 12	27.7 ± 1.3	13.1 ± 1.1

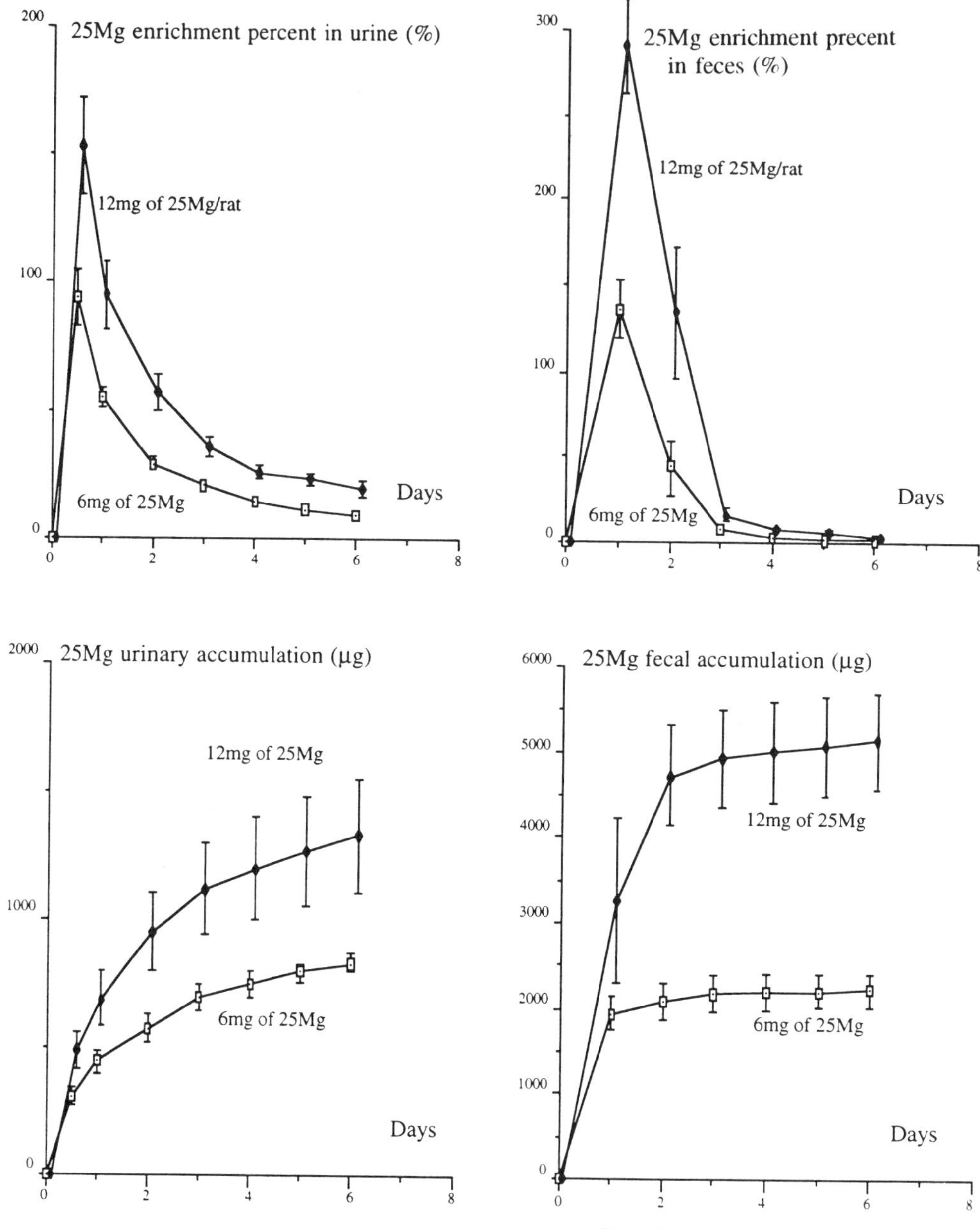

Figure 1. Time course evolution of isotopic enrichment percent of ^{25}Mg/^{26}Mg (%) in urine and faeces in rats.

suitable. The alternative to radioisotopes is the use of stable isotopes. Mg has three natural stable isotopes; ^{24}Mg (78.99%), ^{25}Mg (10.00%) and ^{26}Mg (11.00%) (10), two of which could be used in bioavailability studies. Recent studies demonstrated that the absorption efficiency of extrinsically labeled Mg is similar to the dietary absorption efficiency of ^{26}Mg intrinsically incorporated into vegetables (5) or into milk (11). Although TIMS is still the reference technique for stable isotope studies, inductively coupled argon plasma mass spectrometry has many advantages and could supplant TIMS in many cases (12). In the present work, the feasibility of using stable isotope and ICP/MS technique to study Mg absorption and metabolism was explored in adult rats and an appropriate dosage of an Mg isotope was investigated.

The results of the present study show that the mean daily Mg intake was about 22 mg/rat, the mean of the balance studies, which corresponds to an apparent absorption of Mg of about 50%. Previous works in rats reported similar apparent Mg absorption (40–60% depending on diet composition and Mg level) (13,14,15). Furthermore, as expected, true (net) Mg absorption, obtained from the isotopic studies was higher than apparent absorption by more than 10% when the dosage of ^{25}Mg was 6 mg. This difference is essentially due to the fact that balance studies do not take into account the endogenous fecal excretion of Mg which increases total fecal excretion and causes an apparent reduction in Mg absorption. Isotope administration in liquid form could also be responsible for this increased absorption efficiency in comparison with Mg absorption in the total diet. Moreover, Mg absorption was different depending on the amount of stable isotope administration. Apparent and true absorption was higher in the rats receiving 6 mg of ^{25}Mg than in those receiving 12 mg. This is due in part to the small number of rats in each group but also to the expected reduction of absorption with high level of Mg intake. Indeed, the efficiency of absorption of Mg falls with increasing dose as with other minerals. The time-course fecal ^{25}Mg enrichment, reported in the present study, show that more than 80% or 65% of recovered ^{25}Mg was excreted during 24 hours in rats receiving 6 or 12 mg of ^{25}Mg respectively. In any case, more than 95% of recovered ^{25}Mg was excreted during the first three days in rats receiving either 6 or 12 mg of ^{25}Mg. If the faces of the three first days is pooled for each rat, the expected 25Mg enrichment could reach more than 50% and 100% in rats receiving 6 or 12 mg of ^{25}Mg respectively. The possibility of Dysprosium utilisation as a fecal marker as done by Schuette et al., (16) should permit us in future experiments to collect the faces of the two first days where the expected ^{25}Mg enrichment could attain more than 80% and 200% in rats receiving 6 or 12 mg of ^{25}Mg respectively.

The stable isotopic enrichment in blood traces the appearance and disappearance of ^{25}Mg in plasma and red blood cells in the two groups receiving 6 or 12 mg of ^{25}Mg. There are only a few studies that have investigated blood levels after oral ingestion of tracer Mg. In this study, the peak of plasma ^{25}Mg enrichment was observed around 4 hours reaching 84% and 144% in rats receiving 6 or 12 mg of ^{25}Mg respectively. Others have also observed a peak of 28Mg 4 hours after isotope administration in rats (17). Changes in red blood cell Mg ratios were also observed with a peak at around 24 hours, after ^{25}Mg administration. Maximal red blood cell enrichments were about 25% and 50% for rats receiving 6 or 12 mg of ^{25}Mg respectively. These enrichment levels are sufficiently high to permit their use to calculate several kinetic constants of Mg metabolism, in particular, if combined with the intravenous administration of a second Mg stable isotope (^{26}Mg).

In conclusion, these results indicate that the use of 5–6 mg or less of ^{25}Mg stable isotope per rat permit meaningful investigations of Mg bioavailability. Furthermore, Mg possesses two stable isotopes and thus using a double-label stable isotope technique to explore the metabolism of Mg could be envisaged.

5. REFERENCES

1. G.A. Quamme, *Miner Electrolyte Metab,* **19,** 218–225 (1993).
2. R. Civitelli, L.V. Avioli, In; Johnson LR, *Physiology of the gastrointestinal tract,* third edition, Raven Press, New York, pp. 2173–2181 (1994).

3.   W. Mertz, *J Nutr,* **117,** 1811–1813 (1987).

4.   M. Janghorbani, B.T.G. Ting, *J Nutr Biochem*, **1,** 4–19 (1990).

5.   R. Schwartz, D.L. Grunes, R.A. Wentworth, E.M. Wien, *J Nutr,* **110,** 1365–1371 (1980).

6.   D.D. Gallaher, P.E. Johnson, J.R. Hunt, G. Lykken, M.J. Marchello, *Am J Clin Nutr.* **48,** 350–354 (1988).

7.   C.B. Egan, F.G. Smith, R.S. Houk, R.E. Serfass, *Am J Clin Nutr*, **53,** 547–553 (1991).

8.   E.E. Cary, R.J. Wood, R. Schwartz, *J Micronutrient analysis*, **8,** 13–22 (1990).

9.   R.J. Wood, P.M. Suter, R.M. Russell, *Am J Clin Nutr,* **62,** 493–505 (1995).

10. P. De Bièvre, P.D.P. Taylor, *Inter J Mass Spectro Ion Proc.* **123,** 149–166 (1993).

11. Y.M. Liu, P. Neal, J. Ernst, C. Weaver, K. Rickard, D.L. Smith, J. Lemons, *Pediatr Res,* **25,** 496–502 (1989).

12. B. Sandström, S. Fairweather-Tait, R. Hurrell, W. Van Dokkum, *Nutr Res Rew,* **6,** 71–95 (1993).

13. E.J. Brink, P.R. Dekker, E.C. Van Beresteijn, A.C. Beynen, *J Nutr*, **121,** 1374–1381 (1991).

14. E.J. Brink, A.C. Beynen, *Prog Food Nutr Sci,* **16,** 125–162 (1992).

15. M.J.F. Verbeek, G.J. Van Den Berg, A.G. Lemmens, A.C. Beynen, *J Nutr,* **123,** 1880–1887 (1993).

16. S.A. Schuette, M. Janghorbani, V.R. Young, C.M. Weaver, *Am J Coll Nutr,* **12,** 307–312 (1993).

17. J.K. Aikawa, In ; *Magnesium: its biological significance.* by Aikawa JK, Ed., CRC Press Inc., pp. 43–56 (1981).

12

ESSENTIAL TRACE ELEMENTS IN THE NUTRITION OF INFANTS

P. Brätter

Department of Trace Elements in Health and Nutrition
Hahn-Meitner Institute Berlin
D-14109 Berlin, Germany

1. INTRODUCTION

The newborn child has to develop its own regulation of the metabolic processes. Included in the anabolic structure are the digestive system, the respiratory chain and the endogenous defense system. In addition, the newborn has to develop the ability to synthesize essential compounds from the nutritive fluid, as well as the ability to concentrate waste material in the urine. Apart from other nutrients, an adequate supply of essential trace elements is required to ensure optimal development of all metabolic functions. About 17 of the 90 naturally occurring elements are classified as trace elements. Due to their known physiological importance, 10 of these trace elements (chromium, cobalt, copper, fluorine, iron, iodine, manganese, molybdenum, selenium and zinc) are regarded today as being essential in the nutrition of infants.

In early infancy breast milk or cow's-milk based formulas are the only source of essential trace elements. Mother's milk provides an adequate supply for the full-term infant and its composition is therefore used as a reference. Special attention must be paid to premature infants because they are born with lower levels of essential micronutrients. Trace elements are added, therefore, to pre-term infants' formulas to satisfy their higher dietary requirements.

In recent years several highly informative reviews of trace elements in infant nutrition have been published (1–7). Various factors considered to be of importance in the nutrition of infants are: difficulties in analytical determination, nutritional essentiality, requirements during growth, storage in fetal liver, mean concentration in breast milk, decline with progression of lactation, chemical binding in the nutritional fluid, interactions with other nutrients in fortified formulas, and relation to maternal nutritional status and dietary intake.

The present paper will discuss on the following points: trace element supply of breast-fed and formula-fed infants, from birth up to the age of 6 m, binding pattern of trace elements in breast milk in relation to formulas, and the breast milk level of selenium and iodine in relation to maternal intake.

2. DIFFICULTIES IN ANALYTICAL DETERMINATIONS

All aspects of the discussion of infant nutrition depend on reliable analytical determination of trace elements in breast milk and formulas. Reliable data are available for the concentration of copper, iron and zinc in breast milk but for the other essential trace elements one finds a wide range of values in the literature. This can be due to longitudinal variation in the trace element content and regional dietary conditions; but analytical difficulties can also lead to significant differences. Element concentrations in the range below 10 µg/l must be regarded very critically with respect not only to contamination in the preanalytical steps but in the instrumental analytical determination as well. This includes the elements manganese, cobalt, chromium and molybdenum. It is astonishing that despite the known essentiality of selenium and the availability of advanced analytical methods for determining its concentration no data on the selenium content in infant formulas are normally given by formula producers.

3. DAILY TRACE ELEMENT INTAKES OF INFANTS FROM MATURE BREAST MILK AND FORMULAS IN COMPARISON TO RECOMMENDED VALUES

Comparison of the recommended dietary allowances of trace elements for neonates and young infants as published by the US National Research Council in 1989 (8) with the real average daily intake of fully breast-fed infants reveals striking differences. In Table 1 the intake levels were calculated from milk values which were confirmed by quality control measurements. For formulas, the values given by the producers are used. Most of the trace element intake values of breast-fed infants are far too low, whereas infant formula levels seem to be adequate or even higher. The question has to be asked whether this means that

Table 1. Average daily trace element intakes of infants (0 - 6 months) from mature breast milk and formulas. Comparison with RDA

			Human breast milk (700ml)			Formulas (700ml)	
Element	Unit	RDA[a] (0–6m) (1989)	WHO/IAEA range of six countries[b] 3 months (n = 516)	Casey et al. (25) (USA) 4 weeks (n = 11)	Brätter (this work) (FRG) 3 weeks (n = 44)	Soy based formulas (n = 3)	Cow's milk based formulas (n = 8)
Iron	mg	6.0	0.22 - 0.46	n.d.	0.54	8.3- 8.5	5 - 9
Copper	mg	0.4 - 0.6	0.12 - 0.22	0.25	0.63	0.32 - 0.42	0.36 - 1.4
Zinc	mg	5.0	0.5 - 1.7	2.0	2.0	3.5 - 3.8	2.5 - 9.0
Manganese	µg	300 - 600	2.1 - 2.8	2.0	4.0	140 - 180	50 - 400
Selenium	µg	10	9 - 17	n.d.	10.6	?	3 - 7
Iodine	µg	40	38 - 45	n.d.	n.d.	31 - 70	20 - 105
Fluoride	µg	100 - 500	5 - 12	n.d.	n.d.	?	20 - 150
Chromium	µg	10 - 40	0.5 - 1.0	0.15	3.0	?	7 - 15
Molybdenum	µg	15 - 30	0.3 - 10	n.d.	n.d.	?	20 - 50
Cobalt	µg	0.3 B_{12}	0,1 - 0,25	n.d.	0,8	?	?

[a]US National Research Council 1989 (8).

[b]Minor and Trace Elements in breast milk of Guatemala, Hungary, Nigeria, Philippines, Sweden and Zaire. Calculated from Tab.33 Report of a joint WHO/IAEA collaborative study, WHO Geneva (1989).

n.d. = not determined.

most of the healthy, breast-fed term infants with intakes below the RDA values are at risk of deficiency during this period of life. The problem would seem, however, to lie more with the RDA values. One has to keep in mind that consensus and statistical evaluation provide the basis for the RDA for a given element and that the RDA values are designed to maintain good nutrition for all healthy people in the United States (8). Furthermore, these recommendations have to be regarded only as a guideline for the dietary intake needed to prevent deficiency symptoms, whereas no claim is made that they optimize the parameters of growth, development, health or biological activity (9)

The RDA of nutritionally essential trace elements for infants from birth until 6 months of age cannot be used generally; in addition, separate recommendations must be developed for breast- and formula-fed infants taking into consideration as well the preterm and very low birth weight status. In order to optimize the trace element supply in infant formulas with respect to quantity and quality, improved knowledge of trace element bioavailibility from human milk, soy formula and cow's-milk formula is necessary. As a first approximation for the relevant investigations, human milk may be taken as an adequate source.

4. TRACE ELEMENT CONCENTRATION IN BREAST MILK IN RELATION TO THE MATERNAL NUTRITIONAL STATUS AND NUTRIENT INTAKE

When human milk is used as a standard for adequate trace element supply, the influence of the maternal nutritional status on the element concentrations in the milk must be taken into consideration. Calcium, iron, copper and zinc apparently are not influenced by short-term variations in maternal status provided the mother is well-nourished (3, 10). No significant correlation was observed between dietary zinc intake and zinc concentration in human milk (10, 11); furthermore, zinc supplementation of a zinc adequate diet does not affect the zinc concentration in human milk (12).

Whereas the concentration of divalent metal ions in breast milk seemed not to be influenced by short-term variations in maternal intake, for the anionic species of the essential elements selenium, iodine and fluorine a wide variation in the values reported in the literature can be found due to regional food-chain conditions. Because of the significant responsiveness to maternal dietary intake iodine and selenium are of special interest.

4.1. Iodine

Newborns are particularly sensitive to the effects of both iodine deficiency and iodine excess because of the risk of thyroid impairment (3). A sufficient iodine supply is important for optimal metabolic activities of the developing organism. A range of breast milk iodine content of 29 - 490 µg/l has been reported in the literature (4). An average daily breast milk intake of 700 ml provides about 56 µg of iodine in Europe and 112 µg in the USA. The US National Research Council recommended a dietary allowance of 40 µg per day for neonates and young infants. This corresponds to a daily iodine intake of about 8 µg/kg. In recent iodine balance studies conducted with infants aged 1 month, Delange (13) showed that the RDA values are probably too low. In order to achieve a positive iodine balance the daily intake has to be at least 15 µg/kg in full-terms and 30 µg/kg in preterms.

Human breast milk concentration responds sensitively to maternal dietary iodine intake. The iodine content of cow's milk depends on the geographical area, the local pasturage, and whether or not mineral supplements have been used. It is further influenced by seasonal variations. In the literature values are reported within the range 21 - 970 µg/liter. Thus, the contribution of iodine from cow's milk in formulas varies and has to be taken into consideration in any supplementation plan. Because of the importance of iodine in early infancy we recently checked the reliability of the iodine concentration given by a producer for a formula batch (116 µg/liter). By means of RNAA we obtained only 72 µg/l, which is about 40 % lower than the value given. Reliable values are needed to estimate the iodine supply in early infancy when formula is the only dietary source of iodine. This example shows the need for reliable quality control in the trace element analysis of infants' formulas.

4.2. Chemical Binding of Trace Elements in Infant Nutrition

The feeding practice adopted may be critical to the infant's well-being and development. In the course of the first few months of life, breast milk, cow's-milk-based and soy-based formulas are the only dietary sources of essential trace elements. In breast milk, most of the essential trace elements are bound to specific proteins, which may explain their high bioavailibility. Compared to breast milk, trace elements in formulas show a significantly different binding pattern, since the trace elements are added in the form of inorganic compounds (Table 2). Furthermore, most of the formulas contain much higher trace element levels than occur in human breast milk. This might be necessary to compensate for the lower bioavailibility of the chemical binding forms present after the dry matter is dissolved in water.

There is a need to determine the binding forms of the essential trace elements and to clarify the adequacy of their levels, which means that speciation studies of these dietary sources have to be carried out. From the analytical point of view it is possible to perform speciation of trace elements in infant dietary sources by the on-line combination of liquid chromatography techniques with analytical methods of high detection power such as ICP-MS or ICP-AES.

4.3. Iron

The iron supply in early infancy is currently a matter of discussion, with regard to adequate supplementation (1, 14). In order to obtain more information on this topic we have undertaken speciation studies. As an example, the iron binding patterns obtained by means of a combination of size exclusion chromatography and ICP-MS speciation analysis are shown in Figure. Breast milk is compared with two different formulas. Compared to breast-fed infants the iron supply of formula-fed infants is much higher but the binding patterns are quite dif-

Table 2. Binding form of added trace elements in formulas in comparison to breast milk

Element	Cow's milk based	Soy formula	Breast milk
Iron	(II) sulfate lactate	sulfate	lactoferrin
Zinc	(II) sulfate oxide	sulfate	casein
Copper	(II) sulfate	sulfate	serum albumin
Manganese	sulfate	sulfate	lactoferrin
Molybdenum	Na molybdate	—	xanthine oxidase
Iodine	iodide	K - iodide	inorganic
		Na- iodide	
Chromium		—	?

Table 3. Trace elements in serum of healthy breast-fed and formula fed infants (age 12 - 14 w)

Feeding	n	Mean (+/- SD)	Range	Median
Selenium (ng/g wet)				
Human milk	45	59 (11)	38 - 85	58,0
Pre-Aptamil	29	31,4 (5,4)	21 - 45	31,0
Humana	33	30,1 (5,9)	13 - 48	30,0
Multival	27	34,4 (7,4)	10 - 49	36,0
Pre-Beba	6	19,8 (2,8)	16 - 23	18,5
Zinc (µg/g wet)				
Human milk	44	0.80 (0,13)	0.56 - 1,16	0,81
Pre-Aptamil	30	0,74 (0,22)	0,52 - 1,72	0,71
Humana	33	0,71 (0,12)	0,43 - 0,96	0,72
Multival	28	0,80 (0,12)	0,62 - 1,20	0,79
Pre-Beba	6	0.81 (0,15)	0,59 - 1,01	0,85
Iron (µg/g wet)				
Human milk	31	1,35 (0,38)	0,72 - 2,1	1,37
Pre-Aptamil	19	1,24 (0,45)	0,74 - 2,2	1,14
Humana	27	1,29 (0,42)	0,63 - 2,1	1,26
Multival	8	1,06 (0,34)	0,64 - 1,7	1,13
Pre-Beba	6	1,32 (0,43)	0.80 - 1,8	1,48

ferent. Analysis of the iron in serum of healthy term infants fed for 12 weeks solely on breast milk or on formulas showed comparable iron concentrations (Table 3). Different bioavailibility of iron may explain this finding. Concerning the level of iron fortification and the use of iron-supplemented formula during the first 3 months of life, the various national Committees on Nutrition differ as to their recommendations (1). Iron fortification of all formulas is recommended by the American Academy of Pediatrics, whereas the European ESPGAN has recommended the use of non-iron supplemented formulas before the third month of life.

Most of the infant formulas are fortified with iron, up to 12.7 mg/l in the USA and 7–8 mg/l in Europe. In contrast to the formulas, the iron breast milk level is only in the range 0.2 - 0.5 mg/l (Table 1) . Thus, in infant formulas the ratios of the other trace elements relative to iron are significantly different and the possibility, in particular, of antagonistic interactions between Fe, on the one hand, and Cu and Zn, on the other, in intestinal uptake and transfer must be taken into account (5, 15, 16). Lönnerdal et.al. (5) found a lowered serum copper concentration for high iron supplementation (6.9 mg/l) in the form of ferrous sulfate in a study of infants and drew the conclusion from the results that 4 mg Fe/ liter were adequate for infants up to 6 months of age.

4.4. Zinc

Zinc occurs in human breast milk bound to various whey-proteins. The major low-molecular zinc-binding ligand has been identified as citrate, which is thought to facilitate Zn-absorption (17). In cow's milk casein is the main zinc-binding compound, with its accessibility for the infant being much lower than that of citrate. In fortified formulas zinc is added as sulfate and/or oxide. Speciation analysis of breast milk and cow's-milk-based formulas showed quite different binding patterns (Figure 1); further, there are considerable differences between cow's-milk- and soy-based formulas (not shown here). Zinc intake is higher via cow's-milk formulas than via breast milk but the serum levels of fully breast-fed and solely

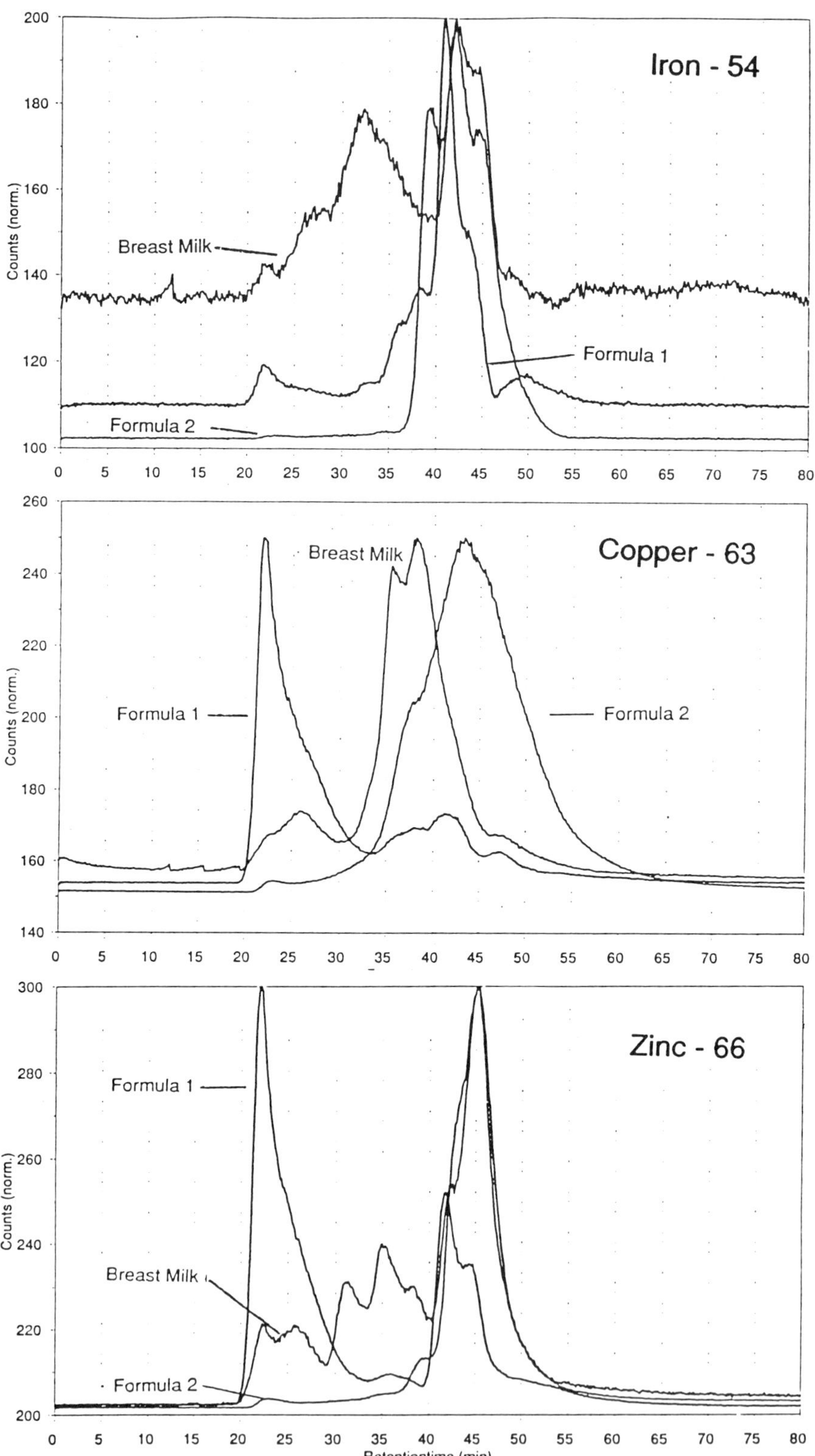

Figure 1. Distribution profiles of Fe, Cu and Zn in breast milk (Berlin, FRG) and formulas (1 = cow milk based Aptamil1, 2 = hypoallergen H.A. 1) by HPLC/ICP-MS.

formula-fed infants measured 3 months after birth are comparable (Table 3). This indicates a better zinc absorption from breast milk. The advantage of human milk in infant feeding has been shown in the treatment of acrodermatitis enteropathica, where human milk but not cow's milk has a therapeutic effect (18).

4.5. Copper

Today, infant formulas are normally supplemented with copper in the range 0.4 - 0.9 mg/liter. As with zinc, significantly different binding patterns were also obtained for copper in human breast milk and formulas. Copper is presumably bound to the whey protein serum-albumin, which is not present in cow's milk-based formulas (Figure 1).

4.6. Selenium

According to data reported from various countries, mature human milk contains selenium in the range 2.6 to 200 µg/l and the individual daily selenium supply of infants via breast milk lies in the range 2 - 140 µg, based on a 700 ml daily milk intake. This wide range reflects the influence of local geochemical conditions on the maternal dietary intake of selenium. At present, the producers of formulas for infants do not give any information on the selenium content of their products. Analyses of various cow's-milk-based formulas distributed in Germany have showed that the selenium content is much lower (one third) as compared with mature human milk.

A comparison of solely breast-fed or formula-fed infants yielded significant differences in the serum selenium levels depending on the different selenium supply (Figure 2, Table 3). The selenium-dependent glutathione peroxidase activities were significantly higher in breast-fed infants than in formula fed groups (5). This may indicate, firstly, the necessity for analyzing the selenium content of formulas and, secondly, the need to substitute selenium in low-level formulas. Much lower selenium values were found in serum of infants undergoing total parenteral nutrition (19) (Figure 2). Infusion solutions are made up of chemically pure components and they can be regarded as selenium-free. Despite the fact that selenium deficiency symptoms have been documented in children undergoing long-term TPN (20), at present selenium supplementation is not routinely used in infusion programs.

Selenium levels in breast milk are strongly correlated with maternal selenium intake. This level might become crucial with respect to selenium supply of the infant in regions of endemic selenium deficiency, so that maternal supplementation with selenium might be of benefit to the suckling infant. With the aim of increasing the selenium content in breast milk, the effectiveness of supplementation of the mother was experimentally investigated by Kumpulainen et al. (21). The results have to be discussed in the context of the finding that the breast milk selenium content is 6 to 7 times lower than the maternal serum concentration, independent of the maternal dietary intake level (22). According to the data obtained by Kumpulainen the daily selenium intake of a breast-fed infant can be increased by about 3µg when its mother is given a daily dose of 100 µg of organic-bound selenium, or increased by only about 1 µg when 100 µg selenium as selenite is used. Thus, a substantial increase in the infant's supply via supplementation of the mother requires a relatively high maternal selenium intake. The quantity to be administered, however, has to be considered in relation to possible interaction with other trace element-containing components during the secretion of breast milk. It has also to be viewed critically with respect to possible effects on the mother's health.

Group	Subject	Status	Type of Nutrition	Age	N
1	Premature and New-Born	Healthy	Oral Milk Formula	3 D	33
2	Infants	Healthy	Breast-Milk	12 - 14 W	35
3			Milk Formula		26
4	Premature and New-Born,	Hospitalized (Operation,	Infusion	1 - 3 D	22
5				2 - 3 W	24
6	Infants	Artificial Respiration, etc.)		5 - 8 W	22

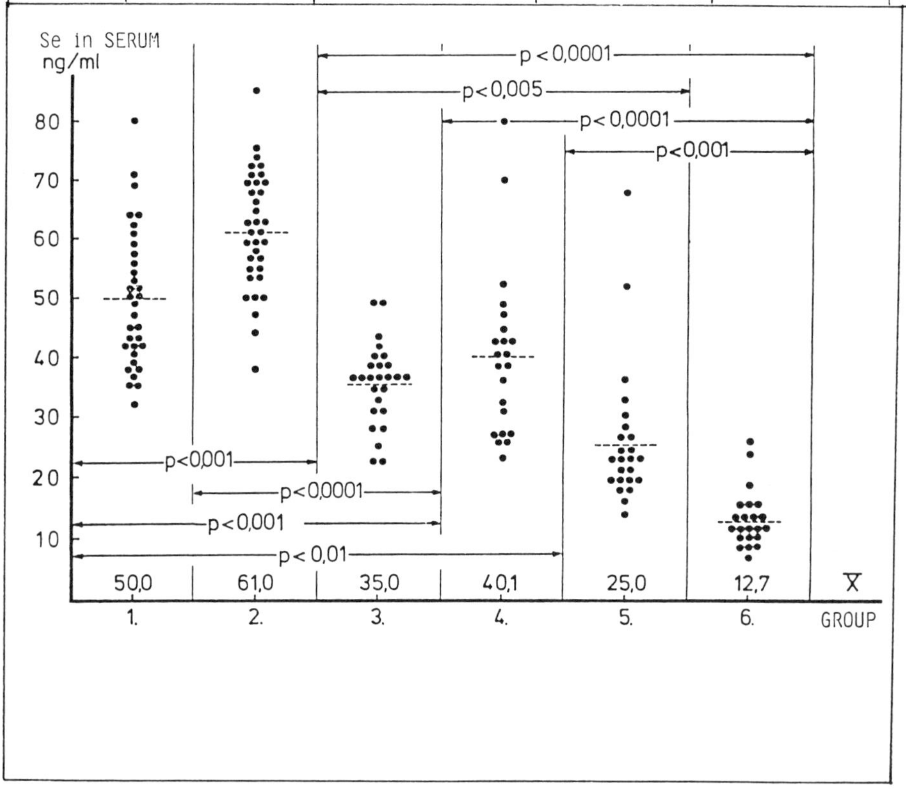

Figure 2. Selenium in serum of infants. Comparison of breast-feeding, formula-feeding and total parenteral nutrition.

We studied the effects of high maternal dietary selenium intake (range 150 - 900 µg/d) on the composition of mature breast milk in seleniferous areas of Venezuela. Analysis showed that selenium in the breast milk was in the range of 50 - 200 µg/l. The results indicated an inverse relationship between zinc in breast milk and the dietary selenium intake level (22, 23). By means of speciation studies significant changes in the zinc-binding pattern were observed. Citrate and the citrate-bound zinc fraction were found to decrease with increasing dietary selenium intake (23). The resulting mean daily zinc intake of about 1.6 mg/d in the seleniferous region was much lower than that of infants in the control region and anthropometric examinations suggested that the growth retardation observed in the seleniferous areas might be associated with the low zinc intake during infancy (24).

4.7. Manganese

Manganese is another essential element which is found in much higher concentrations in most formulas than in breast milk. The average daily intake of formula-fed infants has been reported to be 100 - 1000 times higher than that for breast-fed infants. There is very little information available on the manganese bioavailibility in breast milk and formulas. In breast milk, manganese is mainly bound (67 %) to lactoferrin (26) In cow's milk it is mainly associated with casein. It has been shown that high levels of manganese adversely affect intestinal iron absorption (15, 26, 27). Given the relatively high manganese level of cow's-milk-based formulas there is no reason for supplementation.

5. REFERENCES

1. R.Chierici, C. Gamboni, V. Vigi, *Acta Paediatr. Suppl* **402**: 50–60 (1994)
2. C.J. Bates, A. Prentice, *Pharmac. Ther.* **62**, 193–220 (1994)
3. P.J. Aggett, S.M. Barclay, In „*Principles of perinatal-neonatal metabolism"* Chapter **27**, R.M. Cowett, ed., 500–530 (1991)
4. A. Flynn, *Advances in food and nutrition research*, *36*, 209–252 (1992)
5. B. Lönnerdal, O. Hernell, *.Acta Paediatr.* **83**, 367–73 (1994)
6. L. Davidsson, *Acta Paediatr. Suppl.***395**, 38–42 (1994)
7. J.A. Milner, *J. Pediatr.* **117**, 147–155 (1990)
8. Foods and Nutrition Board. National Research Council. Recommended dietary allowances. 10th ed. Washington,DC, National Academy Press (1989)
9. P. Saltman, in *Encyclopedia of anorganic chemistry,* R.Scott,ed., J.Wiley Co., 2721–2726 (1995)
10. E. Vuori, S.M. Makinen, R. Kara , *Am. J. Clin. Nutr.* **33**, 227–31 (1980)
11. P.B. Moser, R.D. Reynolds, *Am. J. Clin. Nutr.* **38**, 101–108 (1983)
12. M. V. Karra, A. Kirskey, O. Galal, N.S. Bassily, G.G. Harrison, N.W.Jerome, *Nutr.Res.*.**9**, 471–478 (1989)
13. F. Delange, *Annales Nestlé* , **52**, 81–93 (1994)
14. S.H. Zlotkin, *J. Pediatr. Gastroenterol. Nutr.* 16, 1–3 (1993)
15. L.S. Hurley, C.L. Keen, B. Lönnerdal, *Fed. Proc.* **42**,No.6, 1735 (1983)
16. P.J. Aggett, Proc. Nutr. Soc. **47**, 21–25 (1988)
17. B. Sandström, A. Cederblad, B. Lönnerdal, *Am. J. Dis. Child.* **137**, 726–729, (1983)
18. P.A. Walravens, K.M. Hambidge, K. H. Neldner, *J. Pediatr.* **93**,71–73 (1978)
19. P. Brätter, V.E. Negretti de Brätter, U. Rösick, H.B. v. Stockhausen, in *Trace Element Analytical Chemistry in Medicine and Biology- 4.* P.Brätter,P.Schrammel eds., Walter de Gruyter, 133–143 (1987)
20. H.-J. Gramm, A. Kopf., P. Brätter, *J.Trace Elements Med. Biol.* **9** (1995) 1–12 (1995)
21. N. Kumpulainen, E. Vuori, M.A. Siimes, *Intern.J.Vit.Nutr.Res.* **54**, 41–43 (1984)
22. P. Brätter, V.E. Negretti de Brätter, U. Rösick, H.B.von Stockhausen, in *Trace elements in the nutrition of children-II.* Ranjit K. Chandra,ed., Nestlé Nutrition Workshop Series, **23**, Nestec Ltd. Vevey/Raven Press, Ltd., New York, 79–90 (1991)
23. P. Brätter, V.E. Negretti de Brätter, S. Recknagel, R. Brunetto, *J.Trace Elements Med. Biol.* (in press)
24. P. Brätter , V.E. Negretti de Brätter, W.G. Jaffé, H. Mendez-Castellano, *J.Trace Elem.   Electrolytes Health Dis.* **5** , 269–270 (1991)
25. C.E. Casey, K.M. Hambidge, M.C. Neville *Am. J. Clin. Nut.* **41**, 1193–1200 (1985)
26. B. Lönnerdal, C.L. Keen, L.S. Hurley, Am. J. Clin. Nutr. **41**, 550–559 (1985)
27. B. Lönnerdal, *J Nutr. Suppl.,***119**, 1839–1844 (1989)

13

THE EXPANDING FIELD OF ZINC SUPPLEMENTATION IN CHILDREN

P. Walravens

University of Colorado Health Sciences Center
Barbara Davis Center for Childhood Diseases
Denver, Colorado

1. PARAMETERS OF ZINC STATUS

The determination of zinc deficiency still remains elusive, as there is no single, sensitive index that will provide an accurate answer. Therefore , as suggested by Aggett (1), consideration of known factors that lead to zinc deficiency, a judicious selection of laboratory tests and the clinical or biochemical response to zinc supplements remain the basis of the diagnosis. In controlled studies during the childhood years, a positive growth response to zinc supplements is considered the correction of a pre-existing deficiency state. Thus simple, careful measurements may confirm an existing deficiency and are sometimes more useful than various biochemical assays which have been measured in the quest of defining zinc deficiency. When the many factors that influence plasma zinc concentrations and the limitations of hair zinc levels became known, a broad spectrum of functional indices were examined to include a host of zinc metalloenzymes in plasma, erythrocytes and leukocytes, apothymulin levels in serum and white blood cell chemotaxis. Variations in taste acuity and adaptation to darkness, and measurements of cutaneous hypersensitivity are additional functional indices which are sometimes used to monitor responses to zinc supplements.

2. ZINC SUPPLEMENTATION STUDIES

Among the conditions that predispose to zinc deficiency (Table 1), inadequate intake and intestinal diseases such as frequent diarrheal episodes or parasitism put children in developing countries at high risk of zinc deficiency. Thus growing numbers of controlled supplementation studies are being performed in Africa, Asia and Central and South America to assess the effects of zinc supplements on growth, general immunity, resistance to illness and some aspects of cognitive development. In chronic inadequate intake states, deficits in energy, protein and undoubtedly other micronutrients accompany those in zinc. Hence supplementation with a single nutrient whose deficiency seems well documented may not always

Therapeutic Uses of Trace Elements, edited by Nève et al.
Plenum Press, New York, 1996

Table 1. Factors predisposing to zinc deficiency

Low intake or availability
Malnutrition
Vegeterians
Synthetic diets
Infecions
Nutrient interactions

Malabsorption and digestion
Intestinal immaturity
Acrodermatitis enteropathica
Gastro-intestinal surgery
Enteropathies, IBD
Exocrine pancreatic insufficiency
Liver and biliary diseases

Increased losses
Catabolic states
Protein-losing enteropathies
Renal failure and dialysis
Chronic hemolysis
Chelating agents
Exfoliative dermatoses

Increased utilization
Rapid tissue synthesis
Convalescence
Neoplasms
Corrected anemias

lead to improved growth. This was noted in a study from Gambia (2), where zinc supplements did not correct growth faltering during the rainy season, when energy and protein intakes remained at inadequate levels. A small increase in mid-upper arm circumference and a decrease in intestinal permeability were the only findings associated with supplemental zinc. Among older children in Zimbabwe (3), an early effect of zinc supplements on weight and arm muscle area was noted but later disappeared after 12 months of observation. The latter half of the study was however, because of drought, one of very low energy and animal protein intake. Parasitic reinfection with Schistosoma mansoni was significantly decreased in the supplemented group, suggesting a positive effect on immune function.

The planned duration of supplementation may also influence the effects on statural or ponderal changes. In a study of Guatemala City young schoolchildren, the zinc supplement was given, because mainly of absenteeism, on average a total of 90 days during the six months of observation (4). Changes in body composition characterised by increased fat deposition in presence of less muscle loss were the main findings in the study with the effects most marked in children with lower levels of hair zinc content. In non-selected population studies of older children at reasonable risk of zinc deficiency, 12 month studies seem necessary and the merits of preselecting participants with known biochemical markers, for example, low hair zinc levels in mild zinc deficiency, should be discussed.

Breast-fed infants represent another population at risk, as the zinc content of breast-milk declines from 40 to 10 micromoles per liter over the first 6 months of lactation. In a study of immigrant breastfed children in Paris, whose median age was 5.6 months, the daily provision of 5 mg elemental zinc for 3 months led to a 30% increase in length gains of the supplemented boys and a similar difference in weight gains (5). Improved growth, but mainly in females this time, is found in an ongoing study of zinc supplements of younger breastfed infants in Denver (6). The supplements are started at 2 months of age, at a dose of 5 mg daily, and the mean daily weight gains between 4 and 7 months of age, were 3 grams greater in the supplemented group. In small for gestational age (SGA) infants, zinc supplementation starting in the first month of life led to dramatic improvements in mean weight for age Z-scores, from -2.05 at onset to -0.24 at 6 months (7). Differences in weight gains between the treated and placebo groups were significant after two months of supplementation and the catch-up growth was most marked in the SGA girls. This study was prompted by observations that many malnourished children in Chili had prior histories of fetal growth retardation and that catch-up growth in SGA infants, and the concomitant increased tissue synthesis, would lead to higher zinc requirements. The same reasons prompted the use of cow milk formulas supplemented to 15 mg zinc per liter to treat children recuperating from malnutrition (8).

The gender differences in response to zinc supplements remain mysterious. In a study of low income Chilean preschool children, boys, as is often the case, showed significant increments in length gains (9) and a similar effect was noticed in a year long trial of zinc treatment of children and adolescents with short stature (10). In rural Guatemalan infants however, gender was not the discriminating factor, but moderate stunting defined the population whose growth improved with added zinc (11).

Various aspects of immunity and cognition have also been studied in supplementation projects. In China, a three armed study of zinc, zinc and micronutrients, or micronutrients alone, showed significant increases in knee height with the latter two interventions (12). The participants were also administered a series of computer based tests of psychomotor and cognitive function. Zinc alone and with micronutrients improved dexterity and eye-hand co-ordination. Perception, memory and concept formation improved most with the zinc and micronutrient supplement. Attention, search (attention and perception) and serial object recognition were not affected in the study (13). It has been estimated that 20 - 30% of Chinese children have suboptimal zinc nutrition, therefore these growth and cognitive function changes have important, practical implications for better health and educational achievement of the largest uninational childhood group in the world.

In addition to cognitive benefits, zinc supplementation studies have demonstrated decreases in morbidity which have major public health consequences. Diarrhea is a regular finding in severe zinc deficiency and large losses of zinc have been demonstrated in diarrheal stools. Hence, the idea to provide zinc supplements to a large cohort of 937 children, 6 -35 months of age in New Delhi, India (14). The children receiving the zinc supplement had reductions in the risk of continued diarrhea, in the number of watery stools per day and in the duration of watery diarrhea. These effects on duration and severity were greater in children with stunted growth who probably had more severe zinc deficiency. Nearly two-thirds of these children were given zinc for an additional 180 days and they had significantly greater length gains and weight gains in the younger participants than the multivitamin treated control group. Moreover there was a significant reduction in the number of subsequent episodes of diarrhea, the number of days with diarrhea and in the number of days with acute respiratory illnesses (S. Sazawal and R. Black, personal communication). These effects were most marked in the group of children who had low levels of plasma zinc upon entry into the study. Confirmation of the beneficial effects of zinc on diarrheal and respiratory illnesses is also found in a preliminary report from studies of zinc and iron supplementation in Mexico (15) and for diarrheal disease only, in the Guatemalan infant study (16). A very large study in SGA infants is starting in India, and while cumbersome and slow, controlled supplementation studies will continue to demonstrate the importance of zinc as a nutrient.

3. REFERENCES

1. P.J. Aggett, in *Trace Elements in Infancy and Childhood. Annales Nestlé*, **52**, 94–106 (1994).
2. C.J. Bates, P.H. Evans, M. Dardenne, A. Prentice et al., *Br. J. Nutr.*, **69**, 243–255 (1993).
3. H. Friis, T. Ndhloyu,K. Mduluza, K. Kaondera et al., *Faseb J.* **A164** (1984).
4. K.R. Cavan, R.S. Gibson, C.F. Grazioso, A.M. Isalgue AM et al., *Am. J. Clin. Nutr.* **37**, 344–352 (1993).
5. P.A. Walravens, A. Chakar, R. Mokni, J. Denise, D. Lemonnier, *Lancet* **340**, 683–685 (1993).
6. N.F. Krebs, J.E. Westcott, N. Butler-Simon, K.M. Hambidge, *Faseb J.* **10** (1996), abstract.
7. C. Castillo-Duran, A. Rodriguez, G. Venegas, P. Alvarez, G. Icaza, *J. Pediatr.* **127**, 206–211 (1995).
8. L. Schlesinger, M. Arevala, S. Arredondo, M. Diaz et al., *Am. J. Clin. Nutr.* **56**, 491–498 (1992).
9. M. Ruz, C. Castillo-Duran, X. Lara, A. Rebolledo et al., *Faseb J.*, **9**, A736 (1995).
10. C. Castillo-Duran, H. Garcia, P. Venegas, I. Torrealba et al., *Acta Paediatr.* **83**, 833–837 (1994).
11. J. Rivera, K.H. Brown, M.C. Santizo, M. Ruel, B. Lonnerdal, *Faseb J.* **9**, A164 (1995).
12. C. Chen, J. Yang, J. Li, F. Zhao, H. Dayal, H. Sandstead. Faseb J. **9**, A481(1995).
13. J.G. Penland, H.H. Sandstead, X. Chen, J. Li, *Faseb J.*, **10** (1996), abstract.
14. S. Sazwal, R.E. Black, M.K. Bhan, N. Bhandari et al., *N. Engl. J. Med*.**333**, 839–844 (1995).
15. J.L. Dosado, L.H. Allen, P. Lopez, H. Martinez, *Faseb J.*, **9**, A157 (1995).
16. M.T. Ruel, J. Rivera, K. Brown, M.C. Santizo, B. Lonnerdal, *Faseb J.* **9**, A157 (1995).

14

RELEVANCE OF TRACE ELEMENT SUPPLEMENTS IN WOMEN OF DIFFERENT AGES

A. Favier

GREPO Research Group on Oxidative Pathologies
Faculty of Pharmacy
University of Grenoble
F-38700 La Tronche, France

1. INTRODUCTION

Trace element supplementation projects in women bring up specific problems. These are a reflection of the action of hormones on the metabolism of trace elements. Conversely the hormonal metabolism is dependent, at many steps, on various trace elements and hence can be modified by nutritional deficiencies. The adequacy of trace element intake is most important during the major hormonal changes that rhythm women's life.

2. RELATIONSHIP BETWEEN TRACE ELEMENTS AND FEMALE HORMONES

Trace elements can act at all levels of production, action and regulation of hormones (1). In the hypothalamic-pituitary axis, zinc facilitates the synthesis and activation and copper is involved in the terminal amidation of peptides. Zinc also regulates the secretion of prolactin by the pituitary gland (2) and copper in a chelated form, acts on a peptidyl amidase to modulate the release of LHRH (3). The administration of copper to the rat stimulates the release of GnRH and LH (4). Trace elements act on several steps in the synthesis of estrogens. Steroid hormones synthesis is dependent on zinc at the level of 3 and 17 hydroxy-dehydrogenases, and on iron at the level of hydroxylases. Zinc by desaturases, iron by oxygenase and selenium modulate the synthesis of prostaglandins, that have a major influence on the onset of labor (5).

Finally trace elements are necessary for the activity of peripheral receptors that allow the expression of hormonal message. Zinc modulates the form of the zinc-finger proteins of the estrogen receptors, which are cytosolic proteins migrating to the nucleus of the cell under

Therapeutic Uses of Trace Elements, edited by Nève et al.
Plenum Press, New York, 1996

the influence of the hormones to act as transcription factors. Zinc deficient rats present reduced estrogen sensitivity in the absence of any modification in the number of receptors (6). It is worth mentioning that copper also promotes estrogen binding to the protein receptor in the cytosolic compartment (7). The same effect occurs with progesterone, whose receptor functions according to the same model. Zinc gives also an active form to the LH receptors inside the cell membrane .

Conversely female hormones influence the metabolism of trace elements. Estrogens, and or progesterone strongly modify copper metabolism, increasing copper concentrations in serum, liver and kidneys. Estrogen raises serum transferrin and iron levels, by increasing the transcription of the transferrin gene. There are conflicting results on the effects of sex hormones on zinc metabolism, but most studies demonstrate a decrease in serum zinc induced by estrogen administration (8). The increases in serum copper and decreases in serum zinc observed during oral contraception or pregnancy are linked to this estrogen effect. The differences in manganese absorption observed between males and females result certainly from an hormonal action (9).

3. DIETARY ALLOWANCE, INTAKE, AND STATUS OF WOMEN

Trace element requirements for women vary largely with hormonal physiological changes and women's needs are also lower than in men according to their body weight. The onset of menstruation during puberty results in increased iron losses hence iron requirements are higher in women than men until menopause. In France, daily needs are estimated at 21.4 mg of iron for menstruating teen-agers and 18.09 mg for mature women (10). During pregnancy women produce new tissues for the fetal-placental unit and these tissues are rich in trace elements, thus creating new needs for iron or zinc. Lactation induces increased needs for many trace elements that have to be present in milk to ensure adequate growth of the newborn. With the exception of iron, iodine and zinc whose recommended intake during pregnancy or lactation are increased, recommended allowances (11) do not sufficiently take known facts into account. For other trace elements, such as chromium or manganese, the recommended dietary allowances during pregnancy are unchanged thus ignoring women's real needs.

Actual dietary intakes of women are generally far lower than recommended ones. In a large nutritional study performed in France we observed an inadequacy of intakes for zinc, iron, copper (12). Variations in the discrepancies between needs and intakes during life result in important changes in levels of trace elements in biological fluids. Thus, during the Val de Marne study in 1989, we observed a progressive decline of serum zinc with age, while copper and iron increased immediately after puberty (Figure 1). In the same population the calculation of deficit risk, i.e., the frequency of abnormally low values, shows that iron deficits become major when women menstruate or become pregnant, then disappears after the menopause. The risk of zinc deficiency increases later and becomes frequent in the elderly woman.

4. EFFECTS OF TRACE ELEMENTS SUPPLEMENTATION

4.1. During Sexual Maturation

In females, zinc deficiency triggers sexual malfunctions responsible for reproductive disorders. Abnormalities of oocytes , estrogen cycle and ovulation have been observed in

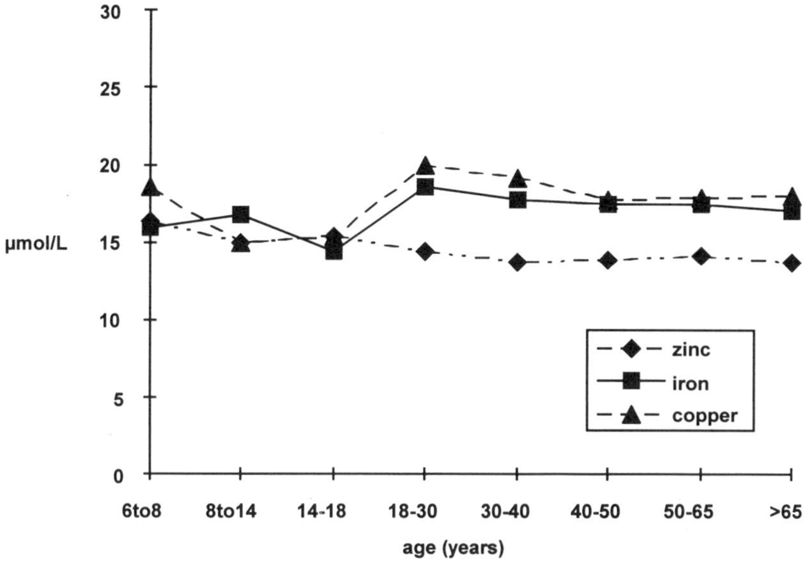

Figure 1. Evolution of serum trace elements (mean) in French normal women according to age (Val de Marne study).

mice (13), rabbits (14), rats (15), or Rhesus monkeys (16) fed zinc-deficient diets. Pregnancy cannot occur in deficient animals. In Iran, young women, 19 and 20 years old, suffering from dwarfism with delayed sexual maturation had correction of symptoms by zinc treatment (17). Similar symptoms were observed in 7 Turkish women practising geophagia (18). Infertility in women secondary to celiac disease was found to be related to zinc deficiency (19). However, measured zincemia in 48 infertile women was found quite normal (20). Selenium supplementation in 6 infertile women whose magnesium blood level was not restored to normal after magnesium supplements, improved their magnesium levels and restored fertility (21).

In pre-menstrual syndrome, an unusual and painful condition that is improved by exercise and diet, supplementation with tocopherol, vitamin B6, and zinc is recommended. A double blind trial with magnesium (22) or with a vitamin and mineral supplement showed beneficial effects (23).

4.2. During Pregnancy

Pregnancy results in a dramatic increase in the need for trace elements. These nutrients are necessary not only for synthesis of fetal and placental tissues, but, during the last month of gestation, they contribute to hepatic stores that will be used by the newborn during the lactating period. An oxidative stress of unknown origin occurs in mothers during pregnancy, as demonstrated by the progressive increase in conjugated dienes or in MDA in blood (24). Unfortunately antioxidant vitamin and mineral supplements failed to modify this increase of peroxides in blood (25).

4.2.1. Benefit for the Mother. Muscle weakness is common after delivery and results from efforts exerted but also from anemia with impaired tissue oxygenation. Iron deficiency is implicated to a considerable extent in the origin of these two disorders (26). Ane-

Table 1. Recommended supplementation in pregnant women

	Demonstrated effect	Supposed effect in industrial countries
Iron	–decreases anemia	–increases birthweight
	–decreases prematurity	–decreases fetal mortality
Zinc	–increases birthweight	–decreased spina bifida
	–reduces pre and postmaturity	–improves delivery
		–protects of gestational diabetes
Copper		–decreases bone malformation
Selenium	–decreases peroxides	–protects from toxemia
		–protects of gestational diabetes
Iodine		–improves brain development
Chromium		–protects of gestational diabetes
Fluorine	–increases birthweight	

mia is very frequent and hemoglobin levels less that 11 g/dL is found in 30–50 % of pregnant women in the third world and in 2–30 % in industrialised countries (27). Many studies have evaluated the effect of iron supplements during pregnancy on maternal iron status, the latter generally estimated by assaying serum ferritin. The doses used should be adjusted in function of the state of reserves and are sometimes very high, from 40 to 250 mg, especially in third world countries. Iron supplements result in higher ferritinemia and hemoglobin levels, but cannot totally prevent the drop in these parameters during pregnancy.

Even if customary, iron supplementation is rarely done correctly during pregnancy. Few pregnant women are supplemented for the duration of gestation. For instance in a survey in Jakarta, the prevalence of anemia did not decrease during supplementation with 300 mg iron sulphate but only 64 % of women claimed to have taken all the tablets . When supplemental intake was checked by iron analysis in stool, only 25 % had a positive test (28). Supplementation must start as early as possible and must last at least 12 weeks; high iron doses that create zinc deficiency must be avoided (29). A supplementation with 45 mg zinc during pregnancy slightly decreases maternal serum copper but increases hemoglobin (30).

4.2.2. Effect of Supplementation on Pregnancy and Delivery.

4.2.2.1. Pre- and post maturity. Premature delivery is considered more common in mildly anemic mothers than in non- anemic mothers (31). In a Kenyan study, the rate of premature births reached 42 % in anemic mothers *compared to* 7 % of children from non-anemic mothers (32). Women giving birth before the 36th week have abnormally low circulating copper levels (33). Several studies in pregnant women have shown a link between low serum zinc levels, or decreased zinc in amniotic liquid, and an abnormal increase in the length of pregnancy (34). Supplementing pregnant women with zinc reduced the prematurity rates (35) and decreased the postmaturity (36).

Zinc deficiency causes a number of parturition disturbances in animals, excessive blood losses, and increased stress (37). A decreased serum zinc was observed at the beginning of pregnancy in 18 women who subsequently experienced a complication during labor or atonic bleeding after delivery (38). Increased bleeding and a very significant increase in the pre-labor phase, especially in the dilatation period was observed in zinc-deficient women. Effective supplementation with 45 mg/d of zinc decreases the frequency of abnormally low labor times.

4.2.2.2. Gestational diabetes. Diabetes whether long standing or of gestational origin alters indices of maternal selenium metabolism during pregnancy, as demonstrated by decreases in plasma selenium and glutathione peroxidase (GPx) levels (39). Pregnant diabetic women have higher lipid peroxidation accompanied by a decrease in GPx and SOD activities than non-diabetic pregnant women (40). Oxidative stress may be implied in the origin of gestational diabetes as it decreases peripheral sensitivity to insulin (41). Zinc supplementation decreases lipid peroxidation in diabetic patients (42). A disturbance in chromium metabolism is observed in pregnant diabetic women who have a transient increase in hair chromium (43). Chromium supplementation has been beneficial for glucose homeostasis in diabetes.

4.2.2.3. Toxemia Gravidis. Toxemia gravidis is one of the most frequent complications of pregnancy. It is characterised by arterial hypertension (pre-eclampsia), associated with proteinuria and edema. It can evolve into an eclamptic crisis and can be complicated by retroplacental hematomas, arterial hypertension, fetal distress, death in-utero, or disseminated intravascular coagulation that can be fatal for the mother or child. Two theories coexist to explain this state; one of placental ischemia and an immuno-genetic theory. Many anomalies of trace elements have been observed: a significant decrease in zincemia (44–46), an increase of placental (47) and serum copper (48–50), an increase in serum iron and ferritin (51). Selenium status seems unchanged (52–53). The increase in serum copper occurs as early as the forth month of pregnancy, when no clinical signs exist (54). Several authors found a decline of the magnesium status. For Kisters (55), plasma magnesium is not lowered while erythrocyte magnesium is low as compared to normal mothers; serum zinc levels decreased while serum copper remained normal. An oxidative stress can originate or participate in toxemic damage (56–57). This oxidative stress can result from placental ischemia or from anomalies of the endothelial cells. Supplementation with zinc (58) or magnesium (59–61) give discordant results. Only supplementation with calcium (62–64) have shown a net beneficial effect. Recently a supplementation with 100 mg of selenium improved blood pressure (65). It is important to notice that some trials with zinc caused an increase in biological signs of pre-eclampsia.

4.2.3. Effect on the Foetus and Newborn.

4.2.3.1. Fetal mortality. Experimental zinc, copper and magnesium deficiencies in animals are responsible for fetal resorptions and stillbirths. The analysis of trace element status of women aborting has revealed low copper (66) or zinc levels (67). A link between the risk of placental abruption and decreased circulating copper levels has been reported (68) and could be explained by the importance of copper for the maturation of collagen and elastin. The levels of fetal mortality and morbidity are higher as anemia is more severe, and there is a link between anemia and a decreased urinary estrogen secretion.

4.2.3.2. Growth retardation. Growth retardation induced zinc deficiency has been reported by various authors and in different animal species: rats, sheep and Rhesus monkeys. Many studies in humans attempted to ascertain a link between the maternal zinc status expressed by serum zinc or leukocyte zinc and the weight of the newborn at birth or the zinc status of the newborn. The latter was expressed by zinc in the amniotic fluid, in tissues, in umbilical cord blood or in certain cases in fetal tissues (69). Birthweight is very often correlated with maternal iron status, and sometimes with copper or chromium status (see revue of Thauvin 1992). Few supplementation studies have evaluated effects on birthweight. Zinc supplements have beneficial results on birthweight (70), particularly when given to pregnant women with a borderline zinc status such as teenagers (71), women at risk of having small-for dates babies (72), or women from low income families (73). In a region where drinking water fluorine levels ranged from 0.02 to 0.16 mg/l, newborns, whose mothers received a fluorine supplement, had a higher birthweight (74).

4.2.3.3. Malformations. Experimental deficiencies of copper, zinc, and manganese during gestation in animals are teratogenic. In humans, only zinc deficiency is associated with a risk of a defective closure of the neural tube, resulting in anomalies such as spina bifida or anencephaly. A number of studies have shown disturbances in zinc status in mothers of children with neural tube defects, when compared to mothers of healthy children. We have done several investigations that have all shown this relationship, in particular when assessing the nutritional status of 25 mothers that had just given birth to a baby with a neural tube defect. Only serum zinc was decreased, and no deficiencies of iron, copper manganese, vitamins B1, B2, B6, C or A, were found, nor were there decreases in folates, which nevertheless are often considered responsible for this abnormality (75).

During gestation, vanadium, molybdenum, nickel and arsenic deficiencies caused a significant increase in abortions and intrauterine death in goats. The newborn of goats deficient in vanadium also suffer from joint and skeletal disorders, while the newborn of mothers deficient in molybdenum and nursed by these deficient mothers have a growth retardation (76). These deficiencies will probably not be encountered in humans, whose needs for these inorganic elements are apparently largely satisfied.

In spite of the recognised relationship between an altered maternal trace element status and the birth of malformed babies, systematic supplementation studies of pregnant women with trace elements have not yet been undertaken. Only a comparison of the effects of a randomised multivitamin and mineral supplement given to 2471 women, versus a mixture of trace-elements (Mn, Zn, Cu) given to 2391 women, showed that only the vitamin and mineral supplement suppressed the appearance of spina bifida (77).

4.2.3.4. Postnatal effects of supplementation during pregnancy. Other consequences of trace element deficiencies, such as decreased immunity, may be manifested after birth. Newborn rats whose mothers were iron deficient during gestation show decreased cytotoxicity of peritoneal NK cells (78). The newborn from mice deficient in zinc during the last two-thirds of gestation have a default in production of IgA and IgG2 which persists up to 6 months after birth in spite of normal postnatal feeding (79).

4.3. In Elderly Women

After menopause women undergo great changes in hormonal metabolism. Deficits, notably in zinc, become more frequent and can decrease the efficiency of hormonal substitution. Using stable isotopes, we observed a decrease with age in zinc and selenium pool size in elderly women, and supplementation with selenium improved selenium status and decreased MDA in women over 65 years of age (80). But during the Mineral-Vitamin-Antioxidant study we observed an increase in MDA in elderly women that remained unchanged by a supplementation with selenium, zinc and antioxidant vitamins (81). Many age-related skin disorders could be improved by supplementation with zinc, copper and selenium. Osteoporosis, one of the most frequent complications of menopause, is decreased by fluorine supplementation (82). Bone density, as measured by photon absorption, was related to intakes of zinc, magnesium and iron (83). Copper has also been suspected to play a role in this

Table 2. Recommended supplements in postmenopausal women

	Demonstrated	Supposed
Zinc	–improves skin, immunity	
Selenium	–decreases peroxides	
Copper		–improves osteoporosis
Fluoride	–improves osteoporosis	
Silicon		–improves osteoporosis
Strontium		–improves osteoporosis
Boron	–improves calcium status	
	–increases SOD, ceruloplasmin	

disorder (84). Supplementation with boron (3 mg/d) decreases the urinary elimination of calcium, magnesium and phosphate, and increases the concentrations in serum of 17 beta-estradiol and testosterone in menopausal women (85). In another trial, boron (4 mg/d) increased SOD and ceruloplasmin activity regardless of coexisting treatment or not with estrogens (86). Supplementation with silicon and strontium may also be beneficial in osteoporosis.

Iron requirements decrease with the end of menstrual cycles and the ferritin levels of postmenopausal women approach those of males, and iron deficiency becomes very unusual (87). Thus iron supplements seem unnecessary after menopause. Even more recent hypotheses suggest that the longer life span of women results from their lower iron status (88). Iron overload favours free radical production and may be responsible for increases in cancer and cardiovascular diseases (89). Some large prospective studies found people with high iron status at increased risk for colorectal cancer (90), but no relationship to cardiovascular risk (91,92). These newer aspects of iron nutrition are of particular importance and deserve further epidemiological investigations.

5. REFERENCES

1. J. Nève, A. Peretz, in *Handbook of Metal-Ligand in Biological Fluids*, Berthon G., ed, Marcel Dekker, New-York , part 2, chapter 2, pp 128–135 (1995).
2. M. Koppelman, *Medical Hypothesis* **25**, 65 (1988).
3. A. Bamea, G. Cho, B. Katz , *Brain Research* **541**, 93 (1991).
4. E. Hazum, *Biochem. Biophys. Res. Comm.* **112**, 1983 (1983).
5. J. Schwabe, D. Rhodes , *TIBS* **16**, 291–296 (1991).
6. G. E. Bunce , M. Vessal, *Steroid Biochem.* **26**, 303–308 (1987).
7. H. Fischman, J. Fischman, *Biochem. Biophys. Res. Comm.* **144**, 505–511 (1987).
8. G. Leblondel , P. Allain, in *Handbook of metal-ligand in biological fluids*, Berthon G. ed, Marcel Dekker, New-York , part 2, pp. 484–494 (1995).
9. J. W. Finley , P. E. Johnson , L. K.Johnson, *Am. J. Clin. Nutr.* **60**, 949–955 (1994).
10. L. Hallberg , L. Rossander-Hulten , *Am. J. Clin. Nutr.* **55**, 1047–1058 (1992).
11. National Research Council, *Recommended Dietary Allowances,* 10th National Academic Press, Whashington (1989).
12. A. Favier, S. Hercberg, J. Arnaud, P. Preziosi, P. Galand, *Trace Elements in Man and Animals VII*, pp. 13–5 to 13–7., Momcilovic B., ed., IMI, Zagreb (1991).
13. T. Wanatabe , F. Sato, A. Endo , *Yamagata Med. J.* **1**, 13–20 (1983).
14. N. A. Shaw, H. C. Dickey, H. M. Brugman, D. L. Blamberg, J. F.Witter, *Lab. Anim.* **8**, 1–7 (1974).
15. H. Swenerton and L.S. Hurley, J. Nutr. **95**, 8–18 (1968).
16. H. Swenerton and L.S. Hurley, *J. Nutr.* **110**, 575–583 (1980).
17. H.A. Ronaghy , J. A. Halsted , *Am. J. Clin. Nutr.* **28**, 831–836 (1975).
18. A. O. Cavdar, A. Arcassoy , S. Cin , E. Babacar, S. Gozdasaglu, *Zinc deficiency in human subjects,* Alan R. Liss, New York, pp. 71–97 (1983).
19. S. Jameson , *Acta Med. Scand.* **593** (suppl), 3–89 (1976).

20. M. H. Soltan, D. M. Jenkins, *Br. J. Obstet. Gynecol.* **90**, 457–459 (1983).
21. J. M. Howard , S. Davies , A. Hunnisett , *Magnes. Res.* **7**, 49–57 (1994).
22. F. Facchinetti, P. Borella, G. Snes, L. Fioroni, E. Nappi, A. Genazzani, *Obstet. Gynecol.* **78**, 177–181 (1991).
23. S . Bradley, N. Chiamori, *J. Am. Coll. Nutr.* **10**, 494–499 (1991).
24. R. Couderc, J.P. Bonnardot, S. Kerisit, D.N Brault, P. Deligne, P. Laruelle, *Biologie Prospective. Proceedings of the 76th colloqium of* Pont à Mousson (France), pp. 577–580 (1989).
25. E. Thauvin. Thesis of the University of Grenoble (1992).
26. F.E. Viteri, B. Torun, *Clin. Haematol.* **3**, 609 (1974).
27. P. Galan, S. Hercberg, Y. Soustre, H. Dupin, *Hum. Nutr. Clin. Nutr.* **39**, 279 (1989).
28. W. Schultink , M. Van Der Ree, P. Matulessi, R. Gross, *Am. J. Clin. Nutr.* **57**, 135–139 (1993).
29. N. R. Williams, D. L. Bloxam , Y. Morardji , P. M. Pattinson-Green, in *Nutrient Availability,* D. Southgate, ed., Royal Society of Chemistry, Cambridge, pp. 220–225 (1989).
30. K. Garg, C. Singhal, Z. Arshad, *Indian J. Physiol. Pharmacol.* **38**, 272–276 (1994).
31. G.J. Ratten , N.A. Beischer , *J. Obst. Gynecol. Br. Commonw.* **79**, 228–237 (1972).
32. M. MaC Gregor, *Scot. Med. J.* **8**, 134–140 (1963).
33. S. Friedman, C. Bahary, B. Eckerling, B. Gans, *Obstet. Gynecol.* **33**, 189–193 (1969).
34. A. Favier, *Biol. Trace Element Res.* **32**, 363–382 (1992).
35. F. F. Cherry, H. H. Sandstead , P. Rojas , L. Johnson, H. K. Batson , X. B. Wang, *Am. J. Clin. Nutr.* **21**, 739–742 (1989).
36. S. Jameson, in *Recent Avances in Zinc Metabolism*, Alan R Liss, New York, pp 32–45 (1982).
37. J. Apgar, *Ann. Rev. Nutr.* **5**, 43–68 (1985).
38. S. Jameson , *Acta Med. Scand.* **593**, 3–89 (1976).
39. A. Farhat, M. Frances Picciano, C. Lammi-Keefe, H. Desilva, *J. Trace Elements Exp. Med.* **8**, 29–39 (1995).
40. K. Twardowska-Saucha , W. Grzeszczak , B. Lacka , J. Froehlich , D. Krywult, *Pol. Arch. Med. Wewn.* **92**, 313–21 (1994).
41. P. Faure, J. L. Lafond, C. Coudray, E. Rossini, S. Halimi, A. Favier, D. Blache, *Biochem. Bioshys. Acta.* **1209**, 260–264 (1994).
42. P. Faure , P. Y. Benhamou, A. Perard, S. Halimi, A. M. Roussel, *Eur. J. Nutr.* **49**, 282–288 (1995).
43. A. Aharoni , B. Tesler, Y. Paltieli, J. Tal, Z. Dori , M. Sharf, *Am. J. Clin. Nutr.* **55**, 104–107 (1992).
44. A. W. Zimmerman, B. S.Dunham, D. J. Nochimson, B. Kaplan, J. M. Clive, S. L. Kunkel, *Am. J. Obstet. Gynecol.* **149**, 523–529 (1984).
45. N. Lazebnick, B. R. Kuhnert, P. M. Kuhnert, K. Thompson , *Am. J. Obstet. Gynecol.* **161**, 437–440 (1989).
46. F. A. Adenyyi , *Acta Obst. Gynecol. Scand.* **66**, 579–582 (1987).
47. M. H. Brophy, N. F. Harris, I. L. Crawford, *Clin. Chim. Acta* **145**, 107–112 (1985).
48. J. A. O'leary, G. S. Novalis, G. J. Vosburgh, *Obstet. Gynecol.* **28**, 112–117 (1966).
49. M. M. Fattah, F. K. Ibrahim, M. A. Ramadan, M. B. Sammour, *Acta Obstet. Gynecol. Scand.* **55**, 383–385 (1976).
50. D. M. Campbell , *Proceedings of the Nutrition Society* **47**, 45–53 (1988).
51. S. S. Entsman , L. D. Richarson , A. P. Killam, Am. *J. Obstet. Gynecol.* **144**, 418–422 (1982).
52. M. Hyvonen, P. Dabek, P. Nikkinen-Vilkki, J.T. Dabek , *Clin. Chem.* **30**, 529–533 (1984).
53. A. C. Roy, S. S. Ratman, R. Karunanithy, *Gynecol. Obstet. Invest.* **28**, 161–162, (1989).
54. J. Badin , M. A. Denne , P. Morin , *Nouv. Presse Med.* **11**, 2763–2766 (1982).
55. K. Kisters, C. Spieker, I. Fafera , C. Muller, K. Rahn , W. Zidek , *Trace El. Med.* **10**, 158–162 (1993).
56. J. Uotila , R. Tuimala , K. Pyykoo, *Gynecol. Obstet. Invest.* **29**, 259–262 (1990).
57. W.F. O'Brien, *Obstet. Gynecol.* **75**, 445–452 (1990).
58. I. F. Hunt, N. J. Murphy, A. E. Cleaver, *Am. J. Clin. Nutr.* **40**, 508–521 (1984).
59. L. Spatling, G. Spatling, *Brit. J. Obstet. Gynecol.* **95**, 120–125 (1988).
60. A. Conradt, H. Weldinger, *Magn. Bull.* **6**, 68–76 (1984).
61. B. M. Sibai, M. A. Villar, E. Bray, *Am. J. Obstet. Gynecol.* **161**, 115–119 (1989).
62. P. Lopez-Jaramillo, M. Narvaez, R. M. Weigel, R. Yepez , *Brit. J.Obstet. Gynecol.* **96**, 648–655 (1989).
63. J. Villar, J. T. Repke, *Am. J. Obstet. Gynecol.* **163**, 1124–1131 (1990).
64. N. Kawasaki , K. Matsui , M. Ito , *Am. J. Obstet. Gynecol.* **153**, 576–582 (1985).
65. L. Han , S. M. Zhou , *Chin. Med. J. Engl.* **107**, 870–871 (1994).
66. S. Friedman, C. Bahary, B. Eckerling, B.Gans, *Obst. Gynecol.* **33**, 189–193(1969).
67. M. W. Breskin, B. Worthington-Roberts, R. Knopp, Z. Brown, B. Plovie, K. Mottet, J. Mills, *Am. J. Clin. Nutr.* **38**, 943–953 (1983).
68. R. Artal, R. Burgeson, F.J. Fernandez, C.J. Obel, *Obstet. Gynecol.*, **33**, 189–193 (1979).

69. A. Favier, *Biol. Trace Element Res.* **32**, 383–398 (1992).

70. G. Kynast , E. Saling , *Gynecol. Obst. Invest.* **21**, 117–223 (1986).

71. F. F. Cherry , H. H. Sandstead , P. Rojas L. Johnson, H. K. Batson, X. B. Wang, *Am. J. Clin. Nutr.* **21**, 739–742 (1989).

72. K. Simmer , L. Lort-Phillips, C. James, R. P. Thompson, *Eur. J. Clin. Nutr.* **45**, 139–144 (1991).

73. R. Goldenberg, T. Tamura, Y. Naggers , R. Copper , K. Johnston , M. Dubard , J. Hauth, *JAMA*, **274**, 463–468 (1995).

74. R. Bergmann, K. Bergmann, in *Trace elements in nutrition*, Nestlé nutrition Workshop Series, vol. 23, Raven Press, New York, pp 105–118 (1991).

75. A. Favier , M. Favier, *Rev. Fr. Gyn. Obst.* **85**, 49–55 (1990).

76. M. Anke, in *Trace elements in nutrition*, Nestlé nutrition Workshop Series, vol. 23, Raven Press, New York, pp 1119–144 (1991).

77. E. Czeizel, I. Dudas, *Orv. Hetil.* **135**, 2313–2317 (1994).

78. N. Hallquist , L. Mc Neil, J. Lockwood A. Sherman, *Am. J. Clin. Nutr.* **55**, 741–746 (1992).

79. R. S. Beach, E. Gershwin , L. S. Hurley, *Am. J. Clin. Nutr.* **38**, 579–590 (1993).

80. A. Bortoli, G. Fazzin, M. Marchiori, F. Mello, R. Brugiolo, F. Martelli, *J. Trace. Elem. Electrolytes Health Dis.* **5**, 19–21 (1991).

81. A.L. Monget, Thesis of the University of Paris VII (1995).

82. M. A. Dambacher, J. Ittner, P. Ruegsegger, *Bone* **7**, 199–205 (1986).

83. R. Angus, P. Sambrook, H. Pocock, J. Eisman, *Bone Miner.* **4**, 265–277 (1988).

84. J. Strain, *Med. Hypoth.* **27**, 333–338 (1988).

85. F. H. Nielsen, C. Hunt, L. Mullen , J. Hunt, *Faseb J.* **1**, 394–397 (1987).

86. F. H. Nielsen, *Environ. Health Perspect.* **102** (S 7), 59–63 (1994).

87. J. Cook, B. Skikne, S. Lynch, M. Reusser, *Blood* **68**, 726–731 (1986).

88. J. Sullivan, *Am. Heart J.* **117**, 1177–1188 (1989).

89. R. Lauffer, *Med Hypoth.* **35**, 95–102 (1990).

90. P. Knekt, A. Reunanen , H. Takkunen , A. Aromaa, M. Heliovaara, T. Hakulinen, *Int. J. Cancer* **56**, 379–382 (1994).

91. D. M. Baer, I. S. Tekawa, L. B. Hurley, *Circulation* **89**, 2915–2918 (1994).

92. Y. Liao, R. S. Cooper , D. L. McGee, *Am. J. Epidemiol.* **139**, 704–12 (1994).

15

DOUBLE-BLIND SUPPLEMENTATION WITH TRACE ELEMENTS, MAGNESIUM AND VITAMINS DURING PREGNANCY IN A RANDOMLY SELECTED POPULATION

H. Faure,[1] M. Favier,[2] E. Thauvin,[1] J. Arnaud,[1] M. Fusselier,[3] and
A. Favier[1]

[1] Laboratoire de Biochimie C
[2] Service de Gynécologie-Obstétrique Sud
C.H.U.G.
BP 217, 38043 Grenoble Cedex 9 France
[3] Laboratoires Aguettant
Rue A. Fleming, 69007 Lyon, France

1. INTRODUCTION

During pregnancy, metabolic changes are observed in women. These changes affect all the nutrients, but they become crucial for certain trace elements and vitamins, and for magnesium, that tend to decrease towards the deficiency area. These changes are caused by an alteration of volume and distribution of body fluids (1), the coverage of the foetus needs, and an important tissue synthesis (2). This have long been known and several supplementation studies are reported in the literature. However, these supplementation trials generally involved a few trace elements and vitamins - often zinc, iron, ascorbate, or folate - and this can be criticised as interactions between these nutrients occur. Indeed, a supplementation with zinc may alter iron and copper metabolism, and the opposite is also true. Folate has been shown to increase fecal excretion of zinc (3).

In the present study, we report the results for iron and zinc parameters in a double-blind randomised study, which involved 102 pregnant women and their newborns. One group received a placebo and another received a supplementation with copper, zinc, iron, chromium, magnesium and vitamins C, B_1, B_2, B_6, B_9, B_{12}, and E. To our knowledge, a global supplementation trial with all these micronutrients has never been reported.

Therapeutic Uses of Trace Elements, edited by Nève et al.
Plenum Press, New York, 1996

2. MATERIAL AND METHODS

2.1. Subjects

One hundred and thirty two pregnant women from the Maternity of the Grenoble University Hospital were initially included, but after beginning the study 6 refused to take their capsules. The women were included in the study at the third month of pregnancy and they all gave their informed consent. Women were examined by a gynaecologist at 3, 6, and 9 months of pregnancy. Some also agreed to be examined 2 months after delivery. Were excluded from the study: diabetic women, and those who suffered from toxoplasmosis, or chronic renal failure, women who had taken a trace element/vitamin supplementation less than six months before the study, deeply anaemic women at the third month of pregnancy (blood haemoglobin <10 g/dl), women who did not speak French fluently, and women addicted to drugs, alcohol, or tobacco. Were also excluded 21 women in whom the second sample could not be taken. Finally, this study included 53 women who took the polyvitamin-mineral supplementation and 49 women who received the placebo. This study was approved by the ethical committee of the Grenoble University Hospital.

2.2. Supplementation

Half of the women randomly received the placebo, and the other half received the supplementation. This supplementation consisted in 50 mg elemental iron and 30 mg elemental zinc daily absorbed *per os* by women. The supplement additionally contained copper (2 m g/d), chromium (50 µg/d), magnesium (200 mg/d), ascorbate (50 mg/d), thiamine (2mg/d) riboflavin (2 mg/d), pyridoxin (4 mg/d), folic acid (1mg/d), cyanocobalamin (2.5 mg/d) and α-tocopherol (15 mg/d). The six capsules (two before every meal) were taken daily from the third month of pregnancy till the delivery.

2.3. Samples, Biological Determinations, and Statistical Analysis

Blood samples were taken at each clinical examination. Cord blood was also taken at delivery and a sample was taken two months after delivery in women who accepted.

Serum iron (ferrozine), ferritine (ELISA), plasma zinc (EAAS), monocyte and polymorphonuclear zinc (ficoll-hypaque separation followed by EAAS) were measured in each sample.

Hemograms were obtained by cell counting. Serum albumin (bromocresol green), prealbumin (radial immunodiffusion) were also determined.

All data were treated using PCSM statistical software (Deltasoft, Meylan, France). The first step of the statistical analysis consisted in verifying the normality of distribution in each group. For normally distributed variables two-ways variance analysis was used for inter- and intra-group variations (4). For non-normally distributed variables, Friedman tests were used for intra-group comparisons, Mann-Whitney U test for intergroup comparisons and Wilcoxon T test to detect significant differences between two times in the same group.

Table 1. Serum ferritin in the supplemented and placebo groups

Month	Medians (n)		
	Supplemented ng/ml	Placebo ng/ml	Inter-group comparisons
3	20 (50)	20 (47)	NS
6	10.5 (44)	3 (42)	p < 0.001
8	11 (43)	3 (45)	p < 0.001
9	10 (28)	3 (28)	p < 0.001
Post-partum	32 (21)	9.5 (8)	p < 0.01
Intra-group comparisons (3rd to 9th month)	p < 0.05	p < 0.001	

3. RESULTS

3.1. Iron Status and Iron Related Indexes

The mothers in the placebo group quickly developed symptoms of iron deficiency: plasma ferritin (Table 1) decreased below 12 ng/ml and mean erythrocyte volume (Table 2) below 80 fl. These haematological parameters show that 9.5% of women in this group were iron deficient (5) at the 8th month, and 25% at the 9th month. Concerning mothers in the supplemented group, mean cell volume increased significantly, and no decrease in plasma folate and vitamin B12 was observed. Till the 8th month, no woman developped iron deficiency. At 9 months, the median of plasma ferritin was 10 ng/ml, but mean erythrocyte volume remained above 80 fl and no sign of iron deficiency was noted.

As shown in Tables 1, 2 and 3, plasma ferritin, mean cell volume and total blood haemoglobin (TBH) were significantly higher in the supplemented group than in the placebo group. At ninth month, 35.7% of women in the placebo group were anaemic (Hb < 11 g/dl) whereas in the supplemented group only 3.8% were below this threshold. Erythrocyte haemoglobin content (p < 0.0001) and haemoglobin concentration (p < 0.05) were significantly higher in the supplemented group than in the placebos.

In newborns, plasma ferritin (118 vs. 89 ng/ml), blood haemoglobin (15.7 ± 2.5 vs. 15.4 ± 3.0 g/dl) and mean erythrocyte volume (106.5 vs. 107.7 fl) were not significantly different in babies from supplemented mothers as compared to the placebo group.

Table 2. Mean erythrocyte volume in the supplemented and placebo groups

Month	Means ±SD (n)		
	Supplemented fl	Placebo fl	Inter-group comparisons
3	88.3 ± 5.6 (50)	88.9 ± 5.3 (44)	NS
6	91.4 ± 4.0 (44)	89.1 ± 6.4 (43)	
8	92.4 ± 3.8 (43)	85.6 ± 11.0 (43)	p < 0.01
9	91.6 ± 4.3 (20)	74.6 ± 8.4 (28)	
Post-partum	88.7 ± 3.5 (21)	85.6 ± 7.6 (8)	NS
Intra-group comparisons (3rd to 9th month)	p < 0.001	p < 0.05	

Table 3. Total blood haemoglobin in the supplemented and placebo groups

Month	Means ±SD (n)		Inter-group comparisons
	Supplemented g/dl	Placebo g/dl	
3	12.5 ± 1.1 (50)	12.5 ± 0.9 (45)	NS
6	12,0 ± 1.0 (44)	11.5 ± 1.1 (43)	
8	12.4 ± 0.9 (43)	11.3 ± 0.9 (45)	$p < 0.001$
9	12.8 ± 1.0 (28)	11.4 ± 1.1 (28)	
Post-partum	13.1 ± 1.0 (20)	12.3 ± 1.0 (8)	$p = 0.05$
Intra-group comparisons (3rd to 9th month)	$p < 0.001$	$p < 0.001$	

3.2. Zinc

Plasma zinc in mothers was low at 3 months (Table 4) and decreased significantly till the 8th month. However this decrease less pronouced in the supplemented group, and a significantly higher concentration of serum zinc was observed in the supplemented group when compared to the placebo ($p < 0.01$). A 10.7 µmol/l deficiency cutoff may be used to calculate the prevalence of zinc deficiency in pregnant women (6). These calculations show that less supplemented women than placebo are zinc deficient (43% vs. 71% at 6 months, 65% vs. 82% at 8 months, and 57% vs. 79% at 9 months). The plasma zinc/serum albumin ratios showed the same variations as plasma zinc. In monocytes and in polymorphonuclear leukocytes, no significant difference in zinc concentration occured within the groups as well as between the two groups at each period of time.

Plasma zinc was not different in newborns from supplemented mothers when compared to the placebos (16.1 ± 4.2 vs. 15.3 ± 3.3 µmol/l)

3.3. Clinical Observations

In mothers, clinical examinations showed that vergetures appeared more frequently ($p < 0.05$) in the placebo than in the supplemented group. Supplementation also decreased the vomiting frequency at the third trimester ($p < 0.05$). Muscular cramps ($p < 0.05$) and insom-

Table 4. Serum zinc in the supplemented and placebo groups

Month	Means ± SD (n)		Inter-group comparisons
	Supplemented µmol/l	Placebo µmol/l	
3	11.7 ± 2.4 (50)	11.4 ± 2.2 (47)	NS
6	11.0 ± 2.3 (44)	9.2 ± 1.7 (42)	
8	10.3 ± 2.6 (43)	9.2 ± 1.4 (45)	$p < 0.001$
9	10.5 ± 2.8 (28)	9.6 ± 1.9 (28)	
Post-partum	12.1 ± 1.4 (21)	12.6 ± 2.58 (8)	NS
Intra-group comparisons (3rd to 9th month)	$p < 0.001$	$p < 0.001$	

nia (p < 0.05) occurred less frequently in the supplemented group. However supplementation induced an increase in threat of premature birth at 8 months (p = 0.0431).

The weight of the placenta was significantly higher in the supplemented group (636 ± 156 vs. 563 ± 103 g) than in the placebo. However children's weight, and other anthropometric indices showed not significant difference.

4. DISCUSSION

In pregnant women, anemia is defined by a TBH lower than 11 g/l. In the placebo group, TBH underwent a 1 g/dl decrease which generated a high prevalence of anemia: indeed, 1 woman out of 3 was anaemic at the end of pregnancy. These results agree with those previously published (7,8,9,10). Iron deficiency, which is defined by a mean cell volume smaller than 80 fl and a serum ferritin concentration lower than 12 ng/ml, appeared in 25% of placebo women at the end of pregnancy. Serum ferritin represents iron stores in the body. Below 12 ng/ml these stores can be considered as insufficient. In this study, all placebo women presented low iron stores (median concentration: 3 ng/ml) from the sixth month to the end of pregnancy. However, this prevalence can be explained also by the relatively high percentage of iron deficient women at the beginning of the study (43% exhibited serum ferritin concentrations lower than 12 ng/ml). Based on serum ferritin concentration (11), iron stores may be evaluated around 24 mg in placebo women. Serum ferritin began to decrease early in the pregnancy (as soon as the third month). Two months after delivery, serum ferritin concentrations increased, but could not reach the third trimester levels. Supplementation prevented anemia in most women as 9% of them had THB lower than 11 g/l at 8 month (33% in the placebo group), and 7% at the 9th month (36% in the placebo women). Serum ferritin decreased in the supplemented group but its median level never decreased below 10 ng/ml. Iron deficiency criteria show that no woman was iron deficient in the supplemented group. Therefore, iron supplementation was efficient in preventing iron deficiency and decreasing largely anemia prevalence.

Plasma zinc decrease during pregnancy, which was observed in the placebo group, has been previously reported by several authors (12, 13, 14). This decrease occurred in the two groups, nevertheless it was significantly smaller in the supplemented women as compared to the placebo. At 8 months, respectively 79% and 98% of women in the supplemented and placebo group showed plasma zinc concentrations lower than 12.5 μmol/l, the lower limit of normal values. These values can be explained by the low level of plasma zinc at the beginning of pregnancy, and perhaps by a competition between iron and zinc in the supplements. Indeed, non heme iron has been shown to compete with zinc in duodenum (15) and iron content of the supplement capsules is 66% higher than zinc content. Moreover, ascorbic acid is known to favour iron duodenal absorption for it keeps this metal at the reduced state. Leukocyte zinc showed no significant variations, in agreement with Simmer (16) but others reported a decrease in cellular zinc during pregnancy (17, 18).

Serum albumin was significantly higher in the supplemented group as compared with the placebo (p < 0.05). Albumin is an important long-term index of nutritional status, which in our study shows the benefit of the supplementation.

Several clinical improvements and one alteration have been noted. These clinical effects are due to the whole content of supplement capsules. Premature threat of delivery was slightly significatively increased in the supplemented group, but these threats did not result in a significant increase in the prematurity percentage. Premature threat of delivery is multifactorial and depends on numerous other factors such as socio-economical level, corporal design, or number of premature births in previous pregnancies. Obtaining a definite conviction would require a much larger population and the study of all favouriting factors.

5. REFERENCES

1. F.E. Hytten In: *Clinical physiology in obstetrics.* F.E. Hytten and G. Chamberlain eds. Blackwell Oxford U.K. pp 482–483 (1980)
2. D. Bouglé , A. Favier , F. Bureau and P. Walravens In: Les oligoéléments en nutrition et en thérapeutique. Tec & Doc Lavoisier- EM Inter, Paris pp 73–94 (1995)
3. D.B. Milne, W.K. Canfield , J.R. Mahalko and H.H. Sandstead *Am. J. Clin. Nutr.* **40** 535–540 (1984)
4. B.J. Winer , D.R. Brown and K.M. Michels In: *Statistical principles in experimental design.*, 3rd ed. Mc Graw-Hill, Inc. New-York (1991)
5. E. Thauvin *Effets nutritionnels et cliniques d'un apport en vitamines, magnesium et oligoéléments au cours de la grossesse.* Thèse de l'Université Joseph Fourier. Grenoble. (1992)
6. S.M. Pilch , F.R. Senti *Assessment of the zinc status of the U.S. population based on data collected in the 2nd natl. health and nutrition examination survey.* Life Sci. Res. Office + FASEB Bethesda ML (1984)
7. G.J. Lewis and D.F. Rowe *Br. J. Clin. Pract.* **40**: 15–16 (1986)
8. C. de Benaze, P. Galan , R. Wainer and S. Hercberg *Rev. Epid. Santé Pub.* **37**: 109–118 (1989)
9. W. Heng , C. Xuencun , W. Wenguang. *Nutr. Res.* **10**: 493–502 (1990)
10. E.M. Knight, B.G. Spurlock , A.A. Johnson. *Nutr. Res.* **11**: 1357–1375 (1991)
11. G.O. Walters , F.M. Miller , M. Worwood. *Clin. Path.* **26**: 770–772 (1973)
12. J. Argemi , J. Serrano , M.C. Gutierrez. *Ann. Nutr. Metab.* **32**: 121–126 (1988)
13. I.J. Hinks , A. Ogilvy-Stuart, K.M. Hambidge and V. Walker. *Br. J. Obst. Gynecol.* **96**: 61–66 (1989)
14. G. Lockhitch , B. Jacobson , G. Quigley. *Clin. Chem.* **36**/6: 973- (1990)
15. N.W. Solomons and R.A. Jacob. *Am. J. Clin. Nutr.* **34**: 474–482 (1981)
16. K. Simmer and R.P. Thompson *Clin. Sci.* **68**: 395–399 (1985)
17. N.J. Meadows , N.J. Ruse , M. Smith. *Lancet,* **ii**, 1135–1137 (1981)
18. K. Raja, P. Leach, G. Smith , D. Mc Carthy and T. Peters. *Clin. Chem. Acta* **123**: 19–26 (1982)

TRACE ELEMENTS AND AGING

M. Ferry[1] and A. M. Roussel[2]

[1] Centre Hospitalier
F-29953 Valence Cedex 9, France
[2] GREPO
Université Joseph Fourier
F-38700 La Tronche, France

1. INTRODUCTION

Elderly people are a rapidly growing segment of the population in both developed and developing countries. This rapid increase is one of the actual demographic trends (1). At the beginning of this century, 4 of 10 people in France were aged > 65 y. Today, it is estimated that 4 in 5 of the population will be elderly in the year 2000. Then, people > 60 y will represent 26.8% of the French population (2). Thus, the maintenance of health with age appears to be one of the main public health challenge for the next 20 years.

Which definition for "elderly"? The life span in different species is fixed in a narrow range. However, the individual duration of life depends on exogenous factors involving, besides the aging process itself, lifestyle, nutritional habits of adulthood and the incidence of diseases. A combination of these factors is considered as a continuum with different rates among individuals, and an universal definition of "elderly" cannot be put forward as yet. It is therefore important to emphasize the relation between nutrition and health in elderly, particularly regarding the possible linkages between impaired nutrition and chronic diseases, and to define specific nutrient requirements for aging people.

2. NUTRITION AND AGING

Nutrition may act in different benefit ways in elderly to promote health and prevent diseases, to improve treatments, and even to support rehabilitation. Numerous clinical and epidemiological data focused on the importance of nutritional status in maintaining performance capacities and quality of life in aging population (3–5). Nutritional status influences the functional decline of organs, of immunity and the incidence of chronic illness (6). Low dietary intakes and, consequently, a significant decrease in body weight, have been reported to precede hospitalization (7) and to increase mortality (8). The follow up of 4116 elderly subjects during 3 years showed low albumin levels as an independent risk factor of death (9).

Conversely, Keller observed in a retrospective study, 5% of gain weight as the best predictor of vital prognosis (10). Several factors (social as isolation, poverty, physical as disabilities, immobility, or dental problems, changes in taste, numerous medications) tend to limit the quality of elderly's intake and to increase the risk of nutrition-related diseases (11). Recently, Roberts reported that elderly exhibited an impaired control of food intake (12), while Rolls described an altered energy regulation (13).

These low total caloric intakes, associated to changes in gastrointestinal, renal, and endocrine functions result in a proportional reduction of essential nutrients, including trace elements (14–15), since it is known that caloric intakes below 1500 kcal/day cannot satisfy the biological needs in micronutrients. Another age- related change with respect to trace element nutriture to be considered is the decline in the lean body mass (16). Trace elements are distributed mainly in the lean body mass, and their concentrations are quite stable along the adulthood. While the body mass is lowered, steady trace element concentrations indicate a decreasing size of the metabolic pool of these elements (17). So, aging may potentially increase the risk of mineral deficiency in elderly people. Assessment of trace element nutriture of the elderly group is difficult. As underlined by Wood (18), difficulties are present in defining appropriate population samples that represent various strata of the elderly from 60 to 100 y. In addition, except for calcium requirement which is well documented (19) especially in postmenopausal women (20), the database of analyzed intakes or controlled metabolic studies are quite limited for most of the trace elements. Thus, because of the limitations of food-intake-assessment methods in elderly people, interpretation of trace element intake data will be done with caution.

3. ASSESSMENT OF TRACE ELEMENT REQUIREMENTS IN ELDERLY

Our knowledge of requirements of trace elements is based almost entirely on data obtained from younger population. There is no agreement as to the trace element dietary recommendations for the aging part of the population (21). Specific trace element recommendations (Iron, Zinc, Iodine, Selenium) have been divided into 2 categories : < 51 y and > 51y. ESSADIs for Copper, Manganese, Fluoride, Chromium and Molybdenum have been set for all adults, although it seems evident that the needs of a 51 y subject are not the same as a 80 y one, and depend on other numerous factors, evocated former, such as diseases and medications. However, until now, most of the nutritional recommendations for the elderly are derived by extrapolation from data of younger adults, and the requirements for trace elements would frequently not be satisfied in aging, especially when the intake is marginal in adulthood. Therefore, the requirements for the elderly should be linked to their health status : healthy elderly for whom the trace element requirements are similar to middle aged, acutely ill elderly with increased requirements, and chronically ill elderly who have low dietary intakes and an increased need for specific trace elements. For this last group, as mentioned later, an oral supplementation should be encouraged (22).

4. ANTIOXIDANT TRACE ELEMENTS AND AGING PROCESS

Considering that aging may be described as a phenomenon which results from randomly accumulated damaging effects to tissues, the role for free radicals in the aging process is a scientific center of interest (23–24) since Harman, first in 1956, promulgated the free radical theory of aging. Iron and copper, as free transition metals, are the most effective promoters of free-radical formation. Carrier or storage proteins effectively complex the free ionic forms and prevent from damaging concentrations, but if some of these defenses are missing, an iron overload could have deleterious effects. On the other hand, zinc, selenium and copper (when bound) are the main antioxidant trace elements involved in antioxidant defenses. Copper, together with zinc, is a constituent of the cytosolic Cu-Zn SOD, and selenium is the catalytic center of glutathione peroxidase. They are essential for function and stability of the enzymatic chain against the deleterious effects of reactive oxygen species. Moreover, zinc has been described as a biological antioxidant (25) acting by competitive effect with iron in Fenton reaction and preventing the thiol groups from oxidation. Thus, it can be hypothesized that an antioxidant trace element deficiency might result, through an imbalanced oxidant/antioxidant equilibrium, in accelerated aging. The pathological consequences of antioxidant trace element deficiencies are numerous, particularly concerning cardiovascular diseases, diabetes mellitus, or degenerative brain diseases. As a matter of fact, institutionalized people must be particularly cared with respect to their antioxidant trace element status (26).

Aging is also associated to impaired immune functions. The immunological theory of aging suggests that the progressive failure of the immune system with age is a causal factor contributing to increase morbidity and mortality in relation with an increased oxidative stress. A link between immunity, nutrition and diseases has been described in elderly people (27). Among the micronutrients modulating the immune response, zinc is essential for host - defense mechanisms (lymphocyte maturation and cell mediated immunity, functional capacity of phagocytes). Selenium is also implicated in immune defense, and selenium deficiencies result in an impairment of both humoral and cellular immunity. These effects are reversible by supplementation which will be discussed later.

5. INDIVIDUAL ELEMENTS

5.1. Iron

Changes in dietary patterns in elderly people can alter the iron bioavailability. In contrast with heme iron, the absorption of non heme iron can be enhanced (ascorbic acid) or inhibited (fibers) by various substances found in food. A reduction in meat (heme iron) consumption leads to a decrease in heme iron absorption. Age-associated hypochlorhydria or acid-lowering medications decrease also iron bioavailability and influence the iron status. However, several recent studies (NHANES II, Boston Nutritional Status Survey, Baltimore Longitudinal Study of Aging and Euronut Seneca) have reported an appropriate intake of iron in elderly population. In the epidemiological French part of Euronut Seneca Study, the prevalence of iron deficiency was low (< 4%) and correlated with the prevalence of anemia following WHO criteria. This result confirms that the identification of iron deficiency in elderly implicates a clinical screening in order to better identify the etiology (28).

Controversial results (29) reported an increase in iron stores in elderly, since serum ferritin concentrations, as index of body stores, increased with age. But, ferritin concentrations can also be influenced by the presence of inflammation. In healthy elderly people not to be suffering from inflammation, no relation between age and ferritin was shown (30) and, thus, aging alone did not seem to result in marked accumulation of iron in the body. Iron intakes appear adequate in elderly and, considering the potential pro-oxidant role of iron evocated former, the use of iron supplementation is not only unnecessary but potentially hazardous, especially when combined with other supplements that enhance iron absorption, such as ascorbic acid (14).

5.2. Zinc

The biological role of zinc has been described in numerous recent reviews (31- 32). Zinc is a component of metalloenzymes, and vital to the biochemical process of cell growth and cell division. Zinc acts in maintenance of membrane integrity, in relation with its antioxidant role, and is essential for immune functions. Because zinc nutriture affects oxidative stress and immunity (33), it seems very important to prevent zinc deficiency in elderly people. Zinc dietary intakes decrease in aging, simultaneous with the decline in energy consumption (34). Several studies demonstrated it well, in healthy free living elderly (35–36) and, with a higher prevalence, in institutionalized patients (37). Reduced absorptive efficiency, interactions with iron or calcium, increased loss as a result of underlying diseases, and drug therapy (antibiotics, antihypertensive drugs, diuretics) enhance the risk of zinc deficiencies in elderly and, in house bound patients, a negative balance was showed (38). The possibility of a subclinical zinc deficient state in the elderly must be taken in consideration. Several associations of clinical changes in aging could be related to such a mild deficiency : modified taste acuity, dermatitis, impaired wound healing, glucose intolerance and altered immune function. The beneficial effect of zinc supplementation on the immune response is well documented (31–32), but excessive intake of zinc can also impair the immune response. However, many studies put forward the possibilities for improving the immune parameters such as T lymphocytes proliferative response to mitogenes, delayed cutaneous hypersensitivity reactions, increased serum thymulin, with daily zinc doses in a range of 50 to 100 mg/day during few months.

A preventive supplementation, besides the treatment of existing disorders, would improve the maintenance of health in elderly. As suggested by Neve (39), in case of clinical signs of zinc subdeficient state, a daily supply (20 to 50 mg/day) could be given. However, excessive intake of zinc may be harmful (40–41), and there is no justification for higher doses, not well tolerated or interacting with other nutrients.

5.3. Selenium

In recent years, evidence has accumulated that selenium plays a key role in protecting cells against accelerated aging (42). This powerful property is mainly due to its antioxidant effect as constitutive trace element of the active site of glutathione peroxidases. Selenium can also detoxify heavy metals (mercury, cadmium, or lead) and xenobiotics such as carcinogenic substances. As well as zinc, selenium is implicated in immune defense and its protective effect in elderly by modulating the immune system offered a recent field of research. Suboptimal intakes in selenium are frequent in general European populations (43), particularly in elderly subjects (44). Using the duplicate diet method, a recent French study in hospitalized elderly subjects reported that 96% of the patients had intakes below two-third the RDA (45). An influence of aging on selenium status has been described (46) and lower plasma and erythrocyte GPx activities in elderly subjects compared to younger controls (47) confirmed a selenium deficient state in elderly population. Institutionalization and hospitalization might represent an aggravating factor, related for an important part to very low selenium intakes, but also due to intercurrent illnesses. Simonoff (48) reported a significant lowered selenium status in hospitalized elderly patients with cancer or cardiovascular diseases.

Certain conditions, observed in the elderly, like impaired immune function or oxidative pathological states associated with aging, could still benefit from an appropriate selenium supplementation. Recently (1990–1995), selenium supplies as organic or inorganic forms (60 µg/day to 125 µg/day during 1 to 4 months) restored the selenium status (49), decreased the lipid peroxidation (50) and induced a clear improvement of the lymphocyte proliferative response to mitogens (51).

Since in elderly people, such as younger adults, selenium intake is largely inadequate, selenium supplementation should be encouraged. However, it is uncertain whether selenium intakes that are much more higher than the current RDA would provide any beneficial effect in the prevention of chronic diseases.

5.4. Zinc and Selenium Pools in Elderly

In the French epidemiological part of the Euronut Seneca Study in free living healthy elderly people, we found decreased zinc and selenium blood levels (52). Consequently, a study was performed by our group (V. Ducros and Coll. submitted) using stable isotopes as tracers to document zinc and selenium study. It appeared that to be ill or institutionalized is a more important criteria than the age itself concerning zinc and selenium alteration status, since we observed a decreased selenium pool size in institutionalized elderly compared to healthy free living elderly subjects. The decrease of zinc pool size seemed also more marked in hospitalized people, while plasma zinc was normal.

5.6. Chromium

Chromium functions as a cofactor for insulin and is required for glucose homeostasis and lipid metabolism (53). Chromium acts by increasing insulin efficiency, but is not a substitute for insulin. Aging is associated with elevated blood glucose and insulin, decreased insulin efficiency, elevated cholesterol and triglycerides, decreased HDL cholesterol and decreased lean body mass. All these changes are also observed in chromium deficiency (54) and have been shown to improve following chromium supplementation. The gradual impairment of glucose tolerance with age is paralleled by the decline of tissue chromium concentrations, although it is not known whether that decline is a physiological phenomenon or due to insufficient dietary intake (14). Dietary chromium intake for most people and especially for elderly people appears to be suboptimal (55) and to diminish with age. This marginal intake is worsen by the increased consumption of simple sugars which also increase basal chromium losses. In elderly, variable responses to supplemental chromium have been reviewed (56), depending on age and degree of glucose intolerance. Subjects with marginally impaired glucose tolerance respond within 3 weeks to 200 µg of inorganic chromium, but older subjects with more severe glucose intolerance may require higher amounts of chromium.

Chromium supplementation of elderly subjects resulted also in significants improvement in lipids (total cholesterol / HDL ratio, triglycerides) (60). Several studies, reviewed by Anderson (57), have reported beneficial effects of chromium supplementation on lipoprotein pattern. This effect is particularly of interest for age-associated cardiovascular diseases, since subjects with CAD have been reported to have lower serum chromium than subjects free of diseases.

Decreased caloric intake, selection of high sugar foods and decreased physiological processes associated with aging contribute to a decline chromium status in elderly. Clinical evidence for a role of chromium in aging and degenerative diseases has been demonstrated by several supplementation studies, and improved chromium nutriture in elderly has been shown to regulate lipid parameters and to optimize glucose tolerance. However, the knowledge of exact requirements of chromium in age-associated diseases requires extensive amounts of additional research.

6. CONCLUSIONS

Trace element intakes and status are frequently inadequate in elderly, resulting in nutrition-related diseases. Considering the essential protective role of zinc and selenium against oxidative damage and decline of immune functions, or the beneficial effect of chromium nutrition on cardiovascular risk and glucose tolerance, there is now increasing evidence that appropriate supplementations, restoring marginal trace element status, can reduce the risk of several common age-related degenerative diseases. Surprisingly, little information is available concerning the real mineral requirements of elderly people. Additional research also need to be done to evaluate the efficiency of preventive supplementations associating several complementary micronutrients. In any case, many authors agree that more attention should be devoted to elderly people who represent the best target of preventive intervention, since the challenge is to maintain in a better state the growing older population.

7. REFERENCES

1. D. Mc Fayden, in *Improving the health of older people, a world view*, R.L. Kane and J.G. Evans, D. Mc Fayden ed., Oxford University Press for WHO, pp. 19–29 (1990).
2. J. M. Robine, *INSEE*, n° 281 (1993).
3. S.C. Hartz, I.H. Rosenberg and R.M. Russel, *Nutrition in the elderly*, Smith - Gordon ed., London (1992).
4. L.C. De Groot, W.J. Van Staveren and J.G.A.J. Hautvast. *Eur. J. Clin. Nutr.*, **45** (suppl 3), 1–196 (1991).
5. M. Schroll, K. Bjornsbo, M. Ferry and B. Livingstone, *Eur. J. Clin. Nutr.* (in press).
6. M. Mowe, T. Bohmer and E. Kindt, *Am. J. Clin. Nutr.*, **59**, 317–324 (1994).
7. J.C. Cornoni-Huntley, T.B. Harris and D.F. Everett, *J. Clin. Epidemiol.*, **44**, 743–753 (1991).
8. J.L. Wallace, R.S. Schwartz and A.Z. Lacroix, *J. Am. Geriatr. Soc.*, **43**, 329–337 (1995).
9. M.C. Corti, J.M. Guralnik and M.E. Salive, *JAMA*, **272**, 1036–1039 (1995).
10. H.H. Keller, *J. Am. Geriatr. Soc.*, **43**, 165–169 (1995).
11. J.E. Morley and A.J. Silver, *Neurobiol. Aging*, **9**, 9–16 (1988).
12. S.B. Roberts, P. Fuss and M.B. Heyman, *JAMA* , **272**, 1601–1606 (1994).
13. B.J. Rolls, K.A. Dimeo and D.J. Shide, *Am. J. Clin. Nutr.*, **62**, 923–931 (1995).
14. W. Mertz, *Prog. Clin. Biol. Res.*, **326**, 229–240 (1990).
15. N.W. Solomons, in *Mineral homeostasis in the elderly*, Alan R Liss ed., New York (1989).
16. H.N. Munro, *Drug Nutr. Interact.*, **4**, 55–74 (1985).
17. W. Mertz, E.R. Morris and J. Cecil Smith, in *Nutrition, Aging and the Elderly*, H.N. Munro and D.E. Danford ed., Plenum Press, New York, pp. 195–233 (1989).
18. R.J. Wood, P.M. Suter and R.M. Russel, *Am. J. Clin. Nutr.*, **62**, 493–505 (1995).
19. R.P. Heaney, J.C. Gallacher and C.C. Johnston, *Am. J. Clin. Nutr.*, **36**, 986–1013 (1982).
20. B. Dawson-Hughes, G. Dallal and E.A. Krall, *N. Engl. J. Med*, **323**, 878–883 (1990).
21. R.J. Wood, *Age*, **14**, 120–128 (1991).
22. Y. Guigoz, *Facts and Res. in Gerontology*, 265–276 (1992).
23. Y. A. Barnett, *Br.J. Biomed. Sci.*, **51**, 278–287 (1994).
24. H. Nohl, *Brit. Med. Bull.*, **49**, 653–667 (1993).
25. T.M. Bray and W. Bettger, *Free Radic. Biol. Med.*, **8**, 281–291 (1990).
26. P. Faure, A.M. Roussel and A.E. Favier, *Age et Nutrition*, **7**, 10–15 (1996).
27. B. Lesourd, E. Alix, M. Fare, and B. Figord, *La Revue de Gériatrie*, **17**, 537–542 (1992).
28. J.L. Schilienger, M. Ferry and B. Lesourd ; *La Revue de Gériatrie*, **15**, 502–505 (1990).
29. G. Casale, C. Bonora and A. Migliavalla, *Age and Ageing*, **10**, 119–122 (1981).
30. R. Yip and P.R. Dallman, *Am. J. Clin. Nutr.*, **48**, 1295–1300 (1988).

31. J. Nève, *De Natura Rerum*, **5**, 100–107 (1991).

32. A.E. Favier, *Age et Nutrition*, **5**, 48–64 (1994).

33. R. Chandra and L.D. McBean, *Nutrition*, **10**, 79–80 (1994).

34. A.S. Prasad, J.T. Fitzgerald and J.W. Hess, *Nutrition*, **9**, 218–224 (1993).

35. Euronut Seneca investigators, *Eur. J. Clin. Nutr.*, **45**, 121–138 (1991).

36. H. Payette and K. Gray-Donald, *Am. J. Clin. Nutr.*, **54**, 478–488 (1991).

37. W. Stafford, R.G. Smith and S.J. Lewis, *Age and Ageing*, **17**, 42–48 (1988).

38. V.W. Bunker, L.J. Hinks and M.F. Stansfield, *Am J Clin Nutr.*, **46**, 353–359 (1987).

39. J. Neve, *Ärztl. Lab.*, **36**, 305–310 (1990).

40. H. Sanstead, *Am. J. Clin. Nutr.*, **61**, 621S-4S (1995).

41. R.K. Chandra, *Lancet*, **340**, 1124–1127 (1992).

42. J. Neve, *Experientia*, **47**, 187–193 (1991).

43. H.J. Robberecht, P. Hendrix, R. Van Cauwenbergh, and H. Deelstra, *Unters. Forsch.*, **199**, 251–254 (1994).

44. A.B. Lehman, *Age and Ageing*, **18**, 339–353 (1989).

45. A. Schmuck, A.M. Roussel, J. Arnaud, V. Ducros, A.E. Favier, and A.Franco, *Am. Coll. Nutr.*, in press (1996).

46. O. Olivieri, A.M. Stanzial, D. Girelli, M.T. Trevisan, P. Guarini, M. Terzi, S. Caffi, F. Fontana, M. Casaril, S. Ferrari and R. Corrocher, *Am.J. Clin. Nutr.*; **60**, 510–517 (1994).

47. D. Campbell, V.W. Bunker, A.J. Thomas and B.E. Clayton, *Br. J. Nutr.*, **62** : 221–227 (1989).

48. M. Simonoff, C. Conric, B. Fleury, B. Berden, P. Moretto, G. Ducloux and Y. Llabador, *Trace Elem. Med.* , **5** : 64–69 (1988).

49. K. O Lassen and M. Horder, *Scand. J. Clin. Lab. Invest.*, **54**, 585–590 (1994).

50. A. Bortoli, G. Fazzin, M. Marchiori, F. Mello, R. Brugiolo and F. Martelli, *J. Trace Elem. Electrolytes Health Dis.*, **5**, 19–21 (1991).

51. A.Peretz, J. Neve, J. Desmedt, J. Duchateau, M. Dramaix and J.P. Famacy, *Am. J. Clin. Nutr.*, **53**, 1323–1328 (1991).

52. B. Lesourd, M. Ferry, J.L.Schlienger, in *Des années à savourer*, Ch. Rapin ed., Payot, Lausanne, (1993).

53. R.A. Anderson, *Biol.Trace Elem. Res.*, **32**, 19–24 (1992).

54. R.A. Anderson, in *Essential and Toxic Trace Elements in Human Health and Diseases*, A.S. Prasad ed., Wiley Liss inc., N.Y., pp. 221–234 (1993).

55. R.A. Anderson, N.A. Bryden and M.M. Polansky, *Biol. Trace Elem. Res.*, **32**, 117–121 (1992).

56. R.A. Anderson, in *Handbook of Nutrition in the Aged*, R.R. Watson ed., CRC Press, Boca Raton PL, pp. 385–392 (1994).

57. R.A. Anderson, in *Risk Assessment of Essential Elements*, W. Mertz, C.O. Abernathy, S.S. Olin ed., ILSL Press, Washington DC, pp. 187–196 (1994).

17

EFFECTS OF NUTRITIONAL DOSES OF ANTIOXIDANT TRACE ELEMENTS AND/OR VITAMINS ON THE METABOLISM OF FREE RADICALS IN ELDERLY

Longitudinal Study

M. J. Richard,[1] P. Preziosi,[2] J. Arnaud,[1] A. L. Monget,[2] P. Galan,[2] A. Favier,[1] and S. Hercberg[2]

[1] GREPO, CHU
BP217X, 38043 Grenoble, France
[2] Institut Scientifique et Technique de la Nutrition et de l'Alimentation
CNAM
75003 Paris, France

1. INTRODUCTION

Interest in the biological effects of reactive oxygen species (ROS) was stimulated by results suggesting that ROS are involved in the pathogenesis of various disorders including photoaging, vascular diseases, cataract, impaired immune functions, or cancers which are all relevant in aging. On the other hand, investigations in laboratory animal and human epidemiological studies corroborated the relationship between protection against free radical related pathologies and nutritional status (1–5).

Some micronutrient defiencies modify the endogenous antioxidant status. Selenium (Se) deficiency is associated with a reduced synthesis of the antioxidant enzyme the glutathione peroxidase (GPX) (6). Inadequate zinc (Zn) intake modifies the stability of copper-zinc superoxide dismutase (CuZn SOD) (7) and the optimal intracellular levels of metallothionein which exerts antioxidant functions. Zn deficiency also modifies the expression of some proteins by the way of transcription factors like. P53 or SP1. In addition, laboratory investigations in cells and animals identified tocopherol (8) and ascorbic acid (9) as antioxidant vitamins for lipophilic and hydrophilic phases respectively. More recently, carotenoids generated a particular interest as defenses against degenerative diseases (10).

Inadequate intakes of micronutrients were described in elderly specially in institutionalized persons. Nevertheless little information exists on antioxidant intakes and in vivo oxidant-antioxidant balance in such people. Should such a relationship be found it could be the

Therapeutic Uses of Trace Elements, edited by Nève et al.
Plenum Press, New York, 1996

bases for strategies to prevent "oxygen radical diseases" of ageing. There is strong evidence that isolated deficiencies are rare in elderly people. As antioxidant nutrients exert synergistic effects, our intervention trial should therefore not be limited to one antioxidant, but instead should examine a combination of agents that help to protect against oxygen-radical injury. Indeed, improving antioxidant intake will not always provide a beneficial effect and an excess of reductants (i.e. high doses of antioxidants) could increase the reduction of oxygen to superoxide and paradoxically, increase oxidative stress (11). The aim of this study was to investigate the effects of nutritional (and not pharmacological) doses of antioxidant vitamins or/and trace elements on the metabolism of free radicals in elderly.

2. MATERIAL AND METHODS

The participants (n = 756; 563 women and 193 men), aged 65–103 years (mean 83.5 ± 7.6) were enrolled in 26 French geriatric centers and randomly assigned to one of the four supplements: vitamins (VIT); minerals (MIN), vitamins and minerals (VIT-MIN), or placebo (P). Subjects with acute illness, history of cancer or taking medication that might interfere with nutritional status were excluded from enrollment. The research protocol was approved by the committee on human experimentation of the Cochin hospital, Paris, France and all subjects gave written informed consent. All subjects had a correct score value using the Minimal Mental Status. The daily treatment received each day over one year was a capsule containing one of the following preparation: vitamins (β-carotene, 6 mg; vitamin C, 120 mg; and dl α tocopherol, 15 mg); trace elements (zinc, 20 mg as zinc sulfate; selenium, 100 µg as sodium selenite); both (VIT-MIN) or only excipients (dicalcium phosphate and cellulose). The supplements and the placebo were identical in appearance and taste and were especially prepared for this study.

Vitamins (carotenoids, tocopherol, ascorbic acid), trace elements (Zn, Se), antioxidant metalloenzymes and indicators of oxidative stress were determined at baseline and after 6 and 12 months supplementation. Tests were carried out on blood collected after overnight fasting in trace element-free BD vacutainer tubes (Becton & Dickinson, Meylan, France). Samples for glutathione determination were immediately processed. Then blood samples were centrifuged for 10 min at 1600 g. Plasma and red blood cells were separated and frozen at -80 °C until analysis. Trace elements were measured by atomic absorption spectrometry using Seronorm Trace Element (Nycomed) as internal quality control. Thiobarbituric acid reactants (TBARs) used as an indicator of lipid peroxidation were evaluated in plasma by fluorescence with a malondialdehyde-kit (Sobioda, Grenoble, France) as previously described (12). Red blood cells GPX was evaluated by the modified method of Gunzler (13) using ter-butyl hydroperoxide as substrat (14). Erythrocyte Cu-Zn SOD activity was measured after hemoglobin precipitation by monitoring the autoxidation of pyrogallol according to the technique described by Marklund and Marklund (15). For the assay of reduced (GSH) and oxidized glutathione (GSSG) 400 µl of whole blood were transferred into a tube containing 3600 µl 6 % (v/v) metaphosphoric acid in water. The solution was mixed and centrifuged for 10 min at 4 °C. The acid protein-free supernatants were stored at -80 °C until analysis. Total GSH was determined by the modified method of Akerboom (16). To assay GSSG, GSH was masked by 2-vinyl-pyridine. An human home-made control was used as internal quality controls for GPX, SOD, TBARs, oxidized and reduced GSH.

Statistical analysis was conducted using SPSS software. Results were presented as mean ± SD. Mean comparisons were performed by two-way analysis of variance testing for a trace element effect, a vitamin effect, and a trace element-vitamin interaction. When normal

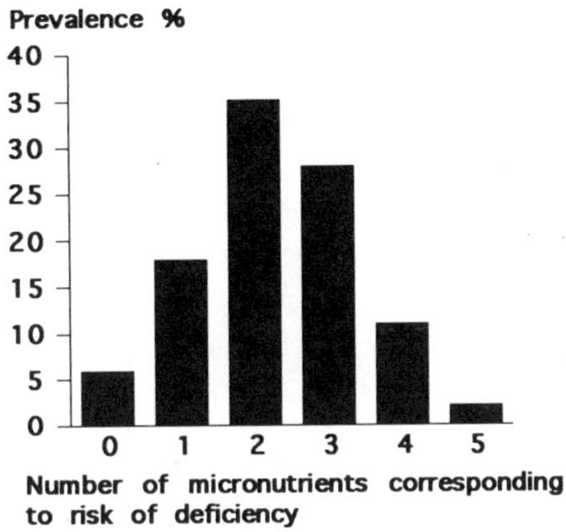

Figure 1. Percentage of subjects with associated deficiencies.

distribution was not proved, logarithmic or square-transformation were used to improve normality.

3. RESULTS

As reported in Figure 1 prevalence of deficiencies in micronutrients according to the cut-offs points retained (17) was very high. 76.2 % of subjects presented 2 or more deficiencies and only 5.9 % presented a normal biological status for all the micronutrients tested. Simple regression analysis showed that most of vitamins and trace elements were negatively correlated with age and bilateral correlations adjusted for age and gender were observed between these different nutrients (17).

Significant response in GPX activity (Figure 2-A) was observed in the subjects taking the MIN or VIT-MIN supplements. In this trial, a positive effect of VIT was observed on this enzyme activity. Table 1 shows the simple correlation coefficients between erythrocyte GPX and blood antioxidant nutrients levels at inclusion before the supplementation. On the contrary no correlation was shown between Cu-Zn SOD and the nutrients tested nor between

Table 1. Correlation between erythrocyte glutathione peroxidase (Se-GPX) and essential antioxidants in plasma at inclusion before treatment

Nutrients	r	p
Se	0.360	< 0.001
β cryptoxanthine	0.243	< 0.001
β carotene	0.227	< 0.001
Vit C	0.218	< 0.001
lycopene	0.188	< 0.001
α carotene	0.170	< 0.001
α tocopherol*	0.103	< 0.01

Erythrocytes SOD (U/mg Hb)
(B)
1.25
1.2
1.15
1.1
1.05
0
2
4
6
8
10
12
Months

Erythrocytes GPX (U/g Hb)
(A)
b****
b****
a****
a****
50
45
40
35
30
0
2
4
6
8
10
12
Months

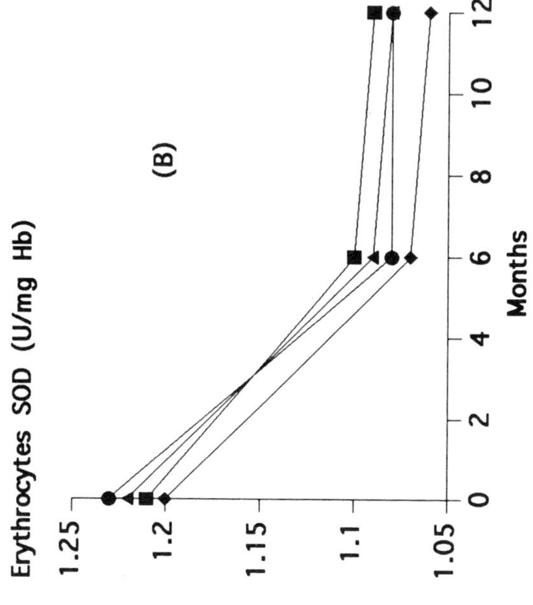

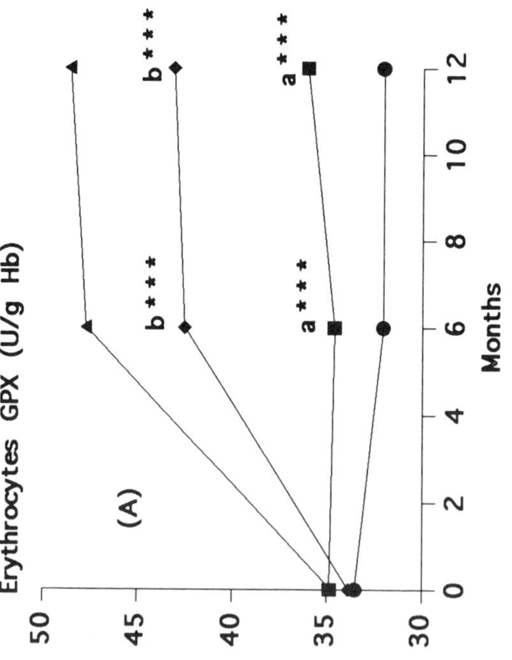

Figure 2.

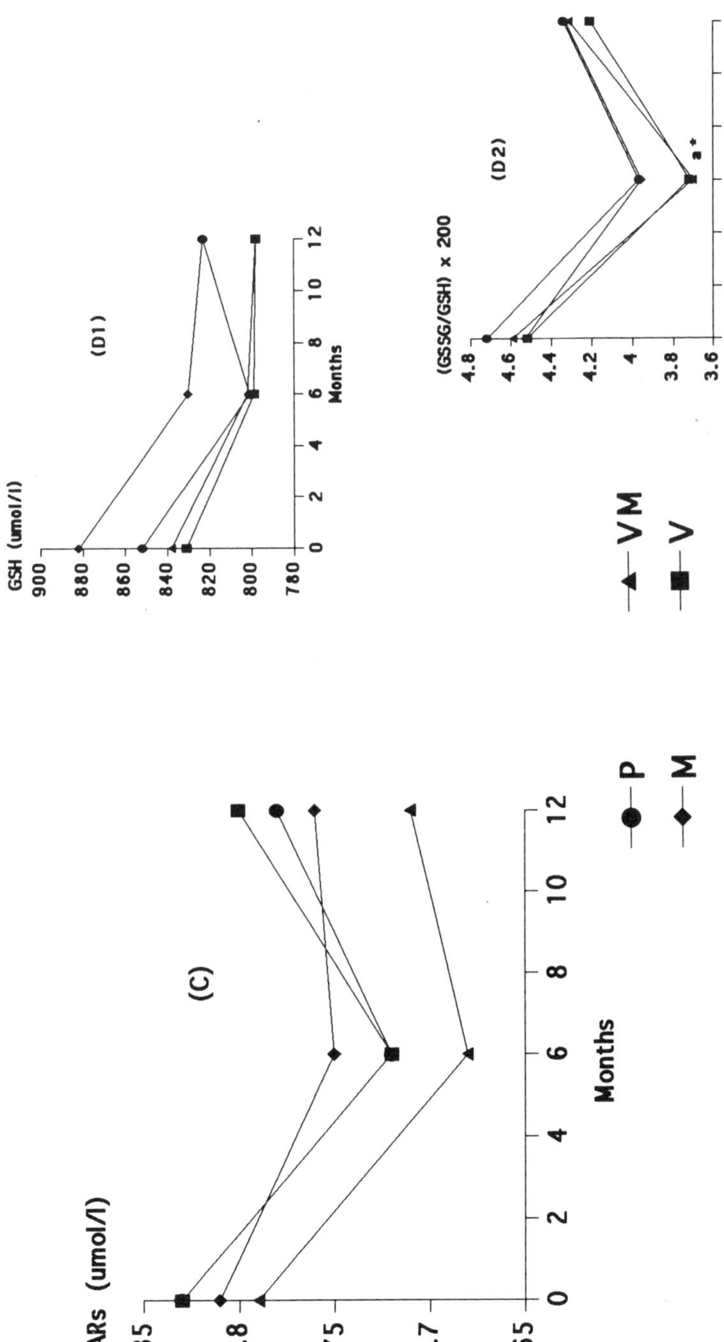

Figure 2. Effect of nutritional doses of antioxidant mixtures (vitamins, minerals or both) on erythrocytes glutathione peroxidase (A), copper-zinc superoxide dismutase (B), plasma lipid peroxidation (C), on intracellular reduced glutathione (D1) and the ratio oxidized glutathione / reduced glutathione (D2). Mean value within each blood taking were compared [two-way analysis of variance testing for a vitamin effect (a) , a trace element effect (b) and a trace element-vitamin interaction (c)]. Abbreviations used: P, placebo group; Vit, vitamin group; Min, trace element group; Vit-Min, Vitamin-Trace element group.

                                                                    M. J. Richard et al.

Table 2. Correlation between the ratio lipid peroxidation/cholesterol
and antioxidant micronutrients and metalloenzymes

	Antioxidants	r	p
plasma	α tocopherol	- 0.498	0.001
	β carotene	- 0.362	0.001
	lycopene	- 0.323	0.001
	α carotene	- 0.185	0.001
	β cryptoxanthine	- 0.247	0.001
	selenium	- 0.185	0.001
	retinol	- 0.101	0.05
	zinc	- 0.097	0.05
erythrocyte	Se-GPX	- 0.124	0.01
total blood	GSH	- 0.09	0.05

SOD and GPX even though in vitro experiments demonstrated a reciproqual protective effect. No modification of the SOD activity was observed during treatment (Figure 2-B).

GSH is an important intracellular antioxidant whose metabolism was strongly linked with α tocopherol and vitamin C. The intracellular levels of its reduced and oxidized forms were determined and the ratio GSSG/GSH was calculated as an index of oxidative stress (Figure 2-D).

TBARs were elevated in elderly people compared to usual values of the laboratory determined in French population (12). As this parameter is strongly related to serum cholesterol, the correlation (Table 2) between this lipidoperoxidation index and antioxidant micronutrients blood levels was adjusted to cholesterol. Nethertheless lipid peroxides levels did not show significant changes over the treatment period (Figure 2-C).

4. DISCUSSION

Based on these findings we confirm a marginal, even a deficient status of antioxidant nutrients in French nursing home residents. Each of the antioxidant micronutrients functions individually to lower the free radical burden (18). But vitamins and minerals interact with each other to protect their antioxidant potential. For instance vitamin C can regenerate the antioxidant form of vitamin E (8), tocopherol protects beta carotene from free radical damage (19), and GPX removes peroxides acting synergistically with vitamin E (7,8). The antioxidant micronutrient interactions enhance the cell protection associated with these essential dietary components. This investigation in a group of elderly institutionalized subjects gave the opportunity to provide a new insight into the influence of several antioxidant nutrients on oxidant-antioxidant balance.

In the present study we confirmed a marginal Se deficiency in French institutionalized subjects. After 6 months of supplementation with moderate nutritional doses of selenite, GPX reached a plateau. A steady state in plasma selenium and erythrocyte GPX levels occured after 1 year of supplementation These results were confirmed by those obtained after 2 years of supplementation in a part of the population (unpublished observations). Our data confirmed that GPX level is positively related to plasma selenium but also to carotenoids, α tocopherol and vitamin C. The negative effect of age on serum levels of these nutrients could explain this relationship.

Plasma TBARs were examined in this study as a possible index of in vivo peroxidation. Indeed, this biological marker is liable to be lowered by long term supplementation with antioxidant micronutrients. Basal TBARs values were not statistically different between the four groups and no modification was observed after 1 year supplementation despite the small standard deviation. The studies on the effectiveness of Se supplementation on lipid peroxidation in elderly people are conflicting; Peretz (20) failed to observe any effect on malondialdehyde after 6 months Se supplementation in elderly subjects. Bortoli (21) evaluated the effect of 30 days Se supplementation with enriched tablets in a group of elderly women living in a geriatric institution. He observed significant changes in erythrocyte GPX but not in plasma malonaldehyde during the period of treatment.

This study further demonstrated that institutionalized elderly people presented high risk of biological mild zinc deficiency but no correlation was observed between plasma Zn and CuZn SOD or between each of these variables and lipid peroxidation. This important trace element, whose deficiency has been involved in the loss of activity of more than 200 enzymes, did not appear to be strongly related with peroxidative damages. However the doses used in this trial did not restore normal plasma Zn concentrations after 1 year supplementation. CuZn SOD might be used as an index of Cu status (22,23), but this study confirmed there is no link between CuZn SOD and Zn in the lack of acute oxidative injury. High intakes of Zn caused low SOD activity due to decreased copper bioavailability (24,25). This reduction in SOD activity may limit the antioxidant potential of high dose Zn supplementation and even would counteract any antioxidant effect of Zn supplementation. At the doses used in this trial, no adverse effect of Zn on SOD could be observed.

TBARs in plasma were strongly influenced by the cholesterol content. Schäefer (26) found similar association with total fatty acids and cholesterol. In spite of this lipid peroxidation marker was related to dietary antioxidants (tocopherol, retinol, β carotene, lycopene, ascorbic acid and selenium) it was not influenced by antioxidant supplementation at nutritional doses in elderly people. These results were not in accordance with those of Tolonen (27) who observed a significant decrease of TBARs in elderly subjects after 3–12 months supplementation with several antioxidants (β carotene, pyridoxal, vitamin C, d-α tocopherol acetate, Zn, organic and inorganic Se) but these results were similar to those of Schäefer (26) who showed that TBARs is not influenced by antioxidant supplementation (α-tocopherol and inorganic selenium) in healthy elderly women. Wartanowicz (28) studied the effects of a daily supplementation with vitamin E or C or both on serum lipid peroxide levels and observed a significant decrease. He postulated that ascorbic acid acts synergistically with vitamin E and Se through maintenance in a reduced state of sulfhydryl compounds such as glutathione. Our results showed a significant decrease of the ratio GSSG/GSH after 6 months of supplementation which suggested a protective effect of vitamins on the reduced state of thiol group.

In conclusion, excessive lipid peroxidation occurs during the ageing process. This increase is correlated with dietary antioxidant levels whose deficiencies could contribute to the changes associated with ageing. Despite the effectiveness of the treatment on blood antioxidant levels, we did not observe any change in the lipid peroxide markers. However we cannot exclude the possibility that the TBA-test is insensitive to detect slight changes in the concentration of peroxidation products during aging and antioxidant supplementation .

ACKNOWLEDGMENTS

We thank Roche Laboratory (Basle, Switzerland) and Labcatal Laboratory (Paris, France) for support of the study.

5. REFERENCES

1. J.E. Manson , J.M. Gaziano , M. Jonas, C. Hennekens. *J .Am .Coll .Nutr.* **12**, 426–432 (1993).
2. M. Garland , W. Willett, J. Manson, D. Hunter. *J. Am. Coll. Nutr.* **12**, 400–411 (1993).
3. B. Ames, M. Shigenaga, T. Hagen. *Proc. Natl . Acad . Sci . USA.* **90**, 7915–7922 (1993).
4. A. Taylor. *J. Am. Coll. Nutr.* **12**, 138–146 (1993).
5. L. Machlin, A. Bendich. *FASEB J.* **1**, 441–445 (1987) .
6. J. Spallholz, M. Boylan. *in Peroxidases in chemistry and biology.* vol.1, J. Everse, K. Everse, M. Grisham, eds, CRC Press, Boston, pp 259–291 (1991).
7. J.A. Tainer, E.D. Getzoff, J.S. Ricardson. *Nature.* **306**, 284–287 (1983).
8. C.K. Chow. *Free Rad. Biol. Med.* **11**, 215–232 (1991).
9. E. Stadtman. *Am. J. Clin. Nutr.* **54**, 1125S-8S (1991).
10. P. Di Mascio, M. Murphy, H.Sies. *Am. J. Clin. Nutr.* **53**, 194S-200 (1991).
11. J. Bland. *J. Nutr. Env. Med.* **5**, 255–280 (1995).
12. M.J. Richard, B. Portal, J. Méo, C. Coudray, A. Hadjian, A. Favier. *Clin. Chem.* **38**, 704–709 (1992).
13. W.A. Gunzler, H. Kremers, L. Flohe. *Z. Klin. Chem. Klin. Biochem.* **12**, 444–448 (1974).
14. B. Portal, M.J. Richard, C. Coudray, J. Arnaud, A. Favier. *Clin. Chem. Acta.* **234**, 137–146 (1995).
15. S. Marklund, G. Marklund. *Eur. J. Biochem.* **47**, 469- 474 (1974).
16. T. Akerboom and H. Sies. in *Methods in enzymology*, vol 77, Academic Press, Inc, New York, pp 373–382 (1981).
17. A.L. Monjet, P. Galan, P. Preziosi, H. Keller, C. Bourgeois, J. Arnaud, A. Favier, S. Hercberg and the geriatrie/MIN.VIT.AOX. network. *J. Vit Nutr. Res.* (in press).
18. A. Bendich. *Ann. NY Acad. Sci. in Micronutrients and immune functions.* A. Bendich and R.K.Chandra, ed. **587**, 168–180 (1990).
19. P. Palozza and N. Krinsky. *Arch. Biochem. Biophys.* **297**, 184–187 (1992).
20. A. Peretz, J. Neve, J. Desmedt, J. Duchateau, M. Dramaix, J.P.Famaey. *Am. J. Clin. Nutr.* **53**, 1323–1328 (1991).
21. A. Bortoli, G. Fazzin, M. Marchiori, F. Mello, R. Brugiolo, F. Martelli. *J. Trace Elem. Electrolytes Health Dis.* **5**, 1–21 (1991).
22. S. Reiser, J.C. Smith, W. Mertz, J.T. Holbrook, D.J. Scholfield, H.S. Powell et al. *Am. J. Clin. Nutr.* **42**, 242–251 (1985) .
23. P.W.F. Fisher, A. Giroux, M.R. L'Abbe. *Am. J. Clin. Nutr.* **40**, 743–746 (1984).
24. S.M. Abdallah, S. Samman. *Eur. J. Clin. Nutr.* **47**, 327–332 (1993).
25. S. Samman. *Free Rad. Biol. Med.* **14**, 95–97 (1993).
26. L. Schäefer, E.B. Thorling. *Scand. J. Clin. Lab. Invest.* **50**, 69–75 (1990).
27. M. Tolonen, S. Sarna, M. Halme, S. Tuominen, T. Westermarck, U. Nordberg, M. Keinonen, J. Schrijver. *Biol. Trace Elem. Res.* **17**, 221–228 (1988).
28. M. Wartanowicz, B. Panczenko-Kresowska, S. Ziemlanski, M. Kowalska, G. Okolska. *Ann. Nutr. Metab.* **28**, 186–191 (1984).

18

COPPER AND ZINC IN THE PATHOPHYSIOLOGY AND TREATMENT OF INFLAMMATORY DISORDERS

Roberto Milanino, Giampaolo Velo, and Mauro Marrella

Istituto Farmacologia
Universita Verona
Policlinico Borgo Roma
I-37134, Verona, Italy

1. INTRODUCTION

The history of copper and zinc in medicine can be said to be as long as our own history. Copper was known and mentioned, especially for its anti-septic and anti-inflammatory properties, in many "medical" books belonging (this is probably the most interesting part of the story) to almost all the more significant ancient cultures such as Egyptian (papyri of Ebers and Smith) (1,3), Chinese (the "medical" book *Nei Ching*) (4), Indian (manuscripts of the Brahamanic age) (4), Roman (the scientific books *De medicina* and *Naturalis Historia*) (3, 4), and the Pre-Columbian American cultures, Aztec and Incaic in particular (see different manuscripts collected by the Invaders) (4). On the other hand, although its use was less widespread than that of copper, also zinc was registered in the manuscripts of the Brahamanic age (ancient India) (4) as a drug endowed with generic therapeutic qualities, and in the Europe of XVIII century it was used as anti-convulsant especially in the treatment of epilepsy (the book *Adversariorum Varii Argumenti*) (5). However, it may be important to recall that the above mentioned books, even the oldest ones, were mostly based upon a previous body of knowledge as well as on "medical" experiences acquired hundreds or thousands years before.

In the early decades of our century, first copper (6) and then zinc (7) were shown to be essential to life. Subsequently some researchers began to suspect that both metals were somehow involved in inflammation. This was because copper was found increased in the serum of rheumatoid patients (8), acutely inflamed laboratory animals (9), and adjuvant arthritic rats (10); moreover, a dramatic accumulation of ceruloplasmin within human inflamed periodontal tissue was also observed (11). On the other hand, a significant decrease in serum zinc levels was measured in subjects suffering from rheumatoid arthritis (12).

It is undisputed that an extraordinary impulse towards a more deep evaluation of the role of copper in inflammation came from the results obtained, at the end of the seventies, by

Therapeutic Uses of Trace Elements, edited by Nève et al.
Plenum Press, New York, 1996

John Sorenson (13) who showed that copper has a remarkable anti-inflammatory potential in laboratory animals. Likewise, although not so conclusively, zinc was also found to be of some interest as a therapeutic agent in human chronic inflammation (14). Nowadays, we may, with a reasonable level of reliability, say that the involvement of both copper and zinc in inflammation as well as their anti-inflammatory properties are not a forgettable myth but a significant, though still not fully understood and certainly underestimate, reality.

This statement will be supported by summarizing the body of experimental evidence obtained in the past two decades, that clearly is in favour of such an hypothesis.

2. COPPER AND ZINC DEFICIENCIES IN RELATION TO THE INFLAMMATORY PROCESS

2.1. Acute Inflammation

Feeding rats on a diet deficient in copper causes a significant enhancement of the acute inflammation. This observation, which probably represents the more direct piece of evidence about the role of endogenous copper on the inflammatory reaction, was first obtained in our laboratory by studying the development of carrageenan-induced paw oedema and pleurisy (15, 16), and was later confirmed by Denko (17) and by Kishore et al. (18) (Table 1). It is important to stress that the preparatory feeding of our female rats, although promoting a quick and dramatic decrease in the serum copper concentration, did not induce significant changes, in comparison with normal animals, as far as the haematology, blood chemistry and histology of the major organs (liver, kidney, spleen, etc.) and tissues (skeletal and cardiac muscle, connectives, bones, etc.) were concerned (unpublished observation). In our work we were also able to show that the extent of the pro-inflammatory effect induced by the copper deficient-diet was not correlated to the level of copper present in the serum of the rats at the end of the preparatory feeding, but it was nevertheless somehow related to the copper deficiency itself

Table 1. Main examples of the pro-inflammatory effect induced by nutritional copper or zinc deficiencies on the acute models of inflammation

Diet								
Cu (ppm)		Zn (ppm)		Days of preliminary		Acute		
Norm.	Def.	Norm.	Def.	treatment	Sex	inflammation[1]	Effect	(Ref.)
10	0.2		–	30	F	C-pO	+ 48% **	(15)
10	0.4	–	–	30	F	C-pO	+ 5%	(16)
10	0.4	–	–	90	F	C-pO	+ 29% **	(16)
10	0.6	–	–	150	F	C-pO	+ 3%	(15)
10	0.2	–	–	30	F	C-P	+ 80% **	(15)
10	0.6	–	–	150	F	C-P	+ 5%	(15)
13	0.6	–	–	30	M	C-pO	+ 31% **	(17)
10	0.5	–	–	20	M	C-pO	+ 8%	(18)
10	0.5	–	–	40	M	C-pO	+ 23% **	(18)
10	0.5	–	–	60	M	C-pO	+ 38% **	(18)
–	–	70	8	60	M	NaU-pO	+ 22% **	(19)

[1]C-pO: Carrageenan-induced paw oedema; C-P: Carrageenan-induced pleurisy; NaU-pO: Na Urate-induced paw oedema.

**p < 0.01; Student's "t" test.

being apparently dependent on both the amount of copper present in the diet and the length of the preliminary feeding (Table 1). Later on, Kishore et al. (18) better defined these crude observations and showed that keeping constant the level of copper in the deficient diet, both the appearance and the extent of the pro-inflammatory effect were clearly linked to the length of the preparatory treatment (Table 1). Moreover, these authors confirmed that the pro-inflammatory effect promoted by copper-deficiency was not related to the amount of total serum copper. They also demonstrated that the residual ceruloplasmin-dependent serum antioxidant activity was not involved as well; however, a significant inverse correlation existed between the biological result measured and the liver metal concentration (18).

Feeding rats on a zinc deficient-diet also causes a significant enhancement of acute inflammation (19). Unfortunately, at least to our knowledge, this is perhaps all we nowadays know about the effect of a dietary zinc deficiency on the development of an acute inflammatory process. However, it may be of interest to note that while a reduction ranging between 95% to 98% of the copper contained in the diet is necessary to promote a pro-inflammatory effect in the acute models of inflammation, the dietary restriction of zinc needed to obtain a similar result is of "only" about 88% (Table 1).

2.2. Chronic Inflammation

At this point of the game, the obvious hope was that examining the effect of copper or zinc deficiencies on the development of chronic inflammation the exciting results obtained studying the acute inflammatory process would have been confirmed, but this was not the case. As a matter of fact, both copper (20, 21) and zinc (21) alimentary deficiencies not only did not exacerbate the severity of the experimental chronic inflammation (the adjuvant arthritis of the rat) but they actually strongly inhibited its normal development.

Considering that the adjuvant arthritis is an experimental disorder, the aetiogenesis of which deeply involves immunity, and that copper (as well as zinc) is an important modulator of immune functions (22, and references therein), we speculated that the above result could be explained on the basis of a lower immunocompetence developed by the deficient rats (20). This hypothesis was subsequently examined by Kishore et al. (23) who observed that rats maintained on a copper-deficient diet capable of promoting an "acute" pro-inflammatory effect were somehow protected from the injury induced by the injection of complete Freund's adjuvant, but they turned out to be in a state of immunosuppression as demonstrated by impaired responsiveness to the T-cell-dependent contact-sensitizing antigen oxazolone and diminished capacity to respond to the T-cell-independent antigen Type III pneumococcal polysaccharide.

However, from our point of view, although some significant results have been achieved, the work with the trace element-deficient diets cannot be considered concluded, and more information about the role that copper and, especially, zinc can play on the development and control of both acute and chronic inflammatory processes may come from further studies with these experimental models.

3. COPPER AND ZINC METABOLISM DURING INFLAMMATION

3.1. Copper and Zinc Concentration in Blood

The appearance of either acute or chronic inflammations causes a remarkable rise in total serum (or plasma) copper (Table 2) accompanied by an increase in ceruloplasmin con-

Table 2. Changes of copper and zinc concentration induced by acute and chronic inflammatory processes in some representative body compartments of laboratory animals

Acute inflammation				Chronic inflammation			
Compart.	Copper	Zinc	(Ref.)	Compart.	Copper	Zinc	(Ref.)
Plasma	⇧	⇔	(25,26)	Plasma	⇧	⇩	(27,28)
	⇧	⇧	(29)		⇧	⇔	(30)
Red cells	⇔	⇔	(26)	Red cells	⇧	⇔	(27)
	⇔	–	(9)		⇔	⇔	(28)
Liver	⇔	⇧	(26)	Liver	⇧	⇧	(27,28,30)
	⇔	–	(9)				
Kidneys	⇔	⇔	(31)	Kidneys	⇩	⇧	(30)
					⇩	⇔	(27)
Infl. area	⇧	⇩	(26)	Infl. area	⇧	⇩	(27)
				Bone	⇩	⇩	(28)
				Muscle	⇔	⇔	(30)
				Heart	⇔	⇧	(30)
				Brain	⇧	⇔	(30)
				Spleen	⇧	⇔	(27)
				Pancreas	⇧	⇧	(30)

⇧ Increased, ⇩ Decreased, ⇔ Unchanged, – Not determined.

centration which, in many but not all the cases, is significantly correlated to that of copper (24, and references therein). However, in both laboratory animals and man, the rise of copper and ceruloplasmin appears to be a sort of "all or nothing" phenomenon , usually rather poorly correlated to the severity of the pathology considered (24, 32), and often unable to discriminate the acute from the chronic processes (24).

Contrary to copper, plasma zinc concentration may be reduced by inflammation (24). However, and again in contrast with copper, zinc level is not modified by the presence of all inflammatory processes, but its decrease seems to be directly correlated to the severity of the disease and is, usually, clearly measurable only in chronic inflammations (Table 2) (24, 32).

Finally, as far as the copper and zinc content of erythrocytes is concerned, the data nowadays available seem to suggest that, in this compartment, the concentration of both trace elements is not significantly influenced, in laboratory animals and man, by either acute or chronic inflammatory processes (24, 32), although some copper increase, most likely devoid of any biological significance (24), has been occasionally observed in the erythrocytes of the adjuvant arthritic rat (Table 2).

3.2. Changes of Copper and Zinc Concentrations in Solid Tissues and Inflamed Areas and Fluids

In Table 2 the changes of copper and zinc concentrations measured by different authors in many body compartments of laboratory animals affected by either acute or chronic inflammations are reported. In acute inflammation, most of the data seem to suggest that the major variation in solid tissues is represented by an increase in liver zinc concentration, while, in the inflamed area, an increase in copper and a decrease in zinc concentration was measured. However, this latter result (i.e. the decrease in zinc) is most likely related to the leakage of the inflammatory exudate within the inflamed area (paw) the zinc concentration of which is over 30 times higher than that normally observed in plasma (26).

The changes of copper and zinc concentrations induced by the chronic process seem to involve much more compartments than those interested by the acute one, and, with few exceptions, the majority of the tissues showed an increase, sometimes very remarkable, in copper, or zinc, or in both metals concentrations was measured (Table 2). In the chronically inflamed areas of laboratory animals (hind paws) the situation observed is much similar to that described above in the case of acutely inflamed tissues (Table 2).

Finally, it is important to remind that in humans, in addition to the previously mentioned accumulation of ceruloplasmin found in the inflamed periodontal tissue (11), both copper and zinc dramatically accumulate in the joint exudate of rheumatoid arthritis patients as it has recently been shown by Peretz and co-workers (33).

3.3. Changes of Copper and Zinc Total Amounts in Solid Tissues and Inflamed Areas

Our own experience taught us that the changes of metalloelements in any body compartment can be properly described only if their total amounts are taken into account with their concentrations. As a matter of fact, the above two parameters give both important but not necessarily consistent information.

A striking example is actually provided by the evidence that the concentration of zinc significantly decreases in the inflamed area (paw) of acutely and chronically inflamed rats, but, in both cases, owing to the oedema and joint hypertrophy which have been forming, the total amount of the metal in the inflamed paws results to be significantly increased (26, 27). Thus, considering the total amount of copper and zinc contained in the body compartments which are known to be the major stores for both metals and/or are directly involved in inflammation (i.e. the blood, the liver, the kidneys, the spleen, and the inflamed area), we found that the experimental acute inflammation causes an increase in liver zinc, in blood copper, and in both metals in the inflamed paws of the animals examined (26, 31). On the other hand, in the adjuvant arthritic rat, together with an accumulation of copper in the whole blood, liver, spleen, and inflamed paws, and of zinc in liver and inflamed paws, a decrease in total kidney copper and total plasma zinc of challenged animals was observed (27). However, and eminently important, if a global balance is done (Figure 1) it seems possible to conclude that both acute and chronic inflammatory processes induce a net accumulation of copper and zinc in the body, and this accumulation of trace elements is often significantly correlated to the severity of the experimental pathology studied.

4. COPPER AND ZINC AS ANTI-INFLAMMATORY AND ANTI-ARTHRITIC AGENTS

4.1. The Theoretical Background

The biological significance of the accumulation of copper and zinc which the inflammatory process promotes was not interpreted in the past in a univocal way by researchers (34, and references therein), however, nowadays the prevailing opinion appears to consider the role of both metals in inflammation as a positive one. The pieces of evidence that may sustain such a view have been outlined in the previous paragraphs and can be summarized as follows: a) copper (11, 35, 36) and zinc (31) naturally contained in the body may act as part of the physiologic "anti-inflammatory response" of the organism oriented to keep inflammation under proper control; b) the inflammatory process is a condition in which more copper and zinc are demanded by the organism that tends to accumulate both metals to better face the *noxa* which challenged it (24); it seems to us especially relevant to recall that acute inflammations in the rat seem to increase the amount of dietary copper required to maintain hepatic Cu-Zn superoxide dismutase levels equal to those of non-stressed animals (37), and also induce a significantly increased absorption and retention of zinc (38); c) although inflammation is *per se* able to cause an increase in the amount of both "endogenous" copper and zinc, this natural effect could be somehow insufficient to effectively control the inflammatory reaction as clearly suggested by the fact that copper and zinc preparations administered to either laboratory animals or man have remarkable anti-inflammatory and anti-arthritic properties (22, 39).

To our knowledge, the copper-containing compounds which have shown a therapeutic value in inflammation are over 100 (39), while only less than half a dozen zinc-containing molecules have been so far successfully tested (22). A detailed description and discussion on all these experimental drugs certainly goes far beyond both the scope and the space given to this presentation. However, we would like to devote a few paragraphs commenting some recent results, obtained in our laboratory and Rheumatologic Clinic, which further highlight the importance of copper and zinc as anti-rheumatic agents.

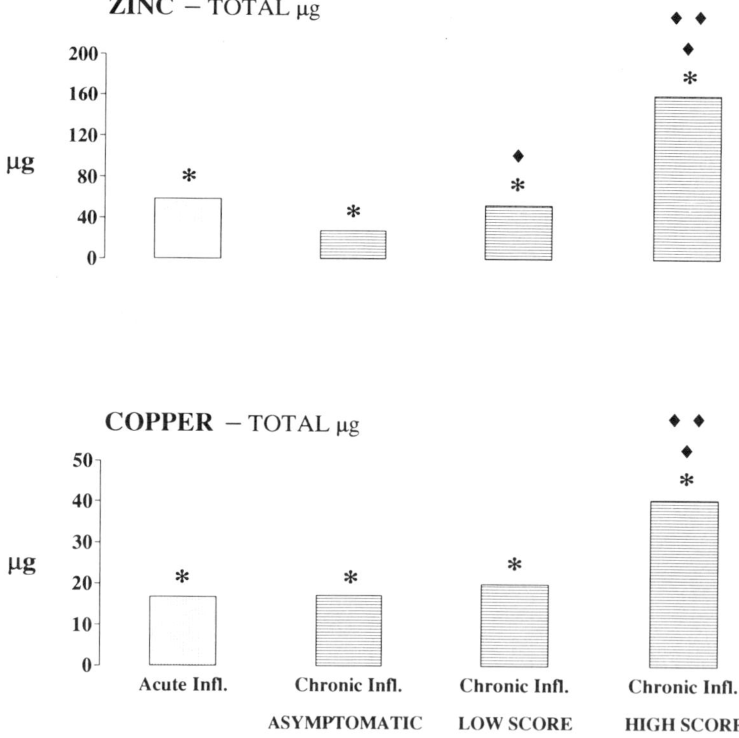

Figure 1. Accumulation (expressed as the differences of total micrograms between inflamed and non-inflamed animals) of copper and zinc in whole blood, liver, kidneys, spleen (considered only in the case of chronically inflamed rats), and inflamed hind paws of acutely and chronically inflamed animals. The data were recalculated from those reported in references (26), (27), and (31). *p << 0.001 (Student's "*t*"-test): acute- and chronic-inflamed rats versus non-inflamed controls. ♦p << 0.001 (Student's "*t*"-test): low- and high-score (28th day) versus asymptomatic (7th day) chronic-inflamed rats. ♦♦p << 0.001 (Student's "*t*"-test): high-score chronic-inflamed (28th day) versus both low-score chronic-inflamed (28th day), and acutely inflamed rats. See text for details.

4.2. The "Copper-Regulated Rat Adjuvant-Arthritis Model"

Rats fed a diet containing 200 ppm of copper (about 40 times the standard requirements) for one month and then tail-injected with heat-killed *Mycobacterium butyricum* suspended in liquid paraffin developed, while still eating the enriched diet, a chronic arthritis that, although similar to that developed by pair-feed controls 14 and 21 days after the inoculum, showed, at day 28, a statistically significant decrease of its systemic involvement (Table 3) (40). Most interesting, this protection against the experimentally induced chronic inflammation was obtained without significantly modifying the main haematochemical markers and the histopathological profiles of the animals subjected to the copper-enriched feeding (41); in particular, although the total amount of copper was found increased in the liver and in the kindneys of the copper supplemented rats, their hepatic and renal functions (Table 3) and histology (41) were found to be completely normal.

The above experimental protocol, which schedules 58 days of copper-enriched feeding and the adjuvant arthritis induction at the 30th day, and which we have called "Copper-Regulated Rat Adjuvant-Arthritis Model", could be, in our opinion, quite suitable to study, both *in vivo* and *ex vivo*, the mechanisms through which copper may exert its control on the inflammatory reaction, following, for example, the production, expression, or activity of histamine, cytokines, nitric oxide, arachidonic acid derivatives, oxygen free radicals, superoxide dismutase, adhesion molecules, and histocompatibility complex proteins, as well as the behaviour of many cell types such as neutrophils, monocytes, lymphocytes, fibroblasts and chondrocytes.

Table 3. Copper-regulated rat adjuvant-arthritis model. Anti-arthritic effect of the copper loading, and some important toxicological parameters measured in the non-inflamed rat after 58 days on the enriched diet (the data were recalculated from those reported in references 40 and 41)

Parameter (units)	Cu 200 ppm (58 days of feeding)	
	% Variation	p [1]
Arthritic score (diet 58 days; arthritis 28th day)	- 30	< 0.01
Plasma Cu (µg/ml)	- 3	NS
Plasma Zn (µg/ml)	+ 27	NS
Blood cell Cu (µg/ml of cells)	- 7	NS
Blood cell Zn (µg/ml of cells)	- 7	NS
Liver Cu (total µg)	+ 454	< 0.01
Liver Zn (total µg)	+ 7	NS
Kidney Cu (total µg)	+ 29	< 0.05
Kidney Zn (total µg)	- 12	NS
Brain Cu (total µg)	+ 1	NS
Brain Zn (total µg)	- 6	NS
Red cells (106/mm3)	0	NS
Hemoglobin (g/dl)	0	NS
Hematocrit (%)	+ 3	NS
Serum AST (U/l)	+ 6	NS
Serum ALT (U/l)	- 2	NS
Serum ALP (U/l)	- 43	< 0.05
Serum creatinine (mmol/l)	+ 6	NS
Serum urea (mmol/l)	- 3	NS

[1] Student's "*t*"-test; $n = 10$ rats per group; NS = statistically non significant.

4.3. The Use of Oral Zinc as a "Disease Modifying Anti-Rheumatic Drug" in Psoriatic Arthritis

After the first and, so far, unique success obtained using oral zinc in the therapy of rheumatoid arthritis (14), and in spite of the fact that, differently from rheumatoid patients, the plasma zinc concentration of the subjects affected by psoriatic arthritis was within normal range, Clemmensen and co-workers (42) attempted to use oral zinc in psoriatic arthritis and their results were so interesting that brought us to repeat their attempt with 18 psoriatic arthritis patients enrolled in a 6-month open trial during which they assumed, *per os*, 200 mg of zinc sulphate (i.e. 40 mg of elemental zinc) three times daily (43). Also in our hands the treatment gave very comforting results the most significant of which are summarized in Table 4. We observed a remarkable decrease in many clinical as well as laboratory markers of the disease, especially in the spontaneous daily consumption of symptomatic drugs that after 4–6 months of zinc therapy was reduced by nearly 90% ($p < 0.001$). Moreover, no side effects were observed and in particular, as judged by the concentration of copper in the erythrocytes (Table 4) and urine (data not shown) and by the concentration of blood iron and haemoglobin, red blood cell count and haematocrit (data not shown), the oral zinc treatment did not induce a clinically significant copper depletion in the treated patients (43).

Following this preliminary observation the Rheumatological Unit of our University decided to open a double-blind cross-over 1-year trial versus placebo aimed at verifying more strictly the efficacy of oral zinc as a "Disease Modifying Anti-Rheumatic Drug" in psoriatic arthritis. This trial has been recently closed and we are, at present, collecting and evaluating the results obtained.

Table 4. Major clinical and laboratory parameters of psoriatic arthritis patients treated with oral zinc sulphate (200 mg three times a day = 120 mg/day of elemental zinc) (the data were recalculated from those reported in reference 43)

Parameter (units)	Basal values	Months of therapy: 4 to 6	
		% Variation	p[1]
Tender joints (n)	13.4	- 46	< 0.01
Swollen joints (n)	4.3	- 56	< 0.05
Ritchie's index	12.4	- 56	< 0.01
Morning stiffness (min)	35.8	- 58	NS
Grip strength (mmHg)	165	+ 12	NS
NSAIDs consumption (tablets: n/week)	14.3	- 87	< 0.001
Erythrocyte sedimentation rate (mm/h)	36.9	- 73	< 0.001
C-reactive protein (mg/dl)	2.3	- 83	< 0.05
α_2-globulins (%)	10.3	- 15	< 0.05
γ-globulins (%)	18.6	- 6	NS
Fibrinogen (mg/dl)	432	- 19	NS
Plasma Cu (μg/dl)	141	- 21	< 0.001
Plasma Zn (μg/dl)	101	+ 50	< 0.001
Blood cell Cu ($\mu\gamma$/dl)	39.5	- 5	NS
Blood cell Zn (μg/dl)	621	+ 4	NS

[1] Student's test; n = 18 patients; NS = statistically non significant.

5. CONCLUSIONS

In its essence the inflammatory process represents the physiologic reaction of the organism directed to defend the organism itself against exogenous and sometimes (auto-immunity) endogenous attacks. An extremely complex network of responses characterizes this process in which different types of cells and mediators are simultaneously involved. Copper as well as zinc may, directly or indirectly, up- or down-regulate some of the biological and biochemical pathways which participate in the above network (Figure 2), exerting, in many but not all the cases, an anti-inflammatory effect.

Copper (39, and references therein) and zinc (44, and references therein) may reduce the release and/or the activity of histamine, zinc may control the synthesis of prostaglandins and leukotrienes acting on phospholipase A_2 (44), and copper, acting on cyclooxygenases, may decrease the synthesis of the "pro-inflammatory" PGE_2 and concomitantly increase the production of the "anti-inflammatory" $PGF_{2\alpha}$ (39). Both metals may also function protecting the articular cartilages; for instance, copper enhances the *in vitro* synthesis of proteoglycans in cartilage-synovium co-cultures (45) and may facilitate the physiologic process of tissue repair by modulating the activity of lysyl oxidases (39), while zinc can regulate the syntheses of nucleic acids and proteins thereby controlling connective tissue metabolism (44). Moreover, copper and zinc may respectively inhibit (46) or regulate (47) granulocyte migration as well as other important responses (such as free radical generation, adhesion, etc.) of these cells (22). The oxygen free radical production is crucial for the development of acute and, probably, chronic (O_2 radicals may favour the formation of endogenous antigens) inflammatory processes (48). The enzyme superoxide dismutase is a copper-zinc protein, and numerous other copper and zinc containing molecules are effective free radicals scavengers (22, 39, 48); however, in the case of copper, it should not be forgotten that this metal may also act as a Fenton-catalyst and, referring to the free radical metabolism and inflammation, its scavenging (i.e. anti-inflammatory) potential could requires a suitable environment and the presence of proper ligands to be expressed (49). In the past few years the importance of cytokines (50), adhesion phenomena (51), and, probably, nitric oxide (52) on inflammation has been highlighted by many researches. Also in these pathways copper and zinc may play a significant role. In particular, copper could down-regulate the secretion of IL-1 and IL-6 by human leukocytes *in vitro* (53) and copper-deficiency may enhance the *ex vivo* production of IL-1 by mouse splenocytes (54); moreover, *in vitro* copper could interfere with the activity of the chondrocyte adhesion-molecule ICAM-1 (55), and it may modulate the activity of nitric oxide synthetase (56). On the other hand, *in vitro* zinc has been shown to up-regulate the expression of the β_2-integrin LAF-1 by peripheral blood mononuclear cells, which has to be probably interpreted as a "pro-inflammatory" effect (55). Finally, last but not certainly least, both copper and zinc are well known modulators of the immune response in animals and man (22).

The above mentioned examples attempt to summarize what we know on the mechanism of action of copper and zinc as "endogenous" modulator of the inflammatory process as well as "exogenous" anti-inflammatory agents. From there, it does not emerge any certainty except, perhaps, the facts that the beneficial effect of both trace elements on inflammation has a multifaceted biological base, as well as that further research conducted in laboratory animals and man is definitely needed.

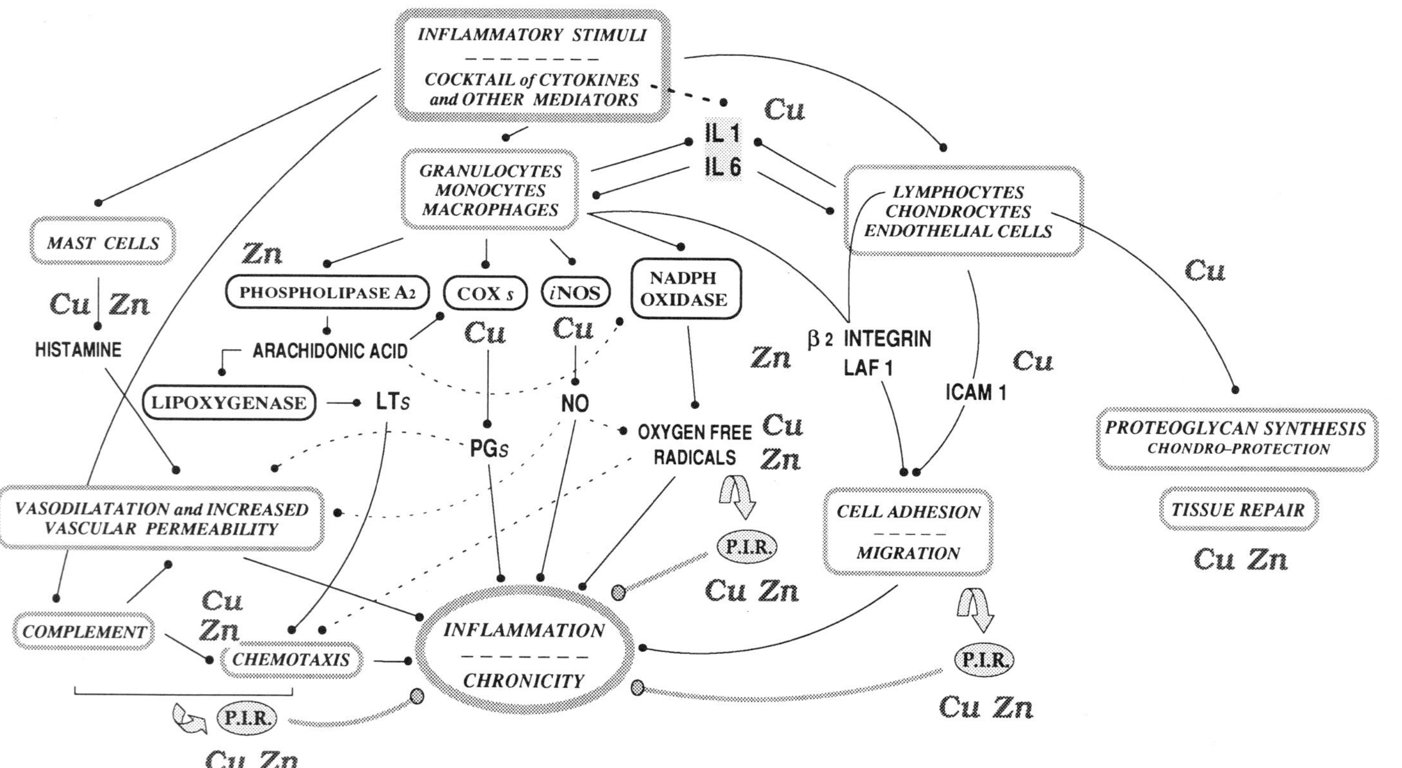

Figure 2. Important biological and biochemical pathways involved in inflammation which may be influenced (direct or indirect up- or down- regulations) by copper and zinc. The inflammatory stimulus elicits the reaction of different types of both resident and circulating cells as well as the activation of some components of the complement. These cells and biomolecules cause, in turn, a cascade of events that may lead to chronicity inducing a Pathologic Immune Reaction (PIR). The continuous lines indicate the main link between the different events, while the dotted lines indicate secondary or indirect connections. IL-1 = interleukin 1; IL-6 = interleukin 6; COXs = cyclooxygenases; iNOS = inducible nitric oxide synthetase; LAF-1 = lymphocyte function antigen 1; ICAM-1 = intercellular adhesion molecule 1; LTs = leukotrienes; PGs = prostaglandins. See text for details.

ACKNOWLEDGMENT

We are most grateful to Prof. Paolo Bellavite (Istituto di Chimica e Microscopia Clinica, Univ. di Verona) for his helpful suggestions in assembling the schema of the inflammatory process which is the basis of Figure 2.

REFERENCES

1. G. Rossi Osmida; *Archeo* **58**, 62–111 (1989).
2. E. David, *I costruttori delle piramidi*, Einaudi, Torino (1989).
3. M. Neuberger, *History of Medicine*, Bedford Medical Publications, London (1910).
4. L. Sterpellone, *Dagli dei al DNA*, Antonio Delfino Editore, Torino (1988).
5. T.U. Hoogenraad, *Trace Elements in Medicine* **1**, 47–49 (1984).
6. E.B. Hart, H. Steinbock, J. Waddell and C.A. Elvehjem, *J. Biol. Chem.* **77**, 797–811 (1928).
7. W.R. Todd, C.A. Elvehjem and E.B. Hart, *Am. J. Physiol.* **107**, 146–158 (1934).
8. L. Helmeyer and G. Stuwe, *Klin. Wochenschr.* **17**, 295–299 (1938).
9. M.M. Wintrobe, G.E. Cartwright and C.J. Gubler, *J. Nutr.* **50**, 395–419 (1953).
10. D.S. Karabelas, *Dissertation University Microfilm Ann Arbor*, MI, Order No 72–31,092. Diss. Abstr. Int. B 1972 (33)6, 2776 (1972).
11. S.C. Sweeney, *J. Dental Res.* **46**, 1171–1176 (1967).
12. W. Niedermeier and J.H. Griggs, *J. Chron. Dis.* **23**, 527–536 (1971).
13. J.R.J. Sorenson, *J. Med. Chem.* **19**, 135–148 (1976).
14. P.A. Simkin, *Lancet* **II**, 539–542 (1976).
15. R. Milanino, A. Conforti, M.E. Fracasso, L. Franco, R. Leone, E. Passarella, G. Tarter and G.P. Velo, *Agents and Actions* **9**, 581–588 (1979).16. R. Milanino, S. Mazzoli, E. Passarella, G. Tarter and G.P. Velo, *Agents and Actions* **8**, 618–622 (1978).
17. C.W. Denko, *Agents and Actions* **9**, 333–336 (1979).
18. V. Kishore, B. Wokocha and L. Fourcade, *Biol. Trace Element Res.* **23**, 97 105 (1990).
19. C.W. Denko, M. Petricevic and M.W. Whitehouse, *Int. J. Tiss. Reac.* **3**, 73–76 (1981).
20. R. Milanino, E. Passarella and G.P. Velo, *Agents and Actions* **8**, 623–628 (1978).
21. G.B. West, *Int. Archs. Allergy appl. Immunol.* **63**, 347–350 (1980).
22. R. Milanino, U. Moretti, M. Marrella, M. Pasqualicchio, R. Gasperini and G.P. Velo, in *Handbook of Metal-Ligand Interactions in Biological Fluids: Bioinorganic Medicine*, vol. 2, G. Berthon ed., Marcel Dekker Inc., New York, pp. 886–899 (1995).
23. V. Kishore, N. Latman, D.W. Roberts, J.B. Barnett and J.R.J. Sorenson, *Agents and Actions* **14**, 274–282 (1984).
24. R. Milanino, M. Marrella, R. Gasperini, M. Pasqualicchio and G.P. Velo, *Agents and Actions* **39**, 195–209 (1993).
25. G.L. Fisher, *Am J. Vet. Res.* **38**, 935–940 (1977).
26. R. Milanino, M. Marrella, U. Moretti, E. Concari and G.P. Velo, *Agents and Actions* **24**, 356–364 (1988).
27. R. Milanino, U. Moretti, E. Concari, M. Marrella and G.P. Velo, *Agents and Actions* **24**, 365–376 (1988).
28. J. Néve, J. Fontaine, A. Peretz and J.P. Famey, *Agents and Actions* **25**, 146–155 (1988).
29. B.F. Feldman, C.L. Keen, J.J. Kaneko and T.B. Farver, *Am J. Vet. Res.* **42**, 1114–1117 (1973).
30. V. Kishore, *Res. Commun. Chem. Pathol. Pharmacol.* **63**, 153–156 (1989).
31. R. Milanino, A. Cassini, A. Conforti, L. Franco, M. Marrella, U. Moretti and G.P. Velo, *Agents and Actions* **19**, 215–223 (1986).
32. R. Milanino, A. Frigo, L.M. Bambara, M. Marrella, U. Moretti, M. Pasqualicchio, D. Biasi, R. Gasperini, L. Mainenti and G.P. Velo, *Clin. Exp. Rheumatol.* **11**, 271–281 (199).
33. A. Peretz, J. Nève and J.P. Famaey, *J. Trace Element Electrolytes Health. Dis.* **3**, 103–108 (1989).
34. R. Milanino, M. Marrella and G.P. Velo, *La Chimica e l'Industria* **76**, 138–143 (1994).
35. J.R.J. Sorenson, *J. Pharm. Pharmacol.* **29**, 450–452 (1977).
36. R. Milanino, E. Passssarella and G.P. Velo, in *Advances in Inflammation Research*, vol. 1, G. Weissmann, B. Samuelsson and R. Paoletti eds, Raven Press, New York, pp. 281–291 (1979).
37. R.A. Di Silvestro and J.T. Marten, *J. Nutr.* **120**, 1223–1227 (1990).
38. R.S. Pekarek and G.W. Evans, *Proc. Soc. Exp. Biol. Med.* **150**, 755–758 (1975).

39. J.R.J. Sorenson, in *Progress in Medicinal Chemistry* 26, G.P. Ellis and G.B. West eds, Elsevier, Amsterdam, pp. 437–568 (1989).
40. U. Moretti, M. Marrella, M. Pasqualicchio, R. Milanino and G.P. Velo, *Pharmacol. Res.* **22**(suppl. 2), 345 (1990).
41. P. Cristofori, A. Terron, M. Marrella, U. Moretti, M. Pasqualicchio, G.P. Velo and R. Milanino, *Agents Actions (Special Conference Issue)*, C118-C120 (1992).
42. O.J. Clemmensen, J. Siggaard-Andersen, A.M. Worm, D. Stahl, F. Frost and I. Bloch, *Br. J. Dermatol.* **103**, 411–415 (1980).
43. A. Frigo, L.M. Bambara, E. Concari, M. Marrella, U. Moretti, C. Tambalo, G.P. Velo and R. Milanino, in *Copper and Zinc in Inflammation*, R. Milanino, K.D. Rainsford and G.P. Velo eds, Kluwer Academic Publisher, Dordrecht, pp. 133–142 (1989).
44. A.S. Prasad, *Clin. Gastroenterol.* **12**, 713–741 (1983).
45. M. Pasqualicchio, M.E. Davies, R. Milanino and G.P. Velo, *Br. J. Pharmacol.* **111** (suppl.), 292P (1994).
46. M. Roch-Arveiller, D. Pham Huy, L. Maman, J-P Giroud and J.R.J. Sorenson, *Biochem. Pharmacol.* **39**, 569–574 (1990)
47. V. Leibovici, M. Statter, L. Weinrauch, E. Tzfoni and Y.Matzner, *Isr. J. Med. Sci.* **26**, 306–309 (1990).
48. R. Milanino and G.P. Velo, in *Trace Elements in the Pathogenesis and Treatment of Inflammation*, K.D. Rainsford, K. Brune and M.W. Whithouse eds, Birkähuser Verlag, Basel, pp. 209–230 (1981).
49. G. Berthon, *Agents and Actions* **39**, 210–217 (1993).
50. C.A. Dinarello, *Blood* **77**, 1627–1652 (1991).
51. B.N. Cronstein and G. Weissmann, *Arthr. Rheum.* **36**, 147–157 (1993).
52. C.H. Evans, *Agents and Actions Suppl.* **47**, 107–116 (1995).
53. P. Scuderi, *Cell. Immun.* **126**, 391–405 (1990).
54. O.A. Lukasewycz and J.R. Prohaska, *Ann. N.Y. Acad. Sci.* **587**, 147–159 (1990).
55. M. Pasqualicchio, M.E. Davies and G.P. Velo, *Inflammopharmacol.* **3**, 35–48 (1995).
56. J.R.J. Sorenson, in this symposium.

19

LOW PLASMA SELENIUM IN PATIENTS ADMITTED IN AN INTENSIVE CARE UNIT IS RELATED TO SYSTEMIC INFLAMMATORY RESPONSE SYNDROME AND SEPSIS

D. Vitoux,[1] X. Forceville,[2] R. Gauzit,[2] P. Lahilaire,[2] A. Combes,[2] and P. Chappuis[1]

[1] Hôpital Lariboisière
Laboratoire Central de Biochimie
Paris, France
[2] Centre Hospitalier de Meaux
Service de Réanimation Polyvalente
Meaux, France

1. INTRODUCTION

The biological role of selenium (Se) has been established since the discovery that Se is a structural component of the active center of the enzyme glutathione peroxidase (GSH-Px) which is the first recognized selenoprotein. GSH-Px catalyzes the reduction of hydrogen peroxide and a variety of organic hydroperoxides by glutathione and thus protects the cells from oxidative damage. Therefore Se is a part of a system that provides a defense against the accumulation of lipid peroxides and free radicals that damage cell membranes and macro molecules. However, evidence has now accumulated that this is not the only function of Se (1). There are several other biologically active selenoproteins (2, 3) and Se seems to play a direct role in the regulation of the inflammatory process (4, 5).

The acute-phase response to injury and infection is a complex sequence of events involving systemic physiological and biochemical alterations. The biochemical changes involve increased oxidation of fat and carbohydrate, increased transfer of amino acids from skeletal muscles to the liver with the synthesis of hepatic acute-phase proteins, and alteration in trace-element metabolism, particularly iron, copper and zinc. However Se metabolism was poorly studied in the acute phase response to injury and infection. A transient decrease in serum Se was described in patients with major trauma (6) and in patients with burns covering 26 to 46 % of body surface (7).

We studied the plasma Se status in surgical and medical patients hospitalized in an Intensive Care Unit (ICU). Plasma selenium was analysed in 192 patients at the day of their admission in ICU and examined as a function of their inflammatory state (presence or not of a Systemic Inflammatory Response Syndrome, SIRS) with a special attention to sepsis.

2. MATERIALS AND METHODS

192 patients admitted to the ICU were studied. There were 56 women (29 %) and 136 men (71 %) with a mean age of 54 ± 18 years. Ventilated patients were 109 (55 %) and mortality rate was 17 %. Patients with acquired immunodeficiency syndrome, pregnancy or polytrauma were excluded. SAPS II is a score that estimates the risk of death without having to specify a primary diagnosis; it includes 17 variables: 12 physiological variables, age, type of admission (scheduled surgical, unscheduled surgical, or medical, or medical), and three underlying disease variables (acquired immunodeficiency syndrome, metastatic cancer, and hematologic malignancy). Three groups were built as a function of SAPS II. The first group included patients with a low risk of mortality (SAPS II < 15), the second group included patients with a middle risk of mortality (SAPS II ≥ 15 and ≤ 30), the third group included patients with a high risk of mortality (SAPS II > 30). SIRS was defined by the presence of one of these clinical features: sepsis, pancreatitis, ischemia, sshock, emergency surgery. Sepsis was defined as conditions where the patient met the criteria for SIRS and also presented with a documented infection. Documented infection was diagnosed using standard definitions.

Heparinized blood sample was drawn at the day of the admission and immediately processed. Samples were centrifuged for 15 min at 1000g and the plasma removed. Plasma was aliquoted and stored at -80°C until analysis. Se was measured by Electrothermal Atomic Absorption Spectrometry (ETAAS). A Perkin-Elmer 5100 Pc spectrometer with Zeeman-effect background correction, equipped with a graphite furnace and autosampler, was used for atomic absorption measurements. Selenium absorption was measured at 196.0 nm. Atomization was made in Perkin-Elmer pyrolytic graphite coated tubes with platforms. Ni (30 µg in the tube) was used as modifier.

3. RESULTS

Plasma selenium concentration on admission in the ICU was strongly decreased in patients presenting a SIRS or a high mortality risk (Figure 1). In patients presenting a SIRS, low plasma selenium was related to the severity of the sepsis (Figure 2). Interestingly, we did not find any increase in the selenium urinary excretion in patients presenting a SIRS (0,18 ± 0,10 µmol/l; n = 21) compared with patients presenting no SIRS (0,21 ± 0,15 µmol/l; n = 15). The percentage of patients who developed more than one new Organ System Failure (OSF) or a nosocomial pneumonia during their stay in ICU was significantly higher when their plasma selenium was less than 0.60 µmol/l on admission (Figure 3). Among the patients hospitalized in ICU, the risk to develop a sepsis and more than one OSF was 4 times higher when plasma selenium was less than 0.60 µmol/l on admission. 22.5 % of patients with plasma Se < 0.60 µmol/l died and the mortality rate was significantly lower in patients with plasma Se > 0.60 µmol/l (14,5%) on admission in ICU. The time-course for plasma selenium to reach normal values with a daily supplementation of 40 µg of sodium selenite was rather different in patients with SIRS (7 weeks) compared to patients without SIRS on admission (2 weeks).

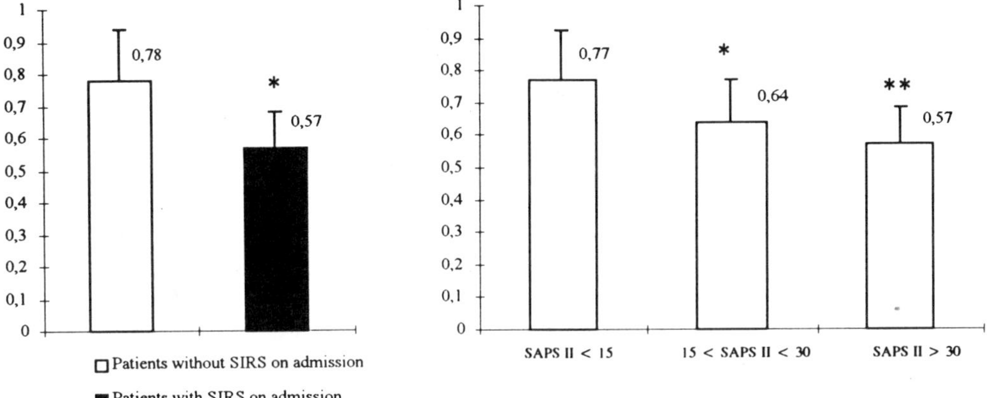

Figure 1. Low plasma selenium (µmol/l) is related to the presence of SIRS (n = 106, *p < 0.001) and high mortality risk (15 < SAPS II < 30, n = 66, *p < 0.05; SAPS II > 30, n = 68, **p < 0.01) in critically ill patients admitted in ICU.

4. DISCUSSION

The metabolic rate after surgery may increase by 10–20 %, whereas in patients with SIRS or severe sepsis, the metabolic rate may increase by up to 50 %. Because essential trace elements are involved in enzyme-catalysed reactions, many of which are central to intermediary metabolism, there is a higher requirement for such trace elements. Fe and Cu are required for the activity of the cytochrome system, Zn plays an important role in protein synthesis and Se is essential for activity of the GSH-Px, one of the key enzyme in patways for free-radical scavenging.

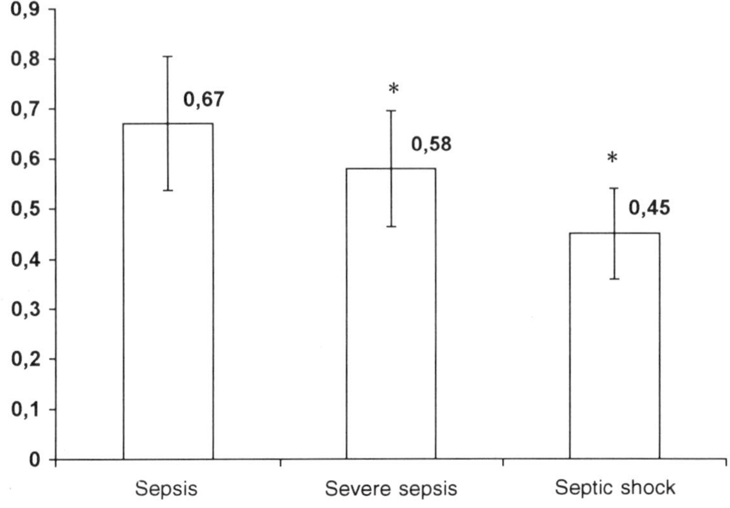

Figure 2. Low plasma selenium (µmol/l) on admission in ICU is related to the severity of the sepsis (sepsis, n = 8; severe sepsis, n = 23, *p < 0.05; septic shock, n = 9, *p < 0.05).

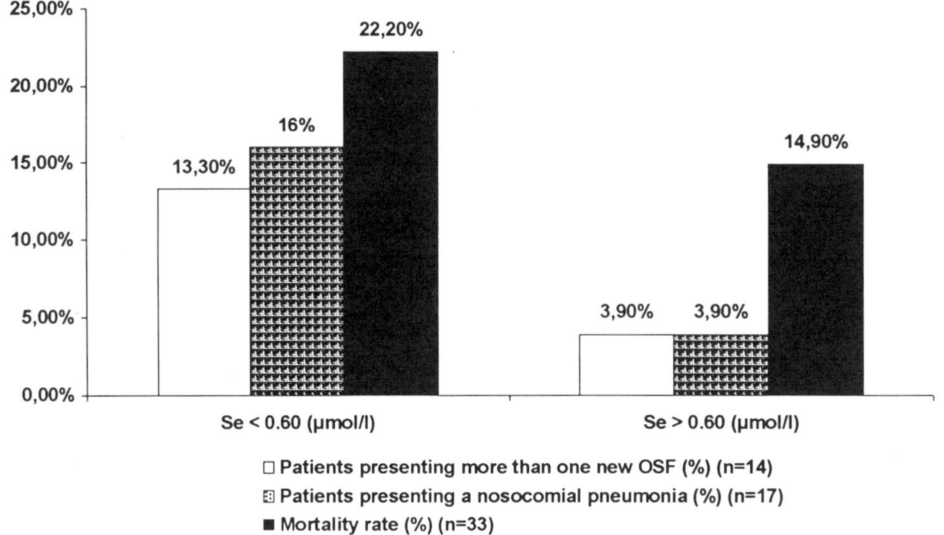

Figure 3. Prevalence of new OSF, nosocomial pneumonia and mortality rate in patients in ICU according to their plasma selenium values.

Little is known about the regulation of selenium metabolism in critically ill patients presenting SIRS together with a sepsis or not. In our patients, low plasma selenium values were not related to an increased urinary loss. Plasma selenium decrease could therefore reflect a defense mechanism against agression requiring a rapid mobilization of circulating selenium. The mechanism by which plasma Se is required is not clear. The fact that plasma GSH-Px activity did not decrease in several patients with low plasma Se concentration (data not shown) suggest that at least renal tissue take and use a part of plasma Se to synthetize the plasma GSH-Px. It is known that Se is necessary for the activity of the GSH-Px and plays, by this way, an important protective role against oxidative damages. Patients with severe sepsis are exposed to increased amounts of free radicals and peroxidation. Some mechanisms occuring in sepsis as the activation of phagocytes in response to infection or tissues reperfusion produce free radicals and peroxides. One of the major protective mechanism against these compounds is GSH-Px and requirement for this enzyme may increase to maintain the adaptative response to the oxidative stress .

However Se plays also an other important role in the modulation of inflammatory and immune responses (8, 9). Some properties of phagocytes such as chemotaxis, migration, ingestion and fungicidal activity are dependent on Se level. Incubation of human neutrophils with Se stimulates phagocytosis and bactericidal activity. Indirect evidences suggest that Se is an important co-factor in immune modulation (4, 10, 11). Deficient animals are more prone to infections and their antibody titers are low. Retarded immunity in deficient animals is improved by Se supplementation. In deficient subjects, the proliferative response of lymphocytes to antigens is impaired. These effects are explained through modulation either of GSH-Px activity or of leukotriene synthesis (12). So, rapid requirement of Se in critical ill patients who have SIRS and/or severe sepsis could reflect the involvement of Se in inflammatory and immune response. On the other hand, these effects in immunity modulation could explain why the risk to develop nosocomial infections during hopitalization is significantly higher when plasma selenium is lower than 0.60 µmol/l.

Our findings may also have a therapeutic interest. Although the decrease in plasma Se is beneficial, the progressive loss of Se could lead to weakness and increase the suceptibility to infection in critically ill patients. This appears particularly important in patients presenting SIRS on admission for which the time to reach normal plasma Se values is about 7 weeks. In these patients optimum supplementation should minimize these changes and perhaps decrease the high incidence of nosocomial infections but more studies are necessary to determine the possible benefical effects of Se supplementation on sepsis prevention in ICU.

5. REFERENCES

1. J. Nève, *Experientia* **47**, 187–193 (1991).
2. B.A. Zachara, *J. Trace Elem. Electrolytes Health Dis.* **6**, 137–151 (1992).
3. D. Behne, C. Weissnowak, M. Kalcklosch, C. Westphal, H. Gessner and A. Kyriakopoulos, *Analyst* **120**, 823–825 (1995).
4. M.L. Handel, C.K.W. Watts, A. Defazio, R.O. Day and R.L. Sutherland, *Proc. Natl. Acad. Sci. USA* **92**, 4497–4501 (1995).
5. E.W. Taylor, *Biol. Trace Element Res.* **49**, 85–95 (1995).
6. M. Berger, T. Lemarchand-Béraud, C. Cavadini and R. Chioléro, *Clin. Intensive Care* **5**, 76 (1994)
7. M. Berger, C. Cavadini, R. Chioléro, S. Guinchard, S. Krupp and H. Dirren, *Nutrition* **10**, 327–334 (1994)
8. A. Peretz, In *Selenium in Medicine and Biology*, J. Nève and A. Favier, ed., W. de Gruyter, Berlin, pp. 235–246 (1988)
9. L. Kiremidjian-Schumacher, M. Roy, H.I. Wishe, M.W. Cohen and G. Stotzky, *Biol. Trace Element Res.* **46**, 183 (1994).
10. M. Roy, L. Kiremidjian-Schumacher, H.I. Wishe, M.W. Cohen and G. Stotzky, *P.S.E.B.M.* **202**, 295–301 (1993).
11. G. Spyrou, M. Bjornstedt, S. Kumar and A. Holmgren, *FEBS Lett.* **368**, 59–63 (1995).
12. F. Weitzel and A. Wendel, *J. Biol. Chem.* **268**, 6288–6292 (1993).

20

DOWN-REGULATION OF NITRIC OXIDE SYNTHASE MAY ACCOUNT FOR THE ANTIINFLAMMATORY ACTIVITIES OF COPPER CHELATES

John R. J. Sorenson

Department of Medicinal Chemistry
College of Pharmacy
University of Arkansas for Medical Sciences Campus
4301 West Markham, Little Rock, Arkansas 72205

1. INTRODUCTION

Roles of nitric oxide ($\cdot$N=O, NO) in decreasing vascular tone, decreasing platelet adhesion and aggregation, increasing the cytotoxic response of macrophages and neutrophils, and serving as a retrograde neurotransmitter in normal biochemically-mediated physiological responses have been reviewed by Moncada, Palmer, and Higgs (1), Lancaster (2). These reviews also present pathological consequences of excess synthesis of NO by Nitric Oxide Synthase (NOS) which include: pain, inflammations, gastrointestinal ulcers, seizures, neoplasias, diabetes, ischemia-reperfusion injuries, and radiation injuries. Synthesis of NO involves an NADPH-dependent diaphorase which supplies reducing equivalents for the conversion L-Arginine to NO.

Down-Regulation of P-450 mono-oxygenase by copper $(II)_2$(3,5-diisopropylsalicylate)$_4$ [$Cu(II)_2$(3,5-DIPS)$_4$] was originally reported by Weser et al. (3,4) which was subsequently suggested to be due to down-regulation of the NADPH-dependent cytochrome P-450 reductase (5,6). Since $Cu(II)_2$(3,5-DIPS)$_4$ is capable of accepting reducing equivalents required for oxygen activation by P-450, it occurred to us that this complex might down-regulate NOS and decrease the synthesis of NO.

To examine this question in a system which is less complicated than NOS but relevant to this enzyme system, $Cu(II)_2$(3,5-DIPS)$_4$ was studied using Porcine Heart Diaphorase (PHD). These studies revealed that $Cu(II)_2$(3,5-DIPS)$_4$ does down-regulate PHD. To examine the relevance of this observation with regard to down-regulation of tissue NOS, the effect of $Cu(II)_2$(3,5-DIPS)$_4$ on reduction of Nitroblue Tetrazolium (NBT) by rat brain NOS was examined using whole brain sections incubated in medium containing an NADPH-generat-

Therapeutic Uses of Trace Elements, edited by Nève et al.
Plenum Press, New York, 1996

ing system (7) and $Cu(II)_2(3,5\text{-}DIPS)_4$. Addition of $Cu(II)_2(3,5\text{-}DIPS)_4$ to this incubation medium decreased reduction of NBT in these brain sections.

We are reporting down-regulation of PHD by $Cu(II)_2(3,5\text{-}DIPS)_4$ wherein this complex does not inhibit PHD but serves as an electron acceptor in decreasing dye reduction. The decrease in reduction of NBT in brain sections by $Cu(II)_2(3,5\text{-}DIPS)_4$ confers additional relevance to the use of PHD as a model in vitro system of NOS down-regulation. In addition, these results are discussed in terms of down-regulation of NOS in accounting for pharmacological activities of $Cu(II)_2(3,5\text{-}DIPS)_4$ and other copper complexes. It is plausible that antiinflammatory, analgesic, antiulcer, anticonvulsant, anticancer, antidiabetic, radiation protection, and radiation recovery activities of copper complexes as well as their reduction of ischemia-reperfusion injury (8,9) may be due to down-regulation of NOS.

2. MATERIALS AND METHODS

Details concerning materials and methods used in these studies have been published (10).

3. RESULTS

Addition of 1 unit of PHD to an equimolar mixture of NADPH and DCPIP lead to the oxidation of NADPH and reduction of the obligate electron acceptor Dichlorophenolindophenol (DCPIP) (10). Addition of 1 to 8 mM $Cu(II)_2(3,5\text{-}DIPS)_4$ produced a concentration related decrease in the initial rate of reduction of DCPIP with little change in the initial rate of oxidation of NADPH. The decrease in reduction of DCPIP was significant at the 99% confidence level when the concentration of $Cu(II)_2(3,5\text{-}DIPS)_4$ was 2 µM. The IC_{50} for $Cu(II)_2(3,5\text{-}DIPS)_4$ -mediated decrease in DCPIP reduction in this PHD system was 1.5 µM. These results demonstrated that PHD oxidation of NADPH occurred in the absence of the reduction of DCPIP, the obligate electron acceptor.

Oxidation of NADPH by PHD in the absence of DCPIP reduction suggested that $Cu(II)_2(3,5\text{-}DIPS)_4$ served as the obligate electron acceptor in down-regulating the NADPH-DCPIP PHD system. This suggestion was examined using the PHD enzyme system without DCPIP and with the addition of increasing concentrations of 5 to 25 µM $Cu(II)_2(3,5\text{-}DIPS)_4$ to a solution of NADPH (10). Recovery of the initial rate of enzymatic activity was directly related to increasing concentration of added $Cu(II)_2(3,5\text{-}DIPS)_4$. The concentration of $Cu(II)_2(3,5\text{-}DIPS)_4$ required to produce 50 percent recovery of the initial rate of PHD activity in this system was 16 µM. Since the concentration of $Cu(II)_2(3,5\text{-}DIPS)_4$ required for 50 percent recovery of the unimpeded initial rate of enzyme activity was an order of magnitude lower than the concentration of NADPH, the role of $Cu(II)_2(3,5\text{-}DIPS)_4$ is clearly catalytic. These demonstrations of enzyme activity recovery are consistent with the suggestion that $Cu(II)_2(3,5\text{-}DIPS)_4$ down-regulates PHD activity and suggested that this complex might down-regulate NOS.

To further examine the possibility that $Cu(II)_2(3,5\text{-}DIPS)_4$ might down-regulate NOS, this complex was added to medium used to stain brain sections for NOS activity (7). Brain sections incubated in medium containing 10 µM $Cu(II)_2(3,5\text{-}DIPS)_4$ were less intensely stained than sections incubated in medium containing no copper complex (10). Densitometric measurements of staining intensity for sections incubated in medium containing $Cu(II)_2(3,5\text{-}DIPS)_4$ were significantly ($P \leq 0.005$, n=10) less intensely stained (integrated volume of color intensity = 6,908 ± 494) than sections incubated in medium containing no Cu(II)$_2$(3,5-DIPS)$_4$ (8,217 ± 275). These data are also consistent with and support the suggestion that this copper complex down-regulates NOS.

4. DISCUSSION

The following discussion has been cited in detail elsewhere (10). Space limitations do not permit citation of all relevant references supporting issues addressed.

Continuous oxidation of NADPH by PHD requires the presence of DCPIP as an obligate electron acceptor. The decrease in reduction of DCPIP in the presence of Cu(II)$_2$(3,5-DIPS)$_4$ with continued oxidation of NADPH suggested that Cu(II)$_2$(3,5-DIPS)$_4$ served as an electron acceptor in down-regulating the reduction of DCPIP. Evidence for this down-regulation was obtained when the deletion of DCPIP and addition of increasing concentration of Cu(II)$_2$(3,5-DIPS)$_4$ caused recovery of NADPH oxidation by PHD.

It is suggested that Cu(II)(3,5-DIPS)$_2$ functions in a catalytic role as the electron acceptor in recovery of PHD activity. As shown in Figure 1, oxidation of Cu(I)(3,5-DIPS)$_2$, the product of PHD reduction of Cu(II)(3,5-DIPS)$_2$, to Cu(II)(3,5-DIPS)$_2$ by dissolved oxygen and disproportionation of the resultant superoxide (11) and/or hydrogen peroxide (12) by Cu(II)(3,5-DIPS)$_2$, which are fast reactions compared to the slow rate of oxidation of NADPH and reduction of DCPIP by PHD, offer a rationale for its catalytic role in down-regulating NOS. The net consumption of oxygen in these reactions would also offer an additional accounting for down-regulation of NOS which requires oxygen for the oxidation of L-Arginine. Disproportionation of superoxide by Cu(II)(3,5-DIPS)$_2$ may also account for the down-regulation of NOS when NOS functions to generate superoxide (13).

Non-Toxic doses of copper complexes, which would supply less than the daily required intake of copper, have been shown to have beneficial pharmacological effects in animal models of disease and injury (8). These small doses of copper complexes including Cu(II)$_2$(3,5-DIPS)$_4$ have antiinflammatory, analgesic, anticonvulsant, antiulcer, anticancer, anticarcinogenic, antimutagenic, antidiabetic, radiation protection, and radiation recovery ac-

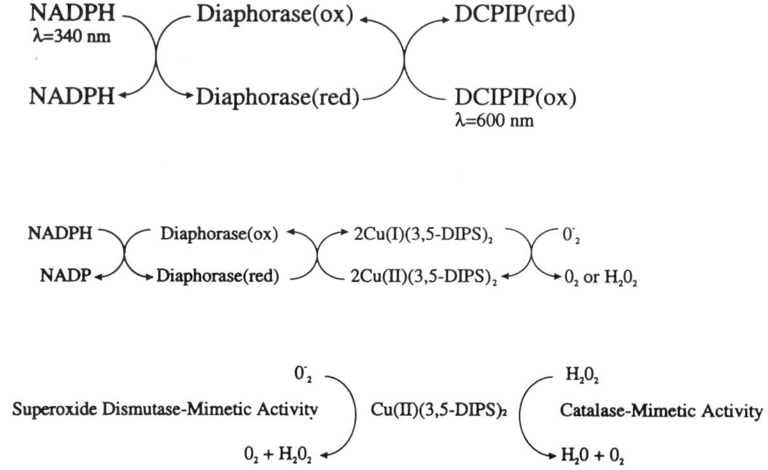

Figure 1. Schemes for the oxidation of NADPH by PHD in the presence of DCPIP and in the absence of DCPIP with the presence of Cu(II)$_2$(3,5-DIPS)$_4$.

tivities and decrease ischemia-reperfusion injury (8,9). In addition to data presented in this report, there is a growing body of literature suggesting that pharmacological activities of copper complexes may be due, at least in part, to the down-regulation of NOS.

Down-Regulation of antigen- or irritant-mediated and cytokine-induced NOS in endothelial cells, macrophages, and polymorphonuclear leukocytes (PMNLs) may partially account for roles of plasma copper complexes in modulating NO-mediated inflammatory responses in animal models of inflammation and may account for the observation that many copper complexes including $Cu(II)_2(3,5\text{-DIPS})_4$ have antiinflammatory activity in animal models of inflammation and man (8). Synthesis of NO plays a key role in mediating inflammation in a variety of animal models of inflammation including: synovial arthritic inflammation caused by intraperitoneal injection of streptococcal cell wall fragments (14), intraarticular injection of carrageenan into rabbit knee joints (15), and intradermal injection of complete Freund's adjuvant in the production of rat polyarthritis (PA) (16) as well as trinitrobenzene sulfonic acid (TNBS)-induced guinea pig ileitis (17). These results are consistent with observations that NO has a role in the immune-mediated response to inflammation such as phorbol-diester induction of inducible NOS in macrophages and NO release (18), streptococcal cell fragment-induced accumulation of PMNLs within synovial tissue and their release of arthritogenic NO, measured as nitrite (14). tubercular antigen-stimulated T-lymphocyte proliferation and release of acid phosphatase by macrophages from PA rats (16), and granulocyte infiltration of guinea pig mucosa associated with TNBS-induced ileitis (17), as well as the anticipated changes in inflamed joint blood flow (15). Synthesis of NO also plays a key role in peripheral, spinal cord, and brain neuronally-mediated nociception and hyperalgesia in various rat models of inflammation including: Carrageenan Paw Inflammation , PA, and Formalin Paw Inflammation (FPI). These results are consistent with the observation that rheumatoid arthritic (RA) and osteoarthritic (OA) patients have elevated synovial and plasma nitrite levels, with RA patients having higher levels, due perhaps to greater disease activity, than OA patients.

Down-Regulation of peripheral, spinal cord, and/or brain neuronal NOS may account for the modulation of nociception by copper complexes in animal models of pain including the writhing mouse and PA rat and may also account for the analgesic activity of $Cu(II)_2(3,5\text{-DIPS})_4$ (10). Results obtained in the Randall-Selitto prostaglandin E_2 [PGE_2]-hyperalgesia model of pain suggested that NO-Mediated analgesic activity while results obtained in the FPI and writhing mouse models of algesia-inflammation and hot plate or heat lamp tail-flick models of nociception demonstrated that down-regulation of NOS produced analgesia. These inconsistencies with regard to the apparent algesic and analgesic effects of NO were clarified with the demonstration that L-Arg exerts a dual role in brain, spinal cord, and peripheral pain processing involving the kyotorphin-met-enkephalin-opioid-receptors and NO-cGMP-glutamate-NMDA-receptor pathways in producing analgesia. Re-Examination of the role of NO in paw pressure nociception demonstrated that in the absence of PGE2-hyperalgesia, down-regulation of NOS produces analgesia. Opioid tolerance and dependence as well as pre-existing dependence mediated via, but not κ receptors were also suggested to be blocked by down-regulation of NOS using the tail-flick model of algesia and by examining behavioral and physical effects of morphine dependence and naloxone or isosorbide dinitrate precipitated withdrawal. These results suggest that $Cu(II)_2(3,5\text{-DIPS})_4$ may also reduce μ-receptor opioid tolerance and dependence as well as pre-existing dependence. Finally, the copper complex of HU 211, a non-psycotropic enantiomer of tetrahydrocannobinol and non-competative inhibitor of the NMDA receptor, may account for the antinociceptive activity of the mixture of this enantiomer and Cu(II)Cl2 in an autotomy pain model.

Down-Regulation of NOS by $Cu(II)_2(3,5\text{-}DIPS)_4$ may account for its anticonvulsant activity in the metrazol and electroshock models of seizure (10). In addition to metrazol-induced and electrically kindled seizures, seizures induced by NMDA-receptor activation, tacrine, and sodium nitroprusside $[Na(I)_2Fe(III)(CN)_5(NO)]$ are mediated via the NO-GC-NMDA pathway. However, pilocarpine-induced seizures were paradoxically facilitated by blockade of the NMDA receptor and inhibition of NOS. Except for these paradoxical results all other results support the hypothesis that NO synthesized by NOS from L-Arg and NO activation of GC leading to NMDA-receptor activation contribute to seizure states and that anticonvulsant activities of $Cu(II)_2(3,5\text{-}DIPS)_4$ are consistent with its down-regulation of NO synthesis.

Down-Regulation of some step in the NOS-cGMP-glutamate-NMDA-receptor pathway accounts for the anesthetic activity of ketamine, kainate, quisqualate, L-Glu, L-Asp, halothane, L-NAME, nitrous oxide, and isoflurane (10). Decreased consciousness and sedation are also produced by down-regulation of this pathway. Down-Regulation of NOS by many copper complexes including $Cu(II)_2(3,5\text{-}DIPS)_4$ may account for its sedative and hypnotic activities found in association with its anticonvulsant activities.

Down-Regulation of NOS may also account for the antiulcer activity of $Cu(II)_2(3,5\text{-}DIPS)_4$ and many other copper complexes (10). This down-regulation accounts for the reduction of alcohol-induced gastric hemorrhage produced by modulation of mucosal blood flow following thyrotropin-releasing hormone administration. Cultured rat gastric mucosal cell secretion of mucin is increased following NO up-regulation, consistent with an irritant induced response, while down-regulation decreased mucin secretion, consistent with the antisecretary activity of copper complexes. Down-regulation of NOS, which did not affect mast cell degranulation, intensified cold-restraint ulceration while administration of hexamethonium or atropine, compounds with cholinergic or anticolinergic activity respectively, strongly antagonized ulcer formation. Both of these compounds antagonized ulcer potentiation by concomitant NOS down-regulation. The observation that down-regulation of acetylcholine and methacholine, both cholinergic agonists, alters blood flow and vascular resistance differently via concentration related or endothelium-derived hyperpolarizing factor release suggests caution in assignment of mechanism of action for these compounds also.

Down-Regulation of NO synthesis by copper complexes may also be useful in decreasing brain ischemia-reperfusion injury (10). Inhibition of cat brain NOS before ischemia decreased caudate volume injury without changing cerebral blood flow and $Cu(II)_2(3,5\text{-}DIPS)_4$ administered before ischemia decreased cat mesentaric ischemia-reperfusion injury.

Down-Regulation of NOS by $Cu(II)_2(3,5\text{-}DIPS)_4$ may also account for the anti-diabetic activity of this complex and $Cu(II)_2(salicylate)_4$ in the streptozotocin-diabetic rat (10). Nitric oxide is suggested to have a role in streptozotocin-induced Type I insulin-dependent diabetes mellitus in mice. Hyperglycemia following repeated treatment with low doses of streptozotocin was significantly reduced by down-regulation of NOS. This down-regulation also prevented steptozotocin-induced and T lymphocyte- and macrophage-mediated inflammatory changes in pancreata and significantly decreased islet β cell destruction in these *in vivo* experiments. These results are consistent with the prevention of interleukin-1 (IL-1) induced destruction of isolated islet cells with down-regulation of NOS in *in vitro* experiments and the observed decrease in formation of Fe-nitrosyl complexes which may serve as NO transporter complexes in transporting NO to and into islet β cells. Down-regulation of IL-1-mediated *c-fos* induction of gene transcription and subsequnt mRNA translation in the synthesis of inducible NOS, formation of Fe-nitrosyl complexes, which may serve as NO sources for the inhibition of mitochondrial electron transport and death of β cells, as well as the induction of cyclooxygenase as observed in in vitro experiments further supports the role of NO in the development of diabetes. However, the seemingly paradoxical observation that IL-1 and tumor necrosis factor α[TNFα] have pronounced antidiabetic activity in *in vivo* experiments may be explained as a role of these cytokines in increasing ceruloplasmin, the principal copper-containing component of plasma, and other plasma low molecular mass copper complexes in down-regulating NOS as having a role in producing the observed antidiabetic effect.

5. REFERENCES

1. S. Moncada, R. M. J. Palmer and E. A. Higgs, *Pharmacol. Rev.* 43, 109–142 (1991).
2. J. R. Lancaster Jr., *Am. Scient.* 80, 248–259 (1992).
3. U. Weser, C. Richter, A. Wendel and M. Younes, *Bioinorg. Chem.* 8, 201–213 (1978).
4. C. Richter, A. Azzi, U. Weser and A. Wendel, *J. Biol. Chem.* 252, 5061–5066 (1977).
5. J. Werringloer, S. Kowano, N. Chacos, and R. W. Estabrook, *J. Biol. Chem.* 254, 11839–11846 (1979).
6. J.J. Reiners, Jr., E. Brott and J. R. J. Sorenson, *Carcinogenesis* 7, 1729–1732 (1986).
7. R. D. Skinner, C. Conrad, V. Henderson, S. A. Gilmore and E. Garcia-Rill, *Exptl. Neurol.* 104, 15–21 (1989).
8. J. R. J. Sorenson, *Prog. Med. Chem.* 26, 437–568 (1989).
9. J.R.J. Sorenson, L.S.F. Soderberg, L.W. Chang, W.M. Willingham, M.L. Baker, J.B. Barnett, H. Salari and K. Bond, *Eur. J. Med. Chem.* 28, 221–229 (1993).
10. J.G.L. Baquial and J.R.J. Sorenson, *J. Inorg. Biochem.* 60, 133–148 (1995).
11. U. Weser, C. Richter, A. Wendel and M. Younes, *Bioinorg. Chem.* 8, 201–213 (1978).
12. G. A. Reed and C. Madhu, in *Biology of Copper Complexes,* J.R.J. Sorenson, ed., Humana Press, Totowa, N.J., pp. 287–298 (1987).
13. S. Pou, W. S. Pou, D. S. Bredt, S. H. Snyder and G. M. Rosen, *J. Biol. Chem.* 267, 24173–24176 (1992).
14. N. McCartney-Francis, J. B. Allen, D. E. Mizel, J. E. Albina, Q. Xie, C. F. Nathan, and S.M. Wahl, *J. Exptl. Med.* 178, 749–754 (1993).
15. H. Najafipour and W. R. Ferrell, *Exptl. Physiol.* 78, 615–624 (1993).
16. A. Ialenti, S. Moncada and M. DiRosa, *Br. J. Pharmacol.* 110, 701–706 (1993).
17. M. J. S. Miller, S. Chotinaruemol, H. Sadowska-Krowicka, J.L. Kakkis, U. K. Munshi, X.-J. Zhang and D. A. Clark, *Agents Actions* 39, C180-C182 (1993).
18. S. Hortelano, A. M. Genaro and L. Bosca, *FEBS* 320, 135–139 (1993).

21

COPPER-HISTIDINE AND -NSAID COMPLEXES IN FENTON CHEMISTRY

Preliminary Results

S. Gaubert, L. Lambs, and G. Berthon

INSERM U305
Equipe "Bioréactifs: Spéciation et Biodisponibilité"
38 rue des Trente-six Ponts, F 31400 Toulouse, France

1. INTRODUCTION

At first sight, the behaviour of copper with respect to inflammation may seem paradoxical. On the one hand, copper exerts antiinflammatory activity. It has long been known that copper complexes of non-steroidal antiinflammatory drugs (NSAIDs) are more effective than parent molecules and that substances inactive by themselves may become antiinflammatory when administered with copper. It has been suggested from this that the active metabolites of NSAIDs are the complexes they form with endogenous copper (1). On the other hand, it is no longer debated that copper, like iron, can act as a catalyst in Fenton-type reactions to produce hydroxyl radicals (2) which are key mediators of the inflammatory process (3).

It has recently been proposed that the above paradox is only apparent. Both aspects of copper activity towards inflammation may actually derive from a single process whose physiological consequences would depend on the nature of the ligand attached to copper ions acting in Fenton chemistry (4). Virtually all the copper present in vivo occurs as complexes with biomolecules, a small but significant fraction of which is labile and likely to undergo redox reactions. Because of the high site specificity of copper-induced hydroxyl radical reactivity (as compared to iron for example), any ligand involved in the copper(II) coordination sphere is attacked in priority to more distant potential targets and may therefore, depending on its nature, aggravate or prevent ˙OH radical damage to these (4). In particular, any copper complex active in Fenton chemistry the ligand of which can form stable metabolites upon ˙OH attack is expected to protect surrounding essential biomolecules (5). In other words, copper pro-oxidant behaviour would in that case result in antiinflammatory activity.

Based on this principle, it has been proposed that copper complexes of ˙OH inactivating ligands (OILs) could be used as lures for the Fenton reaction, and arguments have been presented to support the hypothesis that histidine, which is the predominant low-molar-mass

Therapeutic Uses of Trace Elements, edited by Nève et al.
Plenum Press, New York, 1996

ligand of copper(II) in blood plasma (6), would be the main natural OIL (4). Since then, copper(II)-bound histidine has effectively been detected in ultrafiltrates of rheumatoid synovial fluid (7). Furthermore, histidine has recently been shown to strongly inhibit the apparent production of ˙OH radicals by the Cu(II)/ascorbate system while having virtually no influence on the activity of the Fe(III)/ascorbate system in vitro (8), which suggests a specific role for its coordinating properties in its protective action with copper.

The objective of the present work was therefore to compare the potential OIL activity of histidine to that of two NSAIDs (salicylic and acetylsalicylic acids) and of anthranilate, an inactive substance that becomes antiinflammatory when associated with copper. For each of these substances, true scavenger properties have been distinguished from OIL potentialities by putting inhibitory effects in perspective with copper(II) complex distributions at the outset of the reaction. The combined effect of histidine with each of the other three compounds has also been investigated.

2. MATERIALS AND METHODS

2.1. Reagents and Materials

Salicylic and acetylsalicylic acids were purchased from Aldrich as > 99 % pure Gold Label products. Anthranilic acid, also from Aldrich, was a 98 % pure product. L-Histidine was a Merck biochemical grade reagent (> 99 % pure). L-ascorbic acid was supplied by Sigma as free acid crystals. $CuSO_4$, $5H_2O$ was a Prolabo R.P. Normapur product. 2-deoxy-D-ribose and 2-thiobarbituric acid were purchased from Sigma and hydrogen peroxide from Gifrer (France). EDTA used to sequester copper(II) ions before the last stage of detection was a Fluka puriss. p.a. reagent, and trichloroacetic acid was a Prolabo R.P. Normapur product. All solutions were prepared from triply deionised, distilled and freshly deaerated water. The colorimetric tests used to detect thiobarbituric acid reactive substances (TBARS) were carried out on a PERKIN-ELMER Lambda 2 UV/VIS spectrophotometer.

2.2. ˙OH Radical Production and Detection

All solutions for the Fenton assays were prepared immediately before use under an atmosphere of purified nitrogen. Hydroxyl radicals were generated at room temperature using copper(II)/ascorbate as a redox system in the presence or absence of the different substances tested as potential OILs. Final concentrations of the reactants in mixtures of 2.0 ml were: $CuSO_4$ 0.125 mM, ascorbate 0.156 mM, H_2O_2 3.1 mM. Concentrations of potential OILs in initial solutions were varied so that the ligand-to-copper(II) ratios were successively 0.5, 1, 1.5, 2, 3, 4, 5 and 6. Reactions were initiated by adding hydrogen peroxide. Samples were then incubated in the dark and thiobarbiturate reactivity developed at room temperature. Resulting mixtures were finally heated for 10 min at 100 °C, cooled briefly, and the absorbance measured against appropriate blanks at 532 and 600 nm. The 600 nm absorbance was considered as a non-specific baseline drift and subtracted from A532 (9).

2.3. Speciation Calculations

In order to assess the influence of the coordinating capacity of each ligand in its global inhibitory effect, speciation calculations have been run to determine the distribution of copper(II) complexes at the outset of each experiment. The simulating module of the ESTA program library (10) was used for these calculations, which were based on formation constants recently determined in our group (11,12).

3. RESULTS AND DISCUSSION

3.1. Influence of Histidine and NSAIDs on Apparent 'OH Radical Production

Figure 1 shows TBARS absolute amounts measured in solution for the four substances investigated, expressed as mean absorbances of six determinations for each ligand-to-copper(II) ratio. Variations observed among control values (i.e. relative to ligand-to-copper(II) ratios equal to zero) reflect the degree of reproducibility from one substance to another. Interestingly, standard errors affecting measurements within each ligand series (not shown here for technical reasons) were inferior to these variations.

Among the four substances, ANT induces the most important inhibitory effect at the ligand-to-copper(II) ratio equal to unity. This effect then regularly increases along with the ligand concentration. In contrast, the inhibitory effect of histidine is relatively weak up to ligand-to-copper ratios near 2, but then sharply increases to become by far the most efficient of the four substances at high ligand concentrations (ligand-to-copper ratios ≥ 4). SLA regularly reduces the amount of 'OH radicals detected as its concentration is raised, its inhibitory effect being slightly superior to that of ANT throughout the concentration range investigated. ASA is the less effective inhibitor of the four compounds at high ligand concentrations and is only superior to histidine at low concentrations.

The above influences may be related to the composition of the copper(II) coordination sphere at zero time of the reaction, as can be appreciated from plots of complexed copper(II) global percentages as a function of ligand concentrations (Figure 2). In particular, the separate analysis of copper(II)-histidine complexes of 1:1 and 1:2 metal-to-ligand stoichiometries suggests that the inhibitory effect of histidine is limited by the presence of significant 1:1

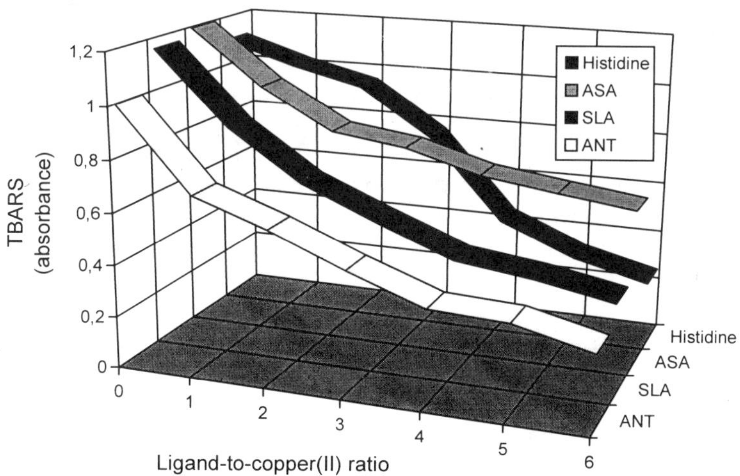

Figure 1. Inhibitory effect of histidine, ASA, SLA and ANT on the apparent production of 'OH radicals.

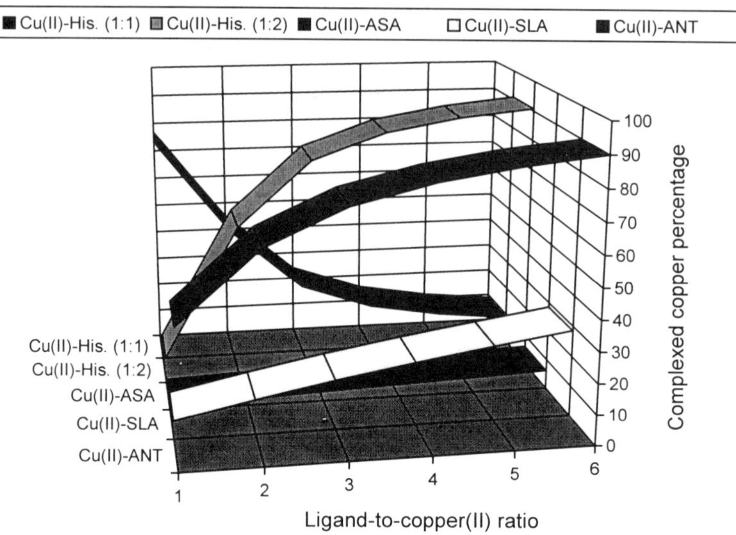

Figure 2. Distribution profiles of copper(II) complexed fractions in the presence of histidine, ASA, SLA and ANT before $^{\cdot}$OH production reaction. (Copper(II) = 0.125 mM)

complexes at ligand-to-copper(II) ratios below ~3, whereas it fully develops as 1:2 complexes become predominant at higher ligand concentrations. It seems logical to infer from this that the initial degree of coordination of copper(II) by histidine conditions the extent of the protection afforded to the detector within the solution. The situation is similar with ANT, the inhibitory effect of which parallels the degree of copper(II) complexation. In contrast, SLA and ASA seem to behave differently. The affinities of these two ligands (especially ASA) for copper(II) are low while their global scavenging activities (notably SLA's) are comparable to those of the other two substances. Relative to the OIL definition (4), histidine and ANT would thus behave more like ligands whereas ASA and SLA would act more as true scavengers even though SLA radical trapping has been said to depend on the degree of Cu(II)-SLA coordination (2).

3.2. Effect of Histidine on the Inhibitory Action of NSAIDs

As the main low-molar-mass ligand of copper(II) in blood plasma, histidine can interact with many copper(II)-mediated processes in vivo, particularly those involving weak ligands. Any investigation of the potential OIL activity of a given substance must therefore take the histidine factor into consideration. For this reason, the influence of histidine on the effect of ASA, SLA and ANT against the apparent production of $^{\cdot}$OH radicals in solution has been investigated.

Figure 3 shows the variations in the inhibitory effect of each substance (ligand-to-copper(II) ratio fixed at 2) induced by increasing concentrations of histidine (ligand-to-copper(II) ratios being varied as in Section 2.2), to be compared to the effect of histidine alone. Figure 4 illustrates the low extent of copper(II) complexation reached by ASA, SLA and ANT in the presence of histidine before the reaction. Given the high stabilities of copper(II)-histidine complexes with respect to those of the other three ligands, the fraction of copper bound to histidine in species of 1:2 metal-to-ligand stoichiometry is about the same in all three ternary

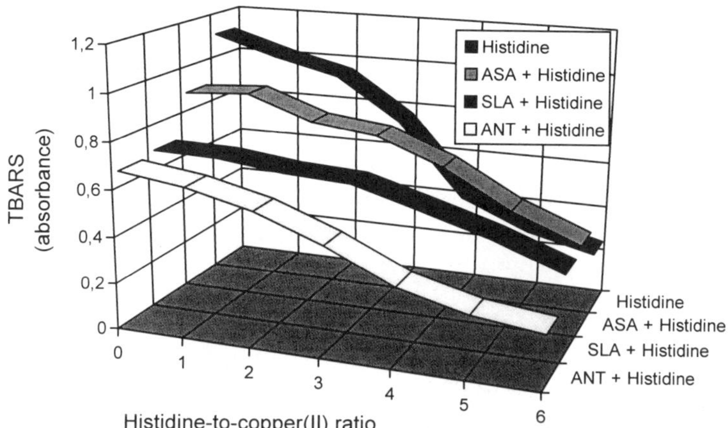

Figure 3. Inhibitory effect of histidine, ASA + histidine, SLA + histidine and ANT + histidine on the apparent production of ˙OH radicals.

mixtures and virtually equal to that shown in Figure 2. The latter is therefore repeated here for comparison. As can be seen on Figure 4, ternary complexation is low for ASA and quite negligible for SLA. It may be significant for ANT, even though no evidence was found for it because of precipitations observed at the higher concentrations required by potentiometry.

The interpretation of the results obtained for ANT is relatively straightforward. Clearly the global inhibitory effect observed progressively identifies with that of histidine alone as ligand substitution takes place to the advantage of histidine in the copper(II) coordination sphere. The contribution of both ligands to binary (and ternary?) complexation induces an effect superior to that of ANT by itself. For ASA and SLA, which bind negligible fractions of copper(II) even at low histidine concentrations, the role of histidine is expected to be pre-

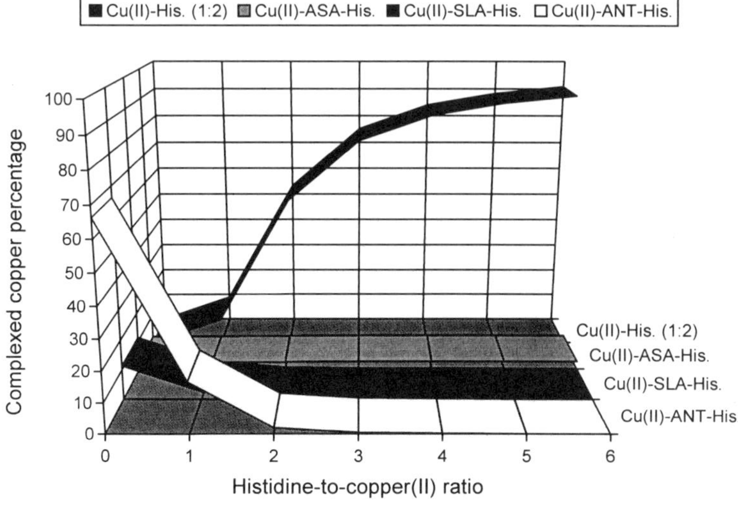

Figure 4. Distribution profiles of copper(II) fractions complexed by histidine (1:2 metal-to-ligand complexes; see text), ASA, SLA and ANT in the ASA + histidine, SLA + histidine and ANT + histidine mixtures (ligand-to-copper ratio = 2) before ˙OH production reaction. (Copper(II) = 0.125 mM)

dominant. Now, the inhibitory effect of histidine is relatively weak up to histidine-to-copper(II) ratios near 2 (see above), and should only develop at ratios around 3 and above. It is exactly what we observe, and the slight advantage of ASA on SLA may be due to the small but significant fraction of copper(II) bound in ternary complexes—hence more "isolated" from the solution. The fact that the inhibitory effects of the three ternary mixtures are globally inferior to that of histidine alone may be due to the influence of other factors (redox potentials, kinetic parameters) on the reaction process itself, and the main result here is that histidine can effectively augment the inhibitory capacity of NSAIDs against the apparent production of ˙OH radicals in vitro.

4. CONCLUSION

In reference to the recently defined notion of OIL (4), this work constitutes the first quantitative attempt to relate the scavenging activity of chemical substances towards copper-induced ˙OH radicals to the capacity of these substances to coordinate the Cu^{2+} ion at the outset of the ˙OH production reaction. Significant results have been found in this preliminary investigation:

i. The previously proposed OIL activity of histidine (4) has been definitely confirmed in vitro.
ii. Combined analyses of scavenging activities with related copper(II) distribution profiles suggest that, as potential OILs, histidine and ANT would be more ligands than true scavengers while ASA and SLA would be more true scavengers than ligands. This provides a beginning of rationale to the observation that some NSAIDs that are antiinflammatory by themselves (e.g. ASA and SLA) are improved in the presence of copper, whereas substances that are devoid of effect by themselves need to be entirely activated by copper (e.g. ANT).
iii. Mixtures of either ASA, SLA or ANT with histidine are more efficient inhibitors than ASA, SLA or ANT by themselves. Whatever the classification of the partner substance as rather true scavenger or ligand in the context of the OIL definition, histidine adds at least part of its intrinsic effect to the protection of ˙OH targets within the solution. However, ternary complexation is expected to induce optimum effects.

5. REFERENCES

1. J.R.J. Sorenson, *Prog. Med. Chem.* **26**, 437–568 (1989).
2. P. Saltman, *Semin. Hematol.* **26**, 249–256 (1989).
3. J.M. McCord, *Science* **185**, 529–530 (1974).
4. G. Berthon, *Agents Actions* **39**, 209–217 (1993).
5. O.I. Aruoma and B. Halliwell, *Xenobiotica* **18**, 459–70 (1988).
6. V. Brumas, N. Alliey and G. Berthon *J. Inorg. Biochem.* **52**, 287–296 (1993).
7. D.P. Naughton, J. Knappitt, K. Fairburn, K. Gaffney, D.R. Blake and M. Grootveld *FEBS Lett.* **361**, 167–172 (1995).
8. J. Randall-Simpson, L. Lambs and G. Berthon, in *Metal Ions in Biology and Medicine*, vol. 4, John Libbey Eurotext, Paris (1996), in press.
9. D.Gelvan and P. Saltman, *Biochim. Biophys. Acta* **1035**, 353–360 (1990).
10. P.M. May, K. Murray and D.R. Williams, *Talanta* **32**, 483–489 (1985).
11. V. Brumas and G. Berthon, *J. Inorg. Biochem.* **43**, 626 (1991).
12. V. Brumas, B. Brumas and G. Berthon, *J. Inorg. Biochem.* **57**, 191–207 (1995).

22

TRACE ELEMENTS AND OTHER ANTIOXIDANTS IN ALCOHOL-RELATED CIRRHOSIS AND CHRONIC PANCREATITIS

André Van Gossum

Department of Hepato-Gastroenterology
Hôpital Erasme
Free University of Brussels
808 route de Lennik, 1070 Brussels, Belgium

1. INTRODUCTION

Excessive alcohol intake is considered to be a major cause for developing cirrhosis and chronic pancreatitis. These two diseases correspond to a degenerative process of tissue parenchyma associated with cell destruction and fibrosis. The cause of both these diseases is probably multifactorial : genetic predisposition, toxic action of alcohol, immune dysregulation, enhancement of fibrosis production and deleterious effect of free radicals. The generation of toxic free radicals from the metabolism of ethanol by hepatic cytochrome P450 II E1 has been implicated in alcoholic liver disease (1).

2. ALCOHOLIC CIRRHOSIS

Several arguments support the hypothesis of an increased lipoperoxidation in alcoholic cirrhosis. Studies have shown that these patients have an increase in conjugated diene production in the liver (2). Situnayake et al. have further described a significant correlation between spectroscopic demonstration of conjugated dienes and the histologic inflammatory process (3). Moreover, Lunel et al. described an increase in chemiluminescence of whole blood in patients with alcoholic hepatitis (4). This test appeared to be of prognostic value. Several groups have also described a significantly increased breath pentane output amongst patients with alcoholic cirrhosis (5,6). On the contrary, increased resistance of erythrocytes to lipid peroxidation has been described in alcoholic cirrhosis (7). This is probably related to changes in membrane lipid composition, because of significant decrease in fatty acids that are susceptible to peroxidation.

The increase in lipid peroxidation may be potentiated by a decrease in antioxidant factors in patients with alcoholic cirrhosis. Indeed, these patients are known to have a decrease in hepatic glutathione (8). Blood vitamin E levels are also lower in severe liver disease (9). Teare et al. have shown that chronic infusion of ethanol in rats leads to progressive decrease in hepatic alpha-tocopherol levels, as it was transformed into alpha-tocopherol quinone (10). On the other hand, Togashi et al. observed a reduced activity of the enzymes Cu, Zn-super-oxide dismutase and catalase within hepatic tissue of patients with alcoholic cirrhosis (11). Selenium levels in the blood (12) and liver (13) are also lower in alcoholic patients. The relationship between impaired selenium status and liver function was examined in 19 hospitalized patients with severe alcoholic cirrhosis. Plasma selenium was found to be significantly lower (mean $\pm$ SD : 54 $\pm$ 13 µg/L) than in healthy controls (83 $\pm$ 11 µg/L). The mean ^{14}C-aminopyrine breath test, an indicator of liver function, was lower than normal (2.7 $\pm$ 19 vs 6.3 $\pm$ 0.9% in controls) and found to be significantly correlated with plasma selenium (r = 0.59, p < 0.05) (14). The lower selenium levels are partly related to insufficient selenium content of diets (15).

We recently studied the effect or oral selenium supplementation on peroxidative parameters in patients with alcohol-related cirrhosis. Seventeen patients with biopsy-proven AC (mean age : 60 y; 7 Fe and 10 Ma) were included in this study. They were classified according to the Pugh-Child Score. They randomly received oral selenium supplementation (200 µg/day) or a placebo during a 2 month-period. The following parameters were measured at baseline and at the end of the supplementation period : Hemoglobin, WBC, platelets count, PTT, GOT, Albumin, plasma selenium (Se), Se-glutathion peroxidase (Gpx), zinc, copper, malondialdehyde production, lag phase of LDL peroxidation and hemolysis of red blood cells, TNF and IL10 production in supernatants of PBMC's cultured for 24 H in the presence or absence of E.coli LPS at 100 µg/ml (Medgenix, Fleurus, Belgium). At baseline, all the parameters were comparable between the 2 groups. They all were blood-selenium deficient (mean : 48 $\pm$ 3 µg/l). Eleven patients completed the supplementation study (6 with selenium and 5 with placebo). Six patients discontinued because of death (n=1), liver-related complications (n=3) or uncompliance (n=2). At the end of supplementation period, plasma selenium (108 $\pm$ 1 vs 44.0 $\pm$ 8 µg/l; p < 0.006) and Se-GPx U.I. (1212 $\pm$ 85 vs 844 $\pm$ 131 U.I.; p < 0.02) were significantly higher in the group receiving selenium than in the placebo group. However, for all the other parameters, there was no significant change after Se supplementation. All the patients were deficient in selenium before supplementation. Although a 2-months oral supplementation in selenium significantly increased plasma selenium and Se-GPx levels, there was no effect on altered peroxidative and immunological parameters in patients with AC.

3. CHRONIC PANCREATITIS

Oxygen-derived species (including free radicals) have been implicated as important mediators in the pathogenesis of several forms of tissue injury (16). Their role has also been debated in acute pancreatitis. Studies in ex vivo models of acute pancreatitis suggested that enzymatic scavengers such as superoxide dismutase and catalase limit the extent of pancreatic injury induced by free fatty acids, ischemia or partial pancreatic ligation (17,18). In several animal models, it has also been demonstrated that these intermediates are involved in the early pathogenesis of acute pancreatitis (19,20,21). A recent study showed that antioxidant treatment reduces tissue damage, biochemical alterations and the final outcome in rats with induced acute hemorragic pancreatitis (22). In case of alcohol-related acute pancreatitis, oxygen species can be excessively released during ischemia-reperfusion injuries (23,24) or produced by inflammatory cells (25). Deleterious effect of such derivatives can be potentiated in case of antioxidant factors deficiency. Such mechanisms could intervene in the development of alcohol related chronic pancreatitis in relation with recurrent episodes of acute pancreatitis and long term exposition to alcohol. Indeed, some of tissue damage occurring in alcohol abusers may be due to generation of free radicals during the metabolism of ethanol and subsequent lipid peroxidation (26). On the other hand, free radical production and deficiency in antioxidant components have been incriminated in carcinogenesis (27,28). Some recent studies suggested a potential benefit of antioxidant supplementation to lower cancer risk in specific populations (29). Interestingly, a higher incidence of pancreatic cancer has been observed in patients suffering from chronic pancreatitis (30).

We recently observed that patients with alcohol-related chronic pancreatitis had low levels in several important antioxidant factors such as selenium, GSH-Px, vitamin E and vitamin A (31). According to the results of the dietary interrogatory, low plasma selenium levels were probably not due to an insufficient oral intake of selenium. Moreover, it seems that intestinal absorptive capacities of the subjects were not altered based on normal levels of other trace elements as zinc, iron and even slightly elevated copper levels. In consideration of normal levels of total proteins, the observed low selenium values could also not be related to changes in plasma proteins. It is therefore likely that the decrease in plasma selenium reflects increased metabolic requirements associated to long-term exposure to factors enhancing lipid peroxidation such as alcohol consumption. Low blood selenium levels have indeed been observed in alcoholics with or without liver damage (32–35). In our patients, plasma selenium levels were not different between drinkers and not drinkers. This finding agrees with the results of Lecomte et al. (36) and Girre et al. (37) who observed no change in plasma selenium of alcoholics after a withdrawal period.

Concerning the origin of modifications in antioxidant vitamins, complex influences can also be evoked. It is known that blood vitamin E concentrations of alcoholics are influenced by intestinal malabsorption, variations in hepatic storage, increased liver metabolism or enhanced degradation due to lipid peroxidation induced by alcohol (38–40). In this study, patients with steatorrhea had lower vitamin E levels than those without steatorrhea. On the other hand, plasma vitamin E levels were higher in current alcoholics than in non-alcoholics. Such an observation confirms results in rats chronically fed ethanol (41). The increase in plasma vitamin E may be related to increase in plasma lipids as a result of chronic ethanol feeding. Indeed, plasma vitamin E is transported by low density lipoproteins the concentration of which is increased after ethanol feeding (42). Vitamin A was especially low in patients with associated diabetes mellitus. It has been recently described that plasma retinol was significantly reduced in the young diabetic subjects as compared to controls, independently from other confounding variables while plasma α-tocopherol was unchanged in diabetic controls (43).

It is also important to notice that all our patients were cigarette smokers. It is indeed documented that cigarette smoking modify to some extent antioxidant balance (44,45). Duthie et al. found an increased peroxidation of smokers' red blood cells when put in contact with hydrogen peroxide as well as an increase in plasma dienes. These processes could be reversed by administration of vitamin E. In cigarette smokers, enhanced breath pentane output (a reflect of increased lipid peroxidation) and decreased GSH-Px activities were also demonstrated. Low vitamin C status was similarly documented in heavy smokers. Chow et al. described an increased concentration of vitamin E within pulmonary tissue of rats exposed to cigarette smoke (46).

4. CONCLUSION

It appears that patients with alcoholic cirrhosis or alcohol-related pancreatitis are deficient in some antioxidant factors due to various causes: decreased nutrients intake in cirrhotic patients, altered capacities in patients with chronic pancreatitis, alcohol consumption, cigarette smoking and probably increased consumption due to oxidative stress. There are several evidences that lipid peroxidation is increased in both conditions. Some experimental arguments support that a increased lipid peroxidation due to toxic free radicals generation is an initial step in the pathogenic process. Prospective studies with antioxidant supplementation in patients with early stages of cirrhosis or chronic pancreatitis are mandatory to definitively establish the theory of free radicals injury in the pathogenesis of these conditions.

5. REFERENCES

1. C. Lieber, *N. Engl. J. Med.*. **319**, 1639–1650 (1988).
2. S. Shaw, K. Rubin, C. Lieber, *Dig. Dis. Sci.* **28**, 585–589 (1983).
3. R. Situnayake, B. Crump, D. Thurnham, J. Davies, J. Gearty, M. Davis, *Gut* **31**, 1311–1317 (1990).
4. F. Lunel, B. Descamps-Latscha, D. Descamps, Y. Le Charpentier, P. Grippon, D. Valla, J.F. Cadranel, J. Trum, P. Opolon, *Hepatology* **12**, 264–272 (1990).
5. S. Moscarella, G. Laffi, G. Buzzeli, R. Mazzanti, L. Caramelli, P. Gentilini, *Hepato. Gastroenterol.* **31**, 60–63 (1984).
6. A. Van Gossum, J. Decuyper, M. Cremer, H. Ooms, *Clin. Nutr.* **9**, 25 (1990).
7. N. Punchard, H. Senturk, J.P. Teare, R. Thompson, *Gut* **35**, 1753–1756 (1994).
8. H. Speisky, A. Mc Donald, G. Giles, H. Orrego H., *Biochem. J.* **255**, 565–572 (1985).
9. S. Moscarella, A. Duchin, G. Buzzeti, *Eur. J. Gastroenterol. Hepatol.* **6**, 633–636 (1994).
10. J. Terae, S. Greenfield, D. Watson, N. Punchard, N. Miller, C. Rice-Evans, R. Thompson, *Gut* **35**, 1644–1647 (1994).
11. H. Togashi, H. Shinzawa, H. Wakabayashi, T. Nakamura, N. Yamada, T. Takahashi, M. Ishikawa, *J. Hepatol.* **11**, 200–205 (1990).
12. J. Aaseth, J. Alexander, Y. Thomassen, J. Blomhof, S. Skrede, *Clin. Biochem.* **15**, 281–283 (1983).
13. B. Dworkin, W. Rosenthal, R. Stahl, N. Panesar, *Dig. Dis. Sci.* **33**, 1213–1217 (1988).
14. A. Van Gossum, J. Neve, *Biol. Trace Elements Res.* **47**, 201–207 (1995).
15. D. Quivy, J. Neve, M. Adler, *Acta Gastroenterol. Belg.* **53**, 286–291 (1990).
16. C. Cross, B. Halliwell, E. Borish, W. Pryor, B. Ames, R. Saul, J. Mc Cord, D. Harman, *Ann. Intern. Med.* **107**, 526–545 (1987).
17. H. Sanfey, G. Bulkley, J. Cameron, *Ann. Surg.* **201**, 633–638 (1985).
18. C. Niederau, M. Niederau, F. Borchard, K. Ude, R. Lüthen, G. Strohmeyer, L. Ferrel, J. Grendell, *Pancreas* **7**, 486–496 (1992).
19. J. Wisner, D. Green, L. Ferrel, I. Renner, *Gut* **29**, 1516–1523 (1988).
20. P. Rutledge, A. Salujy, R. Powers, M. Steer, *Gastroenterology* **93**, 41–47 (1987).
21. M. Schoenberg, M. Büchler, A. Stinner et al, *Gut* **31**, 1138–1143 (1990).
22. M. Schoenberg, M. Büchler, M. Younes, R. Kirchmayer, U. Brückner, H. Beger, *Dig. Dis. Sciences* **39**, 1034–1040 (1994).
23. J. Mc Cord, *N. Engl. J. Med.* **6**, 541–551 (1989).
24. L. Formela, S. Galloway, A. Kingsnorth, *Br. J. Surg.* **82**, 6–13 (1995).
25. H. Rinderknecht, *Int. J. Pancreatol.* **3**, 105–112 (1988).
26. J. Feare, S. Greenfield, N. Punchard, N. Miller, C. Rice-Evans, R. Thompson, *Gut* **35**, 1644–1647 (1994).
27. T. Byers, G. Perry, *Ann. Rev. Nutr.* **12**, 139–159 (1992).
28. G. Block, *J. Natl. Cancer Inst.* **85**, 846–848 (1993).
29. W.J. Blot, J.Y. Li, P.R. Taylor et al, *J. Natl. Cancer Inst.* **85**, 1483–1492 (1993).
30. A. Lowenfels, P. Maisonneuve, G. Cavallini, R. Ammann, P. Lankisch, J. Andersen, E. Dimagno, A. Andren-Sandberg, L. Domelho, *N. Engl. J. Med.* **328**, 1433–1437 (1993).
31. A. Van Gossum, Ph. Closset, E. Noel, C. Mignon, M. Cremer, J. Neve, *Clin. Nutr.* **14**, 19 (1995).
32. A. Tanner, I. Bantock, L. Hinks, B. Lloyd, N. Turner, R. Wright, *Dig. Dis. Sci.* **31**, 1307–1312 (1986).

33. J. Ringstad, S. Knutsen, D. Nilssen, Y. Thomassen, *Biol. Trace Elem. Res.* **36**, 65–71 (1993).

34. A. Van Gossum, J. Neve, *Biol. Trace Elem. Res.* **47**, 201–207 (1995).

35. H. Korpela, J. Kumpulainen, P. Luoma, A. Arranto, E. Sotaniemi, *Am. J. Clin. Nutr.* **42**, 147–151 (1985).

36. E. Lecomte, B. Herbeth, P. Pirollet, Y. Chancerelle, J. Arnaud, N. Musse, F. Paille, Y. Arthur, *Am. J. Clin. Nutr.* **60**, 255–261 (1994).

37. C. Girre, E. Hispard, P. Therond, S. Guedj, R. Bourdon, S. Dally, *Alcohol. Clin. Exp.* **14**, 909–912 (1990).

38. Losowsky M., Leonard P. *Gut* 8, 539–543 (1967)

39. B. Herbeth, L. Didelot-Barthelemy, A. Le Moine, C. Ledevehat, *Am. J. Clin. Nutr.* **43**, 343–344 (1988).

40. S. Jumdar, G. Shaw, A. Thompson, *Drug Alcohol Depend.* **12**, 260–272 (1983).

41. Teare J., Greenfield S., Watson D., Punchard N., Miller N., Rice-Evans C., Thompson R, *Gut* 35, 1644–1647 (1994)

42. T. Ogihara, M. Miki, M. Kitagawa, M. Mino, *Clin. Chim. Acta* **174**, 299–306 (1988).

43. L. Martinoli., M. Di Felice, G. Seghieri, M. Ciuti, L. Di Georgio, A. Fazzini, R. Gori, R. Anichini, F. Franconi, *Int. J. Vitam. Nutr. Res.* **63**, 87–92 (1993).

44. G. Duthie, J. Arthur, P. James, H. Vint, *Ann. NY Acad. Sci.* **570**, 435–438 (1989).

45. E. Hoshino, R. Shariff, A. Van Gossum, J. Allard, C. Pichard, R. Kurian, K.N. Jeejeebhoy, *JPEN* **14**, 300–305 (1990).

46. C. Chow, Airriess G., C. Changcitt, *Ann. N.Y. Acad. Sci.* **570**, 425–427 (1989).

23

ZINC AND DIGESTIVE DISEASES

G. C. Sturniolo,[1] R. D'Inca,[2] C. Mestriner,[2] P. Irato,[3] V. Di Leo,[2]
A. D'Odorico,[2] C. Venturi,[2] G. Longo,[1] and F. Farinati[2]

[1] Department of Internal Medicine
University of Messina
[2] Division of Gastroenterology
[3] Department of Biology
University of Padova, Italy

1. PHYSIOLOGY OF ZINC METABOLISM

Trace element metabolism is regulated by numerous factors which are dependent on availability of nutrients and on the integrity of the gastrointestinal tract and on liver function. It is well known that a diet rich in phytates can reduce zinc bioavailability while a diet rich in animal proteins results in much higher absorption (1). Zinc absorption is regulated homeostatically so that in zinc depleted diets a strong reduction in urinary and fecal zinc excretion is observed. Gastric acidity is essential for zinc absorption as it is for iron. In experiments conducted in healthy volunteers we showed a reduction in zinc absorption when cimetidine (400 mg) or ranitidine (300 mg) were administered (2) (Fig 1). Recently, the same results were obtained after famotidine (40 mg) administration (3). A reduction in zinc absorption has also been observed in the healthy elderly and this may support the evidence of marginal zinc deficiency reported in several studies (4). Zinc uptake appears to be mediated by low molecular weight intracellular ligands able to bind the metal at the brush border and transfer it into the cell. The suggested ligands are picolinic and citric acid and prostaglandins. Deficiency of citric acid has been described in chronic pancreatitis and we observed that zinc absorption reverted to normal after the administration of citric acid in a group of patients with chronic pancreatitis (5). Once in the cell, zinc is stored in metallothioneins and other cysteine-rich molecules which synthesis is stimulated by an excess of zinc in the diet.

2. PATHOLOGICAL CONDITIONS

2.1. Adaptation after Small Bowel Resection

Experimental studies in the rat have demonstrated that zinc deficiency is able to reduce the intestinal DNA levels and mucosal protein content following 70% small bowel resection

Therapeutic Uses of Trace Elements, edited by Nève et al.
Plenum Press, New York, 1996

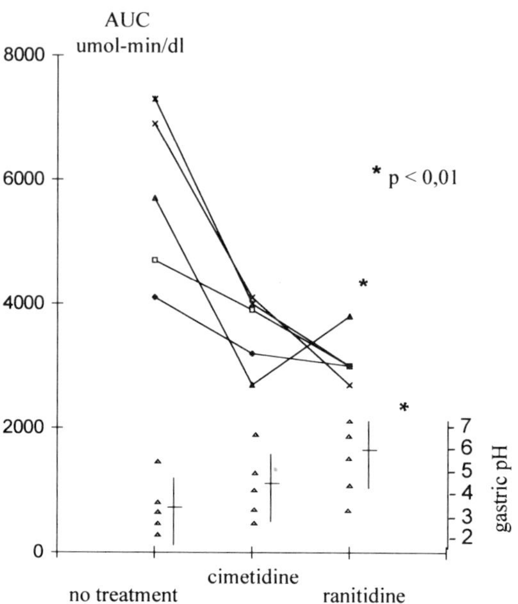

Figure 1. Effect of gastric pH on zinc absorption.

(6). Functional indices, such as maltase, were also affected. Zinc deficiency may therefore play a role also in short bowel syndrome patients but a similar study has not been conducted yet in humans.

2.2. Acute Diarrhoea

Impaired immune function and malnutrition can be associated with zinc deficiency. In developing countries long-lasting and severe diarrhoea is rather common in malnourished children with impaired immune function. In a double blind controlled trial involving 937 children, the daily supplementation with 20 mg elemental zinc in adjunct to oral rehydration resulted in clinically important reduction of duration and severity of diarrhoea (7). The possible mechanisms for the effect of zinc supplementation on diarrhoea include improved absorption of water and electrolytes, enhanced immunological mechanisms, regeneration of gut epithelium or the restoration of its function and increased levels of enterocyte brush border enzymes. The efficacy and effectiveness of zinc supplementation on the intestinal mucosal integrity and function have been demonstrated also in another trial of zinc supplementation in adjunct to antibiotics in 32 children with acute shigellosis (8). Children supplemented with zinc had a significant improvement in intestinal permeability and better nitrogen absorption than unsupplemented children. Fecal protein losses were unchanged. The importance of zinc in this study stays on the stimulation of mucosal regeneration and absorptive function.

2.3. Inflammatory Bowel Disease (IBD)

Zinc metabolism is altered in inflammatory bowel disease: reduced blood zinc levels have been correlated with the severity of acute attacks but the plasma concentration is rather insensitive for detecting zinc deficiency. Deficiency has been referred to inadequate zinc in-

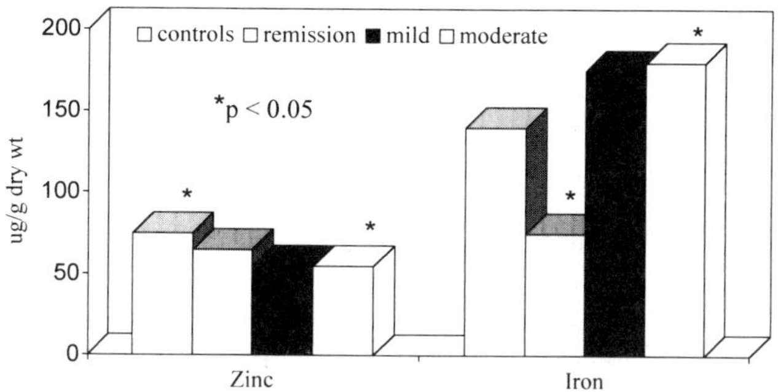

Figure 2. Zinc and iron concentrations in the colonic mucosa of ulcerative colitis and controls patients.

take, increased fecal losses or reduced absorption. In recent studies we observed reduced zinc concentrations also in the colonic mucosa of active ulcerative colitis patients. At the same time iron concentrations were increased (Fig 2). A direct relationship between inflammation and decreased free radical scavengers has been suggested on the basis of a reduced plasma concentration of two antioxidant proteins, metallothionein (MT) and superoxide dismutase (SOD), during disease flare ups. Because IBD is associated with low zinc levels and both SOD and MT contain significant amounts of zinc, it was speculated that treatment with supplemental zinc might increase the mucosal concentrations of these proteins able to protect the tissue against oxygen free radicals mediated damage. High doses of zinc were effective in inducing MT synthesis in small intestine of mice and rats. A randomised double blind placebo controlled cross-over trial involving 14 IBD patients receiving oral zinc supplementation (60 mg elemental zinc) for 4 weeks, however, was not able to effectively induce either SOD or MT (9). We measured another antioxidant protein, glutathione peroxidase, in plasma and colonic tissue of 27 ulcerative colitis patients and found significantly lower plasma levels and significantly higher tissue concentration in moderate activity with respect to mild activity or remission (10). In conclusion, increased iron concentration in the inflamed mucosa of active ulcerative colitis may enhance radical oxygen metabolites production turning on glutathione peroxidase activity and other antioxidants, while mucosal zinc concentration may affect zinc-dependent scavenger enzymes thus delaying ulcer healing.

Physical exercise is known to affect the gastrointestinal tract and the immune system inducing changes in oxygen free radical levels and trace elements status, parameters which are also influenced by Crohn's disease. The influence of mild exercise on lipid peroxidation and plasma and urinary zinc output was evaluated in 6 patients with ileal Crohn's disease in remission. Exercise consisted in cycling at 60% VO_2 max for 1 h. Zinc plasma levels were unaffected by exercise while a significantly increased urinary output was observed. Patients had significantly higher levels of malondialdehyde after exercise (1.77 nmol +/- 0.4) with respect to basal levels (1.62+/-0.4) ($p<0.05$). Zinc losses increase after exercise probably as a result of muscle protein breakdown. This may represent an additional risk factor for zinc deficiency in Crohn's disease.

Zinc status regulates also the intestinal transport of long-chain fatty acids and their esterification into phospholipids. The presence of abnormalities of membrane fatty acid composition was investigated in 20 Crohn's disease patients by Belluzzi et al.(11). The percentage of palmitic, stearic and oleic acids was significantly higher in Crohn's disease, while linoleic arachidonic and n-3 fatty acids were reduced compared to healthy controls. Supplementation with 200 mg of zinc sulphate for 6 weeks resulted in the ability of zinc to abolish the pre-treatment differences in red-cell long-chain acid profiles but did not affect plasma fatty acid values.

Zinc influence on the development and the maintenance of the cellular immune response has been studied in Crohn's disease. The results of the supplementation trials, however, did not offer univocal results. A 3 week-supplementation trial with zinc sulphate performed in 7 Crohn's disease patients showed a reduction in T cell subset percentages. Similar results were reported after zinc supplementation of 20 Crohn's disease patients (12). Natural killer cell activity decreased in treated patients even though no effect was seen on disease activity. No beneficial effect was seen either in an old study involving ulcerative colitis patients(13), or on the humoral and cellular immunity in Crohn's disease (14).

2.4. Liver Disease

Acute and chronic liver diseases are associated with changes in the circulating and tissue concentrations of zinc. In acute viral hepatitis plasma zinc levels are reduced and normalise together with transaminases. Zinc reduction seems to be part of the acute phase response through interleukin stimulation, which redistributes the circulating metal to the inflamed tissues.

Chronic liver disease is associated with altered plasma and urinary zinc levels but the most important risk factor for zinc deficiency is malabsorption. We demonstrated a reduced metal absorption both in alcoholic and in viral cirrhosis. These alterations seem therefore related more to the disease process than to the aetiology of the cirrhosis. Zinc supplementation trials had little efficacy on the amelioration of the liver function except for a recently published open controlled trial in which a mixture of vitamin E, selenium and zinc was administered to 56 patients with acute alcoholic hepatitis (15).

Zinc is undoubtedly an effective treatment for Wilson disease (16) but the exact mechanism of action is still unknown. Possible hypothesis include increased MT synthesis and block of copper absorption at the intestinal level, binding of free copper in the liver cell or even a reduction of lipid peroxidation and oxidative damage. We investigated these possibilities in three different studies. The first involved an experimental model of copper loaded rats. MT concentration was measured in liver, kidney and intestine after zinc supplementation (17). Oral zinc administration in copper loaded rats increased MT contents in all examined organs. MT increased significantly in liver, while the effect on intestine was significant only after supplementation. The second study was performed in Wilson's disease patients. Zinc therapy was able to increase significantly MT concentration in the duodenal mucosa (18). Liver MT are increased even before zinc treatment, therefore we must hypothesise that this cannot be the only mechanism of action of zinc. Copper is known to induce lipid peroxidation and oxidative damage. In the third study we measured malondialdehyde levels and glutathione and cysteine in liver biopsies from patients with Wilson's disease on zinc therapy or on penicillamine and in patients with viral type of hepatitis. We found decreased malondialdehyde levels and increased glutathione turnover in patients treated with zinc compared to those treated with penicillamine confirming the scavenger activity of zinc (19).

3. REFERENCES

1. G.C. Sturniolo , R.D'Incà ., P.E. Lecis et al. *Ital. J. Gastroenterol.* **26**, 247- 250 (1994)
2. G.C.Sturniolo, M.C. Montino, L. Rossetto et al.. *J. Am. Coll. Nutr* **10(4)**, 372–375 (1991)
3. L. Henderson, G. Brewer , J. Dresseman, et al.. *J. Parent. Enter. Nutr.* **19(5)**, 393–397 (1995)

4. G.C. Sturniolo, R. D'Incà, M.C. Montino et al.. *Clin. Nutr.* **13**, 280–285 (1994)
5. G.C. Sturniolo, R. D'Incà, M.C. Montino et al.. *J. Trace El. Exp. Med.* **3**, 267–271 (1990)
6. J. Vanderlhoof , J. Park, C. Grandjean. *Am. J. Clin. Nutr.* **44**, 670–667 (1986)
7. S. Sazawal, R. Blanck, M. Bhan et al. *New. Engl. J. Med.* **333(13)**, 839- 843 (1995)
8. A.N. Alam, S.A. Sarker, M.A. Wahed et al. *Gut* **35**, 1707–1711 (1994)
9. T.P.J. Mulder, A. Van Der Sluys Veer, H.W. Verspaget et al. *J. Gastroenterol. Hepatol.* **9**, 472–477 (1994)
10. G.C. Sturniolo, R. D'Incà, P.E. Lecis, C. Mestriner et al.. *Gut* **W38** (1993)
11. A. Belluzzi, C. Brignola, M. Campieri et al. *Aliment. Pharmacol. Ther.* **8**, 127–130 (1994)
12. Y. Van De Wal, A. Van Der Sluys Veer, H.W. Verspaget et al.. *Aliment. Pharmacol. Ther.* **7**, 281–286 (1993)
13. M.W. Dronfield, J.D.G. Malone, M.J.S. Langman.. *Gut* **18**, 33–36 (1977)
14. A. Animashaun, J. Kelleher, R.V. Heatley et al.. *Clin. Nutr.* **9**, 137–146 (1990)
15. G. Wenzel., B. Kuklinski, C. Ruhlmann et al. *Innere Medicine* **48**, 490–496 (1993)
16. L. Rossaro, G.C. Sturniolo, G. Giacon et al. *Am. J. Gastroenterol.* **85(6)**, 665–668 (1990)
17. P. Irato, G.C. Sturniolo, G. Giacon et al. *Biol.Trace El. Research* **51**, 1–10 (1996)
18. R. D'Incà, C. Mestriner, P.E. Lecis et al.. *Ital. J.Gastroenterol.* **27** suppl.1 , 80 (1995)
19. F. Farinati, R. Cardin, C. Mestriner et al. *Am. J. Gastroenterol* **90(12)**, 2264–65 (1995)

24

SELENIUM AND ANTIOXIDANT FACTORS IN CROHN'S DISEASE

G. Le Moël,[1] T. Gousson,[1] A. Dauvergne,[1] M. Succari,[1] N. Delas,[2]
M. J. Cals,[1] M. Cuer,[1] J. Dumont,[1] F. Callais,[1] M. Bouchoucha,[3]
C. Laureaux,[1] and P. Bernades[4]

[1] Gerbap Groupe Vitamines, Paris (Biochemistry Departments, Hosp.
Bichat/Beaujon/C. Celton/Laënnec/Montfermeil/E. Roux)
[2] Hepato-Gastroenterology Department
Hosp. Montfermeil
[3] Digestive Physiology Lab
Hosp. Laënnec, Paris
[4] Gastroenterology Department
Hosp. Beaujon, Clichy, France

1. INTRODUCTION

Crohn's disease is a chronic pathology of the intestine which develops in sporadic stages of inflammation and affects 60.000 persons in France. The cause of this disease has yet to be ascertained and is probably multifactorious: oxidative stress related to the inflammatory process might be one of the involved factors (1). In fact, several studies have shown a lowering of some factors of antioxidant status such as glutathione peroxydase (GSHPx) and selenium (Se) (2), which is an essential part of the enzyme active site as selenocystein. The aim of the present study was to determine the potential defence against free radicals in patients with Crohn's disease. We have evaluated the activity of major antioxidant enzymes, superoxide dismutase (CU-Zn-SOD), GSHPx and its cofactor Se and three compounds with a biological antioxidant role, retinol, total carotene, and α-tocopherol. In addition, inflammatory and nutritional markers were determined. The results of patients with Crohn's disease were compared with those of colopathic subjects. These later were considered as a control group since they have only functional colonic disorders with a normal coloscopy.

2. MATERIALS AND METHODS

2.1. Subjects

Group 1. 15 patients with active Crohn's disease (CD) (19–63 years), without intestinal resection or nutritional supplement. CD diagnosis included clinical, radiographic, en-

Therapeutic Uses of Trace Elements, edited by Nève et al.
Plenum Press, New York, 1996

doscopic and histological criteria. Disease activity was assessed according to the Crohn's disease activity index (CDAI) (3). All patients had received a treatment with mesalazine and/or corticoïds.

Group 2. 15 control subjects (21–63 years), with chronic idiopathic constipation but without any other digestive disease. The coloscopy and small bowel transit of these subjects were normal. Exclusion criteria for the two groups were parasitosis and all forms of colitis.

2.2. Methods

Albumin, transthyretin, retinol binding protein (RBP), C reactive protein (CRP), α1-glycoprotein acid were determined in serum by nephelometric method (BN 100 Behring, France). Normal range: (35–50) g/l for albumin, (150–400) mg/l for transthyretin, (30–60) mg/l for RBP, (<6) mg/l for CRP, (0.55–1.40) g/l for α1-glycoprotein acid. Cholesterol (CT) and Triglycerides (TG) were assayed in serum by multiparameter equipment using Boehringer reagents (Hitachi 717, Boehringer Mannheim, France). Normal range: (4.4–6.0) mmol/l for CT, (0.50–1.70) mmol/l for TG. Selenium was determined in serum by electrothermic atomic absorption with Zeeman background correction (Varian Spectra AA30, France). Normal range: (0.79–1.39) μmol/l. GSHPx and Cu-Zn-SOD activities were assayed in erythrocytes respectively with Ransel and Ransod reagents (Randox Laboratories, France) (4, 5). Normal range: (14–46) U/g Hb for GSHPx, (0.34–1.22) U/mg Hb for Cu-Zn-SOD. Retinol and α-tocopherol were analysed in serum by high performance liquid chromatography (Waters Company, France) (6). Normal range: (1.7–2.7) μmol/l for retinol, (0.54–1.14) for retinol/RBP, (21–31) μmol/l for α-tocopherol, (4.30–5.75) for α-tocopherol/(CT+TG). Total carotene were measured in serum, after an hexanic extraction step, using a spectrophotometer at 450 nm (Jobin Yvon, France) (7). Normal range: (1.5–3.7) μmol/l and (0.30–0.48) for total carotene/(CT+TG).

2.3. Statistical Analysis

Results were expressed as median and 5–95 percentiles and compared using the Mann-Whitney test. Probabilities <0.05 were considered significant. Correlation coefficients (r) between parameters and between CDAI and each parameter were calculated by regression analysis.

3. RESULTS

For all parameters, the results for the control group were in the normal range, except for three colopathic subjects. The results for CD group were significantly different from those of control group for both inflammatory proteins: α1-glycoprotein acid [1.51; 0.78–2.75 g/l vs 0.58; 0.49–0.90 g/l] p<0.001, CRP [23.2; 2.8–128 mg/l vs <2.6 mg/l (detection limit)] p<0.001. There was no significant difference between CD patients and control group for albumin [39.5; 27.6–51.5 g/l vs 41.6; 36.9–47.4 g/l], transthyretin [235; 88–396 mg/l vs 288; 202–375 mg/l] and RBP [33; 13.5–71 mg/l vs 39; 21–65 mg/l].

Results of antioxidant factors are shown in Table 1. No significant difference was noted for retinol, retinol/RBP, α-tocopherol, α-tocopherol/(CT+TG) and Cu-Zn-SOD activity, between both groups. Concerning the results of Se, E-GSHPx, total carotene and total

Table 1. Comparison of antioxidant factors in Crohn's disease patients and
control subjects (n = 15)

Parameters	CD patients	Control subjects
Se (μmol/l)	0.86 [0.40-1.23] **	0.99 [0.81-1.28]
E-GSHPx (U/g Hb)	20 [12.5-54.1] *	27.5 [20.4-43.4]
Cu-Zn-SOD (U/mg Hb)	1.02 [0.59-1.31]	0.95 [0.68-1.36]
Retinol (μmol/l)	1.40 [0.63-2.56]	1.75 [0.94-2.76]
Retinol/RBP	0.84 [0.60-1.09]	0.92 [0.76-1.13]
α-tocopherol (μmol/l)	24 [16.9-42.8]	26 [15.1-35.1]
α-tocopherol (CT+TG)	3.96 [1.87-6.32]	3.93 [2.87-6.94]
Total carotene (μmol/l)	1.30 [0.56-2.96] ***	2.79 [1.47-3.91]
Total carotene (CT+TG)	0.13 [0.05-0.25] ***	0.26 [0.14-0.38]

Values are expressed as median and 5-95 percentiles, *p < 0.05; **p < 0.01; ***p < 0.001.

carotene/(CT+TG), they were statistically lower in the CD patients than in the control subjects. A highly significant correlation between E-GSHPx activity and serum Se (r=0.69, p<0.01) was found in the CD patients, but not in the control group. Yet no correlation was found between CDAI score and any biological parameters.

4. DISCUSSION

All CD patients had a CDAI higher than 150 (except one) but lower than 450. This shows an active but not a very severe state of the disease. Consequently their nutritional protein status was not modified by the severity of the disease. In the same way, as far as retinol, retinol/RBP, α-tocopherol and α-tocopherol/(CT+TG) values are concerned, our results in the CD group are in the same range as those of the control group. These data are in agreement with Imes et al (8) and Fernandez et al (9). They suggest that our CD patients had an adequate intake and probably no risk of retinol and/or α-tocopherol deficiency. Other authors, Ostrowski et al (10) and Main et al (11) showed that low values for serum retinol were only found in patients with very active disease.

The median values of total carotene and total carotene/(CT+TG) are lower than half of those of the controls. Imes et al (8) also reported low serum carotene values but only in 25 % of the CD patients they studied. No correlation was found by this author and by ourselves between serum total carotene and CDAI. Low serum carotene levels in CD patients could be related to complex events which influence their absorption and transport. Imes (8) showed that low carotene values were associated with steatorrhea for some CD patients; however these low values were not associated with lowering of retinol and α-tocopherol which are also fat-soluble. Another hypothesis for low carotene levels could be their transformation to metabolites during the defence against reactive oxygen species produced by inflammatory processes.

Cu-Zn-SOD activity is not impaired in CD patients, unless an overproduction of free radicals in particular superoxide anion, is involved in the tissue injury and dysfunction associated with Crohn's disease (1). The normal Cu-Zn-SOD activity might be related to the treatment of our patients with mesalazine: this compound is a potent antioxidant and has free-radical-scavenger properties. Consequently it might be implicated as a possible mechanism of defence against free radicals, thus sparing enzymatic defence. Our results showed that both serum selenium and E-GSHPx values are lower in the CD patients than in the control group. In fact, the E-GSHPx activity could be related to the low serum selenium level observed in the group of CD patients. There is a highly significant correlation between these two parameters in the CD group, whereas no correlation was found in the control group as Rannem (2) observed. For Van Dael and Deelstra (13) the correlation between these two parameters exists only below a certain selenium threshold value while above this level it disappears and E-GSHPx activity becomes independent of selenium status as in our control group. Low selenium levels were described in Crohn's disease by Arnaud et al (14), but Rannem (2) observed a significant reduction of selenium only in patients with severe Crohn's disease (large resection of the small bowel). In our patients, a possible explanation for lower selenium rates could be related to increased requirements. Moreover, the large amounts of reactive oxygen species and enhanced levels of lipoperoxides produced in inflammatory bowel disease (1) could modify the antioxidant enzyme defence.

5. CONCLUSION

In Crohn's disease three parameters, ie total carotene, Se and E-GSHPx could be considered as predictive markers of a predeficiency state, whereas denutrition is not shown by any clinical or biological signs. Their decrease may suggest a reduced antioxidant defence against free radicals in Crohn's disease. It would be interesting, to study all antioxidant parameters in a larger group of Crohn's disease patients, to confirm this hypothesis.

6. REFERENCES

1. M.B. Grisham, *Lancet.* **344**, 859–861 (1994).
2. T. Rannem, K. Ladefoged et al., *Am J. Clin. Nutr.* **56**, 933–937 (1992).
3. W.R. Best, J.M. Becktel et al., *Gastroenterology.* **70**, 439–444 (1976).
4. D.E. Paglia, W.N. Valentine, *J. Clin. Lab. Med.* **70**, 158–159 (1967).
5. J. M. McCord, I. Fridovich, *J. Biol. Chem.* **244**, 6049–6055 (1969).
6. G. Catignani, J. Bieri, *Clin. Chem.* **29**, 708–712 (1983).
7. D. Cornwell, F. Kruger, H.B. Robinson, *J. Lipid. Res.* **3**, 65–70 (1962).
8. S. Imes, B.Pinchbeck et al., *Digestion.* **37**, 166–170 (1987).
9. F. Fernandez-Banares, A. Abad-Lacruz et al., *Am J. Gastroenterol.* **84**, 744–748 (1989).
10. J. Ostrowski, P. Janik et al., *Br. J. Cancer.* **55**, 203–205 (1987).
11. A. Main, P.R. Mills et al., *Gut.* **24**, 1169–1175 (1983).
12. J.W. Erdman, T.L. Bierer, E.T. Gugger, *Ann. N.Y. Acad. Sci.* **691**, 76–84 (1993).
13. P. Van Dael, H. Deelstra, *Int. J. Vitam. Nutr. Res.* **63**, 312–316 (1993).
14. J. Arnaud, P. Chappuis et al., *Ann. Biol. Clin.* **51**, 589–604 (1993).

25

TYPE II DIABETES AND CHROMIUM

Richard A. Anderson, Noella A. Bryden, and Marilyn M. Polansky

Nutrient Requirements and Functions Laboratory
Beltsville Human Nutrition Research Center
U.S. Department of Agriculture
ARS, Beltsville, Maryland 20705–2350

1. INTRODUCTION

Chromium was shown to be an essential trace element by Mertz and Schwarz (1) who observed that rats fed a Torula yeast-based diet displayed impaired glucose tolerance. This intolerance was reversed by an insulin potentiating factor that was present in Brewer's yeast, meat and other foods. This component was later shown to be a chromium complex (2). The essentiality of chromium in humans was demonstrated by Jeejeebhoy et al. (3) who reported that a 30-year old female on total parenteral nutrition developed peripheral neuropathy, unexpected weight loss and glucose intolerance that were refractory to the addition of 45 units of insulin per day. However, these symptoms could be alleviated by the inclusion of 250 µg daily of chromium as chromium chloride in the total parenteral nutrition solutions. Following chromium supplementation, no exogenous insulin was required and diabetic-like symptoms were reversed. This work has subsequently been confirmed in several laboratories (see review, 4).

Chromium is involved in carbohydrate and lipid metabolism primarily via its role in the potentiation of insulin action. With only nanogram quantities of chromium, optimal insulin effects can be demonstrated in insulin dependent tissues; in the presence of insufficient levels of dietary chromium, much higher levels of insulin are required.

Increased dietary intake of Cr leads to improvements in glucose intolerance, circulating insulin, insulin receptor number, lipid levels, lean body mass:% fat ratio and in extreme cases nerve and brain disorders (4).

Conversion of Cr to a biologically active form (form that potentiates insulin bioactivity *in vitro*) (5,6) is essential for the utilization of Cr. Tuman et al. (7) reported that biologically active Cr preparations isolated from Brewer's yeast and synthetic biologically active Cr complexes but not simple inorganic Cr compounds, i.e., Cr chloride, lowered plasma glucose and triglycerides of genetic diabetic mice. Unlike control animals, diabetic mice appear to lose the ability to convert Cr to a useable form. Humans, with marginally impaired glucose tolerance, respond with improved glucose metabolism following supplementation with 200 µg of Cr as inorganic chromium chloride (4,8), but diabetics often do not respond unless higher

Therapeutic Uses of Trace Elements, edited by Nève et al.
Plenum Press, New York, 1996

amounts of chromium are given (9). Diabetics appear to convert chromium to a useable form less efficiently since absorption of chromium by diabetics is not impaired but actually higher than that of control or glucose intolerant subjects (10).

A biologically active form of Cr comprised of Cr, nicotinic acid and the amino acids, glycine, cysteine and glutamic acid was isolated from Brewer's yeast (11). Synthetic Cr compounds comprised of these components also possess *in vitro* insulin potentiating activity (11). Naturally occurring organic Cr complexes are postulated to be absorbed better than most inorganic Cr complexes (5). Preformed complexes may not be critical since supplementation of nicotinic acid and Cr to human subjects led to improved glucose tolerance while either component alone was ineffective (12).

Chromium nicotinate (U.S. Patent number 5194615) is also a common human Cr supplement. It does not form stable complexes and with time precipitates from solution. We also studied the incorporation of Cr chloride, Cr acetate, Cr potassium sulfate (ALUM), Cr trihistidine, Cr triglycine, Cr trinicotinic acid (nicotinate) and Cr dinicotinic acid dihistidine (NA-HIS) into rat and pig tissues. Synthesis and sources of the Cr complexes used and methods of Cr analyses have been described (13).

2. CHROMIUM INCORPORATION INTO RAT TISSUES

The incorporation of different forms of Cr into the kidney of weanling rats fed a control low Cr diet and supplemented with various forms of Cr at 5000 ng of Cr per g of diet for three weeks is shown in Figure 1. Raising animals in stainless steel cages with stainless food cups and drinking tubes did not lead to increased kidney Cr compared to animals raised in plastic cages with no direct contact with metal (stainless steel is roughly 18% Cr). Cr chlo-

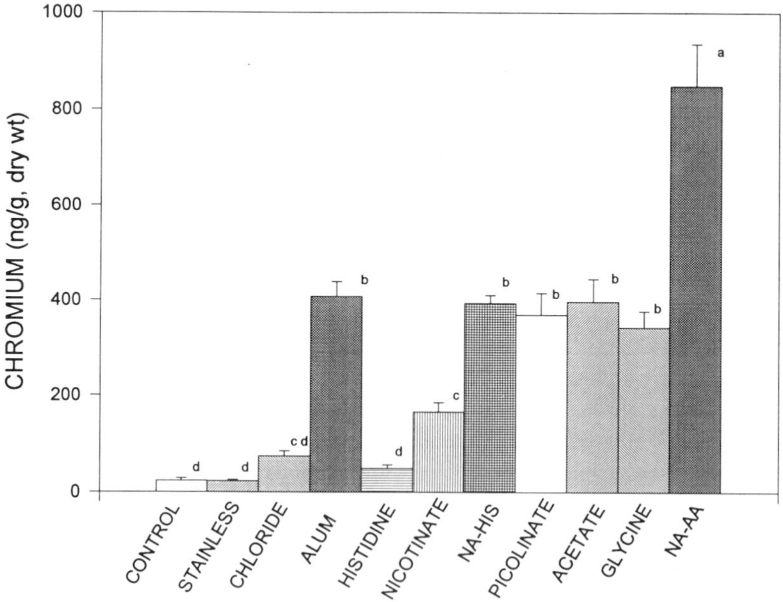

Figure 1. Kidney Cr concentrations of rats consuming a control low Cr diet supplemented with indicated forms of Cr at 5000 ng of Cr per g of diet. Values are mean ± SEM for five animals. Values with different superscripts are significantly different by Duncan's multiple range test at p<0.05. Source: Anderson et al. (13).

ride and Cr histidine did not alter tissue Cr concentrations (Fig. 1). Kidney Cr concentrations for Cr potassium sulfate, Cr nicotinic acid histidine, Cr picolinate, Cr acetate and Cr glycine were similar and all significantly greater than those for Cr chloride, Cr histidine and Cr nicotinate. The complex with the greatest incorporation into the kidney was the Cr nicotinic acid-glycine, cysteine, glutamic acid complex (CrNA-AA) with kidney Cr values significantly greater than those for any other forms of Cr tested. This complex was tested since it appears to have a composition similar to that observed for fractions isolated from yeast (11). The patterns of incorporation of Cr into the liver, spleen, heart, lungs and muscle were similar to that of the kidney for all the Cr complexes but Cr values were significantly lower than those for the kidney (13).

3. CHROMIUM NICOTINIC ACID AMINO ACID COMPLEX AND TYPE II DIABETES

Since the Cr NA-AA appeared to be the most available for the incorporation into tissues, we conducted a study involving 237 type II diabetics supplemented with 0, 200, 400 or 600 µg of Cr as CrNA-AA for six months. Blood glucose, insulin, lipids, hemoglobin A_{1C} and related variables were measured at 0, 3 and 6 months. Subjects continued to take their normal medications and were asked not to change their eating and exercise routines. There were definite improvements in glucose, insulin and lipid variables throughout the study in all groups but Cr effects could not be differentiated from placebo effects. The reasons for the lack of a Cr response are unclear.

4. CHROMIUM INCORPORATION INTO PIG TISSUES

Since the Cr NA-AA complex did not appear to improve the diabetic symptoms of people with type II diabetes, we checked the tissue incorporation of this and other Cr complexes by swine. Pigs were chosen since they are a suitable model for human Cr nutrition studies (14). Two hundred ng of Cr per g of diet of the forms listed (Figure 2) were fed to pigs for 77 days (15). The kidney Cr concentrations of the animals receiving Cr NA-AA were nearly identical to those of the control and Cr chloride animals. There were significantly elevated kidney Cr values for animals supplemented with Cr acetate and kidney Cr of animals receiving picolinate and nicotinate tended to be elevated. This experiment appeared to substantiate the human study in which the Cr NA-AA complex did not show positive effects on people with type II diabetes.

5. FOUR FORMS OF CHROMIUM AND TYPE II DIABETES

We conducted a follow-up human study involving subjects with advanced type II diabetes to determine the form of Cr utilized the most efficiently by people with diabetes. Fifty-seven subjects with overt type II diabetes mellitus were randomized into four groups and supplemented with 200 µg of Cr per day for 3 months in a six month double-blind placebo controlled crossover study. Forms of Cr tested were Cr chloride, Cr nicotinate, Cr picolinate and CrNA-AA. Urinary Cr losses, which are a relative measure of Cr absorption, were 0.41 ± 0.06 µg/d during the placebo period and were not significantly higher during Cr chloride, Cr nicotinate and CrNA-AA supplementation (16). Urinary losses for the diabetics receiving

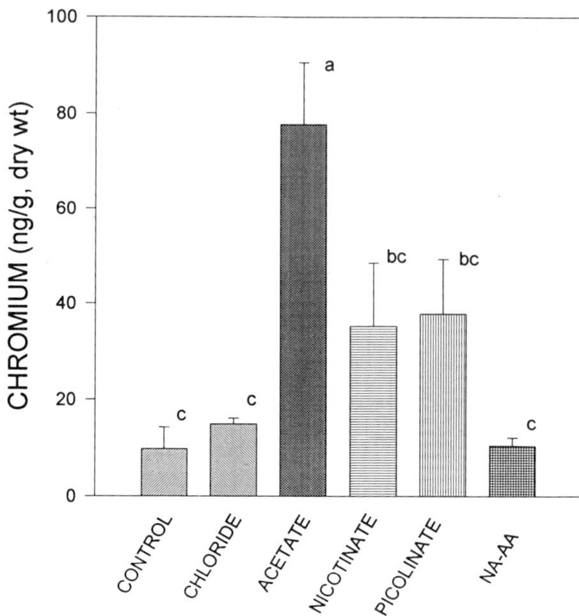

Figure 2. Kidney Cr concentrations of pigs consuming a corn-soybean meal diet supplemented with the indicated forms of Cr at 200 ng of Cr per g of diet. Values are mean ± SEM for four animals. Values with different superscripts are significantly different at p<0.05.

Cr picolinate were 2.65 ± 0.27 µg/d. Urinary Cr losses for control subjects are 0.22 ± 0.02 and increase to 0.99 ± 0.08 µg/d following three months of supplementation with 200 µg of Cr as Cr chloride (8) and larger increases in Cr losses are observed following daily supplementation with 200 µg of Cr as Cr picolinate for control subjects. There were trends in improvements in the glucose and lipid parameters during the Cr supplementation periods but there were no consistent improvements due to any of the supplemental forms of Cr tested. These data confirm that basal urinary losses of people with diabetes mellitus are greater than those of controls and demonstrate that the absorption of Cr chloride, and presumably Cr nicotinate and CrNA-AA, but not Cr picolinate is significantly different for controls than for people with diabetes. Two hundred µg of supplemental Cr does not appear to be adequate to elicit consistent improvements in glucose, insulin and lipid variables of people with diabetes mellitus for the forms of Cr tested.

6. CHROMIUM PICOLINIC ACID AND DIABETES - AMOUNT OF SUPPLEMENTAL CHROMIUM

To determine if people with type II diabetes would respond to higher levels of Cr in the form of Cr picolinate, we conducted a study involving 180 type II diabetics. Adult type II diabetic subjects were divided randomly into three groups and supplemented with: 1) placebo, 2) 100 µg of Cr as Cr picolinate two times per day or 3) 500 µg of Cr as picolinate two times per day. Subjects continued to take their normal medications and were instructed not to change their eating and living habits. Glycated hemoglobin improved significantly after two months in the group receiving 1000 µg of chromium per day and was lower in both Cr groups after four months (17). Two hour meal tolerance glucose values were also significantly lower for the subjects consuming 1000 µg of supplemental Cr after both two and four months. Fasting and 2 hr meal tolerance insulin values decreased significantly in both groups receiving supplemental Cr after 2 and 4 mo. Plasma total cholesterol also decreased after 4 mo in the group receiving 1000 µg of Cr (17). These data demonstrate that supplemental Cr had significant beneficial effects on cholesterol, glycated hemoglobin, glucose and insulin variables of subjects with type II diabetes. The beneficial effects were greater in the group receiving 500 µg two times per day than the group receiving 100 µg two times per day.

While the improvements in the group receiving 100 µg two times per day were significant, there were larger improvements in the group receiving 500 µg two times per day (17). Two hundred µg of supplemental Cr appears to be borderline in observing significant effects in people with diabetes.

7. SUMMARY

This series of studies demonstrate that form of Cr has significant effects on Cr utilization. Chromium chloride, the form used in most animal and human studies until recently, is poorly utilized but appears to be utilized better by people than by rats. Two hundred µg of supplemental Cr per day for type II diabetics appears to be borderline and much larger effects are observed at higher levels. The form of Cr tested that was utilized the best by people with type II diabetes was Cr picolinate.

8. REFERENCES

1. W. Mertz and K. Schwarz, *Arch. Biochem. Biophys.* **58**, 504–508 (1955).
2. K. Schwarz and W. Mertz. *Arch. Biochem. Biophys.* **85**, 292–295 (1959).
3. K.N. Jeejeebhoy, R.C. Chu, E.B. Marliss, G.R. Greenberg and A. Bruce-Robertson, *Am. J. Clin. Nutr.* **30**, 531–538 (1977).
4. R.A. Anderson, in *Essential and Toxic Trace Elements in Human Health and Disease: an Update*, Wiley Liss, Inc., pp. 221–234 (1993).
5. W. Mertz and E.E. Roginski, in *Newer Trace Elements in Nutrition*, Marcel Dekker, N.Y., pp. 123–153 (1971).
6. R.A. Anderson, J.H. Brantner and M.M. Polansky, *J. Agric. Food Chem.* **26**, 1219–1221 (1978).
7. R.W. Tuman, J.T. Bilbo and R.J. Doisy, *Diabetes* **27**, 49–56 (1978).
8. R.A. Anderson, M.M. Polansky, N.A. Bryden, K.Y. Patterson, C. Veillon and W.H. Glinsmann, *J. Nutr.* **113**, 276–281 (1983).
9. R.T. Mossop, *Cent. Afr. J. Med.* **29**, 80–82, (1983).
10. R.J. Doisy, D.H.P. Streeten, J.M. Freiberg and A.J. Schneider, A.J. in *Trace Elements in Human Health and Disease, Vol. II. Essential and Toxic Elements*, Academic Press, N.Y., pp. 79–104 (1976).
11. E.W. Toepfer, W. Mertz, M.M. Polansky, E.E. Roginski and W.R. Wolf, *J. Agric. Food Chem.* **25**, 162–166 (1977).
12. M. Urberg and M.B. Zemmel, *Metabolism* **36**, 896–899 (1987).
13. R.A. Anderson, N.A. Bryden, M.M. Polansky and K. Gautschi, *J. Trace Elem. Exptl. Med.* **9**, (1996) (in press).
14. C.M. Evock-Clover, M.M. Polansky, R.A. Anderson and N.C. Steele, *J. Nutr.* **123**, 1504–1512 (1993).
15. T.L. Ward, L.L. Southern and R.A. Anderson, *J. Anim. Sci.* **(Suppl. 1)**, 189, (1995).
16. R.A. Anderson, M.M. Polansky, N.A. Bryden and J.K. Zawadzki, *FASEB J.* **9**, (1996) (in press).
17. R.A. Anderson, N. Cheng, N. Bryden, M. Polansky, N. Cheng and J. Chi, *Am. Diabetes Assoc.,* San Francisco, CA (Abstract submitted).

26

SERUM SELENIUM, MICRO AND MACROVASCULAR COMPLICATIONS IN DIABETIC PATIENTS

C. da Silva,[1] J. Poupon,[1] P. J. Guillausseau,[2] and P. Chappuis[1]

[1] Laboratoire Central de Biochimie
[2] Service de Médecine B
Hôpital Lariboisière
Paris, France

1. INTRODUCTION

The following study has been undertaken to establish if antiradicalar agents (1) and serum selenium are linked to hypertension and micro and macrovascular complications in diabetic patients. We also addressed the question of any nutritional role of selenium in diabetes, by searching for any relation between selenium and BMI (Body Mass Index) in patients with type 1 insulin dependent (IDDM) and type 2 non insulin dependent (NIDDM).

2. MATERIAL AND METHODS

In accordance to treatment (insulin, or insulin instaured after an oral treatment), age of onset and presence of ketoacidosis, diabetic patients (n=166) were distributed as IDDM patients (n=37; mean age: 38.6 y), NIDDM (n=95; mean age: 53.7 y) and IRDM patients (insulin requiring n=34; mean age: 57.4 y). All of them had given their informed consent before participation in the study. Micro and macrovascular complications were classified using additive scores as follows: no coronary heart disease, absence of carotid or lower limbs artery complications, albumin urinary excretion < 15 mg/24 h, no renal insufficiency (serum creatinine < 150 µmol/l): **score 0**; presence of one or more of these indices: **score 1 to 4**. Subjects presenting a urinary albumin excretion in excess of 300 mg/24 h or serum creatinine above 150 µmol/l were discarded in order to avoid any interference with urinary selenium excretion. Diastolic, systolic blood pressure and BMI were also evaluated.

Serum selenium was determined in fasting subjects by electrothermal Atomic Absorption Spectrometry using a Perkin-Elmer 5100 PC (reproducibility of the method: 2.1 and 3.3% at 1.28 and 0.66 µmol/l respectively). For serum selenium comparisons, control subjects, (n=36; m=1.02 ± 0.15 µmol/l; mean age: 49.4 years old) were included in this study.

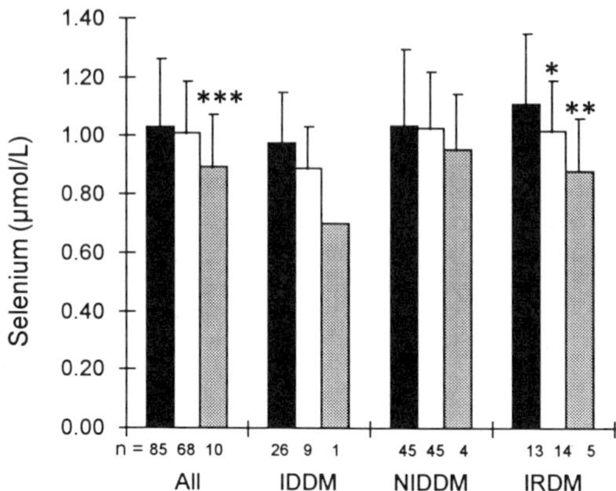

Figure 1. Serum selenium according to the vascular injuries score in the three goups and the whole cohort of diabetics score=0, score=1 or 2, score=3 or 4. n: number of subjects; * p < 0.05, ** p < 0.03, *** p < 0.02 compared to score=0.

Sex was not taken into account as no difference has been reported in mean serum selenium between men and women (2).

3. RESULTS

Mean serum selenium (SD) expressed in μmol/l were 1.02 (0.15) in the control group, 0.95 (0.22) for IDDM patients 1.04 (0.20) for NIDDM patients and 1.02 (0.23) for IRDM patients. Selenium values were lower in IDDM compared to NIDDM patients (p 0.03). This comparison excepted, no difference could be found between each group of diabetic patients, nor between patients and control group.

Distribution of serum selenium (Figure 1) according to the patient's scores indicated that in each diabetic subgroup (IDDM, NIDDM or IRDM), mean selenium decreased according to high scores. This decrease was significant for the whole population and for IRDM patients. In IRDM patients presenting high BMI (Figure 2), low scores were very often associated with high serum selenium values. In patients presenting no diabetes complications, and as evidenced by their slope, regression curves between serum selenium and BMI values showed a positive relation for IRDM and IDDM patients (slope: 0.043 and 0.031, respectively). These relations could not be found in NIDDM patients nor in patients with complications. In uncomplicated NIDDM patients we observed a trend towards an inverse relationship between serum selenium concentration and diastolic blood pressure (p=0.11).

In each diabetes group, the population was divided into 2 subgroups according to serum selenium (hypo to normal and elevated selenium values). The selenium value selected as cut-off was 1.17 μmol/l (mean + 1 SD of controls). In IRDM patients, high selenium subgroup had lower scores (mean=0.3 and median=0) and higher BMI (mean=28.1; SD=3.2) than low selenium subgroup (score: mean=1.3 and median=1; p < 0.01; BMI: mean=25.3 and SD=3.7; p < 0.05). Conversely, no such observations could be evidenced for the 2 other groups of diabetic patients.

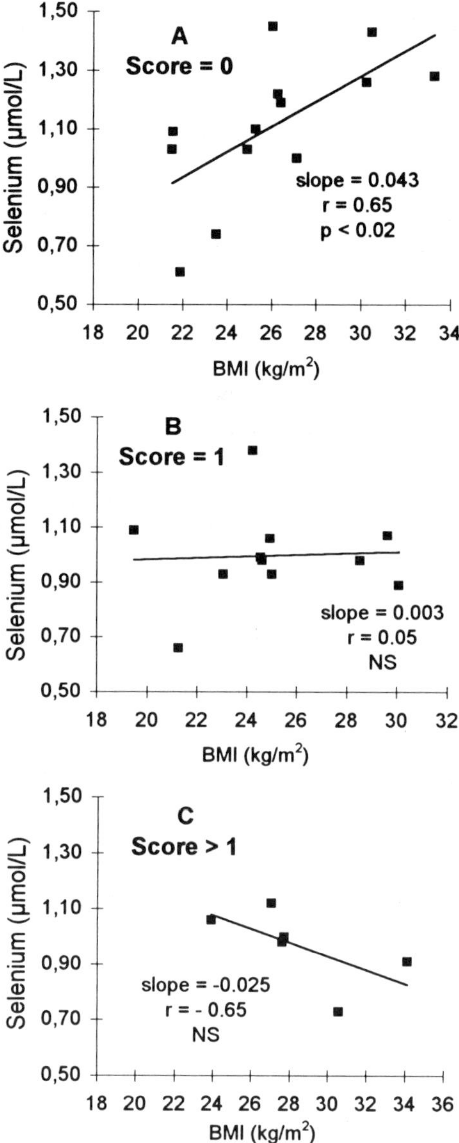

Figure 2. Relation between selenium and BMI in the IRDM classified following the complications score: A: score=0, B: score=1 or 2, C: score=3 or 4.

4. DISCUSSION

Perturbations of mineral status are often encountered in diabetic patients (hypomagnesemia, hyperzinciuria, hypermagnesiuria), as a result of their poor metabolic control. Our study, which was performed in ambulatory patients (with absence of acute illness nor any renal failure) receiving their conventional therapy confirms previous reports mentioning that serum selenium does not vary according to the type of diabetes. This confirms the results obtained by other groups who did not evidence any significant rise in serum glutathione peroxidase activity in diabetic patients of any type (3).

Concerning the influence of complications, the lack of any significant relation with selenium in IDDM and NIDDM patients may be accounted by the limited number of high-scored for IDDM patients. For high-scored subjects, the decrease in serum selenium values observed is to be compared with other mineral modifications such as low serum zinc and high serum copper values already reported. IL1, a cytokine synthesized by macrophages in response to vascular damage, may not only induce liver metallothionein which leads to a zinc accumulation in the liver (decrease in plasma zinc) and to a ceruloplasmin synthesis (increase in plasma copper). IL1 is also a potent immunostimulant and a proinflammatory agent that activates leukocytes. In order to face the free radicals production by leucocytes, serum selenium could participate to cell glutathione peroxidase activity at the expense of selenoprotein P which is known to be the major transporter of plasma selenium. This would explain why low serum selenium is encountered in diabetic patients with complications.

Nutritional status, and especially mineral status is very often impaired in diabetic patients. Serum selenium of subjects presenting no vascular injuries (IRDM and, to a lesser degree, IDDM patients) appears to reflect BMI values, which is related to the patient's nutritional status. Why this happens only for insulinodependent (IDDM and IRDM) subjects remains to be determined. We did not find such relations between selenium and BMI in diabetic patients presenting these vascular complications, perhaps because selenium may then be more involved in the reduction of cell oxidative damages. We have observed a trend towards an inverse correlation between selenium serum levels and blood pressure in uncomplicated NIDDM.

To our knowledge, no relation between serum selenium concentration and blood pressure has been reported so far in diabetic patients. In the general population, Salonen et al. observed an inverse relationship (4) and claimed that selenium could exert its effects on arterial blood pressure through an action against inhibitory effect of prostacyclin synthase (prostacyclin is a highly potent vasodilatating agent). These data should be confirmed by further studies.

5. CONCLUSION

That oxidative stress plays a role in the development of micro and macrocomplications in diabetes mellitus is now recognized. The exact role of selenium in this field remains to be determined, partly because markers of selenium status are still not precise and partly because it may only act when redox insult in diabetes reaches a critical point reflected in this paper by positive scores.

6. REFERENCES

1. D.V. Godin, S.A. Wohaieb, M.E. Garnett and A.D. Goumeniouk, *Mol. Cell. Biol.* **84**, 223–231 (1988).
2. J. Arnaud, P. Chappuis, M.C. Jaudon, J. Bellanger and A. Favier, *Ann. Biol. Clin.* **51**, 589–604 (1993).
3. R.M. Walter, J.Y. Uriu-Hare, K.L. Olin, M.H. Oster, M.S. Bradley, D. Anawalt, J.W. Critchfield and C.L. Keen, *Diabetes Care* **14**, 1050–1056 (1991).
4. J.T. Salonen, R. Salonen and M. Ihanainen, *Am. J. Clin. Nutr.* **48**, 1226–1232 (1988).

27

ZINC AND GROWTH FACTORS

D. Bouglé[1] and F. Bureau[2]

[1] Laboratoire de Physiologie Digestive et Nutritionnelle
Service de Pédiatrie A
[2] Laboratoire de Biochimie A
Centre Hospitalo Universitaire de Caen, France

1. INTRODUCTION

Zinc is essential for growth in humans and animals. The features of Zn deficiency described by Prasad (1) include retarded growth, delayed sexual maturation, skin abnormalities and bone alterations such as open epiphyses. Delays in linear and bone growth and in bone maturation are common and early findings in various conditions associated with Zn deficiency. Zn deprivation causes in a few days decreases in taste acuity and appetite that leads to lower food intake. This anorexia has been related to an inability of the organism to metabolize nitrogen (2, 3). In previously deficient Zn supplemented subjects the increase in growth rate is associated with an increase in food intake and efficiency. Other non specific factors involved in the relationship between Zn and growth include increased activities of enzymes involved in nucleic acid metabolism, protein synthesis and cell division. Zn deficiency causes decrease in protein synthesis within a few days (4, 5); protein utilization is impaired in Zn deficiency (6). Food efficiency is decreased in Zn deficiency (8). Short-life protein levels are increased by Zn supplementation (7) which also leads to an enhancement of immunity and protection against infections. Besides these general effects of Zn on anabolism, protein synthesis, and therefore on growth, it has been observed that Zn deficiency impairs growth even in pair-fed animals on a control diet (4, 8, 9), implying that Zn has specific effects on growth. These effects are mediated by growth hormone and related factors.

2. ZINC AND GROWTH HORMONE

Growth hormone (GH) is a peptide synthetized and secreted by the pituitary. GH-RH (Growth Hormone Releasing Hormone) stimulates the synthesis and secretion of GH, while the inhibitory regulation is mediated by somatostatin (10). Zn seems to be directly involved in the synthesis and release of GH: Zn deprivation usually leads to reduced concentrations of GH, anorexia and a reduction in energy and protein intakes that slow down the weight gains. Malnutrition alone decreases the levels of GH to the same extent as Zn deficiency (11, 12),

Therapeutic Uses of Trace Elements, edited by Nève et al.
Plenum Press, New York, 1996

but the latter impairs growth to a greater extent (9). It has been shown to reduce GH levels even if food intake is normal or slightly impaired (13). Therefore Zn which occurs in high concentrations in the GH secretory granules (14), seems to be directly involved in the regulation of GH synthesis and release. In vitro studies have yield conflicting (either stimulating or inhibitory) results (15- 17). They could explain the conflicting reports on GH levels during Zn deficiency. A short-term deficiency could increase GH levels by inducing the hormone release (18). A prolonged Zn depletion could reduce the storage capacity for GH, leading to unchanged or decreased serum levels (5, 13, 19). In human infants, Zn deficiency has been shown to induce partial GH deficiency, low insulin-like growth factor 1 (IGF-1) levels and growth decline. GH response, IGF-1 values and growth improved after Zn supplementation (11, 12, 20, 21). In some GH-deficient children Zn supplementation improved growth rates and the GH response to stimuli (21–23). It is assumed that the catch-up growth induced by GH therapy increased Zn requirements (24). Differences in Zn status could explain the opposite results of another report which did not show alterations of Zn levels during GH therapy of GH deficient children, nor changes in growth rate during Zn supplementation (25), and increases of basal or stimulated GH levels in apparently normal infants (22).

Zn and GH have independent effects on growth (26): Zn could mediate the binding affinity of GH for its receptor. GH improves bone mineralization and volume and phosphatase activity in rats together with IGF-1 levels (27) but is unable to improve growth and IGF-1 levels of Zn deficient rats (9, 28). Growth effects of GH depend on several associated factors, as clearly shown in the Laron-type dwarfism where growth failure and low levels of IGF-1 occur despite normal GH levels, and by the different evolution of GH and IGF-1 levels during catch-up growth of short children following Zn supplementation (29). On the other hand IGF-1 mediates the growth-promoting effects of GH and its levels are closely related to variations in Zn intakes and status, and to variations of growth related to Zn status.

3. ZINC AND INSULIN-LIKE GROWTH FACTOR 1 (IGF-1)

IGF-1 is a single-chain polypeptide which occurs in plasma at concentrations of 20–80 nM and at lower concentrations in most if not all tissues. The major producers and target organs of IGF-1 are liver and bone osteoblasts. IGF-1 circulating in blood is probably produced by the liver, whereas tissue IGF is produced to a greater part locally. IGF-1 has both endocrine and paracrine functions (30). Several peptide domains of IGF are structural analogs of the insulin chains. They both elicit two types of biological responses, i.e. long-term effects on cell proliferation and short-term metabolic effects such as glucose transport. However they have different potencies, due to the presence of separate insulin and IGF-1 receptors, each of which can crossreact at lower affinity with the heterologous ligand. So far, the mitogenic effects seem to be mediated mainly via the IGF-1 receptor, and the different physiological effects of the two hormones are determined by the distribution of the two receptors on different cells (30). IGF-1 stimulates anabolic processes, protein synthesis, cell division, and thymidine kinase activity in cells. In rats, growth correlates with IGF-1 levels (8, 31), but the relative roles of GH and IGF-1 are still debated. The growth promoting activity of GH could be due on one hand, to direct effects enabling cells to respond to and to produce IGF-1 and, on the other hand, to indirect effects, mostly on the liver, by increasing serum levels of IGF-1 which mimic most but not all effects of GH (30). GH is an efficient inducer of IGF-1 synthesis and secretion, IGF-1 acts as a feedback inhibitor of GH release or GH gene transcription (30). Several nutritional factors such as energy and protein supply also influence IGF-1 levels in both animals (31, 32, 33) and humans (34). IGF-1 levels correlate with nitrogen balance (34). Zn deficiency induces a decrease in serum IGF-1 concentration in rats (5, 8, 9, 18, 28, 31, 35) and human infants (11, 12, 29). This decrease is reversed by Zn repletion (8, 11, 12, 18, 28, 31, 35) and not by GH alone (5, 26), suggesting that Zn is involved in IGF-I production. This effect is distinct from other nutritional factors. The fluctuations of plasma IGF-1 levels correlate with the changes of plasma Zn (32) and an adequate Zn supply is required to reverse the decrease in IGF-1 concentrations observed after energy depletion or feeding a low protein, low Zn diet (31, 32, 28). Only a few reports have given opposite results (9). In human infants serum Zn levels correlated with growth or IGF-I at several periods of follow up during the first year of life (36). In children two to five years old, both Zn and IGF-1 levels correlated with linear growth (37). Zn deficiency could therefore act on IGF-1 synthesis both through direct effects and through malnutrition.

4. ZINC AND BONE

Bone growth retardation is a common finding in Zn deficiency (26, 28) where bone maturation and mineralization are markedly depressed in humans and primates (1, 5, 38, 39). The mechanism of abnormal bone mineralization in the presence of Zn deficiency is unknown. Zn concentration is usually high in bone, suggesting that this element is essential for the calcified matrix. Bone Zn correlates with Zn dietary supply and at intakes of 15 and 45 mg Zn/kg diet it ranges between 51–132 and 185–215 mg/kg dry weight (Bouglé, unpublished data). In pair-fed animals Zn deficiency reduces both serum and bone Zn concentrations (8, 18, 31). Zn is found in proximity to calcification sites and in the layer of preossous tissue where calcification is imminent (40). Zn induces bone formation and anabolism by stimulating bone protein synthesis in vivo and in vitro (41). Zn could play a direct role in osteogenesis: Zn deficiency reduces the number of osteoblasts and chondrocytes of the epiphyseal cartilage in rats (42, 43), and the activity of these cells is enhanced by Zn repletion (43). Zn is a potent inhibitor of osteoclastic bone resorption (44). Zn deficiency reduces the sulfate uptake by glucosaminoglycans (8, 45), a required step of cartilage synthesis and bone growth (46, 47), and synthesis of collagen, which is the most important extracellular bone protein, depends on Zn intakes (45, 48, 49). Wether these effects of Zn on bone matrix are secondary to the general enhancement of protein synthesis or mediated through growth factors is unknown. Significant interactions between Zn and GH and tibial epiphyseal cartilage width and tibial Zn have been displayed (9). Zn stimulates IGF-I production by, and anabolic effects on osteoblasts (50, 51). Bone Zn concentrations correlate with the levels of plasma IGF-1 (8, 32). In short children Zn supplementation increased IGF-1 and osteocalcin levels, and growth (29). IGF-1 in turn has an anabolic effect on bone matrix, by increasing the DNA, protein and collagen content in osteoblastic cells and the number of cells (chondrocytes, fibroblasts, osteoblasts) (51, 52). IGF-1 increases the sulfation of glucosaminoglycans (53), the activities of enzymes involved in the synthesis of bone matrix (54), and favours the differenciation of osteoblasts. Therefore Zn induced IGF-1 mimicks the effects of GH on cartilage; it acts directly to stimulate bone matrix formation more than cell replication and linear skeletal growth (30, 46).

5. OTHER HORMONAL CONSEQUENCES OF ZINC
DEFICIENCIES

Testosterone increases weight, muscle mass, linear growth, and stimulates chondrocyte proliferation. The direct relationship between Zn and testosterone was shown by the delay in sexual development of Zn deficient dwarfs. This clinical picture was reversed by Zn treatment (12, 55, 56): testosterone concentrations were improved by Zn supplementation (11, 12). In experimental human Zn deficiency **thyroid hormones** and TSH decline significantly (57). Thyroid-deficient patients have low Zn absorption, serum and urine levels (58). Zn clearance is increased in **diabetic patients**, and this parameter is inversely correlated with growth rate (59). Zn is involved in the stabilization of insulin and its peripheral action (60). Insulin shares the same cell receptors with IGF-1. Zn deficiency is responsible for a decrease of insulin serum levels in rats (8,18), and sensitivity to insulin (61).

6. CONCLUSIONS

Zinc is involved in every step of growth. It modulates the synthesis, release and activities of most of the growth factors involved in protein synthesis, and more specifically in bone growth. It has also an indirect influence on these factors through nutritional changes induced by variations in food intake and specific enzyme activities.

7. REFERENCES

1. A.S. Prasad, A. Miale, Z. Farid, H.H. Sanstead, A.R. Schulert, W.J. Darby, *Arch. Int. Med.* **111**, 407–28 (1963).
2. B.E. Golden, M.H.N. Golden, *Am. J. Clin. Nutr.* **34**, 892–9 (1981)a.
3. M.H.N. Golden, B.E. Golden, *Am. J. Clin. Nutr.* **34**, 900–8 (1981) b.
4. I. Dorup, T. Clausen, *Br. J. Nutr.* **66**, 493–504 (1991)a.
5. I. Dorup, A. Flyvbjerg, M.E. Everts, T. Clausen, *Br. J. Nutr.* **66**, 505–21 (1991)b.
6. J.M. Hsu, W.L. Anthony, *J. Nutr.* **105**, 26–31 (1975).
7. J. Bates, C.J. McClain, *Am. J. Clin. Nutr.* **34**, 1655–60 (1981).
8. M.S. Bolze, R.D. Reeves, F.E. Lindbeck, M.J. Elders, *Am. J. Physiol.* **252**, E21–6 (1987).
9. D. Dicks, A. Rojhani, Z.T. Cossack, *Nutr. Res.* **13**, 701–13 (1993).
10. M. Berelowitz, M. Szabo, L.A. Frohman, S. Firestone, L. Chu, R.L. Hintz, *Science* **212**, 1279–81 (1981).
11. S.Z. Ghavami-Maibodi, P.J. Collipp, M. Castro-Magana, C. Stewart, S.Y. Chen, *Ann. Nutr. Metab.* 27, 214–9 (1983).
12. Y. Nishi, S. Hatano, K. Aihara, A. Fujie, M. Kihara, *J. Am. Coll. Nutr.* **8**, 93–7 (1989).
13. M. Kirchgeßner, H.P. Roth, *Biol. Tr. Elem. Res.* **7**, 263–8 (1985).
14. O. Thorlacius-Ussing, *Neuroendocrinol.* **45**, 233–42 (1987).
16. M.Y. Lorenson, D.L. Robson, L.S. Jacobs, *J. Biol. Chem.* **258**, 8618–22 (1983).
17. S. Focht, G. Fosmire, W.C. Hymer, *FASEB J.* **5**, A940(Abs) (1991).
18. H.P. Roth, M. Kirchgeßner, *Horm. Metab. Res.* **26**, 404–8 (1994).
19. A.W. Root, G. Duckett, M. Sweetland, E.O. Reiter, *J. Nutr.* **109**, 958–64 (1979).
20. Y.D. Coble Jr, C.W. Bardin, G.T. Ross, W.T. Darby, *J. Clin. Endocrinol. Metab.* 32, 361 (1976).
21. P.J. Collipp, M. Castro-Magana, M. Petrovic, J. Thomas, T. Cheruvansky, S.Y. Chen, H. Sussman, *Ann. Nutr. Metab.* **26**, 287–90 (1982).
22. J. Brandào-Neto, V. Stefan, B.B. Mendoça, W. Bloise, V.B. Castro, *Nutr. Res.* **15**, 335–58 (1995).
23. T. Cheruvanky, M. Castro-Magana, S.Y Chen, P.J. Collipp, S.Z. Ghavami-Maibodi, *Am. J. Clin. Nutr.* **35**, 668–70 (1982).
24. K. Aihara, Y. Nishi, S. Hatano, M. Kihara, M. Ohta, K. Sakoda, T. Uozumi, T. Usui, *J. Pediatr. Gastroenterol. Nutr.* **4**, 610–5 (1985).
25. G.E. Richards, N.M. Marshall, *J. Am. Coll. Nutr.* **2**, 133–40 (1983).
26. A.S. Prasad, D. Oberleas, P. Wolf, J.P. Horwitz, *J. Lab. Clin. Med.* **73**, 486–94 (1969).
27. N.M. Wright, J. Renault, B. Hollis, N.H. Bell, L.L. Key, *J. Bone Mineral Res.* **10**, 127–31 (1995).
28. G. Oner, B. Bhaumick, E.M. Bala, *Endocrinology* **114**, 1860–3 (1984).
29. T. Nakamura, S. Nishiyama, Y. Futagoishi-Suginohara, J. Matsuda, A. Higashi, *J. Pediatr.* **123**, 65–9 (1993).
30. R.E. Humbel, *Eur. J. Biochem.* **190**, 445–62 (1990).
31. Z.T. Cossack, *Br. J. Nutr.* **156**, 163–9 (1986).

32. Z.T. Cossack, *J. Pediatr. Gastroenterol. Nutr.* **7**, 441–5 (1988).

33. S. Takahashi, M. Kajikawa, T. Umezawa, S.I. Takahashi, H. Kato, Y. Miura, T.J. Nam, T. Noguchi, H. Naito, *Br. J. Nutr.* **63**, 521–34 (1990).

34. W.L. Isley, L.E. Underwood, D.R. Clemmons, *J. Clin. Invest.* **71**, 175–82 (1983).

35. E.A. Droke, J.W. Spears, J.D. Armstrong, E.B. Kegley, R.B. Simpson, *J. Nutr.* **123**, 13–9 (1993).

36. K.F. Michaelsen, G. Samuelson, T.W. Graham, B. Lönnerdal, *Acta Paediatr.* **83**, 1115–21 (1994).

37. R. Mokni, A. Chakar, F. Bleiberg-Daniel, J.L. Mahu, P.A. Walravens, P. Chappuis, J. Navarro, D. Lemonnier, *Acta Paediatr.* **82**, 539–43 (1993).

38. H.A. Ronaghi, J.G. Reinhold, M. Mahloudji, P. Ghavami, M.R. Spivey-Fox, J.A. Halsted, *Am. J. Clin. Nutr.* **27**, 112–21 (1974).

39. J.C. Leek, C.L. Keen, J.B. Vogler, M.S. Golub, L.S. Hurley, A.G. Hendrickxs, M.E. Gershwin, *Am. J. Clin. Nutr.* **47**, 889–95 (1988).

40. S. Haumont , *J. Histochem. Cytochem.* **19**, 141–5 (1961).

41. M. Yamaguchi, S. Kishi, M. Hashizume, *Peptides* **15**, 1367–71 (1994)a.

42. A. Suwarnasarn, J.C. Wallwork, L.I Lykken, F.N. Low, H.H. Sandstead, *J. Nutr.* **112**, 1320–8 (1982).

43. M. Yamaguchi, R. Yamaguchi, *Biochem. Pharmacol.* **35**, 773–7 (1986).

44. B.S. Moonga, D.W. Dempster, *J. Bone Mineral Res.* **10**, 453–7(1995).

45. O. Lema, H.H. Sandstead, *Fed. Proc.* **29**, 297 (1970).

46. W.H. Daughaday, L.S. Phillips, A.C. Herington, *Ann. Rev. Physiol.* **37**, 211–44 (1975).

47. C.W. Denko, M. Petricevic, M.W. Whitehouse, *Int. J. Tissue Reac.* **3**, 121–5(1981).

48. F. Fernandez-Madrid, A.S. Prasad, D. Oberleas, *J. Lab. Clin. Med.* **82**, 951–61 (1973).

49. A.S. Prasad, F. Fernandez-Madrid, J.F. Ryan, *Am. J. Physiol.* **236**, E272–5 (1979).

50. M. Yamaguchi, M. Hashizume, *Moll. Cell. Biochem.* **136**, 163–9 (1994)b

51. T. Matsui, M. Yamaguchi, *Peptides* **16**, 1063–68 (1995).

52. J.M. Hock, M. Centrella, E. Canalis, *Endocrinology* **122**, 254–60 (1988).

53. L.S. Phillips, A.T. Orawski, D.C. Belosky, *Endocrinology* **103**, 121–7 (1978).

54. S. Kemp, R.T. Hintz, *Endocrinology* **106**, 744–9 (1980).

55. H.H. Sansdtead, A.S. Prasad, A.R. Schulert, Z. Farid, A. Miale Jr, S. Bassily , W.J. Darby, *Am. J. Clin. Nutr.* **20**, 422–42 (1967).

56. M. Castro-Magana, P.J. Collipp, S.Y. Chen, T. Cheruvanky, V.T. Maddaiah, *Am. J Dis. Child.* **135**, 322–5 (1981).

57. L. Wada, J.C. King, *J. Nutr.* **116**, 1045–53 (1986).

58. W.P. Pimenta, J. Brandào-Neto, P.R. Curi, *Tr. Elem. Med.* **9**, 34–7 (1992).

59. T. Nakamura, A. Higashi, S. Nishiyama, S. Fujimoto, I. Matsuda, *Diabetes Care* **14**, 553–7 (1991).

60. J. Brandào-Neto, J.G.H. Vieira, T. Shuhama, E.M.K. Russo, R.V. Piesco, P.R. Curi, *Biol. Tr. Elem. Res.* **24**, 73–82 (1990).

61. J.Quarterman, C.F. Mills, W.R. Humphries, *Biochem. Biophys Res. Comm.* **25**, 354–8 (1966).

28

IMMUNOSTIMULATING EFFECT OF ZINC SUPPLEMENTS DURING RECOVERY OF SEVERELY MALNOURISHED CHILDREN

P. Chevalier,[1] R. Sevilla,[2] L. Zalles,[2] G. Belmonte,[2] and E. Sejas[2]

[1] LNT, ORSTOM
BP 5045, 34032 Montpellier cedex 1, France
[2] CRIN
Hospital Materno Infantil "German Urquidi"
Cochabamba, Bolivia

1. INTRODUCTION

Pre-school children suffering from protein-energy malnutrition (PEM) show immunodeficiency (1) characterised by impaired cellular immunity (2, 3) and thymic involution (4) with simultaneous altered thymulin content (5) and activity (6). The initial high level of 'null cells' or immature T lymphocytes (7, 8) was reduced by half after incubation *in vitro* with thymulin (9, 10). This indirectly demonstrates the depressed lympho-differentiative capacity of thymic hormones in malnutrition states. In a longitudinal study, clinical and anthropometrical recovery (11) of severely malnourished children was reached in 5 weeks, while thymic recovery required 9 weeks (10). Relapses (12) occurred after "apparently healthy" children were discharged, because they were still immune-depressed. Therefore, along with clinical and nutritional interventions, an immuno-stimulatory treatment was tested to more rapidly restore cellular immunity (13). Although thymulin happened to be the initial candidate, it was discarded because of excessive cost and difficult use and zinc, as a thymulin cofactor (14), was tested first. Compared to previous studies on zinc supplementation in treatment of severely malnourished children (15, 16, 17), we used ultrasonography, a non-invasive technique, to quantify the thymic mass and its restoration, and monoclonal antibodies (MABs) to evaluate the level of immature lymphocytes (9,10).

2. MATERIALS AND METHODS

To test the immunological effects of zinc supplementation, we designed a cohort study of 32 severely malnourished children matched for age, sex, anthropometry and nutritional pathology to malnourished children previously treated without zinc. Children were selected

Table 1. Effect of zinc supplementation on anthropometric recovery (mean ± SD)

	Group	Week 0 (n=32)	p	Week 5 (n=32)	p	Week 9 (n=26)
HAZ	Control	-3.2 ± 1.2	NS	-3.2 ± 1.2	NS	-3.2 ± 1.0
	Zn Sup.	-3.2 ± 1.4	NS	-3.2 ± 1.4	NS	-3.0 ± 1.2
WAZ	Control	-3.7 ± 1.0	***	-2.7 ± 1.2	NS	-2.4 ± 1.1
	Zn Sup.	-3.8 ± 1.1	***	-2.6 ± 0.9	NS	-2.3 ± 1.0
WHZ	Control	-2.4 ± 0.8	***	-1.0 ± 1.1	NS	-0.6 ± 1.0
	Zn Sup.	-2.5 ± 1.2	***	-0.9 ± 1.0	NS	-0.6 ± 0.9
Arm C (cm)	Control	10.9 ± 1.4	***	12.8 ± 1.6	NS	13.8 ± 1.5
	Zn Sup.	11.0 ± 1.6	***	13.0 ± 1.5	NS	13.9 ± 1.4
TST (mm)	Control	4.8 ± 1.9	***	6.8 ± 2.0	*	7.9 ± 1.7
	Zn Sup.	4.6 ± 1.9	***	7.1 ± 2.0	NS	8.0 ± 2.3
UMA (cm^2)	Control	7.1 ± 1.4	***	9.1 ± 2.0	NS	10.3 ± 2.1
	Zn Sup.	7.4 ± 1.8	***	9.4 ± 2.0	*	10.4 ± 1.7
BMI	Control	13.2 ± 1.5	***	15.8 ± 1.8	NS	16.5 ± 1.6
	Zn Sup.	13.0 ± 2.0	***	15.8 ± 1.5	NS	16.3 ± 1.5

HAZ, WAZ, WHZ: z score respectively for height for age, weight for age, weight for height; Arm C : arm circumference, TST : triceps skinfold thickness, UMA : upper arm muscle area, BMI : body mass index. Significance level: * $p < 0.05$, ** $p < 0.01$, *** $p < 0.001$; NS: not significant.

from the Materno-Infantil "German Urquidi" Hospital in Cochabamba (Bolivia). The mean age was 19 months. After parental consent to a 2-month follow-up study, they were admitted to the CRIN (Centro de Rehabilitación Inmuno-Nutricional) and received 2 mg of elemental zinc per kg per day. All of them came from poor Cochabamba suburban areas. Their socio-economic status was characterised by low income, crowded living conditions, lack of sanitation and early weaning. Kwashiorkor, Marasmus and combined PEM diagnoses were based on weight for height (18) and clinical findings: presence of oedema, loss of subcutaneous tissue and diminished muscle mass (19). They were initially treated for respiratory and/or intestinal infections. The same treatment scheme, previously described (10), was applied in both groups: a diet divided in four phases over a 2-month period, daily clinical examination and weekly anthropometric measurements. Thymus size was weekly assessed by mediastinal echographic examination (9, 10) using a portable echo camera (ALOKA SSD-210 DXII, Tokyo) with a 5 MHz linear paediatric probe. Quantification of T lymphocytes was monthly performed using monoclonal antibodies (OKT3, OKT6) (9, 10). Differences between the 2 groups were compared with the Student's *t* test.

3. RESULTS

Zinc supplementation did not change the rate of anthropometric recovery (Table 1). Within each group, after one month, anthropometric parameters were significantly different from these observed on admission levels. Most of them did not change during the second month. During the 2-month period of hospitalisation, in both groups the level of mature lymphocytes (T3 or CD3) increased slightly, meanwhile the level of immature lymphocyte (T6 or CD1) decreased significantly (Table 2). After 5-week hospitalisation, the level of immature lymphocytes in the supplemented group was lower than the levels in the control group at 5 and 9 weeks of hospitalisation. The previously set threshold (10) for the area of the left lobe of the thymus (350 mm^2) was reached by the supplemented and control groups in 5 and

Table 2. Effect of zinc supplementation on immunologic parameters

		n_z, n_c*	Sup zinc	Control	p
Maturelymphocytes (CD3)**	onset	18, 32	60.6 ± 5.3^a	60.9 ± 5.1^c	d
	5 weeks	15, 31	65.1 ± 6.9^d	$65.7 \pm 4.8d$	d
	9 weeks	13, 30	67.8 ± 3.5	65.5 ± 4.0	d
Immature lymphocytes (CD1)**	onset	28, 28	32.5 ± 7.6^c	29.9 ± 5.3^c	d
	5 weeks	26, 26	10.1 ± 3.4^a	15.5 ± 4.0^b	c
	9 weeks	21, 21	8.1 ± 2.7	11.1 ± 4.9	a
Left thymic lobe area***	onset	32, 32	81 ± 7^c	71 ± 10^c	d
	5 weeks	26, 26	362 ± 26^b	229 ± 22^c	a
	9 weeks	23, 23	453 ± 17	387 ± 25	a

*n_z, n_c : number for zinc suplemented and control groups.
** per cent (mean ± SD); *** mm² (mean ± SEM)
Significance: (^{a}) p < 0.05, (^{b}) p < 0.01, (^{c}) p < 0.001; (^{d})not significant.

9 weeks respectively (Table 2). With similar thymic areas at admission, the zinc-supplemented children presented faster thymic recovery than the control children.

4. DISCUSSION

Our results confirmed that children suffering from severe PEM reached anthropometrical and clinical criteria for discharge 5 weeks after admission. Although "apparently healthy", the unsupplemented children were still immune depressed and were kept in the rehabilitation ward for another month until immunological recovery (10). Zinc supplement during the CRIN hospitalisation did not enhance anthropometrical recovery, but significantly reduced the immune recovery period that coincided with anthropometric recovery. Zinc could act as an immune stimulating factor via the biological activation of thymulin (14) which stimulates the maturation and proliferation of thymocytes, or via the inhibition of the endonuclease responsible of apoptosis (20).

Compared to the dietetical cost of inpatient treatment (30 US per month per child), the cost of zinc supplementation (1 $US per month) was one of the cheapest ways to stimulate immunity and reduce the gap between nutritional and immune recoveries. Consequently, children may be discharged after only one month of immuno-nutritional rehabilitation when they are considered sufficiently healthy to face their pathogenic environment.

5. REFERENCES

1. R.K. Chandra, *J. Nutr.* **122**, 597–600 (1992).
2. L. Schlesinger and A. Stekel, *Am. J. Clin. Nutr.* **27**, 615–620 (1974).
3. S. Fakhir et al. *J. Trop. Pediat.* **35**, 175–178 (1989).
4. P.M. Smythe et al. *Lancet.* **2**, 939–944 (1971).
5. B. Jambon et al. *Am. J. Clin. Nutr.* **48**, 335–342 (1988).
6. R.K. Chandra, *Clin. Exp. Immunol.* **38**, 228–230 (1979).
7. R.K. Chandra, *Pediatrics.* **59**, 423–27 (1977).
8. R.K. Chandra, *Acta Paediatr. Scand.* **68**, 841–845 (1979).
9. G. Parent et al. *Am. J. Clin. Nutr.* **60**, 274–8 (1994).

10. Ph. Chevalier et al. *J. Nutr. Immunol.* **3**, 27–39 (1994).

11. J.C. Waterlow, *Protein Energy Malnutrition*, Arnold, London (1992).

12. B. Maire, In *AFTHR*, n°11, T. Marek ed. Banque Mondiale, Washington, pp. 89–102 (1993).

13. S.O. Olusi, G.B. Thurman and A.L. Goldstein, *Clin. Immunol. Immunopathol.* **15**, 687–691 (1980).

14. M. Dardenne et al. *Proc. Natl. Acad. Sci. USA.* **79**, 5370–5373 (1982).

15. M.H.N. Golden, A.A. Jackson and B.E. Golden, *Lancet.* 1057–1059 (1977).

16. C. Castillo-Duran et al. *Am. J. Clin. Nutr.* **45**, 602–8 (1987).

17. L. Schlesinger et al. *Am. J. Clin. Nutr.* **56**, 491–8 (1992).

18. J.C. Waterlow, *Brit. Med. J.* **3**, 566–569 (1972).

19. O. Brunser et al. in *Desnutrición infantil*, F. Monckeberg ed., INTA, Santiago, pp. 13–34 (1988).

20. M.M. Compton and J.A. Cidlowski, *Trends Endocrinol. Metab.* **3**, 17–23 (1992).

29

SERUM COPPER AND PROTEIN-CALORIE MALNUTRITION IN THE FES AREA (MOROCCO)

F-Z. Squali Houssaïni,[1] J. Arnaud,[2] M-J. Richard,[2] J-C. Renversez,[2] B-D. Rossi Hassani,[1] and A. Favier[2]

[1] Faculté des Sciences Dhar Mehraz
BP 1796, Fès, Maroc
[2] CHU de Grenoble
BP 217, 38043 Grenoble cedex 09, France

1. INTRODUCTION

Malnutrition is the result of a negative nutritional balance. Protein-calorie malnutrition (PCM) (1,2) is responsible for a high percentage of deaths among young Moroccan children (3). Copper is an essential micronutrient for growth and participates in the oxidative defence. The copper status is modified in malnutrition (4). The aim of the present study was to determine the copper status in young PCM children living in the Fes area.

2. MATERIALS AND METHODS

Fifty six children, aged 6 to 60 months, were enrolled and divided into 3 groups. Group 1: 20 healthy children (control group) recruited from the out patient clinic of Ain El Qadous under one of the following occasions: second dose of vitamin D at the age of 6.5 months; vaccination against measles at 9 months or growth check-up visit. Group 2: 21 children suffering from severe protein-calorie malnutrition, hospitalised in the paediatric department of Ibn El Katib Hospital in Fes. Group 3: 15 children suffering from moderate PCM recruited from the same hospital. Waterlow's criteria were used to classify malnourished infants.

The serum copper was determined by electrothermal atomic absorption spectrometry, after a 5-fold dilution of the sample in triton X 100, 0.1 % (m/v). Erythrocyte copper-zinc superoxide dismutase activity (SOD) was measured using a clinical biological analyser RA1000 (Bayer, Puteaux, France) by the Marklund and Marlund's method (5). Proteins were analysed by immunephelometry using a Behring Nephelometer (Behringwerke, Marburg, FRG).

Therapeutic Uses of Trace Elements, edited by Nève et al.
Plenum Press, New York, 1996

Table 1. Characteristics of the population: anthropometric parameters and protein markers of nutrition and inflammation

Parameters	Controls	All	Severe	Mild
n	20	36	21	15
Sex, F/M	12/8	16/20	5/16	11/4
Age, months	15 (6–60)	20 (6–60)	18 (6-60)	23 (6–60)
Z-score	0.16	–2.15	–2.15	–2
Weight, SD	(–0.87 – 1.37)	(–2.65 – –1.5)	(–2.65 – –1.7)	(–2.4 – –1.5)
Albumin, g/l	49 (42 – 50)	36 (16 – 52)	31 (16 – 45)	45 (39 – 52)
n, < 35 g/l	0	18	18	0
Orosomucoïd, g/l	1.0 (0.5 – 1.8)	1.8 (0.3 – 1.7)	2.0 (0.9 – 3.7)	1.1 (0.3 – 2.2)
n, > 1,2 g/l	4	20	17	3
CRP mg/l	6 (4 – 10)	19 (5 – 150)	25 (5 – 150)	13 (4 – 46)
n, > 20 mg/l	0	13	10	3
PINI	7.10^{-4}	230.10^{-4}	300.10^{-4}	40.10^{-4}
	$2.10^{-4} - 33.10^{-4}$	$3·10^{-4} - 3808.10^{-4}$	$7·10^{-4} - 3808.10^{-4}$	$2·10^{-4} - 257.10^{-4}$

PINI index: (orosomucoïd g/l x CRP mg/l) /(albumin g/l x prealbumin mg/l). Results expressed as mean (range). Cutoffs were those proposed by Gilson (Albumin and Prealbumin), Cambau (CRP) and Engler (Orosomucoïd). SD: standard deviation, CRP: C reactive protein.

3. RESULTS

The characteristics of the 3 groups of children are summarised in Table 1. The anthropometric parameters were expressed as Z-scores, because of the wide age range of the children (6–60 months). The references used for weight were those published by the National Center of Health Statistics (NCHS) (6,7). All the control group children Z-scores were higher than -1, whereas all the malnourished infants Z-scores were lower than -1. The decrease of albumin, the increase of C reactive protein (CRP) and prognostic inflammatory and nutritional index (PINI index) were related to the malnutrition severity.

The percentage of copper concentrations lower than 12.7 µmol/L was calculated in each group. This value corresponds to the mean minus 2 standard deviation observed in our control group. These percentages were found to be 0 % in control and mild malnutrition groups, 5 % in severe malnutrition group and 3 % in the whole PCM group. Table 2 indicates that ceruloplasmin, such as copper, significantly decrease only in severe malnutrition. On the contrary, SOD activity decrease significantly only in children suffering from moderate malnutrition.

A significant correlation between copper and ceruloplasmin was observed in all groups (r = 0.92, p < 0.001 for the control group and r = 0.82, p < 0.001 for the malnourished group). On the contrary, there was no significant relationship between copper or ceruloplasmin and SOD in both groups. Finally, significant correlations between copper (r = 0.50, p =

Table 2. Copper status in Moroccan healthy and malnourished infants

Parameters	Controls	All	Severe	Mild
n	20	36	21	15
Copper	µmoles/l	24.3 ± 5.8	22.2 ± 6.8	20.6 ± 6.2
SOD	U/g Hb	1.22 ± 0.11	1.19 ± 0.20	1.27 ± 0.20
Ceruloplasmin	mg/l	284 ± 42	272 ± 59	260 ± 66

Results are expressed as mean ± standard deviation.

0.03) or ceruloplasmin ($r = 0.71$, $p = 0.007$) and albumin as well as between copper ($r = 0.66$, $p = 0.003$) or ceruloplasmin ($r = 0.67$, $p = 0.002$) and PINI were observed only in control group.

4. DISCUSSION

The simultaneous decreases of serum copper and ceruloplasmin levels observed only during severe malnutrition agree well with previous studies (8,9). Therefore, serum copper or ceruloplasmin levels could be used as indicators of malnutrition severity. The erythrocyte SOD activity decreases in moderate PCM and returns within the reference range in severe malnutrition. The discrepancy between copper or ceruloplasmin and SOD variations during malnutrition was previously described. As SOD activity decrease was significant in mild PCM, SOD activity could be considered as a more sensitive indicator of copper status than serum copper or ceruloplasmin (4). However, the present results indicate that SOD activity could not be considered as a reliable index of copper nutritional status. Indeed, in severe malnutrition SOD returned to values found in control group. To the best of our knowledge, such variations according to malnutrition severity have never been described. They could be the result of oxidative stress occurring during severe malnutrition. The severe metabolic changes of malnutrition could also explain the absence of significant relationship between copper status parameters (copper and ceruloplasmin) and inflammatory or nutritional indicators (PINI, albumin). This undoubtedly deserves further investigations.

5. REFERENCES

1. A. Briend, B. Maire and J.-F. Desjeux, *Traité de nutrition pédiatrique,* vol. 13, Maloine, Paris, pp 467–512 (1993).
2. M.H.N. Golden and A.A. Jackson, Malnutrition protéino-energétique. *Encycl. Med.,* chir paris nutrition, (1981).
3. Enquête nationale sur la population et la santé 1992 (ENPS 92), Maroc, pp 105–117.
4. C. Castillo-Duran and R. Uauy, *Am. J. Clin. Nutr.* **47**, 710–714 (1988).
5. S. Marklund and G. Marklund, *Eur. J. Biochem.* **47**, 469–474 (1974).
6. P.V.V. Hamill, T.A. Drird, C.L. Johnson et al. *Am J Clin Nutr.* **32**, 607–629 (1979).
7. WHO Working group, Bull WHO **64**, 929–941.
8. M.H.N. Golden, B.E. Golden and F.I. Bennett, *Trace element in nutrition of children*, R.K. Chandra, ed., Raven press, New York, pp 185–207 (1985).
9. H. Heese, A.A. Sive, W.S. Dempster et al, *Trace elements in man and animals,* vol 7, B. Momcilovic, ed., Zagreb, IMI, pp 19.14–19.15 (1991).

30

SERUM TRACE ELEMENTS (Cu, Zn, Se, AND Al) AND THIOBARBITURIC ACID REACTANTS (TBARS) IN HEMODIALYSIS PATIENTS FROM BATNA (ALGERIA)

B. Lachili,[1] J. Arnaud,[2] C. Coudray,[2] N. Zama,[3] A. M. Roussel,[2] C. Benlatreche,[4] and A. Favier[2]

[1] Faculté de Médecine
Université de Batna
Rue Chahid Boukhlouf, 05000 Batna, Algérie
[2] Laboratoire de Biochimie C
CHUG
38043 Grenoble cedex 9, France
[3] Service d'Hémodialyse
CHU Benflis Touhami, Batna, Algérie
[4] INES.SM, Constantine
BP125, Chalet des Pins, 25000 Constantine, Algérie

1. INTRODUCTION

In patients with chronic renal failure (CRF), serum zinc (1,2) and selenium (2–4) concentrations generally decrease, whereas serum copper (1,5) and aluminum (1,6) concentrations increase. These variations could be related to the low dietary intake due to protein restriction or poor appetite, decreased gastrointestinal absorption, decreased bioavailabillity due to drug interactions, inflammatory process and interaction with dialysis membrane (7,8). On the other hand, trace element status variations in CRF patients could explain biochemical and clinical manifestations occuring in these patients. Hypogeusia, taste and smell dysfunctions, hypogonadism, hair-loss, dermatitis, anemia, fatigue, immunodepression, atherosclerosis, osteomalacia and encephalopathy could be in part explained by zinc and selenium deficiencies or aluminum intoxication (6–11). The decrease of serum zinc and selenium concentrations associated to the increase of serum aluminum concentrations in CRF patients could contribute to the high level of thiobarbituric acid reactants (TBARS) reported in hemodialyzed patients (2,12,13). Influence of hemodialysis session have not been extensively investigated (1,5,10). Serum zinc and selenium concentrations increase (1,5,10), whereas serum copper and aluminum concentrations remain unchanged (1,5).

Therapeutic Uses of Trace Elements, edited by Nève et al.
Plenum Press, New York, 1996

The aim of the present study was to determine the concentrations of four trace elements (copper, zinc, selenium and aluminum) as well as TBARS in the serum of CRF patients living in Batna (Algeria) and to evaluate the influence of the hemodialysis treatment.

2. MATERIAL AND METHODS

Fourty five patients (29 males and 16 females) aged 9 to 60 years (median: 39) with chronic renal failure and undergoing periodical hemodialysis treatment were enrolled. Hemodialysis treatment was done either twice a week for 6 hours or thrice a week for 4 hours. Cuprophan were used as dialysis membrane for all the patients. No patients received trace element (copper, zinc and selenium) supplementation. Fifty two healthy individuals (20 males and 32 females) without clinical and biological evidence of renal disease were selected as controls. Their age ranged from 18 to 45 years.

Blood samples were collected using trace element free Vacutainer tubes (Becton Dickinson, Meylan, France) at the beginning (PRE) and the end (POST) of the hemodialysis session. Blood was centrifuged for 10 min at 3000 rpm (2500 g). Centrifugation was performed within 30 to 60 min after blood collection. Serum was put in a trace element free polystyrene (PS) tube. These tubes were previously decontaminated by soaking in a 10 % (W/V) EDTA (Prolabo, Paris, France) solution for 24 hours followed by a thoroughly rinsing with deionized water. Sera were freezed at -80 °C until analysis.

Copper and zinc were determined by flame atomic absorption spectrometry using a Perkin Elmer model 460 (Norwalk, Connecticut, USA). Selenium and aluminum determinations were performed by electrothermal atomic absorption spectrometry using respectively a Perkin Elmer model 5100 and a Hitachi model 8270 (Tokyo, Japan). Seronorm trace element was used as a precision and accuracy internal quality control for trace element analyses. Serum total proteins (TP) were determined using Biuret's method on a clinical biological autoanalyzer Technicon RA1000 (Bayer, Puteaux, France). Seracheck was used as internal quality control for TP analysis.

3. RESULTS

Table 1 indicates the serum concentrations of copper, zinc, selenium, aluminum and TBARS expressed as μmol/l in controls and CRF patients. Aluminum, copper and TBARS concentrations were significantly higher (p<0.001) in patients before hemodialysis than in controls, whereas zinc and selenium concentrations were lower (p<0.001) in patients before hemodialysis than in controls. After hemodialysis, aluminum and copper remained unchanged, zinc and selenium concentrations returned within the reference range (p<0.01) and TBARS concentrations were significantly higher (p<0.05) than before hemodialysis session.

Table 1 also indicates the serum trace element concentrations expressed as μmol/g TP in patients before and after hemodialysis session. Zinc and selenium concentrations increased significantly (p<0.01) whereas copper and aluminum concentrations remained unchanged.

4. DISCUSSION

Serum copper (14), zinc (14), aluminum (1,6) and TBARS (2,12) concentrations in controls were in agreement with those in previous studies. Serum selenium values were in

Table 1. Aluminum, copper, zinc, selenium, and TBARS concentrations in controls and CRF patients before (PRE) and after (POST) hemodialysis session

	Aluminum	Copper	Zinc	Selenium	TBARS
Controls, µmol/l	0.37 ± 0.24	19.3 ± 3.1	13.4 ± 2.0	1.49 ± 0.17	2.71 ± 0.40
CRF patients PRE, µmol/l	0.89 ± 0.60	23.7 ± 7.2	11.1 ± 2.3	1.24 ± 0.28	3.13 ± 0.64
µmol/g TP*	0.014 ± 0.010	0.39 ± 0.13	0.18 ± 0.03	0.020 ± 0.004	
CRF patients POST, µmol/l	1.00 ± 0.75	26.5 ± 8.1	13.1 ± 2.5	1.44 ± 0.36	3.41 ± 0.83
µmol/g TP*	0.015 ± 0.012	0.40 ± 0.11	0.20 ± 0.04	0.022 ± 0.005	

*TP: Total proteins.

the upper part of reference range (14), probably related to high selenium concentrations of the soil or of the typical Algerian diet.

Differences between controls and CRF patients before hemodialysis were in agreement with previous papers concerning aluminum (1,6), copper (1,5,6), zinc (1,2,6), selenium (2–4), and TBARS (2,12,13). Encephalopathy, osteomalacia and microcytic anemia could occur in patients with CRF undergoing long-term hemodialysis due to aluminum accumulation in brain, bone and soft tissues (11,15). The use of either concentrated dialysis solutions with high aluminum content or untreated water for preparing dialysate (15) as well as aluminum-containing phosphate binder treatment have been reported as the main sources of aluminum in hemodialyzed patients in developing countries. The high level of serum aluminum in Algerian hemodialyzed patients could probably explain the high incidence of aluminum intoxication in Algerian CRF patients undergoing long-term hemodialysis. In the present study, 2 patients (one male and one female) had serum aluminum concentration higher than 7.40 µmol/l (16). These 2 patients suffered from encephalophathy and bone disorders. Desferrioxamine (DFO) was used as chelating therapy for six months in these patients and their aluminum-containing phosphate binder treatment was stopped. Serum aluminum concentration was higher than 2.20 µmol/l without clinical signs of aluminum intoxication in 5 patients (16). Clinical signs of zinc and selenium deficiencies were also observed in CRF patients. Five patients suffered from hypogeusia and their serum zinc values were lower than 10.7 µmol/l (7,17). Impotence was observed in 17 patients. Of these 17 patients, 8 had serum zinc lower than 10.7 µmol/l (9,17). Muscular fatigue was observed in 5 patients but without serum selenium deficient levels (18). The decrease of serum selenium and zinc concentrations as well as the increase of serum copper and aluminum concentrations could explain the increase of serum TBARS. The increase of lipid autooxidation could also be explained by the increased oxidative metabolism of neutrophils induced by their interaction with the dialysis membrane (12,13) or by the kidney disease (13).

Differences between the beginning and the end of the hemodialysis session are also in agreement with previous works (1,5,10). These differences are not explained by hemoconcentration as they were also observed when results were expressed in µmol/g TP. Serum copper concentrations were similar before and after dialysis session. Therefore, the cuprophan membranes do not release copper in the blood stream (19). The significant increase in lipid peroxidation products during hemodialysis, expressed in µmol/l was in agreement with a previous paper (13). It could be explained by activation of neutrophils at the surface of the dialyzer membrane (13). However, according to Schettler et al. (13), the elevated TBARS in CRF are not caused by the hemodialysis therapy.

5. REFERENCES

1. P. Allain, H.E. Thebaud, L. Dupouet, P. Coville, M. Pisant, J. Spiesser and P. Alquier, *Nouvelle Presse Méd.* 7, 92–96 (1978).
2. M.J. Richard, J. Arnaud , C. Jurkovitz, T. Hachache, H. Meftahi, F. Laporte, M. Forêt, A. Favier and D. Cordonnier, *Nephron* 57, 10–15 (1991).
3. J.W. Foote, L.J. Hinks and B. Lloyd, *Clin. Chim. Acta* 164, 323–328 (1987).
4. M.D. Saint-Georges, D.J. Bonnefont, B.A. Bourely, M.C. Jaudon, P. Cereze, C. Gard, J.C. Chaumiel and C.L. d'Auzac, *Nouvelle Presse Méd.* 18, 1195–1198 (1989).
5. C. Agenet , C.C. Brugere and J.P. Reynier, *Ann. Biol. Clin.* 47, 493–496 (1989).
6. R.A. Romero, J.A. Navarro, B. Rodriguez-Iturbe, R. Garcia, O.E. Parra and V.A. Granadillo. *Trace Elem. Med.* 7, 176–181 (1990).
7. E. Atkin -Thor, B.W. Goddard, J. O'Nion, R.L Stephen and W.J. Kolff, *Am. J. Clin. Nutr.* 31, 1948–1951 (1978).
8. M.J. Richard and M.C. Jaudon, in *Les oligoéléments en nutrition et thérapeutique*, P. Chappuis and A. Favier, eds., Tec et Doc, Paris, pp 325–341 (1994).
9. L.D. Antoniou, R.J. Shalhoub, T. Sudhakar and J.C. Smith, *Lancet* 2, 895–898 (1977).
10. K. Milly, L. Wit, C. Diskin and R. Tulley, *Nephron* 61, 139–144 (1992).
11. M.R. Wills and J. Savory, *Lancet* 2, 29–33 (1983).
12. J.L. Paul, N.D. Sall, T. Soni, J.L. Poignet, A. Lindenbaum, N.K. Man, N. Moatti and D. Raichvarg, *Nephron* 64, 106–109 (1993).
13. V. Schettler, E. Wieland, R. Verwiebe, P. Schuff-Werner, F. Scheler and M. Oellerich, *Nephron* 67, 42–47 (1994).
15. J.A. Navarro, O.E. Parra, R. Garcia, B. Rodriguez-Iturbe, V.A. Granadillo, D. 14. V. Iyengar and J. Woittiez, *Clin. Chem.* 34, 474–481 (1988).
16. *Off. J. Eur. Comm.* C202, 5–8 (1983).
17. S.M. Pilch and F. R. Senti, *Assessment of the zinc nutritional status of the U.S. population based on data collected in the second national health and nutrition examination survey,* Life Science Research Office, Federation of American Societies for Experimental Biology, Bethesda, (1984).
18. P. Van Dael and H. Deelstra, *Internat. J. Vit. Nutr. Res.* 63, 312–316 (1993).
19. B.H. Barbour, M.Bischel and D.E. Abrams, *Nephron* 8, 455–462 (1971).

31

ABNORMALITIES OF ANTIOXIDANT MICRONUTRIENT STATUS IN HEMODIALYSIS PATIENTS

D. J. M. Malvy,[1] M. J. Richard,[2] J. Pengloan,[3] J. Arnaud,[2] B. Fouquet,[4] H. Nivet,[3] A. Favier,[2] and Ph. Bagros[3]

[1] Laboratory of Public Health
University of Medicine, CHU
Tours, France
[2] Laboratory of Biochemistry C, CHRUG
Grenoble, France
[3] Department of Nephrology, CHU
[4] Department of Reabilitation, CHU
Tours, France

1. INTRODUCTION

Chronic renal failure, which typically results in end-stage renal failure, is commonly associated with metabolic and functional disturbance, leading to calcium phosphate metabolism and nutritional inadequacies (1). Among the causes of malnutrition, abnormal muscle metabolism, endocrine abnormalities, impaired metabolic kidney functions, and disturbances in nutrient metabolism such as amino acids, vitamins and trace elements, have been described in renal failure patients (1–3). Moreover, hemodialysis (HD) performed to compensate for kidney failure appears to be related to metabolic abnormalities. Dialysis membranes may therefore induce the activation of cellular systems leading to production of multiple inflammatory agents (4), with increased and uncontrolled formation of oxidative products (5, 6). Oxygen-activated species cause a wide array of molecular alterations including lipid peroxidation with generation of aldehydes, eg, malondialdehyde, protein and nucleic acid damages, ultimately resulting in cell death. Intracellular antioxidant defence mechanisms primarily involve constitutive and inducible proteins (8, 9). The greater part of the glutathione peroxidase enzyme (GPX), which catalyzes the reduction of all peroxides in the soluble compartment of the cell, is selenium-(Se) independant in humans. Animal and human studies concerning selenium-dependant GPX have shown a close correlation between enzyme activity and Se status. The major hydrophobic membrane antioxidant is alpha-tocopherol, the principal compound of the vitamin E group. Oxidative injury related to dialysis may therefore occur in patients with serious disturbances of the status of micronutrients in-

Therapeutic Uses of Trace Elements, edited by Nève et al.
Plenum Press, New York, 1996

volved in antioxidant defence mechanisms, i.e., selenium, zinc and copper (10). This phenomenon is involved in some pathologic manifestations related to dialysis such as osteoarthropathies (11, 12), or numerous others grouped under the term "accelerated dialysis aging" (13, 14).

The aim of this study was to document the status of antioxidant micronutrients in a population of 76 hemodialyzed patients, and to analyse its relationship with some indicators of lipid peroxidation and the status of protein nutrition. We therefore determined peripheral values for zinc (Zn), selenium (Se) and copper (Cu), vitamins A and E, Se-GPX and lipid peroxide markers such as thiobarbituric acid reactants (TBARs) and organic hydroperoxides (OHP).

2. MATERIALS AND METHODS

2.1. Patients

The patient group included 76 chronic hemodialysis patients, 29 males and 47 females aged 27 to 84 years, mean (58 years), undergoing 2 to 3 times a week dialysis sessions. The number of years of dialysis was from 1 to 15 years, median 5 years. Forty-five patients were dialyzed with a cuprophan membrane, 16 with a polysulfone membrane, and 6 with a polymethylmetacrylate (PMMA) membrane. Blood was sampled before a dialysis session. Rheumatologic data were scored for each patient (0 to 8) after assessment of joints studied for the absence or presence of pain (Clinical Arthropathy Score, CAS, median 2.5), erosive lesions (Erosive Arthropathy Score, EAS, median 2.2), and geodes (Geodic Arthropathy Score, GAS, median 1.4) (11). The joints studied were the cervical spine, dorsal spine, lumbar spine, shoulders, wrist-fingers, hips, knees and feet. The occurrence of an erosive lesion was defined as an irregular or fuzzy radiological aspect of subchondral bone without osteophytosis of adjacent bone (11). Biochemical data were compared with those of 76 healthy sex- and age- matched controls, free from metabolic or renal disease.

2.2. Methods

Venous blood was taken before the dialysis session from fasting individuals, collected on a sodium heparin vacutainer (Trace element-free) and centrifuged promptly at 4 °C. The plasma was stored at -20 °C until analysis. A flame atomic absorption technique was used for plasma zinc determination (15), and an electrothermal atomic absorption spectrometric method was used for measuring plasma selenium (16). Plasma and red blood cell Se-dependant glutathione peroxidase was evaluated by a modified method of Gunzler using terbutyl hydroperoxide as substrate (17). Lipidperoxide evaluation included the determination of thiobarbituric acid reactants (TBARs) using MDA-kit (Sodobia, Grenoble, France) (18), and plasma lipid hydroperoxides (OHP) using the enzymatic technique of Heath and Tappel (19). Serum retinol, carotene and alpha-tocopherol were assessed by high performance liquid chromatography (20). Vitamin A status was obtained by calculation of the molar ratio of retinol to retinol-binding protein in plasma. The vitamin E status was calculated by the ratio of plasma alpha-tocopherol to cholesterol. Nutritional status was assessed by the measurement of plasma concentrations of IgG, IgA, IgM, and 3 related circulating proteins, namely, albumin, transthyretin and retinol-binding protein. Acute phase reactants were determined as for plasma orosomucoid and C-reactive protein. All plasma proteins were measured using a nephelometric immunoassay.

Statistical analysis was conducted after logarithmic transformation of quantitative variables. Statistical differences were calculated using χ^2 and Mann-Whitney U test. Correlations between the variables were estimated by Spearman's rank correlation coefficients. Comparison of paired samples was performed using the paired Wilcoxon rank sum test.

3. RESULTS

Data for biochemical and nutritional markers are shown in Tables 1 and 2. As shown in Table 1, biochemical indicators of nutritional status were not significantly different compared to those of controls, except for plasma RBP which was very elevated in hemodialyzed patients. Levels of plasma acute phase reactants were increased, indicating a chronic inflammatory syndrome in dialyzed patients. This increase was independant of the type of membrane used, but was related to the two radiological rheumatological scores, EAS (r = 0.3, p = 0.03) and GAS (r = 0.19, p = 0.06), although with a borderline statistical significance.

Plasma selenium and GPX were significantly decreased in dialyzed patients (Table 2), although the values of these two variables were statistically independant. Erythrocyte GPX activity was correlated with the selenium plasma levels in patients (r = 0.30, p = 0.01) and dramatically lower than in controls. These disturbances of selenium status were independant of the membrane used and of the number of years of dialysis. Plasma selenium was negatively correlated to the occurrence of inflammatory syndrome, although with low signifance (r = - 0.3, p = 0.03). Plasma zinc was decreased in dialyzed patients. This value was independant of the type of membrane. Plasma copper concentration was not significantly different in patients and controls.

With regard to lipid peroxidation markers, the mean contents of TBARs and OHP showed a significant increase compared to the corresponding levels in the control group. The TBAR values appeared unrelated to the membrane type and were not correlated with values of plasma selenium and erythrocyte GPX activity. Nevertheless, the TBAR value was negatively correlated with plasma GPX (r = - 0.38, p = 0.008).

Table 1. Biochemical plasma parameters in 76 hemodialyzed patients and in 76 controls. Results are expressed as mean ± SD or median (range)

Parameter	Hemodialyzed patients	Controls
Proteins, g/l	62.2 ± 4.5	61.5 ± 3.5
Albumin, g/l	40.0 ± 3.6	43.0 ± 2.5
Transthyretin, mg/dl	34.4 ± 9.1	35 ± 10
IgG, g/l	10.7 ± 3.6	10.5 ± 2.5
IgA, g/l	2.13 ± 0.96	2.20 ± 0.50
IgM, g/l	1.13 ± 0.81	1.10 ± 0.40
RBP, mg/dl	19.4 ± 5.0*	3.6 ± 1.2
Orosomucoid, g/l	1.16 ± 0.39*	0.85 ± 0.20
CRP, mg/l	8.2 ± 0.4*	3.2 ± 1.0
β2-M, mg/l	3.7 ± 1.3*	1.8 ± 0.6
PTHc, ng/ml	5.25 (2.80-7.30)	ND
Osteocalcin, µg/l	3.50 (2.52-6.20)	ND

RBP, retinol-binding protein; CRP, C-reactive protein; ß2-M, ß2-microglobulin; PTHc, parathormone C-terminal fragment; ND, not determined
* $p < 0.001$, comparison between patients and matched controls

Table 2. Plasma micronutrients and indicators of oxidative metabolism (mean ± standard deviation) in 76 hemodialyzed renal failure patients and in 76 healthy controls

Parameter	Hemodialyzed patients	Controls
TBARs, µmol/l	4.85 ± 0.69*	2.55 ± 0.13
OHP, µmol/l	161 ± 64*	120 ± 20
Se, µmol/l	0.77 ± 0.28*	1.07 ± 0.26
Zn, µmol/l	11.5 ± 1.7*	14.3 ± 2.7
Cu, µmol/l	18.7 ± 3.8	16.9 ± 5.2
Plasma GPX, IU/l	130 ± 40*	335 ± 23
Erythrocyte GPX, U/g Hb	35.4 ± 8.6*	49.2 ± 9.2
Retinol, µmol/l	6.0 ± 2.2*	1.60 ± 0.32
Retinol/RBP	0.63 ± 0.12*	0.88 ± 0.30
β-carotene, µmol/l	0.45 ± 0.23*	1.08 ± 0.24
α-tocopherol, µmol/l	22.7 ± 9.6	21.1 ± 5.8
α-T/cholesterol	5.0 ± 2.5*	5.3 ± 1.0
Cholesterol, g/l	1.83 ± 0.60	1.74 ± 0.50

TBARs, thiobarbituric acid reactants; OHP, organic hydroperoxides ; GPX, glutathione peroxidase; retinol / RBP, molar ratio retinol / RBP; α-T/cholesterol, ratio α-tocopherol / cholesterol

$*\ p < 0.001$, comparison between hemodialyzed subjects and controls

Mean plasma retinol was higher in the dialysis patients. This increase was strongly related to those of its carrier protein in the blood, retinol-binding protein ($r = 0.85$, $p < 0.001$). Furthermore, the molar ratio of retinol to retinol-binding protein in plasma was lower in patients, than in controls suggesting disturbance in vitamin A storage or transport. Plasma retinol and ratio values were inversely correlated with circulating parathormone value ($p = 0.02$). Moreover, the plasma retinol level was negatively correlated to joint scores CAS ($r = -0.30$, $p = 0.03$) and GAS ($r = -0.42$, $p = 0.002$), but not to the geodic score GAS ($r = -0.007$, $p = 0.5$). Finally, this parameter was not correlated with cholesterolemia, although with borderline unsignificance. Circulating carotene was decreased in patients. There was no correlation between plasma retinol and carotene levels. Plasma alpha-tocopherol was not different in hemodialyzed patients and controls. However, a decrease was observed when the tocopherol cholesterol ratio was calculated. Moreover the plasma tocopherol to cholesterol ratio was correlated with erythrocyte GPX activity ($r = 0.25$, $p = 0.04$).

4. DISCUSSION

Plasma TBARs and OHP in hemodialyzed patients were higher than in controls, indicating a higher degree of lipid peroxidation. Oxidative stress may result from the association of endogenous over-production of reactive oxygen species by polynuclear leucocytes, caused by activation by the membrane itself or by a compound released by this membrane (21), and failure in antioxidant defence mechanisms (8). The low antioxidative capacity and a continuous oxidative burden in HD patients may thus be related to increased susceptibility of lipid peroxidation in red blood cells in such patients (13).

Abnormalities in indicators for antioxidant systems and lipid peroxidation were associated with decreased levels of plasma selenium and zinc, as previously reported (5, 22). The issue of selenium deficiency in chronic uremic patients is still unresolved. The lack of an effective parameter to assess Se status in such patients may account for the lack of consensus (23). HD patients are reported to have either low plasma, serum, whole blood and erythrocyte Se concentrations (2, 22, 24, 25), or similar concentrations as healthy controls (26). The lack of correlation between plasma selenium and duration of dialysis is consistent with previous reports (22), although a negative link has been found elsewhere (25). Our data do not confirm previous results concerning the influence of dialysis membrane on peripheral selenium value, related to movement of the element across the membrane. Plasma Se concentrations have been found to be lower in Se-deficient HD patients dialyzed with a highly permeable membrane such as polysulfone or polyacrilonitrile, presumably because Se could be lost through the pores of the membrane, compared to cuprophan or cellulose acetate hollow fibres (25). In agreement with prior data (5, 27), a significant correlation was observed between plasma selenium and erythrocyte GPX activity, owing to the role of selenium in the synthesis of this enzyme.

The decrease in plasma GPX was statistically independant of the plasma selenium value. This phenomenon may be explained by the role of endogenous ligands or toxic metals which could inhibit plasma GPX (28). Moreover, in HD patients there is a defect in the system involving glutathione reductase, glucose 6-phosphate dehydrogenase and the pentose shunt, preventing generation of sufficient NADPH to maintain glutathione in a reduced state (28). Decreased dietary and protein intake (22) and increased urinary loss (29) have been suggested as the mechanisms which might modify the trace-element status in uremia. In our study, nutritional status remained in the normal range in HD patients, and no correlation between albumin, transthyretin and zinc or selenium was found. Nevertheless, a strong correlation between blood Se and serum albumin in HD patients has been reported elsewhere (22).

Inadequacy of vitamin E status in HD patients has been previously documented (30, 31). The role of vitamin E deficiency as a predisposing condition to oxidative cell or organ stress has led to supplementation trials to achieve normalization of oxidative status in HD patients (30–32).

Although the plasma carotene and vitamin A values may not be accurate indicators of whole-body status, we can hypothesize that possible carotene deficiency must be considered in patients with chronic renal failure. The results concerning the relation between carotene and retinol suggest that carotene may act independantly from vitamin A, through other mechanisms such as free radical scavenging and singlet oxygen quenching. Hyperretinolemia was accompanied by elevated plasma levels of its binding protein, the retinol-binding protein (RBP), related to a decrease in renal elimination of RBP in chronic renal failure. Functional abnormalities of RBP and disturbances in cellular absorption of retinol have been described in such situations (33). It must be noted that the values of the molar ratio of retinol to RBP in HD patients, reported to be a reliable indicator of liver vitamin A storage (20), was always less than unity, and its mean value lower than in controls. The link between vitamin A status and calcium-phosphate metabolism markers suggests a role of the compound in bone resorption and rheumatologic pathogenesis in uremia and dialysis (34). According to some authors (34), vitamin A could slow down the development of secondary hyperparathyroidism in dialysis patients and might induce a lesser degree of bone resorption.

5. CONCLUSION

This study confirms the occurrence of an oxidant stress associated with hemodialysis. The alterations in the antioxidant systems appeared correlated with zinc and selenium status. The consequence could be aggravated by the concomitant free radical stress resulting from the dialysis procedure. The only reliable method to evaluate nutrient deficiency lies in a positive response to supplementation. Formal proof concerning the link between these biological disorders and some clinical manifestations related to hemodialysis will therefore only be provided by long term supplementation with such micronutrients combined with clinical, dietary and biological follow-up.

6. REFERENCES

1. M.J. Blumenkrantz, J.D. Kopple, R.A. Gutman, Y.K. Chan, G.L. Barbour, Ch. Robert, F.H. Shen, V.C. Ghandi, C.T. Tucker, F.K. Curtis and J.W. Coburn, *Am. J. Clin. Nutr.* **33**, 1567–1585 (1980).

2. J.W. Foote, L.J. Hinks and B. Lloyd. *Clin. Chim. Acta* **164**, 323–328 (1987).

3. R. Cornelis, L. Mees, S. Ringoir, and J. Hoste, *Mineral Electrolyte Me*tab. 2, 88–93 (1979).

4. B. Jahn, M. Betz, R. Deppisch, O. Jannsen, G. Hansch and E. Ritz, *Kidney Int.* **40**, 285–290 (1990).

5. M.J. Richard, J. Arnaud, C. Jurkovitz, T. Hachache, H. Meftahi, F. Laporte, M. Foret and A. Favier, *Nephron* **57**, 10–15 (1991).

6. K. Trznadel, L. Pawlicki, J. Kedziora, M. Luciak, J. Blaszczyk and A. Buczynski, *Free Rad Biol. Med.* **6**, 393–397 (1989).

7. L.A. Videla and V. Fernandez, *Arch. Biol. Med. Exp.* **21**, 85–92 (1988).

8. S.P. Andreoli, *Pediatr. Nephrol.* **5**, 733–742 (1991).

9. Y.R.A. Donatti, D.O. Slosman and B.S. Polla, *Biochem. Pharmacol.* **40**, 2571–2577 (1990).

10. C.J. Diskin, *Nephron* **44**, 155–156 (1986).

11. A.M. Bergemer, B. Fouquet, P. Cotty, D. Blanchier, P. Tauveron, P. Goupille, J. Pengloan and J.P. Vallat, *Rev. Rhum.* **56**, 533–538 (1989).

12. J. Munoz-Gomez, E. Bergada-Baroda, R. Gomez-Perez, E. Lopart, E. Subias, J.K. Rotès-Querol and M. Solé, *Ann. Rheum. Dis.* **44**, 729–733 (1985).

13. S. Schmidtmann, R. Baehr and K. Precht, *Nephrol. Dial. Transplant.* **5**, 600–603 (1990).

14. E.R. Maher, D.G. Wickens, J.F.A. Griffin, P. Kyle, J.R. Curtis, and T.L. Dormandy, *Nephrol. Dial. Transplant.* **3**, 277–283 (1988).

15. J. Arnaud, J. Bellanger, F. Bienvenu, P. Chappuis and A. Favier, *Ann. Biol. Clin.* **44**, 77–87 (1986).

16. J. Nève, S. Chamart and L. Molle, in *Trace Element Analytical Chemistry in Medicine and Biology*, P. Bratter and P. Schramel, eds., Walter de Gruyter, Berlin, pp. 349–359 (1987).

17. W.A. Gunzler, H. Kremers and L. Flohe, *Z. Klin. Chem. Klin. Biochem.* **12**, 444–448 (1974).

18. M.J. Richard, B. Portal, J. Meo, C. Coudray, A. Hadjian and A. Favier, *Clin. Chem.* **38**, 704–709 (1992).

19. R.L. Heath and A.L. Tappel, *Anal. Biochem.* **76**, 184–191 (1976).

20. J.P. Vuilleumier, H.E. Keller, D. Gysel, and F. Hunziker, *Internat. J. Vit. Nutr. Res.* **53**, 265–272 (1983).

21. D. Docci, R. Bilancioni, L. Baldratti, C. Capponcini, F. Turci and C. Feletti, *Clin. Nephrol.* **34**, 88–91 (1990).

22. B. Dworkin, S. Weseley, W. Rosenthal, E. Schwartz and L. Weiss, *Am. J. Med. Sci.* **293**, 6–12 (1987).

23. M. Bonomini, S.K. Mujais, P. Ivanovich and H. Klinkmann, *Nephron* **60**, 385–389 (1992).

24. G. Kallistratos, A. Evangelou, K. Seferiadis, P. Vezyraki and K. Barboutis, *Nephron* **41**, 217–222 (1985).

25. M.D. Saint Georges, D.J. Bonnefont, B.A. Bourely, M.C. Jaudon, P. Cereze, C. Gard, J.C. Chaumiel and C.L. d'Auzac. *Presse Médicale* **18**, 1195–1198 (1989).

26. K. Milly, L. Wit, C. Diskin, and R. Tulley, *Nephron* **61**, 139–144 (1992).

27. M. Kuroda, T. Imura, K. Morikawa and T. Hasegawa, *Trace Elem. Med.* **5**, 197–103 (1988).

28. G.H. Theil, C.E. Brodine and P.D. Doolan, *J. Lab. Clin. Med.* **58**, 736–742 (1961).

29. S.K. Mahajan, *J. Am. Coll. Nutr.* **8**, 296–304 (1989).

30. M. Taccone Galluci, O. Giardini, C. Ausiello, A. Piazza and D. Bandino, *Clin. Nephrol.* **25**, 81–86 (1986).

31. K. Ono, *Nephron* **40**, 440–445 (1985).

32. A.S. Yalçin, M. Yurtkuran, K. Dilek, A. Kilinç, Y. Taga and K. Emerk, *Clin. Chim. Acta.* **185**, 109–112 (1989).

33. E. Delacoux, Th. Evstigneff, M. Leclercq, M.C. Rettori, S. Delons, C. Nazet, and C. Blanchet-Bardon, *Clin. Chim. Acta* **137**, 283–289 (1984).

34. M. Praga, P.D. Rubio, J.M. Morales, F. Canizares, L.M. Ruilope, V. Gutierrez-Millet, J. Nieto and J.L. Rodicio, *Am. J. Nephrol.* **7**, 281–286 (1987).

32

PREVENTION OF ALUMINUM EXPOSURE IN HEMODIALYSIS PATIENTS

Results after the Application of European Guidelines in Aragon (Spain)

P. Nosti,[1] M. D. Zapatero,[2] M. L. Calvo,[1] A. García de Jalón,[1] and J. Escanero[3]

[1] Unidad de Nutrición y Metales
   Servicio de Bioquímica. Hospital Miguel Servet
   Zaragoza, España
[2] Hospital de Alcañiz
   Teruel, España
[3] Facultad de Medicina
   Universidad de Zaragoza, España

1. INTRODUCTION

In the Spanish Autonomous Community of Aragón, with about 1.200.000 inhabitants, there are currently 314 patients suffering from chronic renal failure (CRF). They are undergoing hemodialysis in 7 public hospitals. These centers centralize their controls of the aluminum (Al) levels in the patients' sera and in the fluids used for dialysis in the "Unidad de Metales del Servicio de Bioquímica" of the Miguel Servet Hospital.

Ever since Berlyne et al. published the first report about the probable toxic role played by Al in patients with CRF in 1970, this problem has remained of great importance. These patients are exposed to three different sources of Al intoxication (1): the diluting water used in hemodialysis and hemofiltration, the therapeutical administration of Al hydroxide $[Al(OH)_3]$ as a phosphate binder and the water intended for human consumption, to which Al sulphate is added as a flocculating agent.

In order to limit the risk of osteomalacia and encephalopathy due to Al toxicity in the increasing number of persons who require hemodialysis, the European Union has approved the Directive 86/C184/04, which determines the maximum permitted Al values in hemodialysis concentrates, solutions for peritoneal dialysis, solutions for hemofiltration, diluted hemodialysis solutions and diluting water. Furthermore, it recommends the periodicity of the controls of the Al content in water and in the different dialysis fluids, as well as in patients'

Therapeutic Uses of Trace Elements, edited by Nève et al.
Plenum Press, New York, 1996

serum (2). Another European Directive, (80/778/CEE), established the requirements of the quality of water for human consumption (3).

In the present study, we show the results of the aluminemia of patients undergoing hemodialysis in our region, individualized for each public hospital controlled in Aragón, and we try to associate the inter-center differences found with a possible causal factor.

2. MATERIAL AND METHODS

Al content was measured in sera from 249 patients suffering from CRF. They were undergoing hemodialysis in seven different public hospitals of Aragón, which periodically send us samples. For the purposes of this study we only considered the last measurement of Al of each patient until December 31, 1995. The centers carrying out the hemodialysis also send periodically samples of diluting water for hemodialysis, which were analyzed as well.

The analytical determinations were carried out by atomic absorption spectrometry (Perkin-Elmer 1100-B) equipped with a graphite furnace HGA 700 and an AS-70 autosampler, and a deuterium arc background corrector.

A venous blood sample was drawn with tubes for serum separation with SST gel and clot activator (Vacutainer L42911®). After clotting, tubes were spun for 10 minutes at 3000 r.p.m. until complete serum separation. Serum was immediately transferred to polystyrene tubes (Nunc cryotubes 363452®), to reduce as much as possible the contact of the sample with the glass. Tubes were stored at 4°C until analyzed. The Al measurement was always performed within 48 hours after the sample collection (4). Serum samples were diluted with nitric acid 2 %, and when properly mixed, placed in the autosampler. Dialysis fluids were directly analyzed. The Al measurements in the samples received in the laboratory are under the international quality control program, "Worldwide Interlaboratory Quality Control Aluminium", run by the Societé Française de Biologie Clinique.

A questionnaire was sent to each hospital to collect more information about the criteria applied to decide whether or not to administrate $Al(OH)_3$, the dosage and length of the treatment and the number of patients who received it.

3. RESULTS AND DISCUSSION

The "Unidad de Metales del Servicio de Bioquímica" of the Miguel Servet Hospital carries out aluminemia controls for about 79 percent of the patients in Aragón who are undergoing hemodialysis. Using as reference values the Al levels established in Annex II of the Resolution 86/C14/04 by the European Community Council, the overall results for the region showed on December 31, 1995 the following percent distribution: 91.16 % of patients had Al measurements in serum below 60 µg/l; 7.23 % had aluminemias greater than 60 µg/l and less than 100 µg/l and only 1.61 % exceeded 100 µg/l. By comparing these data with the results published in 1984 about the situation of 1122 hemodialysis patients in Spain (74.2 % < 100 µg/l; 14 % > 100 µg/l-< 200 µg/l and 11.8 % > 200 µg/l) (5), a marked improvement was noted (see Table 1). In Table 2 the total number of patients undergoing hemodialysis and the total number of patients being controlled for their Al level in serum for each public hospital in Aragón is showed. Sorted by unidentified public hospitals, Table 3 shows the percentages of patients included in each of the following subgroups: a) aluminemia < 60 µg/l, b) aluminemia > 60 µg/l-< 100 µg/l and c) aluminemia > 100 µg/l. According to the information received through the questionnaires, hospital 2, which has the highest percentage of patients

Table 1. Comparison of the percent distribution of hemodialysis patients tested for aluminum in serum in Spain in 1984 with patients tested in Aragón in 1995

Hemodialysis patients	< 100 µg/l	100 - 200 µg/l	> 200 µg/l
Spain (n = 1122)	74.2	14.0	11.8
Aragón (n = 249)	98.4	1.6	0.0

Table 2. Total number of patients undergoing dialysis and total number of patients controlled

Hospitals	Patients undergoing dialysis (n = 314)	Patients controlled (n = 249)
Alcañiz	19	19
Barbastro	22	18
Calatayud	25	25
Clínico Universitario *(Zaragoza)*	84	67
Miguel Servet *(Zaragoza)*	85	41
Obispo Polanco *(Teruel)*	37	37
San Jorge *(Huesca)*	42	42

Table 3. Percent distribution of patients tested for aluminum in serum

Hospital[§]	Aluminium measured in serum		
	< 60 µg/l	60 µg/l - 100 µg/l	> 100 µg/l
Hospital 1	72.97	27.03	0
Hospital 2	85.37	9.75	4.88
Hospital 3	95.24	2.38	2.38
Hospital 4	95.52	4.48	0
Hospital 5	96	0	4
Hospital 6	100	0	0
Hospital 7	100	0	0

[§] Numbers randomly assigned to the different hospitals in Aragón

with Al levels exceeding 100 µg/l, reported the highest doses of $Al(OH)_3$, the longest periods with this treatment and the highest percentage of patients treated as well. Conversely, hospital 7, of which no patient had a serum Al level over 60 µg/l, reported the most frequent control in sera and dialysis waters and the lowest dosage of $Al(OH)_3$.

The last measurement of Al content in diluting water of the different hospitals in 1995 was in no case greater than 1,5 µg/l (data from center 4 are lacking). We have to mention one measurement done during the last year in center 1, and which demonstrated an Al contamination problem in the water for human consumption (800 µg/l) and in the diluting water (27 µg/l) as well. Some patients monitored after this increased exposure to Al presented with higher serum Al levels than in previous controls.

4. CONCLUSIONS

In Aragón 91.2 % of the hemodialysis patients presented with aluminemia below 60 µg/l, the maximum recommended level by the European Directive of 1986. 98.4 % of the hemodialysis patients showed aluminemia below 100 µg/l. The highest value registered in our Laboratory is 168 µg/l. The aluminum concentration in dialysis fluids was not greater than 1.5 µg/l. The periodic controls of the serum Al in patients undergoing dialysis and of the dialysis fluids is absolutely necessary in order to find out possible abnormalities in the hemodialysis centers, regarding accidental contaminations and treatments with $Al(OH)_3$. Finally, 10 years after the publication of the European Directive regulating the Al levels in serum and in dialysis fluids, its efficacy has been proved in order to minimize the Al accumulation in patients undergoing hemodialysis.

5. ACKNOWLEDGMENTS

We thank Dr. Angel Calvo for his assistance in the elaboration of the poster. Our gratefulness is expressed to all nurses of the "Unidad de Nutrición y Metales", who contributed to this study.

6. REFERENCES

1. M. González. In *Alteraciones del Aluminio y otros elementos traza (Zn, Cu y Fe) en pacientes hemodializados*. Universidad de Zaragoza. 1992.
2. Council Directive of 16 June 1986 (86/C184/04) relating to the protection of patients undergoing hemodialysis through a maximum decrease to the aluminium exposure.
3. Council Directive of 15 June 1980 (80/778/CEE) relating to the quality of water intended for human consumption.
4. M.D. Zapatero. In *Niveles de Aluminio sérico en la población de Zaragoza: estudio transversal y factores relacionables*. Universidad de Zaragoza. 1995
5. -A. Berlin, S. Challah, N.H. Selwood, G. Mattiello and M. Lai. *Nephrologie*. **6**, 51–56 (1986).

33

USEFULNESS OF PLASMA ZINC PROTOPORPHYRIN (ZPP) DOSAGE IN THE ASSESSMENT OF IRON STATUS IN THE IRON DEFICIENT RAT

M. Boudey,[1] N. Aït-Oukhatar,[1] M. H. Read,[2] M. Mallet,[3] F. Bureau,[4] P. Arhan[1] and D. Bouglé[1]

[1] LPDN Laboratoire de Physiologie Digestive et Nutritionnelle
[2] Laboratoire de Pédiatrie
[3] Laboratoire d'Hématologie et de Cytologie
[4] Laboratoire de Biochimie A
CHU de Caen F14033 Caen, France

1. INTRODUCTION

Iron is a mineral frequently deficient in diets of industrialised countries (1). Assessment of iron status still requires the determination of several parameters of iron metabolism (2). Heme is formed in the developing erythrocytes by insertion of iron into a formed porphyrin ring. In the event of insufficient iron supply or impaired iron utilisation, zinc is substituted for iron into porphyrin IX. The Zinc Protoporphyrin (ZPP) formed in the chelation process is stable and remains in the red cell for its 120 days of life span. Because ZPP concentration increases in iron deficiency, its dosage has been proposed as an early index of iron deficiency using haematofluorometry (3).

This study compares the evolutions of ZPP, blood count and iron liver stores in young rats during iron deficiency and repletion.

2. METHODS

2.1. Animals and Diets

Male Sprague-Dawley rats were housed individually in custom-made plexiglas cages. Weaning rats (D group, n = 8), were fed an iron deficient diet (5 mg/kg, UAR, Villemoisson sur Orge, France) for 4 weeks, followed by a control diet (250 mg Fe/kg as $FeSO_4$) for 2 weeks. Two groups were fed the control diet for 6 weeks, either ad libitum (C group; n = 8), or pair-fed to the D group (PF group; n = 8). The constitution of a pair fed group was necessary because of a decreased food intake during iron deficiency such as previously described (4).

2.2. Methods

Blood was collected on EDTA, for blood count and ZPP determination, using retro-orbital sinus sampling method, at 4 and 6 weeks. At 6 weeks, rats were killed after blood collection, and liver was taken. Using acid digestion by microwave technique (Microdigest 301-Prolabo), 500 mg of desiccated liver sample were digested in 5 ml of 65 % nitric acid (Merck), for 10 min at 120 °C.

Blood count was obtained with a Coulter Counter (Coulter S890 - Coultronics). After washing erythrocytes, RBC-ZPP ratio (μmoles of ZPP per mole of heme) was measured in duplicate with an hematofluorometer (Protofluor, Héléna France). Liver iron concentration was measured by Atomic Absorption Spectrometry (AAS 1100B with Zeeman effect, Perkin-Elmer).

At each period, comparison between blood count value, liver iron and ZPP level were evaluated by analysis of variance. Paired t-test was used to compare blood count value associated with both period of study (4 and 6 weeks). At 6 weeks, iron liver stores within the whole rats were compared with others parameters by multiple regression .

3. RESULTS

Results are shown in Table 1. At 4 weeks, analysis of variance showed low RBC, haemoglobin, hematocrite, MCV, and high ZPP in the deficient D group as compared to C and PF groups (p < 0.001). After the repletion period, D group showed lower RBC, haemoglobin, hematocrite and total liver iron, than in C and PF groups (p < 0.001) but similar MCV and ZPP concentration. No correlation was observed between iron liver and either blood count or ZPP concentration, at 6 weeks in the three groups. For each parameter studied, paired t-tests were not significant between the two periods in control and pair fed groups, but were significant in depleted group (p < 0.001).

Table 1. Iron parameters in response of deficiency in growing rats (mean ± SD)

	4 weeks			Stat		6 weeks			Stat	
Groups	Control C	Deficient D	Pair-fed PF	C/D	PF/D	Control C	Deficient D	Pair-fed PF	C/D	PF/D
RBC 10^6/mm^3	6,6 ± 0,5	3,5 ± 0,6	7,1± 0,6	*	*	7,0 ± 0,4	6,2 ± 0,4	7,4 ± 0,5	*	*
Hb g/100ml	14,2 ± 0,8	5,6± 1,0	15,5± 0,8	*	*	15,3 ± 0,7	13,2± 0,9	15,2 ± 1,0	*	*
Ht%	39,6 ± 2,6	14,3 ± 2,4	42,8 ± 3,3	*	*	41,3 ± 2,4	37,7± 2,3	42,1± 3,0	*	*
MCV μ^3	56 ± 10	41± 1	60± 1	*	*	57± 6	59 ± 3	59± 2	ns	ns
ZPPμmol/mol Heme	53± 3	117± 7	66± 5	*	*	63± 19	78± 14	71± 8	ns	ns
Liver iron mg	-	-	-	-	-	1,77± 0,41	1,34± 0,34	1,97± 0,34	ns	*

RBC, Red Blood Cell; Hb, haemoglobin; Ht, haematocrite; MCV, mean corpuscular volume.
*Statistical analysis: ANOVA 95%: p < 0.001; ANOVA C/PF : ns

4. DISCUSSION

Data at 4 weeks confirmed a severe anaemia in iron deficient D group together with an increase of ZPP level. At 6 weeks, all indices progressed towards normal values but only ZPP concentration and MCV actually reached normal range ($p < 0.001$). ZPP measurement is an important parameter of iron deficiency, as an index of erythropoiesis (3–6). Especially in experimental studies in animal and when compared with immunological determinations such as serum ferritin and transferrin, this method is independent from species, accurate, rapid and reproducible, and has a low cost (5).

Our data show that evolution of ZPP is related to iron intake. Several authors have reported the same results (4–7). Combination of blood count and ZPP determination is a very interesting way to assess iron deficiency (8). But our results show that during repletion, there is no correlation between iron stores and parameters such as blood count and ZPP concentration. ZPP measurement has been proposed to document more precisely the iron deficiency even if the relation between iron stores and peripheral parameters was not always determined. In conclusion, ZPP was a sensitive index of iron intake, but did not seem to improve the assessment iron status.

5. REFERENCES

1. S. Hercberg, P. Preziosi, P. Galan, in *Les oligoéléments en médecine et en biologie*, ed., SFERETE Tech & Doc Lavoisier, pp. 233–346 (1991).
2. M. Worwood, in *Iron metabolism in healthy and disease*, ed., W Saunders Company Ltd, London, **14**, 450–463 (1994).
3. R.F. Labbe et al, *Clin. Chem.* **25**, 87–92 (1979).
4. A. Dhur, P. Galan, S. Hercberg, *Ann. Nutr. Metab.* **34**, 280–287 (1990).
5. J. Hastka, J.J. Lasserre, A. Shwarzbeck, and R. Hehlmann, *Clin. Chem.* **40**, 768–773 (1994).
6. E.E. Langer, R.G. Haining, R.F. Labbe, P. Jacobs, E.F. Crosby, and C.A. Finch, *Blood* **40**, 112–129 (1972).
7. J.R. Hunt, C.A. Zito, J. Erjavec, and L.K. Johnson, *Am. J. Clin. Nutr.* **59**, 413–8 (1994).
8. C. Hershko, A.M. Konijn, G. Link, J. Moreb, F. Grauer and E. Weissenberg, *Clin. Lab. Haemat.* **7**, 259–269 (1985).

34

PLASMA ANTIOXIDANT TRACE ELEMENT LEVELS AND RELATED METALLOENZYMES IN ALGERIAN WOMEN

Impact of Pregnancy

B. Lachili,[1] A. M. Roussel,[2] J. Arnaud,[2] M. J. Richard,[2] C. Benlatreche,[1] and A. Favier[2]

[1] Faculté de Médecine
Université de Batna
DZ-05000 Batna, Algérie
[2] GREPO
UFR de Pharmacie
F38700 La Tronche, France

1. INTRODUCTION

Since its independence, a real demographic burst has been observed in Algeria. The Algerian population which was 8,5 millions in 1962, is now reaching 29 millions of inhabitants. Although the Algerian Government favours the birth control, the fecondity index remains 4.4 per woman and 40 % of the population is younger than 15 y old (1). In these conditions, the biological and clinical survey of Algerian pregnant women represents an important public health challenge. In this study, we have focused on trace element status in Algerian pregnant and non pregnant women as no data were available concerning trace elements in this group of population, except few works about iron deficiency. However, numerous changes have been reported in micronutrients during pregnancy (2) associated to biological changes in mother, or increased needs due to the fetal growth and placental tissue constitution, and sometimes consequently to decreased dietary intakes. Thus, the risk of deficiency especially in antioxidant trace elements has to be considered in relation with a potential enhanced oxidative risk during pregnancy.

The aim of the present study was to measure plasma antioxidant trace element levels (Zn, Se, Cu) and related metalloenzymes (Cu-Zn SOD, Se-GPx) in healthy non pregnant Algerian women and to study the impact of pregnancy on these parameters in relation with oxidative stress.

Therapeutic Uses of Trace Elements, edited by Nève et al.
Plenum Press, New York, 1996

Table 1. Impact of pregnancy on plasma antioxidant trace element levels

	Group		
	I non pregnant	II first pregnancy	III multipares
Age	26.9 ± 7.7	24.5 ± 3.8	33.1 ± 5.2
Zn (µmol/l)	14.1 ± 3.1	$9.1^a \pm 2.2$	$9.1^b \pm 1.9$
Cu (µmol/l)	16.7 ± 4.5	$35.8^a \pm 6.1$	$34.1^b \pm 6.1$
Se (µmol/l)	1.4 ± 0.2	$1.6^a \pm 0.3$	$1.6^b \pm 0.3$

[a] p<0.005 between I and II; b: p<0.005 between I and III.

2. SUBJECTS AND METHODS

Four hundred and sixty women from rural and urban north east Algerian areas participated to the study. They were enrolled in medical centers (Batna, Mila, Constantine). Three groups were formed: group I (healthy non pregnant women, 18–47 y, n = 187, no contraception), group II (first pregnancy, 17–37 y, n = 70) and group III (2 to 6 pregnancies, 22–48 y, n = 203). Blood samples were obtained at delivery in groups II and III and at the begining of the menstrual cycle for group I.

Plasma zinc, copper, selenium (EAAS), and TBARs (fluorimetric method), erythrocyte SOD (Marklund method) and GPx (Gunzler method) were measured according to the usual techniques in our group.

All data were examined with on an IBM PC compatible computer using PCMS statistical software (Deltasoft, Meylan, France). Results were compared using the ANOVA test.

3. RESULTS

In non pregnant women, plasma zinc concentrations were elevated whereas copper levels remained in a low range. Fifty of women younger than 30 years old and 23 % older than 30 years exhibited plasma copper values below than 13,3 µmol/l.

In women at delivery, pregnancy resulted in a significant decrease in plasma zinc concentrations (p < 0.005). Selenium was enhanced and related to increased GPx activity. Plasma lipid peroxidation monitored by TBARs was dramatically higher at delivery than in group I (p < 0.005) while, despite an elevated cupremia, Cu-Zn SOD was low (p < 0.005) (Table 2). However, an aggravating impact of the number of pregnancies on these parameters was not observed.

Table 2. Impact of pregnancy on oxidative stress indexes

	Group		
	I non pregnant	II first pregnancy	III multipares
SOD (U/g Hb)	1.20 ± 0.19	$1.01^a \pm 0.12$	$1.04^b \pm 0.14$
GPx (U/g Hb)	41.7 ± 9.4	$51.5^a \pm 13.2$	$53.1^b \pm 9.1$
MDA (µmol/l)	2.4 ± 0.3	$3.1^a \pm 0.5$	$3.1^b \pm 0.4$
Vit E/Lipids	5.07 ± 1.30	$3.77^a \pm 0.90$	$3.78^b \pm 0.90$

[a] p<0.005 between I and II; b: p<0.005 between I and III.

4. DISCUSSION

The main objective of this work was to contribute to the knowledge of trace element status in Algerian women before and during pregnancy. Compared to levels in young European women as reported by the French Val de Marne Study (3), plasma zinc and selenium concentrations were high in non pregnant women and none of these subjects exhibited values below 10.7 µmol/l for zinc and 0.80 µmol/l for selenium. For this last element, these data might be due to the dietary habits of the population consuming natural foods rich in cereals, fruit and vegetables from selenium rich soils. Similar results were recently reported in Zairian mothers (4).

At delivery, the increase in cupremia was expected and already reported by many authors as an adaptative process during pregnancy in relation with ceruloplasmin increase. Whereas an adequate zinc status was observed before pregnancy, plasma zinc levels significantly decreased in pregnant wowen. This fall was previously described (5,6,7) and seems to be due to increased needs during pregnancy. Surprisingly, the number of pregnancies did not affect the zinc status. However, the decreased Cu-Zn SOD activity might be in relation with the impaired zinc status since zinc is well known to play a structural role in SOD stabilization and efficiency (8). The results concerning the increase in selenium and red cell glutathione peroxidase were interesting. In agreement with previous study (9), the significant higher enzyme activity might be an adaptative response to lipid peroxidation during pregnancy. However, these results are controversial as others showed a decreased selenium status in pregnant woman (10). A raise in lipid peroxidation was recently documented during pregnancy but the exact significance of such an increase is still unknown (11). The biological sources of peroxides could be related to the observed elevation in circulating lipids during pregnancy or with a placental polyunsaturated fatty acid peroxidation. Our data are consistent with others and highlight the need of taking into account the antioxidant status during pregnancy. With respect to the oxidative risk, the pregnant women participating to the study seem to be partly protected by their high selenium status. This beneficial effect should be complemented by restauring zinc status.

5. REFERENCES

1. Ministère de la Santé et de la Population.Office National des Statistiques, Ligue des Etats Arabes. 1994, Alger.ISSN 0184 77 83.
2. D. Bouglé, A. Favier, F. Bureau and P. Walravens, in *Les oligoéléments en Nutrition et en Thérapeutique*,Tec et Doc Lavoisier, Paris, pp. 73–94 (1995).
3. S. Hercberg, P. Preziosi, P. Galan, M. Deheeger, L. Papoz, H. Dupin, *Rev. Epidem. Santé Publ.* **39**, 245–261 (1991).
4. J. Arnaud, P. Preziosi, L. Mashako, P. Galan, C. Nsibu, A. Favier, C. Kapongo and S. Hercberg, *Eur. J. Clin. Nutr.*, **48**, 341–348 (1994).
5. S. Jameson, *Ann. N. Y. Acad Sci.***15**, 178–198 (1993).
6. A. Favier, *Rev. Prat.*, **43**, 146–151 (1993).
7. M. Hambidge, M. Hackshaw, N. Wald, *Br. J. Obstet. Gynecol.*, **100**, 746–749. (1993).
8. T.M. Bray and W.J. Bettger, *Free Radic. Biol. Med.*, **8**, 281–291 (1990).
9. J. Uotila, R. Tuimala, T. Aamio, K. Pykko and M. Ahotupa, *Eur. J. Obstet. Gynecol. Reprod. Biol.*, **42** ,95–100 (1991).
10. S.T. Davidge, C.A. Hubel, R.D. Brayden, E.L. Capelen and M.K. McLaughlin, *Obstet. Gynecol.*, **79**, 897–901 (1992).
11. D. Carone, G. Loverro, P. Greco, F. Capuano, L. Selvaggi, *Eur. J. Obstet. Gynecol. Reprod Biol.* **51**, 103–109 (1993).

35

IMPAIRED ZINC AND COPPER STATUS AND ALTERED FATTY ACID CELL MEMBRANE COMPOSITION IN ESSENTIAL HYPERTENSION

C. Russo, O. Olivieri, D. Girelli, M. Azzini, P. Guarini, A. M. Stanzial, S. Friso, R. Pasqualini, and R. Corrocher

Institute of Internal Medicine
University of Verona
Verona, Italy

1. INTRODUCTION

The apparent antagonism between Zinc (Zn) and Copper (Cu) is a well-known phenomenon and has been documented in various chronic diseases. An impairment in Zn and Cu status has been shown in hypertensive animals (1), suggesting a possible relationship with the pathogenesis of hypertension. Zn and Cu are involved in lipid metabolism, and several studies indicated that a major aspect of Zn-Cu interaction may be related to the opposing effects on essential fatty acids (EFA) metabolism or by differential stimulation of the synthesis of prostaglandins (PGs), whose influence on blood pressure (BP) regulation has been widely documented (2). The conversion of C-18 polyunsaturated fatty acids (PUFA) into the longer-chain metabolites proceeds through desaturation and elongation steps; desaturation processes are catalyzed by the $\Delta 4$-$\Delta 5$-$\Delta 6$ desaturases, which are the rate-limiting enzymes in the pathway, and whose activities specifically require Zn and Cu (2). Thus, Zn and Cu availability may modulate the balance of PGs precursors. Also, since C-20 and C-22 are the major PUFA found in phospholipids, most of the cellular functions (fluidity, permeability or the ion transport systems activity) are in some way related to their metabolism (3).

Several membrane abnormalities have been described in EH, so that a "membrane hypothesis" for its pathogenesis has been purposed; abnormal desaturase activity could impair membrane phospholipid composition and function. Objective of this study was to evaluate plasma Cu and Zn status and red blood cell (RBC) membrane fatty acid profile in essential hypertensive (EH) patients compared with normal subjects. Particular attention was paid to C20:4/C18:2, C20:5/C18:3 and C22:6/C20:5 ratios, which are usually considered expressions of the desaturase enzymes activity (4); Cu-dependent Superoxide Dismutase (SOD) was also measured, as index of Cu status.

Therapeutic Uses of Trace Elements, edited by Nève et al.
Plenum Press, New York, 1996

2. MATERIALS AND METHODS

2.1. Patients

Two groups of 105 EH patients (48 ± 16 years old) and 100 normotensive controls matched for age were selected. EH were recruited among the patients referred to our Hypertension Outpatient Clinic, who had a diastolic BP > 95 mm Hg at 3 visits over 3 months, without any treatment. Patients with secondary hypertension (assessed by clinical, biochemical and instrumental tests) were excluded. Normal subject were selected as previously described (5). In order to avoid confounding effects by any pathological process on the parameters studied, subjects recognized to be affected by any chronic diseases or acute intercurrent illness were excluded. Pregnant women and subjects taking any medication including the contraceptive pill or vitamin preparations, or on particular diets were also excluded.

2.2. RBC Membrane Fatty Acid Analysis

Blood samples (15 ml) were collected after overnight fasting by EDTA-containing vacutainer tubes. RBC were separated by centrifugation at 1000 x g for 15 min (4 °C), the buffy coat was removed and the RBC were washed three times with 154 mmol/l NaCl. Analysis of RBC membrane (250 μl of packed RBC haemolyzed in an equal volume of double-distilled water) fatty acids was performed on total lipids extracted with 4.5 ml of isopropanol/chloroform (11/7, v/v) containing 0.45 mmol/l 2,6-di-ter-p-cresol (BHT) as antioxidant. A gas-chromatographic (Hewlett Packard 5980 chromatograph, Hewlett Packard, Palo Alto CA, USA) method, based on the fatty acid direct transesterification technique, was employed as previously described (6). Analyses were performed in duplicate on each sample. Peak identification and quantification were made with commercially available reference fatty acids (Sigma, St. Louis MO, USA). The areas of the peaks were measured using an automatic plotter-integrator (Hewlett Packard 3392 A). Fatty acid composition data were expressed as g/100 g fatty acid methyl esters. Fatty acids from C 12:0 to C26:0 were measured, unidentified peaks accounting for < 0.5 % of the total.

2.3. Other Parameters

Serum concentration of copper and zinc were determined by flame atomic absorption spectrometry (FAAS) using a Perkin Elmer 372 double-beam spectrophotometer equipped with an air-acetylene flame burner and hollow cathode lamp. All analyses were performed in duplicate. The coefficient of variation for both Cu and Zn was < 5 %. The reference values of our laboratory for zinc and copper were 94 ± 12 μg/dl and 111 ± 19 μg/dl respectively.

Erythrocyte Superoxide Dismutase activity was measured according to the method of Misra and Fridovich (7), which uses the conversion of epinephrine to a coloured product by an autooxidation reaction, inhibited by SOD. The coloured product is measured spectrophotometrically at 480 nm at 25 °C. The enzyme activity is expressed as IU/g Hb. The reference values of our laboratory are 8157 ± 1048 IU/g Hb (5).

Statistical analysis was carried out with the aid of an Apple Macintosh SE/30 computer, using the Systat 5.2.1 program. Differences between the groups were tested by Student's t-test; simple correlations were determined by Pearson's correlation coefficient. Significant values were considered when p < 0.05.

Table 1. Significantly different parameters between essential hypertensive (EH) patients and normal subjects (N)

Parameters	EH n = 105	N n = 100	p *
Zinc (mg/dl)	114 ± 30	97 ± 13	< 0.001
Copper (mg/dl)	100 ± 22	108 ± 15	< 0.01
SOD (IU/g Hb)	7208 ± 2070	8343 ± 2323	< 0.005
RBC Chol (mmol/l cell)	3.2 ± 0.3	3.4 ± 0.3	< 0.05
RBC fatty acid profile (g/100g fatty acid methyl esters)			
C 12:0	0.04 ± 0.02	0.06 ± 0.06	< 0.01
C 14:0	0.3 ± 0.1	0.4 ± 0.1	< 0.001
C 16:0	22.6 ± 1.2	22.4 ± 0.8	NS
C 18:0	17.3 ± 0.9	17.5 ± 0.8	NS
C 18:1	14.6 ± 1.3	14.5 ± 1.2	NS
C 18:2 ω-6	8.8 ± 1.0	10.5 ± 1.4	< 0.001
C 18:3 ω-3	0.08 ± 0.02	0.13 ± 0.06	< 0.001
C 20:0	0.5 ± 0.1	0.5 ± 0.1	NS
C 20:4 ω-6	20.4 ± 1.8	19.9 ± 1.4	< 0.05
C 20:5 ω-3	0.6 ± 0.2	0.7 ± 0.2	NS
C 22:0	1.8 ± 0.2	1.8 ± 0.3	NS
C 22:6 ω-3	6.4 ± 1.2	5.7 ± 1.1	< 0.001
C 24:0	5.2 ± 0.6	4.9 ± 0.7	0.005
C 26:0	0.3 ± 0.1	0.3 ± 0.1	NS
SFA	47.9 ± 2.9	47.9 ± 1.2	NS
MUFA	15 ± 1.4	14.8 ± 1.2	NS
PUFA	36.8 ± 1.5	37.2 ± 1.4	NS
C 20:4 ω-6/C 18:2 ω-6	2.3 ± 0.4	1.9 ± 0.3	< 0.001
C 20:5 ω-3/C 18:3 ω-3	10.0 ± 7.6	5.6 ± 2.3	< 0.001
C 22:6 ω-3/C 20:5 ω-3	9.8 ± 2.5	8.5 ± 2.3	< 0.001

Means ± SD; *= Student t-test.

3. RESULTS

Table 1 and Table 2 summarize the results of the study. EH patients showed lower Cu and higher Zn concentrations and lower SOD activity. With respect to lipid analysis, EH showed a decrease in RBC membrane cholesterol, C18:2 ω-6 and C18:3 ω-3, and an increase in C20:4, C22:6, C20:4/C18:2 ratio, C20:5/C18:3 ratio and C22:6/C20:5 ratio, compared with controls (Table 1). Plasma Zn was significantly and negatively correlated with C18:2 (r = - 0.220) and C18:3 (r = - 0.268) and positively correlated with C20:4/C18:2 ratio (r = 0.231) and C20:5/C18:3 ratio (r = 0.289) (Table 2).

Table 2. Simple correlations of zinc with some of the parameters examined

Parameter	r	p *
C 18:2 ω-6	- 0.220	< 0.05
C 18:3 ω-3	- 0.268	< 0.05
C 20:4 ω-6/C 18:2 ω-6	0.231	< 0.05
C 20:5 ω-3/C 18:3 ω-3	0.289	< 0.05

Means ±SD; *= Pearson's correlation coefficient.

4. DISCUSSION

In our study EH patients exhibit a different cell membrane lipid profile, Zn and Cu status and SOD activity compared with normotensive subjects. To achieve reliable results, particular care was paid to some methological aspects. Very strict criteria were adopted to include only healthy subjects in the control group, so avoiding all the possible pathological influences on the parameters tested. The study was performed on free-living conditions, but people taking preparation containing vitamins or other trace elements were not admitted. Malnutrition could be excluded on the basis of normal clinical and biochemical evaluations, such as body mass index, hemoglobin, serum albumin, etc. Finally, our population was large enough to make reliable the results obtained.

EH has been associated with various functional membrane abnormalities (8–10), suggesting that an intrinsic structural abnormality of cell membranes components may be a key etiological factor. We demonstrated that desaturase activity and long-chain PGs precursors are increased in EH. Arachidonic acid (AA: C 20:4) and docosahexaenoic acid (DHA: C 20:5), which are the major polyunsaturated fatty acids in phospholipids, derived from the precursors linoleic acid (C 18:2 ω-6) and α-linolenic acid (C 18:3 ω3) respectively (2–3). These precursors are metabolized through desaturation-elongation process but desaturase is the main rate-limiting step for this pathway. Δ6 desaturase converts C 18:2 ω-6 to dihomo-γ-linolenic acid (DGLA: C 18:3 ω-6) and C 18:3 ω3 to stearidonic acid; Δ5 desaturase is responsible for the biosynthesis of C 20:4 from DGLA and eicosapentaenoic acid (EPA: C 20:5) from stearidonic acid, which are the major eicosanoid precursors of PGs series 2 and 3. Δ5 desaturase also controls the balance between eicosanoids of series 1 (derived from DGLA) and 2, which have opposite effects on BP. Finally, Δ4 desaturase converts EPA into DHA (2–3). Linoleic and α-linolenic acid are "essential"fatty acids, depending exclusively on dietary intake. Since they cannot be synthesized in vivo, and are metabolized only through the above mentioned process, the C20:4/C18:2, C20:5/C18:3 and C22:6/C20:5 ratios are considered expression of the desaturase-elongase activity (4). Erythrocytes lack desaturase and elongase enzymes, thus RBC membrane fatty acids are representative of the hepatic metabolism. In liver all these enzymes are bound to the microsomial lipid bilayer and require Zn to express their activity. The specific mechanism is not well known, but it has been proposed that Zn bound to the enzymes could influence their configuration. Cu is known to influence Zn levels with antagonistic effect; by this way it can also affect desaturase activity (11). All these parameters were examined in our study. Previous similar works were generally conducted on animals (12–13) and only few data exist on cellular membrane structure and function in EH patients. Two groups have previously demonstrated a decrease in linoleic acid (14) associated with an increase in arachidonic acid (AA) (15), in platelets and erythrocyte membranes respectively of EH patients; nevertheless, the relevance of these preliminary observations was limited by the small sample size of the groups examined. Our data confirm and extend these results, showing higher C20:4/C18:2, C20:5/C18:3 and C22:6/C20:5 ratios in RBC membranes of EH. Considering fatty acids metabolism, this may indicate a more active conversion of linoleic and linolenic acid into their metabolites. This hypothesis is supported also by the differences found in trace elements levels. EH showed higher levels of Zn and lower levels of Cu, as expected on the basis of an increased desaturase activity; the latter data was confirmed by the lower SOD, which is Cu-dependent and thus represent an index of Cu status (1). The positive correlations between Zn and C20:4/C18:2 and C20:5/C18:3 ratios are also in agreement with the mentioned hypothesis. Another possibility is that eicosanoid precursors are preferentially incorporated and retained in membrane phospholipids: this is consistent with the inibitory effect of Zn on phospholipase A$_2$ activity (2–11). Although recent evidence has been provided that Δ5 and Δ6 (and probably Δ4) activity is higher in Spontaneously Hypertensive (SHR) than in normotensive control rats (4), our study first reports similar results in a large sample of humans. This data supports the concept that in EH cell membrane presents an intrinsic structural abnormality, thereby potentially modifying related cellular functions. Whether these observations are of relevance in EH pathogenesis, remains to be confirmed by further appropriate studies.

5. REFERENCES

1. G.D. Allen and L.M. Klevay, *Current Opinion in Lipidology* **5**, 22–28. (1994).
2. S.C. Cunnane, *Prog. Lipid Res.* **21**, 73–90 (1982).
3. D.F. Horrobin, *Semin. Thromb. Hemostasis* **19**, 129–137 (1993).
4. M. Narce, P. Asdrubal, M.C. Delachambre, E. Vericel, M. Lagarde, J.P. Poisson, *Mol. Cell. Biochem.* **141**, 9–13 (1994).
5. O. Olivieri, D. Girelli, A.M. Stanzial et al, *Am. J. Clin. Nutr.* **60**, 510–517 (1994).
6. D. Girelli, M. Azzini, O. Olivieri et al, *Clin. Chim. Acta* **211**, 155–166 (1992).
7. A. Misra and P. Fridovich, *J. Biol. Chem.* **247**, 3170–3175 (1972).
8. M. Canessa, N. Adragna H. Solomon, T.M. Connolly and D.C. Tosteson, *N. Engl. J. Med.* **302**, 772–778 (1980).
9. R.P. Garay, G. Dagher, M.G. Pernollet and M.A. Devynck, *Nature* **284**, 281–283 (1980).
10. G. Bianchi, P. Ferrari , D. Trizio et al, *Hypertension* **7**, 319–325 (1985).
11. S.C. Cunnane, *Prog. Food Nutr. Sci.* **12**, 151–188 (1988).
12. J.J. Strain, *Proc. Nutr. Soc.* **53**, 583–598 (1994).
13. A.F. Dominiczak, Y. McLaren, J.R. Kusel et al, *Am. J. Hypertens.* **6**, 1003–1008 (1993).
14. A.J. Naftilan, V.Y. Dzau, J. Loscalzo, *Hypertension* **8** (suppl II), S19-S24 (1986).
15. J.D. Ollerenshaw, A.M. Heagerty, R.F. Bing and J.D. Swales, *J. Human Hypertens.* **1**, 9–12 (1987).

36

PHARMACOLOGICAL USES OF ZINC AND OTHER TRACE ELEMENTS IN DERMATOLOGY

M. T. Leccia

Service De Dermatologie
C.H.U.
A. Michallon, 38043 Grenoble Cédex 08, France

1. INTRODUCTION

Zinc and four other essential trace elements, copper, manganese, selenium and silicon, exert well known biological functions in the skin. From a purely theorical point of view, these trace elements play interesting structural and/or antioxidant roles in cutaneous cells. However, except for zinc on account of true deficiencies corrected by supplementation, there are few data on possible pharmacological uses of these compounds. It is important to note that skin concentrations of trace elements vary from one person to another and in a same person from one area to another. The epidermis levels for all trace elements are higher than the dermis levels. Futhermore, levels and variations of trace elements in the skin and in the plasma can be independent: plasma levels can only be a partial indication of skin reserves. So, it is very difficult to assess the exact role of a possible cutaneous deficit in trace elements when clinical presentations are suggestive whereas plasma levels are in normal range.

2. EXPERIMENTAL DATA

2.1. Structural Role

In the epidermis, copper, selenium, silicon and zinc participate in the process of keratinization by their capacity to bind sulfhydril groups (1). Keratohyalin grains, rich in histidine, bind metals such as zinc and copper. One clinical illustration of these activities is represented by the pili torti which correspond to hairs showing repetitive twisting present in the Menkes syndrome, a congenital copper deficiency. Zinc as a cofactor of DNA and RNA polymerases is essential for cell divisions and protein synthesis and thus for the normal growth of the epidermis. Defects of healing in zinc deficiency attest of the importance of these functions.

In the dermis, copper, manganese, silicon and zinc are cofactors of key enzymes involved in the synthesis of collagen, elastin fibers, proteoglycans and glycoaminoglycans, components of the extracellular matrix. Copper and zinc are essential cofactors of lysyl-oxi-

Therapeutic Uses of Trace Elements, edited by Nève et al.
Plenum Press, New York, 1996

dases which participate in collagen and elastin synthesis by oxydative desamination of lysine and hydroxylysine residues to tropoelastin and tropocollagen. Like in the epidermis, zinc via metalloenzymes stimulates fibroblasts growth. Manganese regulates the activity of the galactosyl- and glucosyl-transferases for collagen synthesis. With ascorbate, this trace element regulates the activity of prolidase, an enzyme that degrades collagen fibers and regulates collagen levels in the dermis. Silicon is a cofactor of prolyl-hydroxylases in conjonctive tissues. Above all, this element binds to macromolecules such as components of the extracellular matrix allowing their stabilization and their protection against oxydation. Each collagen chain links 3 to 6 silicon atoms.

Copper and zinc are concerned by two more specific activities. Copper is an essential cofactor of tyrosinase, a key enzyme for melanin synthesis, that catalyzes the conversion of tyrosine to dihydroxyphenylalanine. In this way, this element participates in the natural photoprotection of the skin. Zinc, but also copper, can inhibit the activity of type I 5-alpha-reductase, an enzyme present in the skin that catalyzes the reduction of the testosterone into dihydrotestosterone. These two elements have interesting effects in androgen-related pathologies of human skin, already partly demonstrated for zinc.

2.2. Antioxidant Role

The antioxidant role of selenium is linked to glutathione peroxidases, enzymes present in cytoplasm and mitochondria that eliminate hydrogen peroxide and reduce the formation of a large number of hydroperoxides. The addition of selenium in culture medium protects cultured skin cells against ultraviolet radiations (2). When selenium is supplied per os (3) or by percutaneous application (4) in mice, lipid peroxidation is decreased and skin tumors appearance is delayed. In humans, Burke et al. have investigated the protective effect of different concentrations of topical L-selenomethionine on the minimal erythema dose after 15 days of application (5): the active preparation increased the minimal erythema dose (MED) with a dose-dependence (maximal protection for the concentration of 0.05% of L-selenomethionine). In another study, authors have evaluated the efficacy of an oral administration of selenium and copper associated with a vitamin complex on minimal erythema dose and sunburn cell formation in human skin (6). They showed a decrease of sunburn cells only for suberythemal doses without any effect on the MED.

The antioxidant role for copper, manganese, silicon and zinc is only theoretical (1). Copper and zinc are cofactors of the copper-zinc superoxide dismutase, the catalytic activity of the enzyme being essentially governed by copper. Zinc inhibits free radical production and stabilize cell membranes against lipid peroxidation. Zinc stimulates the transcription of the gene of metallothioneins, small proteins with antioxidant activity. Manganese eliminates superoxide anion through the manganese superoxide dismutase. The element could also directly scavenge free radicals. The antioxidant role of silicon by protection against lipid peroxidation has been demonstrated in endothelial cells. It could exert similar properties in the extracellular matrix of the dermis.

3. PHARMACOLOGICAL APPLICATIONS

3.1. Zinc

Zinc salts were used with spectacular clinical results in true deficiency and in an empirical way in several inflammatory cutaneous pathologies.

3.1.1. Zinc Deficiency Diseases. Zinc deficiency was first described in Acrodermatitis Enteropathica, an autosomal recessive disorder with impaired absorption of the trace element from the gastrointestinal tract due to a deficit of the zinc-binding ligand (7,8). The evocative cutaneous symptoms appear in the first months of life after the child is weaned from mother's milk. More recently, acquired zinc deficiencies were reported in premature and in healthy full-term breast-fed infants (9,10). In these cases, the deficit could be associated with a low maternal breast milk zinc concentration. Premature infants are especially susceptible because they have low zinc stores and a decreased intestinal absorption. Dermatological manifestations include pustular and bullous dermatitis with acral and perioral distribution, pustular paronychia, angular stomatitis and glossitis. In some cases, a generalized alopecia can be present. Periorificial lesions are often surinfected such as for impetigo. Similar pictures can been observed in malabsorption syndromes, inflammatory bowel diseases, alcohol-dependent persons, glucagonoma and jejunoileal bypass, affections with low zinc levels (11). In these patients erythematosquamous eruption, particularly localized on the inferior members, are apparent. Low zinc levels with typical cutaneous lesions were also described in total parenteral nutrition (12). In all these cases, parenteral or oral zinc treatment led to a spectacular regression of cutaneous symptoms in few days. Moynahan in 1974 (8) and Kay in 1976 (13) were the first authors to show the efficacy of zinc in Acrodermatitis Enteropathica and in total parenteral nutrition respectively. Zinc gluconate is used at mean doses of 200 mg daily, the length of treatment depending on the involved affection.

3.1.2. Empirical Uses. Zinc salts, especially zinc gluconate, has been used in acne since the first description of Michaëlsson (14). Three months of treatment with 30 mg of zinc metal daily (200 mg of zinc gluconate) are necessary to obtain a significant decrease of inflammatory components such as papules and pustules. The exact mechanisms for the effects of zinc in acne are not completly understood. This element seems to act at four levels. B. Dreno et al. showed that zinc treatment (200 mg/day as zinc gluconate) significantly decreases granulocyte zinc level associated with an inhibition of polymorphonuclear leukocyte chemotaxis (15). Clinical response to zinc therapy is directly correlated to a decrease in granulocyte zinc concentrations with concomittent plasma zinc increase. Studies also demonstrated an inhibitory effect of zinc on type I 5-alpha-reductase, an enzyme involved in androgen-dependent skin pathologies which causes the reduction of testosterone into dihydrotestosterone (16,17). The third target of zinc salts could be keratinocytes: a recent study effectively showed that zinc reduces keratinocyte activation state suggesting that this action is involved in the antiinflammatory effects of zinc treatment (18). Finally, studies showed that zinc participates in the eradication of erythromycin-resistant propionibacteria (19). Zinc is also used as zinc acetate in association with topical solution of erythromycin to treat acne; the preparation is probably active through a antibacterial effect on propionibacterium acnes. Topical preparations with zinc and copper salts are used in order to decrease sebum secretion.

In spite of contradictory results, some authors described efficiency of oral zinc treatment in pathologies such as pustular psoriasis and some microbial dermatitis, when plasma zinc levels are initially low (20). In cutaneous diseases such as atopic eczema, dermatitis herpetiformis, seborrheic dermatitis, Darier's disease, results of skin and plasma zinc measurements are divergent. G. Michaëlsson showed a significant decrease in epidermal and/or dermal zinc contents in dermatitis herpetiformis, acne, psoriasis and Darier's disease, suggesting that these patients have a moderate but chronic zinc deficiency which may negatively influence the clinical course (21). Data have up to now shown that there is no correlation between skin and plasma levels, and that zinc depletion seems to be independant of the extent of the disease. However, zinc therapies have sometimes been proposed and some beneficial effects were observed.

As previously shown, zinc is necessary for migration, proliferation and maturation of the epidermis and for maintaining cutaneous integrity by fibroblasts proliferation and collagen synthesis. So, zinc has been used from the Ancient Egyptians as a skin protectant and as a stimulant of wound healing. This is precisely what justifies its presence in local preparations as zinc oxide or zinc sulfate. The element could be interesting for healing of leg ulcers when serum zinc levels are low (22).

Finally, dry and brittle hairs in zinc deficiency also demonstrate the importance of the element in keratinization. J. Arnaud et al. found no modification in zinc status in telogen defluvium, which confirms previous results in alopecia areata (23,24).

These reports demonstrate contradictory results on skin and serum zinc status in dermatological diseases and the efficiency of zinc therapies. It is however important to remember that marginal zinc deficiency, with atypic clinical signs, may exist in clinical conditions associated with decreased absorption or increased losses of zinc. On the other hand, serum zinc concentrations can be decreased by numerous non-specific factors such as inflammatory reactions or medications, without any cutaneous zinc deficit.

3.2. Copper

The importance of copper in keratinization is illustrated by hair abnormalies called pili torti presented by children affected by the Menkes' syndrome, a X-linked recessive disorder (11). Topical preparations with copper salts have been used for a long time for their antiseptic and antimicrobial properties (copper gluconate or sulfate). As previously shown for zinc, copper may have inhibitory effects on the 5-alpha-reductase present in skin (17). It is thus associated with zinc to treat hyperseborrhoea (25).

3.3. Selenium

Selenium and glutathione peroxidase activity were assessed in skin diseases. Some authors showed a decrease of blood enzyme activity in acne (26), psoriasis (27), atopic dermatitis and dermatitis herpetiformis (28). C. Deffuant et al. showed a significant decrease of plasma selenium levels in cutaneous lymphoma and melanoma, correlated with the severity of the disease (29). It is not known whether the deficit participates in the extent of the disease or is a consequence. Oral selenium treatment corrected the low glutathione peroxidase activity but had no effect on the extent of skin lesions. Controlled studies are needed to evaluate the exact role of this element in cutaneous pathologies. Selenium-containing preparations are also used in seborrheic dermatitis, seborrheic scalp disorders, pityriasis versicolor in order to eradicate Malassezia furfur, the responsable agent of these disorders. Finally, selenium is used in cosmetologic topical preparations for skin aging on the basis of its antioxidant properties.

3.4. Manganese and Silicon

Manganese gluconate is used in association with copper gluconate to treat breast cracks, with two daily applications during a few days (30). Silicon is found in hair lotions to treat hair loss. Silicon is also used as silanols in cosmetology. This element is recommended for skin aging, stretch marks or cellulite. Most of these indications are poorly documented.

4. CONCLUSIONS

Despite interesting investigations on the role of several trace elements in the skin, there are in fact very few data on their biological actions and on their potential or effective applications in dermatology. Further experimental work is necessary to demonstrate the exact role of these molecules in human skin physiology and clinical controlled studies remain to be done to confirm their possible use in cutaneous pathologies.

5. REFERENCES

1. M.T. Leccia, in *Les oligoéléments en nutrition et en thérapeutique*, P. Chappuis, A. Favier, ed., Lavoisier, France, pp. 355–362 (1995).
2. M.T. Leccia, M.J. Richard, J.C. Béani, H. Faure, A.M. Monjo, J. Cadet, P. Amblard, A. Favier, *Photochem. Photobiol.* **58**, 548–553 (1993).
3. E.B. Thorling, K. Overvad, P. Bjerring, *Acta Path. Microbiol. Immunol. Scand. Sect.A.* **91**, 81–83 (1983).
4. R. Cadi, J.C. Béani, M.J. Richard, A. Richard, S. Bélanger, A. Favier, P. Amblard, *Nouvel. Dermatol.* **10**, 266–272, (1991).
5. K.E. Burke, R.G. Bedford, G.F. Combs, I.W. French, D.R. Skeffington, *Photodermatol. Photoimmunol. Photomed.* **9**, 52–57 (1992).
6. G. La Ruche, J.P. Césarini, *Photodermatol. Photoimmunol. Photomed.* **8**, 232–235 (1991).
7. E.J. Moynahan, *Lancet* **2**, 399–400 (1974).
8. N. Danbolt, K. Closs, *Acta Derm. Venereol.* **23**, 127–169 (1942).
9. C.S. Munro, C. Lazaro, C.M. Lawrence, *Br. J. Dermatol.* **121**, 773–778 (1989).
10. L.J. Roberts, C.F. Shadwick, P.R. Bergstresser, *J. Am. Acad. Dermatol.* **16**, 301–304 (1987).
11. S.J. Miller, *J. Am. Acad. Dermatol.* **21**, 1–30 (1989).
12. J. Chevrant-Breton, B. Dreno, *Ann. Dermatol. Venereol.* **106**, 141–150 (1979).
13. R.G. Kay, C. Tasman-Jones, J. Pybus, R. Whiting, H.A. Balck, *Ann. Surg.* **183**, 331–333 (1976).
14. G. Michaëlsson, L. Juhlin, K. Ljunghall, *Arch. Dermatol.* **13**, 31–36 (1977).
15. B. Dreno, M. Trossaert, H.L. Boiteau, P. Litoux, *Acta Derm. Venereol.* **72**, 250–252 (1992).
16. D. Stamatiadis, M.C. Bulteau-Portois, I. Mowszowic, *Br. J. Dermatol.* **119**, 627–632 (1988).
17. Y. Sugimoto, I. Lopez-Solache, F. Labrie, Y. Luu-The, *J. Invest. Dermatol.* **104**, 775–778 (1995).
18. A. Guéniche, J. Viac, G. Lizard, M. Charveron, D. Schmitt, *Acta Derm. Venereol.* **75**, 19–23 (1995).
19. R.A. Bojar, E.A. Eady, C.E. Jones, W.J. Cunliffe, K.T. Holland, *Br. J. Dermatol.* **130**, 329–336 (1994).
20. B. Dreno, M.A. Vandermeeren, J.F. Stalder, H.L. Boiteau, H. Barriere, *Dermatologica* **173**, 209–212 (1986).
21. G. Michaëlsson, K. Ljunghall, *Acta Derm. Venereol.* **70**, 304–308 (1990).
22. A. Phillips, D. Maureen, *Clin. Exp. Dermatol.* **2**, 395 (1977).
23. J. Arnaud, J.C. Béani, A. Favier, P. Amblard, *Acta Derm. Venereol.* **75**, 248–249 (1995).
24. Momcilovic B., Handl S., Oremovic L., Zjacic-Rotkvic V, in *Trace elements in man and animals*, TEMA 5: Commonwealth Agricultural Bureaux, Mills C.F., Bremner I., Chesters J.K., ed., London, pp. 785–788 (1985).
25. A. Frappaz, L. Mostaghini, *Nouv. Dermatol.* **11**, 78–80 (1992).
26. G. Michaëlsson, L.E. Edqvist, *Acta Derm. Venereol.* **64**, 9–14 (1984).
27. G. Michaëlsson, B. Berne, B. Carlmark, A. Strand, *Acta Derm. Venereol.* **69**, 29–34 (1989).
28. L. Juhlin, L.E. Edqvist, L.G. Ekman, K. Ljunghall, M. Olsson, *Acta Dermatovener (Stockholm)* **62**, 211–214 (1982).
29. C. Deffuant, P. Celerier, H.L. Boiteau, P. Litoux, B. Dreno, *Acta Derm. Venereol.* **74**, 90–92 (1994).
30. J. Barrat, G. Crepin, Y. Darbois, S. Sirot, *Rev. Fr. Gynécol. Obstét.* **87**, 1–3 (1992).

37

SELENIUM AS A PHARMACOLOGICAL AGENT AGAINST HEAVY METAL POISONING AND CHEMICAL OR PHYSICAL CARCINOGENESIS

J. Poupon

Laboratoire Central de Biochimie
Hôpital Lariboisière
Paris, France

1. INTRODUCTION

Selenium (Se) is an essential trace element and a constituent of two enzymes, glutathione peroxidase (GPx) and type I iodothyronine 5'-deiodinase. GPx is a component of the antioxidant defense system and therefore plays an important role in the detoxification of xenobiotic-induced free radicals. However, most studies assessing the beneficial role of Se against xenobiotics have been conducted in Se-deficient animals. In these models, Se supplementation at nutritional doses restored normal selenium status, so that the effects eventually observed were attributed to a restoration of GPx activity. Other mechanisms may also be involved when higher Se doses are given. This review will focus on the effects of pharmacological doses of Se given to Se-adequate animals. Other reviews of nutritional and epidemiological studies are already available (1–5).

2. SELENIUM AND HEAVY METALS

Since the first observation by Kar (6) that Se protects animals against cadmium (Cd) toxicity, this property has been confirmed by numerous studies extended to other metals. Se can actually react with many metals to form selenide or protein complexes (3). Although the term "heavy metals" applies to a large number of elements, only mercury (Hg) and Cd will be extensively discussed here. These metals have been thoroughly studied in connection with Se, whereas very few data exist for lead (Pb) and silver (Ag). Special attention will be given here to platinum since the toxicity of this metal (an element of the very powerful anticancer drug cisplatin) is reduced by Se.

2.1. Mercury

Parizek (7) first demonstrated the protective effect of selenite against Hg-induced acute toxicity. Rats were injected with $HgCl_2$ (0.02 mmol/kg, sc) alone or together with sodium selenite (0.03 mmol/kg, sc). Se prevented morphological alterations induced by Hg alone in kidneys and almost completely reduced the mortality of animals. Since this first observation, the interactions between Se and Hg have been intensively studied, particularly for mineral forms (Hg and $HgCl_2$), but also for organic forms (methyl and ethyl Hg). As mineral and organic compounds do not have similar metabolism and toxicity, their interaction with Se may be different.

Injecting selenite into animals protects them against the acute toxicity of $HgCl_2$ and increases the whole-body retention of Hg (8). This increased retention is observed for both elements, and maximum retention of one element is noted when the other is given in excess or in equimolar ratio to the other (9, 10). In these conditions, numerous studies in rats and mice have shown that both the renal content and urinary excretion of Hg are decreased in a dose-dependent manner (11). In comparison with selenite, selenomethionine (SeMet) affects the urinary excretion of Hg less, does not alter the renal concentration of Hg (12) and induces a lower increase of Hg levels in other organs (13). Chronic exposure to selenite or SeMet increases the whole-body retention of Hg following injection of $HgCl_2$. Dietary Se increases the whole-body retention of $HgCl_2$ by 50 % (14). Retention is higher with selenite than with SeMet (15). Se doses of 7.5 µmol/kg increase the fraction of Hg present in kidneys but decrease this percentage when given at 75 µmol Se /kg producing a concomitant increase of Hg content in liver and spleen and thereby indicating a modification of Hg toxicokinetics (15). It is postulated that Hg would preferably bind to renal metallothioneins at low Se level, while an increase in Se would result in the formation of a HgSe complex poorly excreted by the kidneys. Compared to selenite, the distribution of Hg in the gastrointestinal tract and kidneys is different with SeMet, probably because the HgSe complexes formed display different properties (15).

A direct interaction between Hg and Se has been suggested (16), as their administration schedule has a great influence on Se effects. Following exposure to Hg, Se and Hg are found in a molar ratio of 0.96 in human tissues, thus with high retention (thyroid, pituitary gland, kidneys, brain). This relation is also observed in subjects who were exposed to Hg a long time before, suggesting that Se and Hg are strongly bound, thus keeping Hg out of the biological turnover (17). In fact, the Hg:Se ratio approaches 1 as mercury levels increase since free Se (biologically active) becomes negligible (18). As demonstrated for Cd (see below), it is likely that Hg reacts with Se metabolites alone. In the organism, selenate and selenite are converted into hydrogen selenide (H_2Se), enabling them to bind to cations (M) to form MSe which further forms a complex with proteins (19). The formation of such a complex and its uptake by the reticulo-endothelial system account for the lower urinary excretion, increased whole-body retention and Hg-to-Se ratio of around 1 observed in tissues. Cavalli recently reported that, Se and inorganic Hg in dolphins liver are in a 1:1 ratio and that 40 % of Se is in the form of selenocystein, suggesting that Hg is bound to SeCys-rich proteins (20). This Hg-Se complex was also detected in the nuclei of renal tubules (21). In the kidneys, selenite administration reduces mercury binding to metallothioneins (MT) but increases it to high-molecular-mass proteins (22, 23). $HgCl_2$ subcutaneously administered at 5 µmol/kg to rats increases MT, copper (Cu) and zinc (Zn) concentrations in the kidneys as well as the urinary excretion of these two elements. When administered jointly in equimolar amount per os, selenite reduces the renal levels of Hg, MT, Cu and Zn and prevents Cu but not Zn urinary losses. It has been shown that Se removes Hg from MT, leading to increased whole-body retention and a lower kidneys uptake. MT induction is then partly inhibited, and Cu and Zn levels in kidneys and urine are consequently decreased (24). As vitamin E and other antioxidants have a similar effect against Hg-induced toxicity, it has been hypothesized that Se could act, at least in part, by detoxifying free radicals generated by Hg (21).

Methylmercury (MetHg) is the main organic derivative of Hg. The effects of Se on organic Hg are somewhat different than for inorganic Hg. Hg Se-induced retention is less pronounced with MetHg than with $HgCl_2$ (14). As for $HgCl_2$, a Se-MetHg detoxifying complex could be formed (25). Long-term supplementation of mice with selenite increases the elimination of Hg administered orally as MetHg. However, Hg accumulates in brain while cellular repartition in kidneys and liver is not altered (25). In rats, the effect of selenite on the metabolism of ethylmercuric chloride administered per os depends on the molecular ratio of Se to Hg. When Se is in excess (Se:Hg = 10), cumulative and especially urinary Hg excretion is reduced as a consequence of the increase of Hg concentration in most organs (brain, stomach and liver) and the decrease in kidneys and blood. In an equimolar ratio, Se does not affect Hg excretion, so that levels in organs, including kidneys, are increased. Study of the subcellular repartition of Hg does not indicate increased binding to high-molecular-weight proteins. Modifications in kidneys are only observed with low doses of Hg, probably as a consequence of inorganic Hg release through renal metabolism (26). Indeed, an increase of metallothionein-like proteins has only been observed in kidneys and depends on the quantity of inorganic Hg released from Hg mercury (27).

2.2. Cadmium

The protective action of Se against the necrotizing effect of small doses of Cd on the testis was demonstrated by Kar (6). Se exerted a protective effect against acute Cd toxicity in rats (28–32). The increase in GPx activity in liver, heart, kidneys and lung could be linked to this Se-induced protection (33). However, Gasiewicz suggested that Se interacts with plasma Cd at a ratio of 1:1 to form a protein-bound complex. *In vitro*, this complex, which is not formed when Cd and Se are mixed with plasma alone, only appears in the presence of erythrocytes (34). In the kidneys, Se stimulates the incorporation of Cd into high-molecular-mass proteins and therefore almost completely eliminates the binding of Cd to MT (35). The effects of Se on the chronic toxicity of Cd are less well known (31).

2.3. Lead and Silver

As for Hg and Cd, Se prevents Pb toxicity (36–38). Selenite and Pb acetate co-administered to rats during 8 weeks counteracts almost all of the detrimental effects of Pb (increase of urine GOT, ALP, proteins and δ-ALA, increase of G6PH in kidneys, decrease of GOT-GPT and ALP in kidneys, decrease of blood δ-ALA-D, hemoglobin and hematocrit). Pb uptake by blood, liver and kidneys is also reduced by Se (39). The interaction between Se and Ag has recently been described (40).

2.4. Platinum

Cisplatin, *cis*-diaminedichloroplatinum (CDDP), a widely used and powerful antitumor agent, has dose-limiting nephrotoxicity. Numerous compounds have been tested to reduce this toxicity but unfortunately such reduction is generally accompanied by a simultaneous decrease of activity. Since Naganuma (41) demonstrated that selenite can protect against CDDP-induced toxicity without altering its antitumor properties, this effect has been confirmed in rodents by several studies (42–44). The mechanism for this protection is not fully elucidated, but it has been shown that CDDP can react with Se metabolite CH_3-SeH to form Pt-Se-CH_3 (45). In addition, as CDDP toxicity is attributable in part probably to free radical-mediated peroxidation processes (46), Se may also act by stimulating GPx activity. Although selenite and Ebselen (a synthetic selenium compound) are effective, selenomethionine seems ineffective (unpublished observations). Despite the good results obtained in rodents with selenite, no specific clinical trial has been performed in humans.

3. SELENIUM AND CARCINOGENS

Se is a powerful agent for preventing chemically- or radiation-induced cancer. Because numerous carcinogens, including X-rays, UV radiations and many chemicals, generate free radicals, the anticarcinogenic properties of Se could result from its ability to detoxify reactive oxygen species (47).

3.1. Chemicals

Carcinogenic effects of a wide range of compounds are inhibited by Se (48). Several mechanisms may be involved, depending on the dose and chemical form of the carcinogen (3, 49–50): antioxidant effects via GPx activity, implication of Se in the activation/inactivation of growth factor processes, cell-growth inhibition modulated by glutathione, alterations of carcinogen metabolism, direct (reduction of DNA methylation) or indirect (increased DNA repair) protection of DNA, direct detoxification of carcinogens, and stimulation of the immune system. Se can act either at the initiation phase or at the postinitiation phase (promotion), and a prolonged administration of Se is recommended after exposure to the carcinogen (49, 51). Reddy et al. (52) studied the effect of dietary excess of selenite during initiation and postinitiation phases of colon carcinogenesis induced by azoxymethane (AOM) in rats. They showed that Se given at 2.5 ppm significantly reduced the incidence of tumors only when administered during the postinitiation phase. However, a reduction in the number of tumors could also be observed during the initiation phase.

The chemical carcinogens most often studied are dimethylbenz[a]anthracene (49, 53), benzo[a]pyrene (47, 54), and aflatoxin B1 (55–57). After exposure of rats to aflatoxin B1, Se administered 6 ppm in drinking water significantly reduced the formation of oxidized derivatives of DNA such as 8-OHdG (57). A modification of carcinogen metabolism has been shown for some compounds such as 2-acetylaminofluorine or 1,2-dimethylhydrazine (DMH) (51). When fed a Se-rich diet at 5 ppm, rats injected with DMH for 20 weeks displayed fewer distal colon tumors compared to rats fed with a normal diet at 0.2 ppm Se (58), thus demonstrating the role of supra-nutritional doses of Se.

3.2. Radiation and UV

Se supplementation to mice reduces UV-induced skin reactions such as pigmentation and inflammation (59). Dermal application of Se-rich water to mice can delay the appearance of UV-induced cancers (60). In mice, Se given at 0.5 mg/kg in the diet has a beneficial effect on UVB-induced skin carcinogenesis while Se at 0.1 mg/kg shows no effect (61). UVA-induced carcinogenesis results from a mechanism involving free radicals (62). In vitro, selenite (0.1 µg Se/L in the culture medium) increases the survival of human skin fibroblasts exposed to UVA (5 J/cm^2). GPx activity is increased, while TBARS (markers of lipid peroxidation) are reduced (62).

Ionizing radiation generates free radicals (OH$^\bullet$ and O$_2^-$) through water radiolysis leading to peroxidation of biomolecules. Se offers partial protection against lethality, probably by detoxification of the free radicals formed (63). This effect is less marked for high irradiation doses (above LD_{50}) but an association of Se with amifostine (WR 2721) reduces WR 2721 toxicity and increases protection (64). The efficiency of selenomethionine is controversial (63). The mechanism of Se-induced protection against ionizing radiation may not depend exclusively on its antioxidant properties since this protection is also observed when Se is injected just before or even after the irradiation dose. Moreover, the addition of Se to cell cultures increases GPx activity but does not reduce cell sensitivity (63). Se protection against experimental radiation-induced carcinogenesis is less effective and has only been reported in *in vitro* studies (63). In rats, 3 days after skin irradiation, selenite given at 1 or 2 ppm Se in diet had no effect on the time required to develop tumors (65).

4. CONCLUSION

Selenium used at pharmacological doses has proved effective against the toxicity of many xenobiotics, especially heavy metals and carcinogens. However, the seleno-compounds used are mainly selenite and selenomethionine which exert their own toxicity at high dosages. There is increasing evidence that the development of synthetic organoselenium compounds is very promising. Such new compounds have been developed to achieve greater chemoprotection and decreased toxic side effects. The results obtained so far with selenocyanates are encouraging (53, 66). Finally, few human clinical trials have been conducted, so that human applications would seem to deserve further investigation.

5. REFERENCES

1. G.F. Combs, S.B. Combs, *Pharmacol. Ther.* **33**, 303–315 (1987)
2. J.T. Salonen, *Ann. Clin. Res.* **18**, 18–21 (1986)
3. G.N. Schrauzer, *Biol. Trace Elem Res.* **33**, 51–62 (1992)
4. M. Garland, M.J. Stampfer, W.C. Willet, D.J. Hunter, in *Natural antioxidants in human health and disease* , pp. 263–286 (1994)
5. A. Favier, in *Les oligoéléments en nutrition et en thérapeutique*, P. Chappuis, A. Favier (Eds), Lavoisier Tec&Doc, Paris, pp. 135–151 (1995)
6. A.B. Kar, P.R. Das, B. Mukerji, *Proc. Natl. Inst. Sci. India* **26B**, 40–50 (1960)
7. J. Parizek, I. Ostadalova, *Experientia* **23**, 142–143 (1967)
8. V. Eybl, J. Sykora, F. Mertl, *Arch. Toxikol.* **25**, 296–305 (1969)
9. A.E. Moffitt Jr, J.J. Clary, *Res. Commun. Chem. Pathol. Pharmacol.* **7**, 593- (1974)
10. P. Kristensen, J.C. Hansen, *Toxicology* **12**, 101–109 (1979)
11. P. Kristensen, J.C. Hansen, *Toxicology* **16**, 39–47 (1980)
12. L. Magos, T.W. Clarkson, S. Sparrow, A.R. Hudson, *Arch. Toxicol.* **60**, 422–426 (1987)
13. L. Magos, T.W. Clarkson, A.R. Hudson, *Environm. Ther.* **228**, 478–482 (1984)
14. O. Andersen, J.B. Nielsen, *Environ. Health Perspect.* **102 suppl. 3**, 321–324 (1994)
15. J.B. Nielsen, O. Andersen, *J. Trace Elem. Electrolytes Health Dis.* **5**, 245–250 (1991)
16. A. Naganuma, Y. Ishi, N. Imura, *Ecotox. Environ. Safety* **8**, 572–580 (1984)
17. L. Kosta, A.R. Byrne, V. Zelenko, *Nature* **254**, 238–239 (1975)
18. M. Nylander, J. Weiner, *Br. J. Ind. Med.* **48**, 729–734 (1991)
19. L. Magos, M. Webb, *CRC Critical Reviews in Toxicology* **8**, 1–42 (1980)
20. S. Cavalli, N. Cardellicchio, *J. Chromatogr. A* **706**, 429–436 (1995)
21. R.A. Goyer, *Am. J. Clin. Nutr.* **61 suppl.**, 646S-650S (1995)
22. E. Komsta-Szumska, J. Chmielnicka, J.K. Piotrowski, *Biochem. Pharmacol.* **25**, 2539–2540 (1976)
23. J. Chmielnicka, E.A. Brzeznicka, *Bull. Environ. Contam. Toxicol.* **19**, 183–190 (1978)
24. J. Chmielnicka, E.A. Brzeznicka, A. Sniady, *Arch. Toxicol.* **59**, 16–20 (1986)
25. A.W. Glynn, N-G. Ilbäck, D. Brabencova, L. Carlsson, E-C. Enqvist, E. Netzel, A. Oskarsson, *Biol. Trace Elem. Res.* **39**, 91–107 (1993)
26. E.A. Brzeznicka, J. Chmielnicka, *Environ. Health Perspect.* **39**, 131–142 (1981)

27. E.A. Brzeznicka, J. Chmielnicka, *Environ. Health Perspect.* **60**, 423–431 (1985)
28. R.J. Niewenhuis, P.L. Fende, *Biol. Reprod.* **19**, 1–7 (1978)
29. H. Ohta, *Acta Scholae Med. Univ. Gifu* **33**, 931–946 (1985)
30. H. Ohta, S. Imamiya, *Kitasato Arch. Exp. Med.* **59**, 27–36 (1986)
31. Z.Z. Wahba, T.P. Coogan, S.W. Rhodes, M.P. Waalkes, *J. Toxicol. Environ. Health* **38**, 171–182 (1993)
32. B. Wlodarczyk, B. Biernacki, M. Minta, W. Kozaczynski, T. Juszkiewicz, *Bull. Environ. Contam. Toxicol.* **54**, 907–912 (1995)
33. M. Sidhu, M. Sharma, M. Bhatia, Y.C. Awasthi, R. Nath, *Toxicology* **83**, 203–213 (1993)
34. T.A. Gasiewicz, J.C. Smith, *Biochim. Biophys. Acta* **4281**, 113–122 (1976)
35. R.W. Chen, P.D. Whanger, P.H. Weswig, *Bioinorg. Chem.* **4**, 125–133 (1975)
36. S.C. Rastogi, J. Clausen, K.C. Srivastava, *Toxicology* **6**, 377–388 (1976)
37. F.L. Cerklewski, R.M. Forbes, *J. Nutr.* **106**, 778–783 (1976)
38. B. Nehru, A. Iyer, *J. Environ. Pathol. Toxicol. Oncol.* **13**, 265–268 (1994)
39. S.J.S. Flora, S. Singh, S.K. Tandon, *Acta Pharmacol. Toxicol.* **53**, 28–32 (1983)
40. J.P. Berry, P. Galle, *J. Submicrosc. Cytol. Pathol.* **26**, 203–210 (1994)
41. A. Naganuma, M. Satoh, M. Yokoyama, N. Imura, *Res. Commun. Chem. Pathol. Pharmacol.* **42**, 127–134 (1983)
42. J.P. Berry, C. Pauwells, S. Tlouzeau, G. Lespinats, *Cancer Res.* **44**, 2864–2868 (1983)
43. K. Ohkawa, Y. Tsukada, H. Dohzono, K. Koike, Y. Terashima, *Br. J. Cancer* **58**, 38–41 (1988)
44. G. S. Baldew, C.J.A. van den Hamer, G. Los, N.P.E. Vermeulen, J.J.M. de Goeij, J.G. McVie, *Cancer Res.* **49**, 3020–3023 (1989)
45. G. S. Baldew, J.G.J. Mol, F.J.J. de Kanter, B. van Baar, J.J.M. de Goeij, N.P.E. Vermeulen, *Biochem. Pharmacol.* **41**, 1429–1437 (1991)
46. J. Poupon, P. Chappuis, in *Trace Elements and Free Radicals in Oxidative Diseases*, A. E. Favier, J. Nève, P. Faure, eds., AOCS Press, Champaign, Illinois, pp. 261–275 (1994)
47. C. Borek, A. Ong, H. Mason, L. Donahue, J.E. Biaglow, *Proc. Natl. Acad. Sci. USA* **83**, 1490–1494 (1986)
48. P.D. Whanger, *Fundam. Appl. Toxicol.* **3**, 424–430 (1983)
49. H.W. Lane, D. Medina, L.G. Wolfe, *In Vivo* **3**, 151–160 (1989)
50. G. Hocman, *Int. J. Biochem.* **20**, 123–132 (1988)
51. C. Ip, *Ann. Clin. Res.* **18**, 22–29 (1986)
52. B.S. Reddy, S. Sugie, H. Maruyama, P. Marra, *Cancer Res.* **48**, 1777–1780 (1988)
53. K. El-Bayoumy, *Carcinogenesis* **15**, 2395–2420 (1994)
54. P. Prasanna, M. Jacobs, S. Yang, *Mutation Res.* **190**, 101–105 (1984)
55. A. Francis, T. Shetty, R. Bhattacharya, *Mutation Res.* **199**, 85–93 (1988)
56. C.Y. Shi, Y.C. Hew, C.N. Ong, *Hum. Exp. Toxicol.* **14**, 55–60 (1995)
57. H.M. Shen, C.N. Ong, B.L. Lee, C.Y. Shi, *Carcinogenesis* **16**, 419–422 (1995)
58. T.J. McGarrity, L.P. Peiffer, *Carcinogenesis* **14**, 2335–2340 (1993)
59. E.B. Thorling, K. Overvad, P. Bjerring, *Acta Path. Microbiol. Immunol. Scand. A* **91**, 81–83 (1983)
60. R. Cadi, J.C. Beani, S. Belanger, M.J. Richard, A. Richard, A. Favier, P. Amblard, *Nouv. Derm.* **10**, 266–272 (1991)
61. B.C. Pence, E. Delver, D.M. Dunn, *J. Invest. Dermatol.* **102**, 759–761 (1994)
62. M.T. Leccia, M.J. Richard, J.C. Beani, H. Faure, A.M. Monjo, J. Cadet, P. Amblard, A. Favier, *Photochem. Photobiol.* **58**, 548–553 (1993)
63. J.F. Weiss, V. Srinivasan, K.S. Kumar, M.R. Landauer, M.L. Patchen, in *Trace Elements and Free Radicals in Oxidative Diseases*, A.E. Favier, J. Nève, P. Faure, eds., AOCS Press, Champaign, Illinois, pp. 211–222 (1994)
64. P. Bienvenu, F. Herodin, M. Fatome, J.F. Kergonou, in *Selenium in Medicine and Biology*, J. Nève, A. Favier, eds., De Gruyter, Berlin, pp. 129–132 (1988)
65. H.S. Zackheim, K.K. Fu, A.S. Chan, *Cancer Lett.* **70**, 123–127 (1993)
66. J. Nayini, K. El-Bayoumi, S. Sugie, L.A. Cohen, B.S. Reddy, *Carcinogenesis* **10**, 509–512 (1989)

38

THE ANTIOXIDATIVE ROLE OF SELENIUM IN CADMIUM CHRONIC INTOXICATION

Z. Zloch

Medical Faculty of Charles University
Institute of Hygiene
Pilsen, Czech Republic

1. INTRODUCTION

Cadmium is known to be one of the most toxic environmental and industrial pollutants. In the Czech Republic, 10–20 percents of manufactured foodstuffs have been contaminated with cadmium (Cd) - due to polluted soils - in levels surpassing the tolerable concentration. As the Cd - induced toxicosis is partly characterized by the tissue injuring oxidative processes (1), dietary antioxidants are believed to ensure the effective prevention of oxidant damage. Due to biological interaction of Cd with selenium (Se), the Se - compounds are supposed to hinder Cd - oxidant activity. However, in the area of Middle Europe including the region of Czech Republic, the selenium dietary intake has been rather low (2). Nevertheless, some people have been used to take regularly pharmaceuticals containing Se and other trace elements. As a danger of adverse action in case of non adequate Se - intake arises, an examination of oxidant / antioxidant interrelationships of Cd and Se is of interest. Our aim was to clarify some oxidation / reduction aspects of a subchronic Cd - intoxication at subcellular level which was, in parallel, modified by a dietary intervention with selenium.

2. MATERIALS AND METHODS

2.1. Animals

Wistar albino male rats weighing 210 g were fed a commercial pellet diet containing approximately 0.04 mg Cd / kg and less than 0.05 mg Se / kg. The rats had access to drinking water containing less than 0.01 mg Cd and 0.005 mg Se / L and to the diet ad lib. In two experimental groups the drinking water was enriched with cadmium ($CdCl_2$) so that 4 - 6 mg Cd was daily consumed in relation to 1 kg body weight. In the parallel group, drinking water contained the same amount of Cd plus the supplement of selenite referred to a daily intake of

5 - 10 μg Se 1/ kg body weight. The control group was administered the drinking water without minerals addition. After 8 weeks experimental period, the animals were sacrificed.

2.2. Preparation of Mitochondria

To obtain mitochondria, tissues were homogenized in 10 volumes of ice - cold 0.32 M saccharose containing 1 mM EDTA at pH 7.4. The homogenate was centrifuged at 1300 g for 3 min, the supernatant was centrifuged at 17 000 g for 10 min. The sediment was washed with saline buffered (hydrogenphosphate) to pH 7.4.

2.3. Analytical Methods

Activity of enzymes glutathione peroxidase (GSH - Px) and superoxide dismutase (SOD), both in the erythrocytes, and the total antioxidant status in plasma were estimated using sets of Randox Laboratories LTD, U. K. Reduced glutathione (GSH) contents in mitochondria and in whole tissues were measured with a photometric assay with use of Ellman's reagent. Total vitamin C in mitochondria was estimated using 2,4-dinitrophenylhydrazine reagent. Lipid peroxide concentrations in mitochondria and in the whole tissues was assessed as the content of thiobarbituric acid reactive substances (TBARS, expressed as malondialdehyde, MDA). The Student's paired test was used for data comparison between different groups of animals.

3. RESULTS

The antioxidant status in plasma of both experimental groups provided a decrease in comparison with the control group (Table 1). GSH - Px activity in the erythrocytes of Cd - treated animals was significantly depressed. A moderate increase of GSH - Px, not reaching the control values, was noticed in the Se - supplemented group. SOD activity in the erythrocytes was significantly inhibited if the rats were exposed to cadmium only. The animals supplemented with selenium reached a partial restoration of SOD activity . Concerning the GSH contents in mitochondria, a lowered level in all tissues of Cd - intoxicated animals and an abolition of these deficits in rats influenced by an increased intake of Se were confirmed . Vitamin C contents in mitochondria provided different values according to the tissue examined. In the testes and kidneys a decreased concentration of this antioxidant was estimated in group intoxicated with Cd. In mitochondria, the highest oxidative injury expressed as TBARS was determined in cadmium exposed rats. The values in Cd - treated, Se - supplemented group were significantly more favourable and they were nearly the same as in the control group. However, in the liver the differences between experimental and control groups were negligible and in the testes, the greatest lipid peroxidation revealed in Cd + Se exposed rats.

4. DISCUSSION

The exposure of the animals to the mild doses of cadmium was responsible for biochemical changes reflecting both a decrease of the antioxidant potential and an oxidant injury of cell organelles in experimental rats. In kidneys, lung and brain, selenium addition to the diet brought about a benefit for the Cd - damaged intracellular biostructures.

Table 1. The effect of cadmium and selenium on some parameters of antioxidant defense and of oxidant injury

Parameter	Cd	Cd + Se	C
GSH - Px in RBC (U / g Hb)	45 ± 13*	53 ± 15	65 ± 16
SOD in RBC (U / g Hb)	950 ± 186*	1098 ± 372	1256 ± 215
Total Antioxidant Status (µM Trolox)	1.03 ± 0.33	1.01 ± 0.23	1.14 ± 0.24
GSH in Mitochondria (µmol / g Prot.)			
Liver	4.2 ± 0.4*	4.9 ± 0.4	5.3 ± 0.3
Kidneys	0.8 ± 0.1*	1.7 ± 0.2	1.9 ± 0.2
Testes	18.3 ± 1.9**	36.6 ± 3.1*	29.9 ± 2.1
Lung	11.9 ± 1.3*	14.6 ± 1.3	13.8 ± 1.1
Vitamin C in Mitochondria (µmol / g Prot.)			
Liver	17.7 ± 1.9	16.7 ± 1.5	17.7 ± 1.4
Kidneys	14.6 ± 1.8**	41.9 ± 3.7**	27.7 ± 3.8
Testes	10.7 ± 2.1*	14.4 ± 2.2	12.3 ± 1.4
Lung	3.9 ± 0.3	3.9 ± 0.8	4.2 ± 0.6
Lipid peroxidation in Mitochondria (µmol MDA / g Prot.)			
Liver	1.04 ± 0.02*	1.22 ± 0.11*	1.53 ± 0.12
Kidneys	0.070 ± 0.002*	0.050 ± 0.004	0.050 ± 0.004
Testes	1.59 ± 0.01	2.45 ± 0.12**	1.51 ± 0.10
Lung	2.84 ± 0.18*	2.28 ± 0.21	2.12 ± 0.15

Values are means ± SD of 7 - 8 rats
* Significant Effect of cadmium (+ Se resp.) relative to corresponding control at P < 0.05
** Significant Effect of Metals (Cd, Se) relative to corresponding control at P < 0.01

In both experimental groups the antioxidant enzymes GSH - Px and SOD were inhibited, in the group exposed to cadmium alone this change have being to a deciding extent. GSH - Px is a selenoprotein, activity of which is in a relation to the selenium status of organism (3). SOD activity depends on the metal - ions availability and an exposure to Cd induced probably an antagonism of cadmium with the central ions of SOD. The results of total antioxidant status determination confirmed an impairment of the antioxidant defense potential in rats intoxicated with cadmium without any changes after selenium application. In particular, this finding is in agreement with the results of lipid peroxidation analysis, the higher antioxidant status being in relation to the lower oxidant deterioration.

Contents of the reduced glutathione (GSH) in tissues of Cd - intoxicated rats were depressed and a renewal of GSH content in Se - supplemented animals revealed implying a fortified protection of cells against cadmium toxicity. Vitamin C varied in a greater range indicating some deficit of this antioxidant in kidneys and testes. The decreased tissue concentration of vitamin C in rats is indicative for a disturbed biosynthesis and / or an impaired transport of ascorbate (4).

The increased lipid peroxidation in the mitochondria of kidneys, lung and brain (Table 1) confirmed the prevailing oxidant effect of cadmium on intracellular biomembranes. The selenium application contributed to a diminution of this effect. On the other hand, a lowered lipid peroxidation in liver mitochondria of Cd - intoxicated rats - as compared to the Se - supplemented and to the controls - was estimated. A significant increase of peroxidative damage occured surprisingly in the testes of rats treated with cadmium and selenium unlike the Cd - intoxicated animals (without Se) and the controls. Indeed, enhanced doses of selenium can depress the antioxidant defense in some tissues leading to a higher degree of toxic injury (5). Oxidant moiety of Cd ions is believed to be conditioned on their chemical forms -

free or bound in metallothionein - that modify their chemical behavior. Selenium can alter distribution of cadmium in tissues and influence the character of its chemical bounds in biomolecules (6). In this way, selenium may prevent cadmium toxicity. On the contrary, cadmium administration to animals can be linked to a redistribution of selenium in organism inducing the different antioxidant responses to the oxidant challenge (7).

In conclusion, the adverse effects on mitochondria of subchronic cadmium exposure are tissue specific. Kidneys, lung and brain are the main targets of oxidant damage caused by cadmium. The selenium activity in relation to the Cd - induced oxidant injury is rather beneficial. Nevertheless, a deteriorating side - effect in testes of animals treated by cadmium together with selenium has been detected. Appropriate caution should be exercised when applying selenium as a supportive, health promoting factor in cadmium subchronic intoxication.

5. REFERENCES

1. M. Sugiyama, *Cell Biol. Toxicol.* **10**, 1 - 22 (1994).
2. V. Korunová, Z. Škodová, J. Dìdina, Z. Valenta, J. Paøizek, Z. Píša and M. Stýblo, *Biol. Trace Elem. Res.* **37**, 91 - 99 (1993).
3. J. R. Arthur and G. J. Beckett, *Proc. Nutr. Soc.* **53**, 615 - 624 (1994).
4. G. Mozesová and E. Ginter, *Int. J. Vit. Nutr.Res.* **62**, 342 - 343 (1992).
5. A. Albrecht, M. A. Pélissier and M. Boisset, *Toxicol. Lett.* **70**, 291 - 297 (1994).
6. M. P. Waalkes, T. P. Coogan and R. A. Barter, *Crit. Rev. Toxicol.* **22**, 175 - 201 (1992).
7. J. E. Spallholz, *Free Rad. Biol. Med.* **17**, 45 - 64 (1994).

39

METABOLISM OF SELENIUM IN A MODEL OF MESENTERIC ISCHEMIA-REPERFUSION INJURY

Role of N-Acetylcysteine

Ph. Bauer,[1] F. Belleville-Nabet,[2] E. Vauthier,[3] C. Colin,[3] F. Dubois,[2] J. C. Guédenet,[4] O. Bodenreider,[5] B. Dousset,[2] P. Nabet,[2] J. P. Mallié,[3] and A. Larcan[1]

[1] Service de Réanimation Médicale, Hôpital Central
54035 Nancy Cedex
[2] Laboratoire de Biochimie, Hôpital Central
54035 Nancy Cedex
[3] Laboratoire de Néphrologie, Faculté de Médccine
54500 Vandoeuvre
[4] Laboratoire d'Anatomo-Pathologie, Faculté de Médecine
54500 Vandoeuvre
[5] Département D'information Médicale, Hôpital Central
54035 Nancy Cedex, France

1. INTRODUCTION

Selenium is the essential co-factor of glutathione peroxidase involved in free radicals detoxification. In case of ischemia-reperfusion and depending on the model studied, the tissue variations of selenium and reduced glutathione can be related to various mecanisms (1) ; among others, a redistribution has been suggested. The aim of this study was to confirm this phenomena and to study the effect of N-acetylcysteine (NAC), a free radical scavenger (2), in a model of moderate intestinal ischemia-reperfusion injury by using hemodynamic, biochemical and histological data.

2. MATERIAL AND METHODS

Three groups of 10 male adult Wistar rats were randomly allocated to receive either saline (0,75 ml/kg) without (group A) or with clamping-release of the mesenteric artery (group

Therapeutic Uses of Trace Elements, edited by Nève et al.
Plenum Press, New York, 1996

B) or NAC (150 mg/kg) with clamping-release of the mesenteric artery (group C). We used an experimental model previously described which consisted of a transient intestinal ischemia in anesthetized rats by occluding the superior mesenteric artery for 20 minutes and releasing the clamp for 20 minutes (3). Plasma selenium, glutathione peroxidase (GSHPx) activity, intestinal, hepatic and renal selenium and reduced glutahione (GSH) contents were determined as previously described (3). Cyclic GMP was determined by radio-immuno-assay (4). Statistical analysis was performed using parametric and non parametric tests of analysis of variance.

3. RESULTS

After release of the clamp, a drop in blood pressure was observed (B vs. A and C vs. A, p = 0,0001) (Figure 1) with a more marked fall of the diastolic pressure in the group of rats pretreated with NAC (C vs. B). The biological data are shown on Table 1. A significant decrease in bicarbonate and increase in lactate levels were noted in case of ischemia-reperfusion injury (B vs. A and C vs. A). In the same way, hemoglobin and hematocrit levels were increased in case of ischemia-reperfusion injury (B vs. A and C vs. A). No change was noted regarding plasma selenium or globular GSHPx activity either within a group or between the different groups. In contrast, GSH content was increased in kidney after administration of NAC (A vs. C and B vs. C). On histological studies, a desquamation of the apical portion of the epithelium or the whole villosity was observed with an oedematous chorion and in some cases thrombi of the capillaries (Figure 2). These alterations were more frequently encountered after administration of NAC (C vs. B).

4. DISCUSSION

The small intestine is extremely susceptible to ischemia; a massive burst of damaging oxygen free radicals at reperfusion may generate much of the mucosal injury. Antioxidant

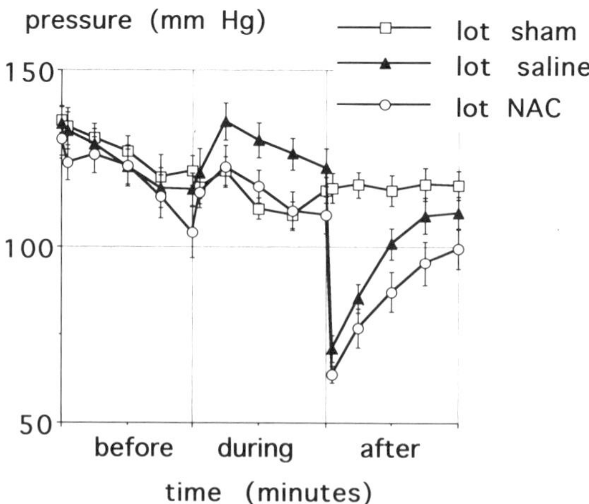

Figure 1. Mean arterial pressure during clamping of the superior mesenteric artery in three groups of rats (mean ± SEM).

Table 1. Biological data in the 3 groups of rats, n=10 in each group
(results are expressed as mean ± SEM)

Group	A (sham)	B (saline)	C (NAC)	ANOVA (p)
Bicarbonate (mmol/l)	22,4* ± 0,3	20,3 ± 0,5	19,7 ± 0,7	0,0023*
Lactate difference (mg/l)	-44* ± 16	44 ± 21	83 ± 32	0,0031*
Hematocrit (%)	46,9* ± 0,8	49,9 ± 1,0	51,9 ± 0,7	0,0010*
Hemoglobin (g/dl)	15,9* ± 0,3	16,9 ± 0,3	17,6 ± 0,2	0,0010*
Plasma selenium (µg/l)	347 ± 16	371 ± 16	402 ± 21	0,1014
globular GSHPx (µmol/gHb/mn)	62,7 ± 8,2	67,2 ± 5,1	61,9 ± 7,9	0,8569
Intestinal selenium (µg/g protein)	3,3 ± 0,4	3,2 ± 0,7	3,5 ± 0,6	0,9098
Hepatic selenium (µg/g protein)	5,6 ± 0,3	5,2 ± 0,3	5,5 ± 0,2	0,7008
Renal selenium (µg/g protein)	7,4 ± 0,3	7,1 ± 0,2	8,1 ± 0,4	0,1443
Intestinal GSH (µg/mg protein)	12,2 ± 1,3	9,5 ± 0,3	11,9 ± 1,3	0,1686
Hepatic GSH (µg/mg protein)	8,8 ± 0,7	9,2 ± 0,5	10,5 ± 0,9	0,2309
Renal GSH (µg/mg protein)	11,8 ± 0,7	10,9 ± 1,0	15,9* ± 1,1	0,0021*
intestinal cGMP (nM/g)	0,134 ± 0,020	0,167 ± 0,013	0,152 ± 0,012	0,3576

therapies directed toward the ablation of free radicals may prevent these injuries (5). NAC is a potent free-radical scavenger which has been widely used in acetaminophen overdose ; it prevents hepatic necrosis and improves the survival of patients with fulminant hepatic failure (6). In septic patients, NAC has provided a transient improvement in responders (7). By enhancing GSH or other sulfhydryls, and, as a result, reducing oxydative stress (8), by inhibit-

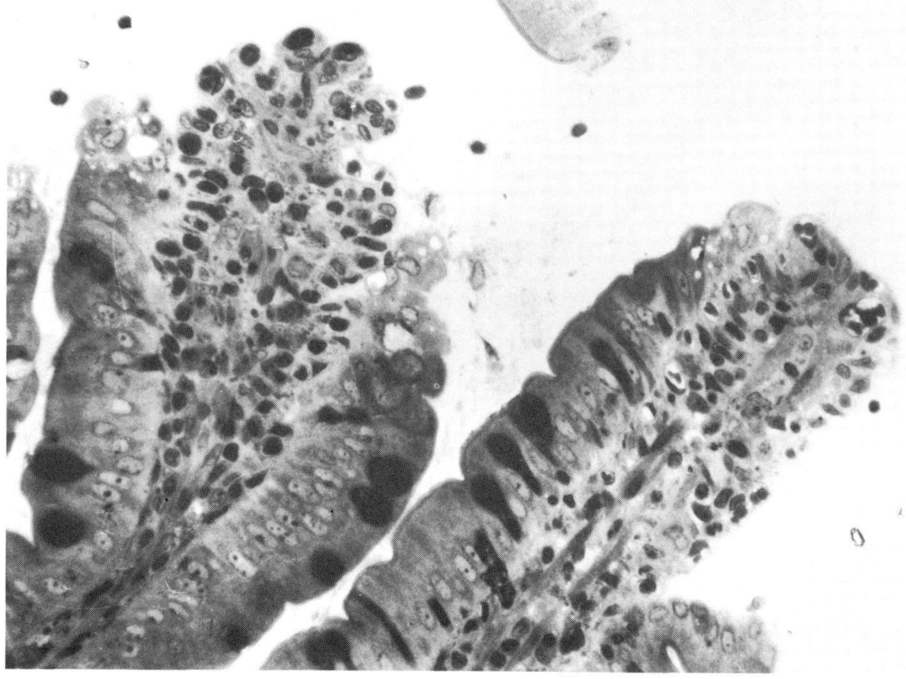

Figure 2. Desquamation of the apical portion of the epithelium (jejunum) with an oedematous chorion after ischemia for 20 minutes and reperfusion for 20 minutes of the small intestine in rat (magnification x 80).

ing tumor necrosis factor, stimulating endothelium-derived relaxing factor and reversing nitrate tolerance, NAC may improve the microcirculation and tissue oxygenation (9).

In this animal model, no significant changes were observed as far as tissue selenium and GSH were concerned, except in renal tissue after administration of NAC. Indeed in the kidney, we did not show any statistical difference between A and B, whereas we observed a difference between A and C and B an C respectively. This can be explained by a) a short period of ischemia limiting the loss of GSH, b) a short period of reperfusion and c) NAC at 150 mg/kg given preinsult effectively replenishes intracellular stores of GSH. However, the hemodynamic and histological changes observed could suggest *a contrario* some beneficial effect of NAC on splanchnic microcirculation. When administered postinsult, NAC has been effective in a model of acute lung injury (10) and may be of interest in case of transient ischemia of the small intestine by modulating the splanchnic nitric oxide response to oxygen free radical stimulation (11, 12). This hypothesis deserves further investigation.

5. REFERENCES

1. D.D. Gibson, D.J. Brackett, R.A. Squires, A.K. Balla, M.R. Lerner, P.B. McCay, L.R. Pennington, *Free Radic. Biol. Med.* **14**, 427–433 (1993).
2. K.R. Knight, K. MacPhadyen, D.A. Lepore, N. Kuwata, P.A. Eadie, B. McO'brien, *Clin.Sci.* **81**, 31–6 (1991).
3. Ph. Bauer, F. Belleville-Nabet, F. Watelet, F. Dubois, A. Larcan, *Biol. Trace Elem.Res.* **47**, 157–163 (1995).
4. F. Raul, M. Galluser, R. Schleiffer, F. Gossé, M. Hasselmann, N. Seiler, *Digestion* **56**, 400–405 (1995).
5. H.J. Schiller, P.M. Reilly, G.B. Bulkley, *Crit. Care Med.* **21**, S92-S102 (1993).
6. P.M. Harrison, J.A. Wendon, A.E.S. Gimson, G.J.M. Alexander, R. Williams, *N. Engl. J. Med.* **324**, 1852–1857 (1991).
7. C.D. Spies, K. Reinhart, I. Witt, A. Meier-Hellmann, L. Hannemann, D.L. Bredle, W. Schaffartzik, *Crit. Care Med.* **22**, 1738–1746 (1994).
8. O.I. Aruoma, B. Halliwell, B.M. Hoey, J. Butler, *J. Free Radicals Biol. Med.* **6**, 593–597 (1989).
9. H. Zhang, H. Spapen, D.N. Nguyen, M. Benlabed, W.A. Buurman, J.-L. Vincent, *Am. J. Physiol.* **266**, H1746-H1754 (1994).
10. J.A. Leff, C.P. Wilke, B.M. Hybertson, P.F. Shanley, C.J. Beehler, J.E. Repine, *Am. J. Physiol.* **265**, L501-L506 (1993).
11. I.R. Hutcheson, B.J.R. Whittle, N.K. Boughton-Smith, *Br. J. Pharmacol.* **101**, 815–820 (1990).
12. S.I. Myers, R. Hernadez, A. Castaneda, *Cardiovasc. Surg.* **3**, 207–210 (1995).

40

RELATIONSHIP BETWEEN ANTIOXIDANT ENZYME ACTIVITIES AND CARDIAC SUSCEPTIBILITY TO ISCHAEMIA AND REPERFUSION DURING AGING IN RATS

Preliminary Study

F. Boucher, S. Tanguy, S. Besse, and J. de Leiris

Physiopathologie Cellulaire Cardiaque, CNRS ESA-5077
Université Joseph Fourier
BP 53X, 38 041 Grenoble Cedex, France

1. INTRODUCTION

Numerous studies have shown that ischaemic damage and the so called reperfusion syndrome of the myocardium are, at least in part, associated to a burst of oxygen-derived free radicals (1). The involvement of reactive oxygen species (ROS) in post-ischaemic contractile dysfunction is supported by several studies underlying the benefit of the use of antioxidant catalytic agents such as superoxide dismutase and/or catalase during experimental cardiac reperfusion (2,3). Endogenous catalytic ROS scavengers, namely superoxide dismutase (SOD), catalase (CAT) glutathione peroxidase (GPx) and glutathione reductase (GRx) play key roles in the cellular defence systems against ROS under pathophysiological conditions (3–5). The myocardial activity of these enzymes is therefore determinant to protect the heart during post-ischaemic reperfusion. It is now well established that ageing is associated to an increase in cardiac susceptibility to ischaemia (6) and reperfusion (7) in rats. The present study was designed to define the relationship between the endogenous antioxidant status and cardiac susceptibility to ischaemia and reperfusion during ageing.

2. MATERIAL AND METHODS

2.1. Animals and Heart Preparations

Two groups of 4 month (adults: n = 9) and 16 month (elderly: n = 7) old male Wistar rats were used. The animals were allowed free access to standard laboratory diet (UAR,

Therapeutic Uses of Trace Elements, edited by Nève et al.
Plenum Press, New York, 1996

Villemoisson-sur-Orge, France) and to tap water prior to experimentation. All experiments were conducted in conformity with the guiding principles in the care and use of animals.

The animals were anaesthetised (pentobarbital : 60 mg/kg b.wt, i.p.) and received heparin (500 IU/kg b.wt, i.v.). Hearts were perfused via the aorta (constant pressure of 9.81 kPa), with a Krebs-Henseleit buffer containing $K^+ = 5.9$ mM, $Ca^{2+} = 1.36$ mM and glucose 11 mM. A water-filled ultra-thin balloon (8) was introduced into the left ventricle and inflated to impose an end-diastolic pressure of 4.0 ± 0.5 mm Hg. The volume of the balloon was kept constant throughout the experiment. The hearts were electrically paced at 5 Hz (300 bpm) *via* a monopolar electrode. After a 20 minute control perfusion period, the hearts were subjected to a 20 minute global, total, normothermic (37 °C) ischaemia followed by 30 minutes of reperfusion. Cardiac function was monitored *via* a pressure transducer. Coronary flow was measured by timed collection every 5 minutes.

2.2. Biochemical Assays and Statistical Analysis

At the end of the perfusion protocol, hearts were freeze-clamped. Hydroxyproline myocardial content was assessed by the method of Prockop and Udenfriend (9). Antioxidant enzyme activities: SOD (10), CAT (11), GPx (12) and GRx (13) were assayed biochemically.

Results are expressed as means $\pm$ standard deviation (SD). Student's t-test was used. The correlation between two variables was investigated using linear regression analysis. A difference was considered statistically significant when $p < 0.05$.

3. RESULTS

As expected, ageing is associated to a significant increase in myocardial hydroxyproline content, taken as an index of fibrosis (Table 1), and to a decrease in normoxic left ventricular developed pressure (LVDevP) (Table 1). Similarly, at the end of 30 minutes of post-ischaemic reperfusion, the recovery of LVDevP, expressed as a percentage of pre-ischaemic values, appeared significantly lowered (Table 1). As shown in Table 2, CAT and GRx were also affected by ageing. Moreover, a positive correlation between normoxic LVDevP and CAT activity was shown in adults but not in elderly (Fig. 1). Interestingly, GPx activity was correlated to the recovery of LVDevP measured after 30 minutes of reperfusion in adults, but this phenomenon was not verified in elderly (Fig. 2).

Table 1. Effect of aging on myocardial fibrosis and heart function. Means $\pm$ SD

Group(age)	Hydroxyproline (μmol/g w.wt)	Pre-ischaemic LVDevP (mmHg/g w.wt)	Recovery of post-ischaemic LVDevP (%)
4 months	5.68 ± 0.90	93 ± 12	50 ± 21
16 months	7.3 ± 1.6 *	64 ± 11 **	24 ± 16 *

LVDevP=left ventricular developed pressure measured at the end of the 20 min pre-ischaemic perfusion period. Post-ischaemic LVDevP is expressed as a percentage of the corresponding pre-ischaemic value. * $p < 0.05$; ** $p < 0.01$

Table 2. Effect of aging on myocardial antioxidant enzyme activities. Means ± SD

Group(age)	SOD(IU/mg protein)	CAT (IU/mg protein)	GPx (IU/g protein)	GRx (IU/g protein)
4 months	13.8 ± 1.9	20.8 ± 5.6	369 ± 60	5.4 ± 0.6
16 months	14.6 ± 2.0	14.6 ± 2.2 *	376 ± 73	6.4 ± 0.5 *

SOD = superoxide dismutase, CAT = catalase, GPx = glutathione peroxidase, GRx = glutathione reductase. * p < 0.05.

4. DISCUSSION

Our results suggest that cardiac function might be related to the cellular ability to reduce peroxides. Indeed, cardiac contractility under normoxic conditions, is negatively correlated to cardiac catalase activity in adult rats. Moreover, cardiac recovery of contractility during post-ischaemic reperfusion in adult rats is positively correlated to GPx activity. Additionally, catalase, which is of one of the major enzymes involved in the defence systems against oxidative stress, decreases with age. Finally, no compensatory increase in glutathione peroxidase activity occurs in old rats, suggesting that ageing might be associated to an alteration of the ability of cardiac cells to reduce peroxides.

5. CONCLUSIONS

These preliminary findings support the hypothesis that the age-dependent increase in cardiac susceptibility to ischaemia and reperfusion might be related to perturbations of the cellular defence systems against oxidative stress, and particularly those implicated in the reduction of peroxides. Further studies are now required to verify whether new pharmacological or nutritional interventions designed to restore the balance between the antioxidant enzymes in the heart can reduce the cardiac consequences of ageing.

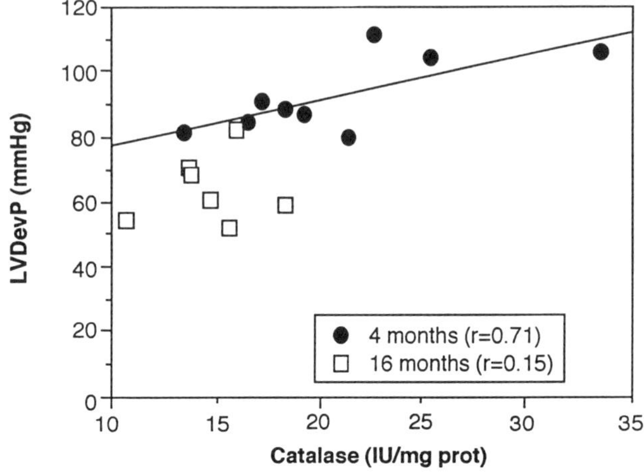

Figure 1. Relationship between pre-ischaemic LVDevP and catalase activity in adult (4 months) and old (16 months) rats. LVDevP = Left ventricular developed pressure measured at the end of the 20 min pre-ischaemic perfusion period.

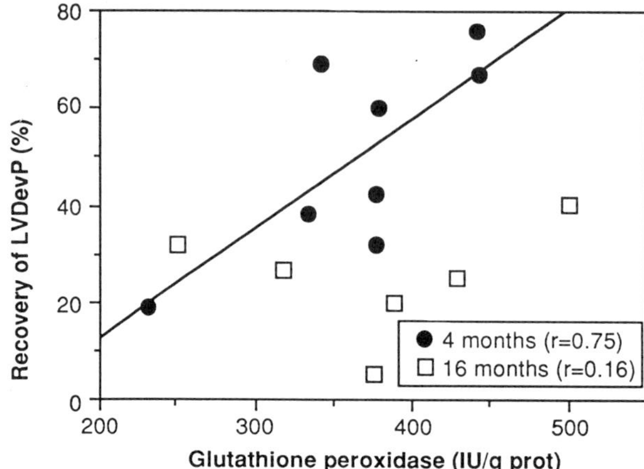

Figure 2. Relationship between cardiac glutathione peroxidase activity aṅd post-ischaemic recovery of LVDevP in adult (4 months) and old (16 months) rats. Recovery of LVDevP is expressed as % of LVDevP measured at the end of the pre-ischaemic perfusion period.

ACKNOWLEDGMENTS

Stéphane Tanguy is supported by Volvic, France. The authors thank Patrice Ghezzi from Collège Saint Paul, France, for his technical assistance.

6. REFERENCES

1. F.R. Boucher, S. Pucheu, C. Coudray, A. Favier and J.G. de Leiris. *FEBS letters*. **302**, 261–264 (1992).
2. M. Bernier, D.J. Hearse, and A.S. Manning. *Circ. Res.* **58**, 331–340 (1989).
3. J.M. Downey, B. Omar, H. Oiwa and J. Mc Cord. *Free Rad. Res. Commun.* **12–13**, 703–720 (1991).
4. W.S. Thayer, *FEBS Lett.* **202**, 137–140 (1986).
5. E. Pigeolet, P. Corbisier, A. Houbion, D. Lambert, C. Michiels, M. Raes, M.D. Zachary and J. Remacle. *Mech. Aging Develop.* **51**, 283–297 (1990).
6. K. Ataka, D. Chen, S. Levitsky, E. Jimenez and H. Feinberg. *Circulation.* **86** (suppl II), 371- 376 (1992).
7. L.H.E.H. Snoeckx, G.J. Vann der Vusse, W.A. Coumans, P.H.M. Willensen and R.S. Reneman. *Cardiovasc. Res.* **27**, 874- 881 (1993).
8. M.J. Curtis, B.A. Macleod and R. Tabritzchi. *J Pharmacol Meth.* **15**, 87–94 (1986).
9. D.J. Prockop and S. Udenfriend. *Anal. Biochem.* **1**, 228–239 (1960).
10. S.L. Marklund. *J. Biol. Chem.* **251**, 7504–7507 (1976).
11. R.F. Beers and I.W. Sizer. *J. Biol. Chem.* **19**, 133–135 (1952).
12. L. Flohe and W.A. Günzler. Assays of glutathione peroxidase. *Meth. Enzymol.* **105**, 114–121 (1984).
13. I. Carlberg and K. Mannervi. *Meth. Enzymol.* **113**, 484–490 (1985).

41

PHARMACOKINETICS OF PLATINUM IN A PATIENT UNDERGOING HEMODIALYSIS AFTER ACUTE RENAL FAILURE DUE TO TREATMENT WITH CARBOPLATIN

H. Jaumain,[1] M. Bret,[1] M. Accominotti,[2] C. Ardiet,[3] E. Chatelut,[4] and J. Moskovtchenko[1]

[1] Service de Réanimation, Pavillon P
[2] Laboratoire de Biochimie et Analyse des Traces
   Hôpital Edouard Herriot
   Place d'Arsonval, 69437 Lyon Cedex 03
[3] Laboratoire de Pharmacocinétique, Centre Léon Bérard
   Lyon
[4] Centre Claudius Regaud
   Toulouse, France

1. INTRODUCTION

Platinum-based therapy is currently used in anticancer regimens. Carboplatin [diamine (1,1-cyclobutanedicarboxylato) platinum], one of the second generation platinum compounds (1), has been introduced in clinical use as a less nephrotoxic alternative to cisplatin. The binding of platinum to plasma proteins is less pronounced than for cisplatin, resulting in a higher proportion and a longer half-time of active free platinum in blood. In addition, renal excretion by glomerular filtration is rapid and, unlike cisplatin, occurs without tubular secretion (2). At the conventional dose of 400 mg/m2, carboplatin dose-limiting toxicity is essentially hematologic (neutropenia and thrombocytopenia). This toxicity can be prevented by administration of cytokines such as GM-CSF, G-CSF or IL-3 allowing a dose escalation over 1000 mg/m2. However, at these doses, other signs of toxicity such as nephrotoxicity appear. In this paper, we described the case of a patient with acute renal failure after carboplatin administration.

Therapeutic Uses of Trace Elements, edited by Nève et al.
Plenum Press, New York, 1996

2. INDIVIDUAL CASE

2.1. Patient History

M P., a 57 year old was admitted to the intensive care unit at the Edouard Herriot Hospital in Lyon in May 1995 and presented bone marrow aplasia and acute renal failure with hyperhydratation (7 kg), pulmonary oedema, oliguria and sodium overload. The background of M P. was a Hodgkin disease in 1976. Eight days before his admittance (D 0), he received an antitumor treatment, for a non Hodgkin malignant lymphoma (holoxan, vepeside and carboplatin : 450 mg/m^2/day, i. e 870 mg/day for 3 days). On D 8, five days after treatment, holoxan and vepeside serum levels were normal. On D 8, 9, 10, the patient was dialysed during 4 hours with a conventional bicarbonate hemodialysis (HD). To improve carboplatin elimination (molecule of low MW: 371,6 Da), a new hemodiafiltration technique, the acetate-free biofiltration (AFB) was used from D 11. On D 14, an autologous bone marrow transplantation was performed: transplant was successful (total remission). Later, the patient presented chronic renal insufficiency with anuria, requiring repeated hemodialysis sessions. Six months later, the patient suffered from lymphoma relapse.

2.2. Methods

Blood samples were taken for platinum determination before and after hemodialysis sessions. Blood was collected into EDTA (K3) tube and immediately centrifuged. The plasma was removed and cooled to 4 °C. The morning after, plasma samples were sent to the laboratory. Total platinum (TP) analysis was performed by electrothermal atomic absorption spectrometry. The others aliquots were frozen at -20 °C until free platinum analysis (48 days after beginning). Free platinum analysis (FP) was made by flameless atomic absorption spectrophotometry after centrifugal ultrafiltration using Amicon ultrafiltration cones. Total and free platinum levels were determined before hemodialysis sessions from D 8 and D 12 to D 30, and after hemodialysis sessions only in few cases.

During the standard bicarbonate HD, the basic depuration mechanism is a diffusive transport of the small molecules. Ultrafiltration (UF) capacity is limited to 1 to 3 l. per session. UF has proved to be more efficient than diffusion in transfering medium-sized and small molecules (peptides, urea, creatininemia). AFB is a hemodiafiltration technique, characterised by buffer-free dialysate and simultaneous infusion in post-dilution mode of isotonic bicarbonate infusion. Thus associating diffusion and a higher ultrafiltration rate (5 to 8 l. per session), must increase the clearance of small molecules like free platinum. AFB is performed with a MONITRAL Bio generator (Hospal, Lyon, France) equiped with an UF automatic control system. Dialysate electrolyte concentrations are constant: sodium (139 mmol/l), potassium (2 mmol/l), chloride (145.5 mmol/l), calcium (1.75 mmol/l) and magnesium (0.5 mmol/l). Should a higher potassium concentration be required, the ion is added to the acid concentrate diluted to 1/35. Infusion of sodium bicarbonate solution (166 mmol/l) is performed by the generator with simultaneous dialytic exchanges and infusion. The prescribed bicarbonate flow solution depends on the predialysis bicarbonatemia to get a postdialysis bicarbonatemia of approximately 25 mmol/l.

2.3. Results

They are shown in Table 1. TP level was very high: 2000 µg/l on D 8, along with increasing creatininemia: 79 µmol/l on D 0, 111 µmol/l on D 3, 540 µmol/l on D 6, 730 µmol/l

Table 1. Evolution of creatininemia, TP and FP (rates before and after dialysis)

day (D)	type of dialysis	events	diuresis ml/24h	creatininemia (µmol/l)	TP (µg/l)	FP (µg/l)
0	-	carboplatin 870 mg	4900	79	-	-
1	-	carboplatin 870 mg	2800	88	-	-
2	-	carboplatin 870 mg	7300	91	-	-
3	-	-	4500	113	-	-
4	-	-	700	171	-	-
5	-	-	1400	385	-	-
6	-	-	2000	540	-	-
7	-	-	2500	665	-	-
8	HD	emergencies	1400	730 - 419	2000 - *	-
9	HD	-	1750	660 - 486	1825 - *	-
10	HD	-	800	719 - 398	1800 - *	-
11	AFB	-	400	721 - 104	1800 - 1300	-
12	AFB	-	150	571 - 241	1320 - 1160	362 - 158
13	AFB	-	200	430 - 222	1300 - 1075	373 - 258
14	AFB	bone marrow trans- plantation	160	470 - 258	900 - *	403 - *
15	AFB	-	450	517 - 342	1200 - 850	273 - 143
20	AFB	-	600	612 - 350	550 - 450	375 - 249
30	AFB	-	1350	630 - 263	375 - 290	102 - 63

* no blood sample collected after hemodialysis session

on D 8. The plasma TP levels before HD were respectively 2000 µg/l, 1825 µg/l.and 1800 µg/l. On D 11, TP levels fell rapidly from 1800 to 1300 µg/l after AFB, decreasing to 900 µg/l on D 14 (autologous bone marrow transplantation). Pharmacokinetic of FP were similar to those obtained for TP (Figure 1). Later, creatininemia remained stabilized (about 600µmol/l) and creatinine clearance remained under 2 ml/mn, in spite of a decrease in TP and FP levels.

3. DISCUSSION

In the present case, the total dose of carboplatin was very high (>1000 mg/m^2) and the renal function was supposed normal. Indeed, on D 0 the creatininemia was normal but the creatinine clearance was not measured and M. P. received many cycles of chemotherapy for a Hodgkin disease. In emergency, the hemodialysis technique was standard with a good clinical result on hyperhydratation (pulmonary and general oedema) but not on renal function, and TP levels did not fall rapidly. Autologous bone marrow transplantation had to be started quickly. Previous studies have shown that AFB may have several advantages over conventional HD in patients with chronic renal failure, i.e improved hemodynamic tolerance and acid-base balance correction, higher ultrafiltration, absence of dialysate backfiltration and decreased risk of endotoxin transfer related complications (7, 8, 9). A study (10) had

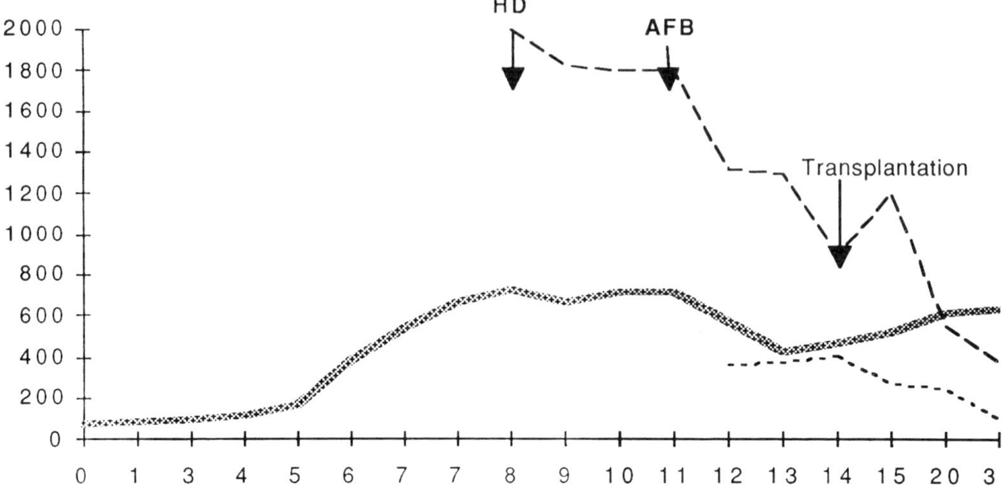

Figure 1. Kinetics of creatininemia, total and free platinum before hemodialysis sessions. *x*-axis: days (D 0 is the beginning of the chemotherapy). *y*-axis: Creatininemia (μmol / l) (—); total platinum (μg / l) (− −); free platinum (μg / l) (- -).

been undertaken to evaluate AFB in isolated acute renal failure patients. A prospective randomized study was carried out in our department in order to evaluate the efficacy and tolerance of AFB technique versus bicarbonate HD technique: there was a better small molecule removal and a better tolerance for the AFB technique in patients showing hyperhydratation with anuria and sodium overload. So, we decided to use AFB technique to obtain a more pronounced decrease of TP levels. TP rate actually decreased from 1800 to 1300 μg/l in one seance and transplantation was realised on D 14, with no myelotoxic levels (TP: 900 mg/1 and FP: 403 mg/l). During the following days, we noted a success of the transplantation (total remission); TP and FP levels were stabilized under 900 mg/l and 400 mg/l. As can be seen from Figure 1, the results were not corrected for free platinum apparent decay; howewer, since the corrective factor of 2,4 % per day evaluated by Erkmen and al (11) rendered irrelevant values in the present study, we decided to present raw concentration data. Even with the under estimation to irreversible platinum binding to plasma proteins, these values seem in accordance with the known spontaneous fate of platinum compounds, spontaneously or under dialysis. Furthermore, the concentrations measured allow the comparison of free and total platinum before and after dialysis.

The evolution of creatininemia, TP and FP levels were similar. Filterable free platinum is considered to be responsible for the toxic as well as anticancer effects. Filterable free platinum is eliminated by hemodialysis. The late beggining of dialysis after platinum overdose is said to be ineffective in clearing platinum. But, platinum found in the daily plasma probably reflects the slow release of platinum from tissues to which carboplatin became initially bound. After each hemodialysis treatment, we found a significant elimination of plasma platinum. This may indicate that an amount of unbound drug is available for dialysis. As we supposed, hemodiafiltration AFB was a better hemodialysis technique than conventional HD to remove platinum. After daily dialysis, it was possible to realise the bone marrow transplantation within the time limit; however the treatment was probably introduced too late for reversing the renal insufficiency.

4. REFERENCES

1. W.J.F. van der Vijgh, *Clin. Pharmacokinet.* **21**, (4), 242–261 (1991).
2. M. Chazard, L. Lenaz, *Bulletin de la Société de Cancérologie Privée* **7**, 77–82 (1988).
3. A.H. Calvert, S.J. Harland, D.R. Newell, *Cancer. Chemother. Pharmacol.* **9**, 140–147 (1982).
4. R. Canetta, K. Bragman, L. Smaldone, M. Rozencweig, *Cancer Treat. Rev.* **15**, (S B), 17–32 (1988).
5. M.J. Egorin, D.A. Van Echo, E.A. Olman et coll., *Cancer Res.* **44**, 5432–5438 (1984).
6. A. Calvert, D.R. Newell, L.A. Gumbrell et coll., *J. Clin. Oncol.* **7**, 1748–1756 (1989).
7. T. Petitclerc, I. Sitavanc, A. Hamani, C. Jacobs, *XVI Journées d'uro-néphrologie de la Pitié Salpétrière,* 145–153 (1990).
8. K. Man, C. Cianconi, B. Perrone, P. Chauveau, C.Jehenne,.*Trans. Am. Soc. Artif. Intern. Organs* **35**, 8–13 (1989).
9. A. Albertazzi, P.F. Palmieri, E. Mastrangelo,.*Kidney Int.* **43**, (S. 41) S 188-S 194 (1993).
10. C. Meloni, M. Morosetti, L. Meschini, M. Taccone-Gallucci, C.U. Casciani, *Artif. Organs* **17**, 188–190 (1993).
11. K. Erkman, M.J. Egorin, L.M. Reyno, R. Morgan, J.H. Doroshow. *Cancer Chemother. Pharmacol.* **35**, 254–256 (1995).

42

THE INFLUENCE OF LITHIUM AND MAGNESIUM ON DIGESTIVE LESIONS INDUCED BY PLATELET ACTIVATING FACTOR IN RATS

M. Nechifor, B. I. Neughebauer, M. Adomnicai, E. Teslariu, and C. Filip

Department of Pharmacology
University of Medicine and Pharmacy
Iasi, Romania

1. INTRODUCTION

The Platelet Activating Factor (PAF) is a lipidic autacoid having the structure 1-O-alkyl-2-acetyl-sn-glyceryl-3'-phosphorylcholine and is widely spread in a great variety of human and animal tissues (1–3). Numerous physiological and pathological involvements of PAF are known. Several authors (1,4–6) have shown that PAF produces ulceration and hemorrhages in the digestive tract. The ulcerogenic effect of PAF was observed after intravenous or topical administration at the gastric mucosa level. The mechanisms through which PAF exerts its lesional activity at the level of gastric mucosa are very complex and not completely identified. Among mechanisms proposed to explain the ulcerogenic effects of PAF are: damages of microcirculation of gastric mucosa (7), increase of free radical production (8), increase of acid secretion, formation of platelets microaggregates, increase of aggregation of neutrophils and release of lysosomal enzymes (9). In this paper, the effects of magnesium, calcium and lithium on the lesions produced by intravenous administration of PAF in rats were investigated.

2. MATERIALS AND METHODS

Five series of 10 male adult Wistar rats, weighing within 180–200 g and bread in normal laboratory conditions were used. Sixteen hours before the experiment, the animals were deprived of food, but received water ad libitum. Series I served as control and received no substance. Series II received PAF 2 µg/kg i.v. in tail veins. Series III received PAF 2 µg/kg i.v. and $MgSO_4$ 0.15 g/kg i.p. 1 hour before PAF administration. Series IV received PAF

Therapeutic Uses of Trace Elements, edited by Nève et al.
Plenum Press, New York, 1996

$2\mu g/kg$ i.v. and Li_2CO_3 0.1 g/kg i.p. 1 hour before PAF administration. Series V received PAF 2 $\mu g/kg$ i.v. and calcium gluconate 0.1 g/kg i.v. 1 hour before PAF administration.

The experiment was carried on at the room temperature (20 °C). All animals were killed by carotid section, 40 minutes after PAF administration. The histological examination of the gastric mucosa of all animals was performed. The gastric mucosa lesions were macroscopically observed and quantified by using the following scale: 0: normal mucosa; 1: congestive mucosa; 2: 1–3 hemorrhagic areas; 3: hemorrhagic areas + 1 to 3 ulcerations; 4: hemorrhagic areas + more than 3 ulcerations; 5: 1 or more perforations. The results were statistically interpreted by the "t" Student test.

3. RESULTS AND DISCUSSION

The experimental data are reported in Table 1. They clearly show that PAF induces a strong lesion at gastric level. The intravenous administration of $MgSO_4$ 1 hour before PAF decreases significantly the intensity of gastric lesions. Li_2CO_3 has no significant activity, while calcium gluconate increases slightly, but not statistically significant the PAF-induced gastric lesions. It is well known that PAF enhances the intracellular concentration of calcium (10). This fact may explain the slight increase of gastric lesions during calcium administration. On the other hand, verapamil, a calcium channel blocker, reduces the binding of PAF to its receptor (11). Thus, it is possible that the calcium excess might play an important role in the regulation of the binding of PAF to its membranous receptors. Nigam et al. (12) have shown an increase of PAF plasma levels in patients with breast cancer and hypercalcemia. It is possible that the slight increase of the lesional score observed after calcium administration, is partially due to an increase of endogenous PAF production. PAF exerts the gastric lesional effects after binding to its receptor. The PAF antagonists, as BN 52021, BN 50727 or kadsurenone inhibit the ulcerogenic effects of PAF (13). In the same time PAF stimulates the synthesis of some icosanoids with ulcerogenic effects, especially leukotrienes and TxA_2 (14,15). This situation has been observed not only at the gastric level but also in leukocytes and platelets from the digestive tract blood flow (16). The synthesis of leukotrienes and TxA_2 is stimulated by calcium ionophores, but is decreased by Mg^{2+}. This might be a possible way through which Mg^{2+} reduces the PAF-induced gastric mucosal lesions. While the PAF synthesis is increased in different experimentally models of stress, it seems possible that the positive effect of Mg^{2+} in stress may be at least partially due to the decrease of digestive lesions of PAF. In correlation with other studies of our team (17), lithium does not significantly influence the gastric activity of PAF. Our data therefore pleads for a possible involvement of some cations in the modulation of the intensity of digestive effects induced by PAF.

Table 1. The influence of $MgSO_4$, Li_2CO_3 and calcium gluconate on PAF-induced lesions of gastric mucosa

Series	Substance	Lesional Score	p[*]
I	Control	0.2 ± 0.1	< 0.01
II	PAF	3.1 ± 0.3	—
III	PAF + $MgSO_4$	1.9 ± 0.2	<0.05
IV	PAF + Li_2CO_3	2.9 ± 0.2	NS
V	PAF + Ca gluconate	3.3 ± 0.3	NS

*p value (compared with series II); NS = nonsignificant

4. REFERENCES

1. P. Braquet, L. Touqui, T.S. Shen, B.B. Vargaftig, *Pharmacological Rev.* **39**, 97–145 (1987).
2. G.E. Plante, R.L. Hebert, in *Ginkgolides - Chemistry, Biology, Pharmacology and Clinical Perspectives*, P. Braquet, ed., J.R. Prous Science, Barcelona, pp. 5775–602 (1988).
3. J. Benveniste, P.M. Hemson, C.G. Cochrane, *J.Exp.Med.* **136**, 1356–1377 (1972).
4. B.J.R. Whittle, T. Morishita, Y. Ohya, F.W. Leung, P.H. Guth, *Am. J. Physiol.* **251**, G772–778 (1986).
5. J.L. Wallace, Whittle B.J.R., *Prostaglandins* **31**, 989–998 (1986).
6. A.C. Rosam, J.L. Wallace, B.J.R. Whittle, *Nature* **319**, 54–56 (1986).
7. J. Bjork, L. Lindbone, B. Gerdin, G. Smedegard, K.E. Arfors, J. Benveniste, *Acta Phisiol. Scand.* **119**, 305–308 (1983).
8. T. Binnaka, T. Yamaguchi, J. Hirohara, *Scand. J. Gastroenterol* **66**, 2629–2634 (1989).
9. J.W. Doebber, M.S. Wu, T.Y. Shan, *Biochem. Biophys. Res.* Commun. **125**, 980–987 (1984).
10. A. Etienne, N. Baraggi, in *Ginkgolides - Chemistry, Biology, Pharmacology and Clinical Perspectives*, P. Braquet, ed., J.R. Prous Science, Barcelona, pp. 115–125 (1988).
11. P.J. Wade, *Thromb. Res.* **41**, 251–262 (1986).
12. S. Nigam, S. Muller, C. Benedetto, *J. Lipid. Mediat.* **12**, 323–328 (1989).
13. M. Koltai, D. Hosford, Ph. Guinot, A. Esanu, P. Braquet, *Drug* **42**, 174–200 (1991).
14. P.V. Peplow, D.D. Mikhailidis, *Prostagland. Leuk. Essent. Fatty Acids* **41**, 71–82 (1990).
15. A. Dembinska-Kiec, B.A. Peskar, M.K. Muller, B.M. Peskar, *Prostaglandins* **37**, 69–91 (1989).
16. S.H. McColl, E. Krump, P.P. McDonad, M. Braquet, P.H. Naccache, *J. Lipid. Mediators* **2**, S119-S128 (1990).
17. M. Nechifor, B.I. Neughebauer, M. Adomnicai, E. Teslaru, C. Filip, A. Negru, *Rom. J. Physiol.* **30**, 151–154 (1993).

43

MANGANESE(II) COMPLEXES WITH OROTIC ACID DERIVATIVES AS SCAVENGERS OF SUPEROXIDE RADICALS

ESR and Voltammetric Studies

J. P. Souchard,[1] P. L. Fabre,[2] J. P. Patau,[1] M. Massol,[1] P. Castan,[2] and F. Nepveu[1]

[1] Laboratoire de Synthèse, Physico-Chimie, et Radiobiologie
[2] Laboratoire de Chimie Inorganique
Université Paul Sabatier
118, route de Narbonne, F-31062 Toulouse Cedex, France

1. INTRODUCTION

The superoxide anion radical, O_2^-, plays a key role in the initiation stage of oxidative damage in biological systems (1). The main line of defense in mammalian organisms for controlling intracellular, and to a lesser extent extracellular, O_2^- radicals are the Cu/Zn and Mn-containing superoxide dismutase (SOD) enzymes and, recently, the application of SOD as a drug has attracted much attention (2). Since various problems are associated with using an enzyme as a drug (cost, bioavailability, stability, immunogenicity), non-toxic and low-mass metal complexes that catalyze the dismutation of O_2^- might be able to substitute for SOD in such applications.

As a part of a programme aimed at evaluating the antioxidative properties of natural or synthetic molecules and in accordance with the reported O_2^- scavenging activity of the Mn^{2+} ion in a number of *in vitro* and *in vivo* systems (3), we previously studied the variation of this activity in manganese complexes with weak ligands such as carboxylic acids (4). Recently, we described the quantitative analysis of such properties by ESR (5). Among the series of Mn(II) complexes investigated by this technique, the Mn(II)bis(lactato)diaquo complex demonstrated a high SOD-like activity and was tested *in vivo* using a classical anti-inflammatory activity evaluation model (6). The experiments showed significant anti-inflammatory activity at doses between 20 to 200 mg/kg after a single oral administration to the rat (7). These studies, which focused on a first series of Mn(II) complexes, have demonstrated the possibility to use ESR to screen drugs with the view of detecting and quantitating SOD-like activities before *in vivo* tests.

Therapeutic Uses of Trace Elements, edited by Nève et al.
Plenum Press, New York, 1996

The present work reports a study of the O_2^- radical scavenging properties of three Mn(II) complexes containing orotic acid (2,6-dioxo-1,2,3,6-tetrahydropyrimidine-4-carboxylic acid, H_3L^1) or one of its substituted derivatives, 3-methyl-orotic acid (H_2L^2) or 5-nitro-orotic acid (H_3L^3). This class of ligands was selected because of the nature of these substituted carboxylic acids as well as their biological role as precursors of pyrimidine bases in nucleic acids of living organisms. The properties of these compounds have been evaluated and compared to those of known antioxidants (vitamin E, trolox, vitamin C, phenol, 2,6-di-tert-butyl-4-methylphenol or BHT) in buffered aqueous and organic solvents by means of cyclic voltammetry and ESR. The aim of this work is to compare the effectiveness of these two methods for screening antioxidant drugs before *in vivo* tests.

2. MATERIALS AND METHODS

2.1. Materials

Manganese(II) complexes were prepared and characterized by X-ray crystallography (8) leading to the following formula and masses: with orotic acid (H_3L^1): $[Mn_2(HL^1)_2(H_2O)_6]$ ($M_r = 526.1$); with 3-methyl orotic acid (H_2L^2): $[Mn_2(L^2)_2(H_2O)_6]$ ($M_r = 554.2$); with 5-nitro-orotic acid (H_3L^3): $[Mn_2(HL^3)_2(H_2O)_5K_2Cl_2]$ ($M_r = 747.2$). Xanthine oxidase (XOD), orotic acid, 5-nitro-orotic acid, 5,5-dimethyl-1-pyrroline-N-oxide (DMPO), dimethylaminoethylamine (DMAEA) and activated charcoal were obtained from Sigma Chemical Co.; acetaldehyde (AC) and diethylenetriaminepentaacetic acid (DTPA) were of analytical grade and purchased from Prolabo; ascorbic acid, D-L-α-tocopherol (vitamin E), BHT and Trolox^R were of the hightest quality available from Aldrich Chemical Co. DMPO was purified and prepared as reported elsewhere (5). 3-methyl-orotic acid was synthezised as previously reported (8).

Electrochemical measurements were carried out by cyclic voltammetry with a home-made potentiostat driven by a PC computer. The electrochemical cell was a conventional three electrode configuration: working electrode, gold disk (diameter 1 mm); reference electrode, SCE; and counter electrode, Pt wire. Experiments were carrried out at room temperature in oxygenated dimethylformamide (DMF) with Bu_4NClO_4 as bulk electrolyte.

The ESR spectra were recorded with a Bruker ER 200 D spectrometer at room temperature by starting a 3 min scan after adding acetaldehyde *in fine*. Recording conditions were: central field, 3500 G; scan range field, 200 G; microwave frequency, 9.71 GHz; frequency modulation, 70 KHz; field modulation intensity, 1 G; time constant 0.5 sec, scan speed, 24 G per min.

2.2. Methods

The cyclic voltammetry method is based on the electrochemical properties of the O_2/O_2^- redox couple which appears to be reversible at a gold electrode in DMF (Figure 1). When an O_2^- scavenger is added, the cyclic voltammogram is modified according to the kinetics of the O_2^- deactivation (9). On the backward sweep, the anodic peak current, Ip_{ox}, which corresponds to O_2^- oxidation, is lower than the initial cathodic peak current, Ip_{red}, of the reduction of O_2 into O_2^- during the forward sweep. The analysis of the peak current ratio RIp (= Ip_{ox}/Ip_{red}) as a function of the potential scan speed and of the concentration of the drug tested allows the determination of the kinetic constants. The method has been standardized with vitamin E. When the concentration of vitamin E is raised, the oxidation peak of

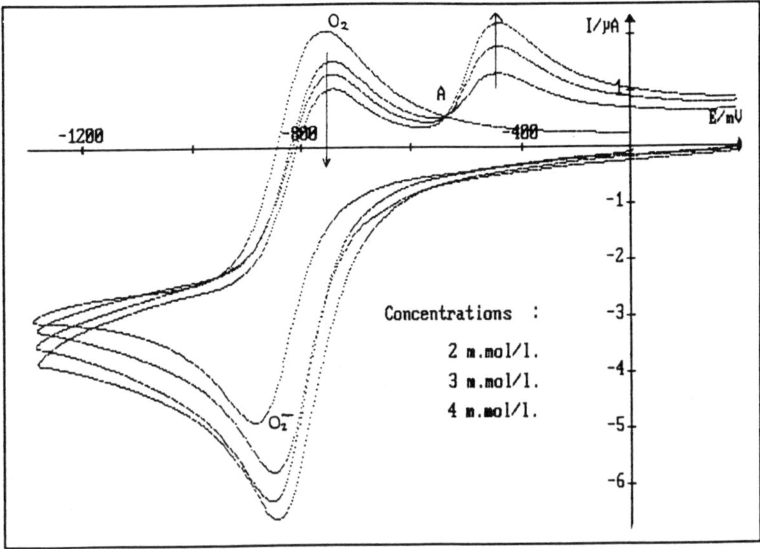

Figure 1. Cyclic voltammogram in DMF of O_2 with or without vitamin E. Potential scan speed 0.1 V/s.

O_2^- vanishes and an oxidation peak appears near -0.4 V which is attributed to an adduct between vitamin E and O_2^-. The peak ratio also is decreased (Figure 2). The three compounds (phenol, BHT, vitamin E) display the same O_2^- scavenging activity as shown by the RIp/concentration curve. From the standardisation of the RIp/concentration curve with vitamin E, activities of the various compounds can be expressed as vitamin E equivalent.

For ESR measurements, a 1 ml solution was prepared in a glass tube by adding the reagents in the following order: DMPO (50 mM), xanthine oxidase (0.018 U/ml), tested compound (x mM) and acetaldhehyde (60 mM) in phosphate or DMAEA buffer pH 7.4 containing 15 µM DTPA (9). DMAEA buffer was used to test manganese complexes (5). For each concentration of compound tested, the percentage of inhibition of the ESR signal com-

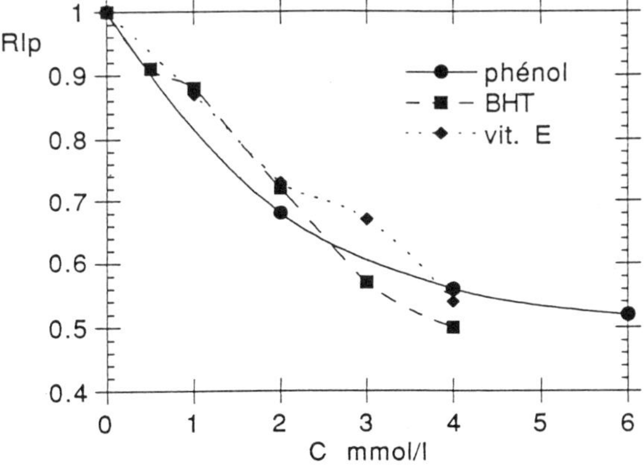

Figure 2. Peak ratio Rip as a function of the concentrations of the compounds tested.

pared to the signal of the blank has been calculated by adding the height of the sextet peaks of the O_2^- dependent AC/XOD carbon-centered signal (10). The O_2^- scavenging activity was quantified by the compound concentration inducing a 50% decrease in the control ESR signal intensity (IC_{50}). The AC / XOD system with DMPO in phosphate or DMAEA buffers produced a sextet ESR signal resulting from the presence of the O_2^- anion in the medium (5, 10) (Figure 3). This signal may be due to a methylcarbonyl-DMPO adduct.

3. RESULTS AND DISCUSSION

In order to compare the two methods, results are expressed with reference to vitamin E used as a standard (as described above) (Figure 4). Due to the higher activity range in aqueous media than in DMF, a logarithmic scale is used to present ESR results. This difference may be explained by the presence of H_3O^+, which plays an important role in the deactivation process through O_2^- protonation (consequently, O_2^- scavenging activities may be related to the acidity of the compounds). Whatever the method used, similar histograms were obtained for the various compounds in both solvents.

The ESR studies reveal that the Mn(II) complexes tested are about 100 times more active than the retinol hydrosoluble group Trolox[R] and the free orotic acid derivatives. The

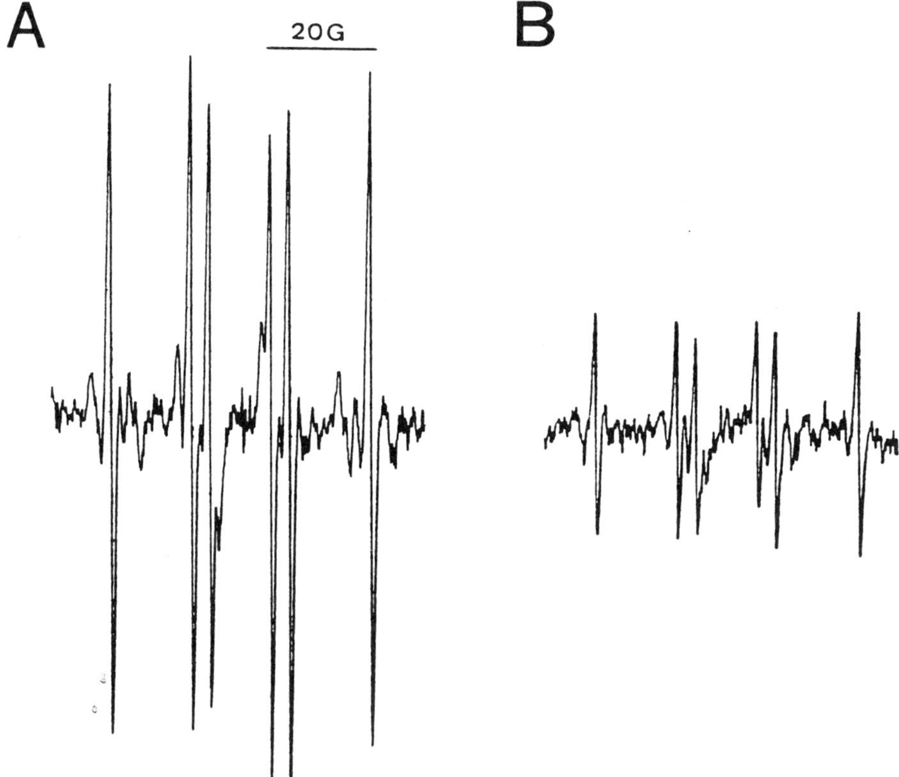

Figure 3. ESR spectra obtained with the AC/XOD superoxide-generating system (A), and inhibition (70%) of the signal intensity by the Mn(II)-5-nitro-orotic acid complex $[Mn_2(HL^3)_2.5H_2O.K_2Cl_2]$ at 7.10^{-7} M (B).

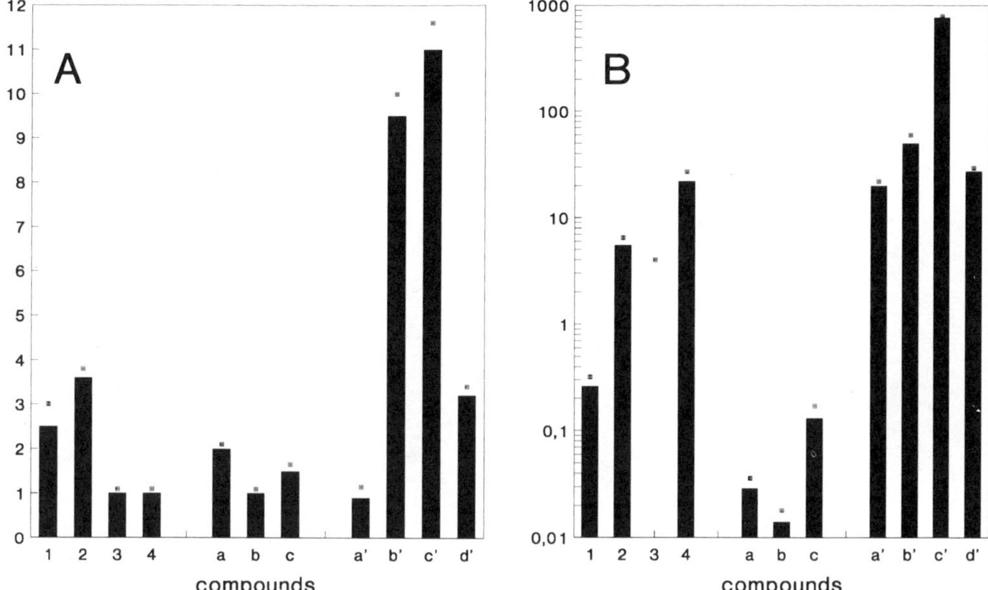

Figure 4. Superoxide scavenging activity of compounds measured by: A) cyclic voltammetric studies (CV); B) Electron Spin Resonance studies (ESR). The relative activities of the compounds is expressed as their equivalent in vitamin E, i.e: A) RIp(compound)/RIp(vitamin E); B) IC50(vitamin E)/IC50(compound). The compounds tested are: 1) trolox[R]; 2) ascorbic acid; 3) BHT; 4) phenol; a) orotic acid (H_3L^1); b) 3-methyl-orotic acid (H_2L^2); c) 5-nitro-orotic acid (H_3L^3); a') $[Mn_2(HL^1)_2.6H_2O]$; b') $[Mn_2(L^2)_2.6H_2O]$; c') $[Mn_2(HL^3)_2.5H_2O.K_2Cl_2]$; d') $MnCl_2$.

Mn(II) complex derived from 5-nitro-orotic acid, $[Mn_2(HL^3)_2(H_2O)_5K_2Cl_2]$, is 28-fold more effective than $MnCl_2$ and the complex with orotic acid, $[Mn_2(HL^1)_2(H_2O)_6]$. Spectrophotometric studies and potentiometric studies (8,11) have shown that orotic acid, (H_3L^1), only forms very weak complexes with the Mn^{2+} ion. The complex isolated and tested, $[Mn_2(HL^1)_2(H_2O)_6]$, which dissociates immediately in aqueous solution, therefore displays about the same activity as $MnCl_2$. However, the activity of Mn^{2+} should not be seen as that of the free hydrated metal ion since the buffer used (DMAEA) contains complexing groups. The electron-withdrawing effect of the 5-nitro-orotic acid and its stronger complexing properties compared to the two other ligands (8, 11) could explain the better activity of its Mn(II) complex.

4. REFERENCES

1. B. Halliwell, J.M.C. Gutteridge, *Free radicals in Biology and Medicine,* 2nd ed., Clarendon Press, Oxford (1989).
2. W.H. Bannister, *Biological and Clinical Aspects of Superoxide and Superoxide Dismutase*, Elsevier North Holland, New York (1980); G. Jadot, *Les Superoxydes Dismutases*, Masson, Paris (1988).
3. F.S. Archibald, I. Fridovich, *Arch. Biochem. Biophys.*, **214**, 452–463 (1982).
4. S. Rivomanana, M. Massol, P. Derache, F. Nepveu, in *Electron Spin Resonance (ESR), Applications in Organic and Bioorganic Materials*, B. Catoire, ed., Springer-Verlag, Berlin, pp. 105-111 (1992).
5. J.P. Souchard, M. Massol, F. Nepveu, *J. Chim. Phys.*, **93**, 214–219 (1996).

6. C. Declume, *J. Ethnopharm.*, **27**, 91–98 (1989).
7. C. Declume, F. Nepveu, *Fund. Clin. Pharmacol.*, **5**(9), 832 (1991).
8. F. Nepveu, N. Gaultier, N. Korber, J. Jaud, P. Castan, *J. Chem. Soc. Dalton Trans.*, 4005–4013 (1995).
9. P. Coffre, D.T. Sawyer, *Anal. Chem.*, **58**, 1057–1061 (1986); C.P. Andrieux, P. Hapiot, J.M. Saveant, *J. Am. Chem. Soc.*, **109**, 3768–3772 (1987).
10. J.P. Souchard, M. Massol, F. Nepveu, *submitted*
11. E.R. Tucci, B.R. Doody, N.C. Li, *J. Phys. Chem.*, **5**, 1570–1576; E.R. Tucci, C.H. Ke, N.C. Li, *J. Inorg. Nucl. Chem.*, **29**, 1657–1664 (1967).

44

ANTIBACTERIAL PROPERTIES OF SOME METAL SALTS AND LANSOPRAZOLE AGAINST *HELICOBACTER PYLORI* USING MIC DETERMINATION, ELECTRON MICROSCOPY AND FLOW CYTOMETRY ANALYSIS

F. Vicari,[1] P. Franck,[3] M. C. Conroy,[2] L. Marchal,[4] A. Lozniewski,[2] M. Joubert-Collin,[5] S. Forestier,[5] B. Foliguet,[4] P. Nabet,[3] and M. Weber[2]

[1] FLEEP Nancy
[2] Laboratoire de Microbiologie - CHU Nancy
[3] Laboratoire de Biochimie - CHU Nancy
[4] Laboratoire de Microscopie Electronique
Faculté de Médecine
Vandoeuvre les Nancy
[5] Laboratoires TAKEDA
Puteaux

1. INTRODUCTION

Helicobacter pylori associated gastritis responds to a variety of triple therapies. The combination of a proton pump inhibitor, amoxicillin and clarithromycin is particulary effective (1,2,3) and is one of the triple therapies recommended by the French conference of consensus. Amoxicillin and bismuth are among the most useful of the antimicrobial drugs that do not induce resistance in *H. pylori*. Currently, the triple therapy of asymptomatic gastritis does not seem to be suitable because of the cost and the risk of developing antibacterial resistance. However, a potential therapy using different metal salts could be envisaged for patients who do not have peptic ulcer. In fact, metals such as bismuth, silver or aluminium have been used in the treatment of gastritis for their topical activity but some of them have also bacteriostatic properties (4).

The aims of this work were to study the bacteriostatic activity of some metal salts and lansoprazole by classical susceptibility tests (agar dilution), to determine the morphological effects and ultrastructural damage after exposition to these metal salts and lansoprazole by transmission microscopy, and finally to characterize the effects on the bacterial external surface and respiratory metabolism by flow cytometry analysis.

Therapeutic Uses of Trace Elements, edited by Nève et al.
Plenum Press, New York, 1996

2. MATERIALS AND METHODS

2.1. Minimum Inhibitory Concentrations (MICs) Determination

Bacterial strains were twelve clinical isolates *H. pylori*, *H pylori* ATCC 49503, *H. pylori* CIP 101260. The MICs were determined by an agar dilution method using the medium : brain-heart infusion agar supplemented with 10 % horse blood and 1 % polyvitex. The inoculum was ajusted to 2 Mc Farland units. Metal salts (bismuth subcitrate, bismuth citrate, cupric sulfate, cupric chloride, silver nitrate, colloïdal silver, gold chloride, aluminium phosphate, aluminium hydroxide and lansoprazole) final concentration range were 1–256 mg/l.

2.2. Electron Microscopy

Bacterial cultures were used to prepare bacterial suspensions which were then centrifuged. The residue was fixed in 2.5 % glutaraldehyde, washed, post-fixed with osmic acid and embedded in Epson resin. Thin sections were mounted on grids and studied by transmission electron microscopy (CM 12, Philips). The cell membrane and internal structures were assessed for damage. The result was given as the percentage of microorganisms with normal ultrastructure.

2.3. Flow Cytometry Analysis

The reference strain of the Institut Pasteur was *H. pylori* CIP 101 260. A home - made selective medium (agar Muller-Hinton supplemented with 10 % horse blood and 10 % polyvitex and antimicrobials agents (vancomycin 2 mg/l, nalidixic acid 8 mg/l, fungizone 12,5 mg/l) was inoculated. The concentrations of bismuth (colloïdal bismuth suspension solution), silver (silver nitrate, Aldrich Chemical Company), and lansoprazole (Takeda Chemical Industries) were 0.25, 0.5, 1 , 2 , 4 , 8 , and 16 mg/l. The plates were incubated at 37 °C in microaerobic atmosphere in a jar with an appropriate gas mixture consisting of 5 % O2 - 10 % CO2 - 85 % N2. 48 hours later, the culture was used to prepare a bacterial suspension in distilled water, which was subdivided, in different fractions. The bacteria were then stained for 5 minutes at 37 °C with ethidium bromide (EB) - (Sigma) at a final concentration of 50 µg/l, for 20 minutes at 37 °C with DCFH-DA (2'-7' - dihydrodichlorofluorescein diacetate) (Molecular Probes) at a final concentration of 1µM, and for 5 minutes at 30 °C with Rhodamine 123 (Molecular Probes) at a final concentration of 2.5 mM. Samples were analysed with a flow cytometer EPICS-C (Coultronics, Margency, France) at 488 nm.

3. RESULTS AND DISCUSSION

3.1. Minimum Inhibitory Concentrations (MICs) Determination

Results are presented in table 1. These results (except for aluminium) suggest that metal salts have bacteriostatic activity against H. pylori.

Table 1. MICs determinations

Antimicrobial agent	Mic range (mg/l)
Bismuth subcitrate	2 - 8
Bismuth citrate	4 - 64
Cupric sulfate	4 - 64
Cupric chloride	4 - 64
Silver nitrate	0.5 - 32
Colloïdal silver	2 - 16
Gold chloride	1 - 32
Aluminium phosphate	>256
Aluminium hydroxide	>256
Lansoprazole	0.25 - 32

3.2. Transmission Electron Microscopy

Figure 1 shows the percentage of microorganisms with normal ultrastructure in the presence of different antimicrobial drugs and metal salts. The results show that bacterial damage occured in the presence of amoxicillin and lansoprazole but silver salt caused similar changes. Bismuth did not seem to alter *H.pylori* morphology.

3.3. Flow Cytometry Analysis

The membrane integrity was monitored by incorporating EB (intercalating dye) which does not normally cross intact membranes. Figure 2 shows the fluorescence histograms produced from *H. pylori* grown in the presence of silver. The ethidium bromide incorporation was increased (increasing red fluorescence) showing an altered membrane integrity. Bismuth and lansoprazole did not alter directly the bacterial membrane.

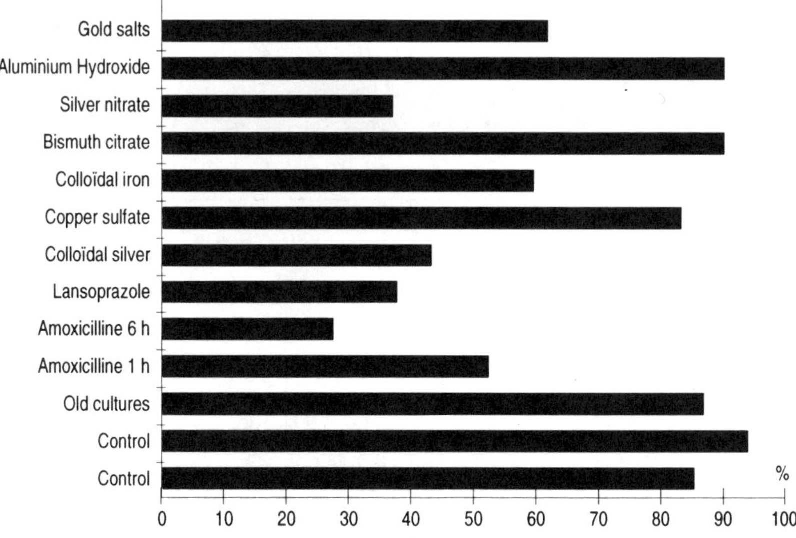

Figure 1. Percentage of microorganisms with normal ultrastructure under different conditions assessed by electron microscopy.

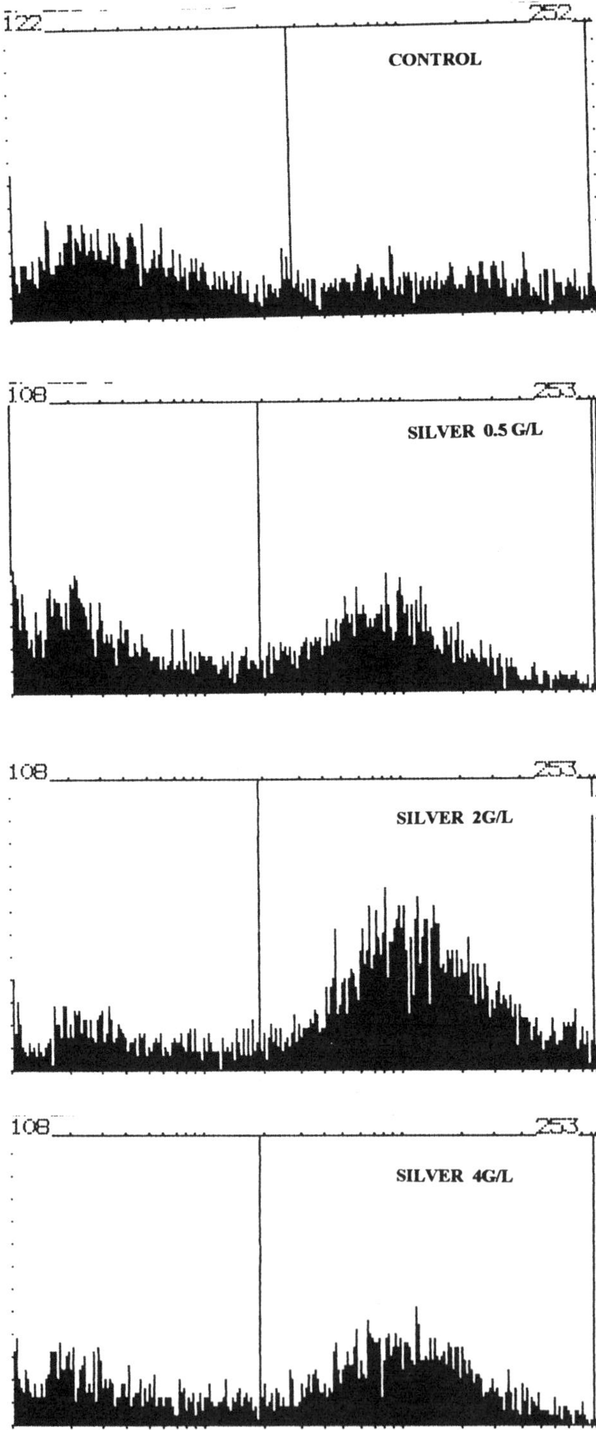

Figure 2. Flow cytometry analysis of membrane integrity : EB does not normally cross intact membranes. The red fluorescence was increased for bacteria in the presence of silver.

The membrane potential was monitored by incorporating potential - sensitive probes like rhodamine 123 (5,6). In bacteria, the cellular apparatus for energy metabolism is localized on the cytoplasmic inner membrane. The potential accross this membrane is dependent on energy metabolism and decreases when the membrane is pertubed by physical or chemical agents (6). Bismuth and lansoprazole induced a large variation of the potential membrane shown by an important decrease of fluorescence. The changes in membrane potential assessed the loss of viability of the bacteria. Nakao et al (3) reported that lansoprazole induces morphological alterations (shown by electron microscopy) but they supposed that antibacterial mechanisms of lansoprazole against *H. pylori* are probably different from those of classical antimicrobial agents. The important decrease of the membrane potential could account for this mechanism and could explain the substantial anti *H pylori* activity of lansoprazole (MIC range of lansoprazole : 0,25 - 32 mg/l).

The results obtained by flow cytometry with bismuth were similar to those of lansoprazole. This could explain the good results obtained with bismuth - based therapies. No significant modification of the oxidative metabolism was showed with the probes used in this study (DCFH and DHR 123).

4. CONCLUSIONS

This preliminary study shows that different metal salts could be of interest for therapy again *H. pilori*. Bismuth, which is largely used, had approximatively the same antibacterial activity than the lansoprazole via an inhibition of the membrane potential. Silver induced a loss in membrane integrity. Other metal salts such as copper or gold were bacteriostatic against *H pylori*.

5. REFERENCES

1. V. Berry, K.Jennings, G.Woodnutt, *Antimicrob. Agents Chemother.* **39**, 1859–1861 (1995).
2. B.J. Marshall, *Am. J. Gastroenterol.* **89**, S116–5128 (1994)
3. M. Nakao, M. Tada, K. Tsuchimori, M. Vekata, *Eur. J. Clin. Microbiol. Infect. Dis.* **14**, 391–399 (1995)
4. F. Vicari, B. Foliguet, M.C. Conroy, A Lozniewski, J.J. Denis , L. Marchal, M. Joubert-Collin, S. Forestier, M. Weber, *Acta Endosc.* **25**, 275–277 (1995)
5. C.Y. Cohen, E. Sahar, *J Clin. Microbiol.* **27**, 1250–1256 (1989)
6. R.I. Jepras, J. Carter, S.C. Pearson, F.E. Paul , M.J. Wilkinson, *Appl. Environ. Microbiol.* **61**, 2696 - 2701 (1995)

45

THE ROLE OF SELENIUM IN COPPER-INDUCED DAMAGE IN COPPER LOADED RATS' LIVERS

I. C. Fuentealba, B. Horney, J. Daley, and A. Tasony-Ferraro

Atlantic Veterinary College
University of Prince Edward Island
Charlottetown, Canada

1. INTRODUCTION

Copper (Cu) toxicosis, associated with increased hepatic Cu content, occurs as a familial disorder (Wilson's disease) in humans (1) and some breeds of dogs (2,3). The Cu-loaded rat has proved to be a useful model in which to investigate Cu-associated diseases (4,5). The mechanisms by which Cu exerts its toxic effect on the cell are still unclear, cell death may occur due to lipid peroxidation of lysosomal membranes (6) or as a consequence of nuclear disorganization (7). It is also possible that copper enters the nucleus as a consequence of lipid peroxidation of nuclear membranes (8). Selenium (Se), an essential trace element, is part of the glutathione peroxidase (GSHPx) and other selenoenzymes involved in removal of hydrogen peroxide and lipid peroxides produced during oxidative processes in cells (9).

The aim of this study was to evaluate, using morphologic and clinico-pathologic methods, the protective role of Se in copper-induced liver damage.

2. MATERIALS AND METHODS

Weanling male Wistar rats (n=32) of uniform age and weight were randomly allocated to four groups of eight rats each as follows: Group A was fed a high Cu diet (1500 ppm); Group B was fed a high copper diet (1500 ppm) and Se supplementation (2 ppm Se); Group C was fed a Se deficient diet and copper excess (1500 ppm); Group D received a normal rodent diet. After six weeks, blood samples were taken for glutathione peroxidase assay, metal analysis and determination of liver enzymes (ALP, GGT, SDH, AST and ALT). Rats were killed and their livers removed for histological examination, determination of glutathione peroxidase activity and metal (Cu and Se) analysis, by atomic absorption spectrophotometry. The data were analyzed using multivariate analysis of variance within general linear models using SAS software (10).

Therapeutic Uses of Trace Elements, edited by Nève et al.
Plenum Press, New York, 1996

Table 1. Cu and Se levels in the livers and serum of rats receiving a high Cu diet (Group A), high Cu diet with Se supplementation (Group B), high Cu/ Se deficient diet (Group C) and normal rodent diet (Group D). (Mean ± SD, N=8)

Group	Liver Cu	Serum Cu	Liver Se	Serum Se
A	581 ± 181	1.7 ± 0.2	1.3 ± 0.2	0.38 ± 0.10
B	503 ± 199	1.8 ± 0.4	8.0 ± 3.7	0.44 ± 0.05
C	820 ± 222	2.0 ± 0.2	0.2 ± 0.01	0.10 ± 0.02
D	4.22 ± 0.15	1.2 ± 0.01	0.9 ± 0.1	0.43 ± 0.09

3. RESULTS

Results of Cu and Se content in serum and liver tissue are presented in Table 1. Hepatic Cu levels were significantly higher ($p < 0.005$) in all the Cu-loaded groups (A,B and C) compared to control rats (4.22 ± 0.15 µg/g wet weight tissue). Moreover, significant differ-

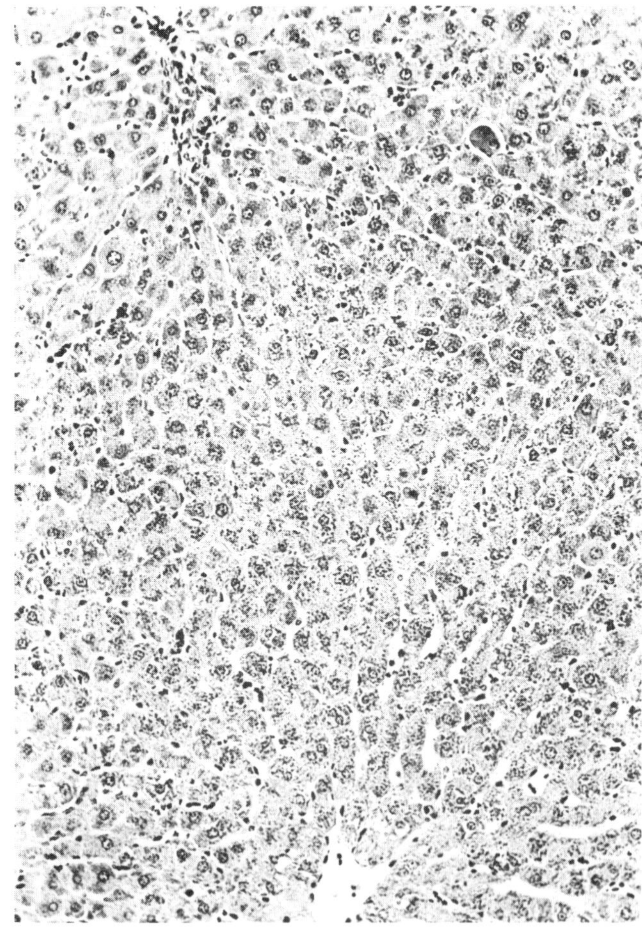

Figure 1. Section of liver from a Cu-loaded rat showing mononuclear inflammatory infiltration in zone 1. HE stain (x 32).

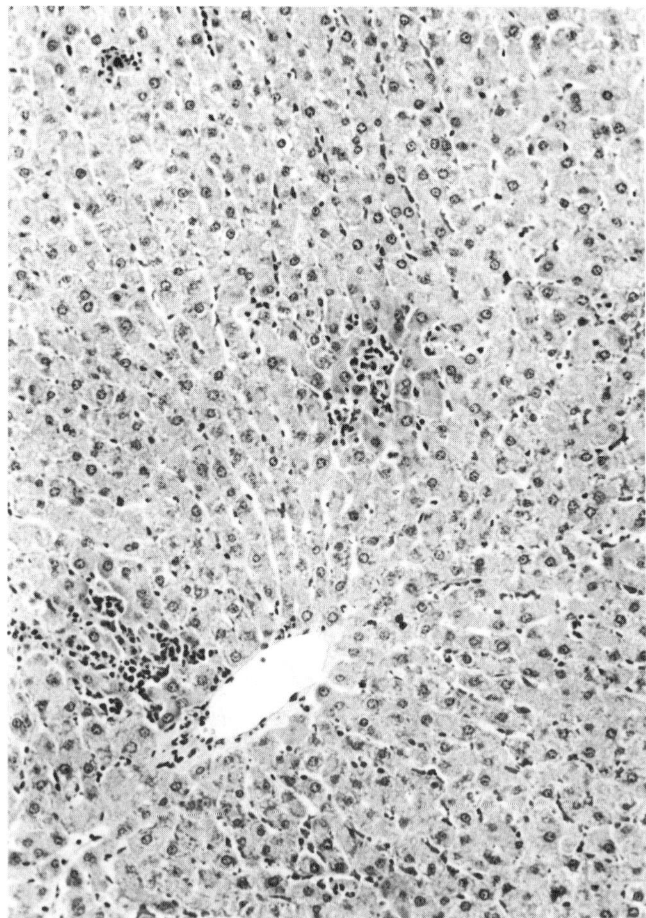

Figure 2. Section of liver from a copper-loaded rat showing increase in the number and size of inflammatory foci affecting zones 1 and 2. HE stain (x 32).

ences were detected among the Cu-loaded groups, as hepatic Cu was significantly higher in the Se-deficient group (820 ± 222 µg/g wet weight tissue), followed by the Cu-loaded groups with normal Se levels (group A) and Cu-loaded with Se supplementation (group B). Serum Cu was lower in the control group (1.24 ± 0.05 µg/g) compared to all Cu-loaded groups, but no statistically significant differences were detected between different Cu-loaded groups (A,B,C). Se levels in the liver reflected selenium-supplementation. A significant ($p < 0.005$) increase of hepatic Se was detected in group B (7.99 ± 3.7 µg/g), whereas the group with Se deficient diet had a significantly lower hepatic Se content (0.20 ± 0.01 µg/g). Serum Se was also lower in the Se deficient diet (0.10 ± 0.02 µg/g) compared groups A, B, and D, but no differences were detected between the Cu-loaded groups. Cu-loading did not affect hepatic Se levels as no statistical difference was detected between the control diet (group D) and the Cu-loaded diet with normal Se levels (group A).

Glutathione peroxidase levels in whole blood and liver tissue correlated well with Se levels. Glutathione peroxidase was markedly decreased in group C in both blood (216 ± 18 U/gHb) and liver (10 ± 3 U/g tissue). Control levels were 577 ± 76 U/gHb) in blood and 166

± 26 U/g tissue in liver. A significant (p < 0.005) increase in GSHPx activity was detected in the blood of Se-supplemented rats but not in their livers.

Histological changes in the livers of Cu-loaded rats were more prominent in zone 1 (Figure 1), and consisted of necrotic hepatocytes and multifocal accumulation of lymphocytes and plasma cells, often surrounding necrotic hepatocytes. In general, histological changes were more severe in rats fed the Se-deficient diet. In this group, necrotic hepatocytes were numerous and there was an increase in the size and number of inflammatory foci, which extended into the midzone (Figure 2). Lesions were less severe in the Se-supplemented rats. Livers from rats in group D were unremarkable (Figure 3). Rubeanic acid detected Cu granules within hepatocytes in zone 1 in the livers of all copper-loaded rats (Figure 4). Cu was not detected in control livers.

Results of liver enzymes activity are presented in Table 2. In general, hepatic enzymes (ALT, SDH) were increased in all Cu-loaded groups. ALT control values were 56 ± 17 (U/L), enzyme activity was higher in Cu-loaded rats with Se supplementation (406 ± 265 U/L), followed by group A and C. ALP was lower in Se deficient rats (group C) and SDH

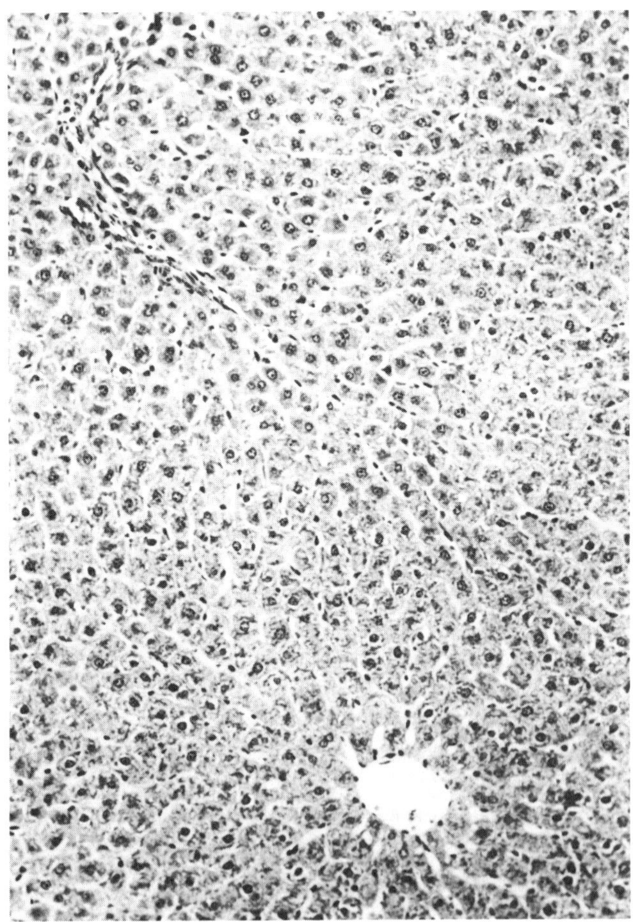

Figure 3. Section of liver from a control rat. HE stain (x 32).

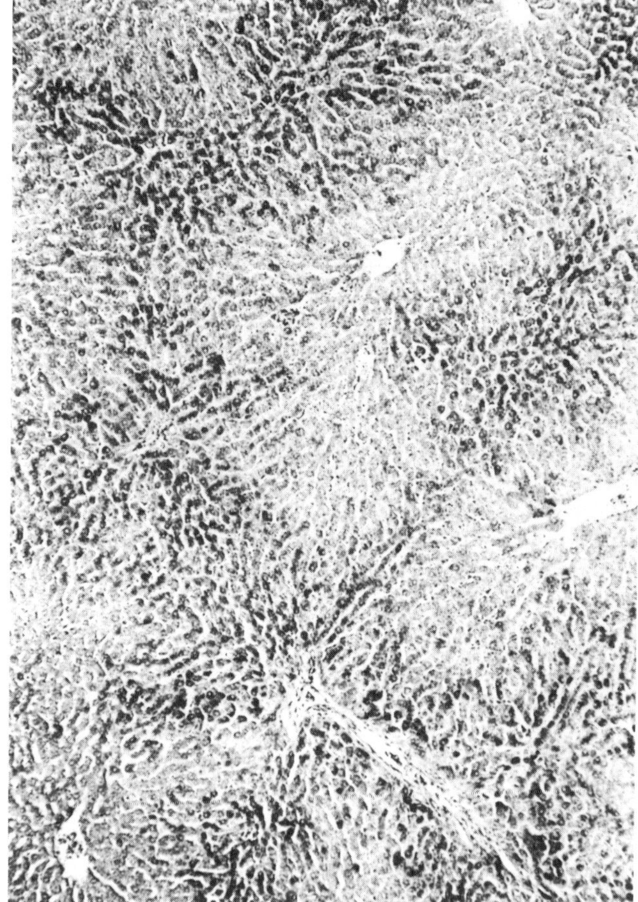

Figure 4. Section of liver from a Cu-loaded rat depicting Cu granules within hepatocytes in zone 1. Rubeanic acid stain (x 13).

was elevated in all Cu-loaded groups. Differences between groups were not detected in total protein, albumin, glucose, cholesterol and total bilirubin.

4. DISCUSSION

Dietary Se deficiency resulted in an increase in the severity of liver lesions caused by Cu. Se supplementation reduced the severity of Cu-induced hepatocellular damage, as evidenced by a decrease in the severity of lesions and the accumulation of Cu within the liver.

The exact mechanism of the protective effect of Se on heavy metal toxicity is not known, but it has been associated to a strong tendency to complex with metals (11) and its antioxidant properties. Se is a component of glutathione peroxidase (12), an antioxidant system capable of reducing lipid hydroperoxides and hydrogen peroxide (10). Se increases the reductive capacity of glutathione peroxidase by acting as a cofactor of the enzyme, thus help-

Table 2. Hepatic enzymes (ALT, ALP, SDH) and glutathione peroxidase activity (GSHPx) in the blood and livers of rats receiving a high Cu diet (Group A), high Cu diet with Se supplementation (Group B), high Cu/ Se deficient diet (Group C) and normal rodent diet (Group D). (Mean ± SD, N=8)

Group	ALT	ALP	SDH	GSHPx Blood	GSHPx Liver
A	237 ± 100	246 ± 44	89 ± 46	477 ± 69	125 ± 16
B	406 ± 265	249 ± 47	173 ± 166	731 ± 65	122 ± 9
C	216 ± 121	156 ± 27	163 ± 100	216 ± 18	10 ± 3
D	56 ± 17	227 ± 38	10 ± 5	577 ± 76	166 ± 26

ing to prevent peroxidation of polyunsaturated fatty acids present in cell membranes (9). Se has been shown to protect against the toxic effects of Cadmium (10).

The presence of Se in liver tissue was a reflection of dietary Se. Similarly, hepatic glutathione peroxidase activity was higher in Se-supplemented rats and markedly reduced in Se deficient rats. This association between dietary Se and Se levels in tissues, and GSHPx in liver tissue has been described by other authors (13).

Results from this study support the concept of a potential use of antioxidants in the treatment of Cu-associated diseases. It is not known whether lipid peroxidation is the primary mechanism or only an epiphenomenon in copper-induced hepatotoxicity. Should lipid peroxidation play any role in Cu toxicity, then the use of antioxidants would prevent or diminish cell injury and may be of value in the prevention and the development of adequate treatment of Cu poisoning.

5. REFERENCES

1. E.J. Underwood, *Trace elements in human and animal nutrition*. 4th ed. Academic Press, New York (1977).
2. R.M. Hardy, J. B. Stevens and C.M. Stowe, *Minnesota Veterinarian* **15**,13–24 (1975).
3. L.P. Thornburg, D. Shaw, M. Dolan, *Veterinary Pathology* **23**,148–154 (1986).
4. S. Haywood, *J. Comp. Pathol.* **90**,217–232 (1980).
5. I. Fuentealba, S. Haywood, and J. Foster, *Exp. Mol. Pathol.* **50**, 26–37 (1989).
6. P. Hochstein, K.S. Kumar and S.J. Forman, *Ann. N.Y. Acad. Sci.* **355**,240–248 (1980).
7. S. Haywood, M. Loughran and R.M. Batt, *Exp. Mol. Path.* **43**,209–219 (1985).
8. D.C. Sorjonen, and D.M. Vaughn, *Comp. Cont. Ed.* **11**,248–255 (1989).
9. J. Gambhir and R. Nath, *Indian J. Exp. Biol.* **30**,597–601 (1992).
10. SAS Institute Inc. SAS/STAT TM Guide for Personal Computers, Version 6 Edition, Cary, NC: SAS Institute Inc., pp 378 (1985).
11. J. Parizek, Elsevier Scientific Publishing Co Amsterdam) pp. 498 (1976)
12. J.T. Rotruck, A.L. Pope, H.E. Ganther, A.B. Swenson, D.G. Hafeman and W.G. Hoekstra *Science*, **179**,588 (1973).
13. S. Osame, T. Ohtani, S. Ichijo, *Jpn. J. Vet. Sci.* **52**,705–719 (1990).

46

COMPARISON OF AN IMMUNOHISTOCHEMICAL AND A HISTOCHEMICAL STAIN IN DETECTION OF COPPER IN RAT TISSUES

J. E. Mullins,[1] R. A. Fredrickson,[2] I. C. Fuentealba,[1] and R. J. F. Markham[1]

[1] Atlantic Veterinary College
University of Prince Edward Island
Charlottetown, PE, Canada, C1A 4P3
[2] Diagnostic Chemicals Ltd.
Charlottetown, PE, Canada, C1E 1B0

1. INTRODUCTION

Routine pathological diagnosis of hepatic copper (Cu) has traditionally depended upon histochemical stains such as rubeanic acid, rhodanine and orcein (1). The sensitivity and specificity of these stains are often unknown. In recent years, metallothionein (MT) has been used in immunohistochemical techniques, to indicate presence and distribution of heavy metals, in particular Cu, within biological tissues (2,3). Metallothionein (MT) is a low molecular weight (6000 D), cysteine-rich, cytoplasmic protein, with a high affinity for cadmium (Cd), Cu, zinc (Zn) and other heavy metals. It is thought to be involved in cellular detoxification due to metal sequestration (4). The purpose of this study was to compare MT immunostaining with rubeanic acid staining of Cu in normal and Cu-loaded rat tissues.

2. MATERIALS AND METHODS

2.1. Preparation of Metallothionein and Immunization

Rabbit MT (Sigma Chemical Co., St Louis, MO) was purified by anion exchange using fast protein liquid chromatography (Pharmacia, Montreal, PQ) (5). Purified MT was polymerized into soluble complexes with glutaraldehyde (6) and then mixed with Ribi's adjuvant in a 1:1 ratio (v/v). Three 15–16 week old Balb/c mice (Charles River, St Constant, PQ) were injected intraperitoneally with 100 µg of polymerized MT on days 0, 21, 35, 45

Therapeutic Uses of Trace Elements, edited by Nève et al.
Plenum Press, New York, 1996

and 55. The mice were anaesthetized with halothane (Wyeth-Ayerst, Montreal, PQ), and exsanguinated by transthoracic cardiac puncture. Serum was stored at -20 °C.

2.2. Experimental Animals

Thirty two male 10-month old Wistar rats (Charles River) were randomly placed in groups of four. Two groups were fed rat chow containing 1500 mg/kg $CuSO_4$ (Teklad Diets, Madison, WI) and two groups were fed rat chow alone (< 10 mg/kg Cu). Food and water were free choice. After 6 weeks rats were killed by CO_2 and cervical dislocation. Kidney, liver and small intestine were removed from each rat and fixed for 24 hours in 10 % neutral buffered formalin.

2.3. Histochemistry and Immunoperoxidase Staining

Paraffin embedded tissues were sectioned at 5 μm thickness and stained for Cu with rubeanic acid (7). Tissue sections (5μm) were applied to 0.1 % w/v poly L-lysine coated slides (Sigma). All reagents were prepared in Dulbecco's phosphate buffered saline (D-PBS) and used at room temperature. Tissues were deparaffinized, hydrated and treated sequentially with 1.5 % hydrogen peroxide for 30 minutes, 1 % normal goat serum (Vector Lab Inc., Burlingame, CA) for 30 minutes; primary antibody (polyclonal sera against rabbit MT diluted 1:400 or monoclonal antibody (MAb) 1E9 against horse MT at 1:12000 (kind gift of Dr. Jasani)) for 60 minutes, horse radish peroxidase-goat anti-mouse IgG (1:200)(Cappel, Organon Teknika Corporation, Durham, NC) for 60 minutes and substrate (18 mL D-PBS, 7 mL of 30 % H_2O_2 and 10 mg 3, 3'- diaminobenzidine (Sigma)) for 4 minutes. Between each step, slides were washed in D-PBS/0.1 % Tween-20 (J. T. Baker Chemical Co, Phillisburg, N.J) for 10 minutes. Slides were counterstained with aqueous 4.7 % haematoxylin (Shandon, Pittsburgh, PA) for 5 seconds, dehydrated, cleared, and mounted with Flo - texx (Lerner Laboratories, Pittsburgh, PA). Specificity of staining reaction was checked in several different control experiments: prior absorption of primary antibody with rabbit MT, substitution of primary antibody with D-PBS or normal mouse serum (1:100), and omission of both primary and secondary antibodies.

2.4. Metal Analysis

Hepatic Cu content was determined by atomic absorption spectrophotometry (AAS) (Perkin-Elmer Zeeman 5100 graphite furnace with Zeeman background corrector) as described elsewhere (8).

3. RESULTS

Mean hepatic Cu content was 4.2 μg/g in control and 581.7 μg/g (wet tissue weight) in Cu-loaded rats. Staining was specific for MT since none of the negative control procedures gave any staining. In control rats the presence of Cu could not be detected using rubeanic acid staining. In control rat livers occasional hepatocytes showed moderate nuclear and cytoplasmic immunoreactivity. Moderate MT immunoreactivity appeared predominantly in the cytoplasm of renal cortical proximal convoluted tubular epithelial cells (PCT) and Paneth cells in the small intestine.

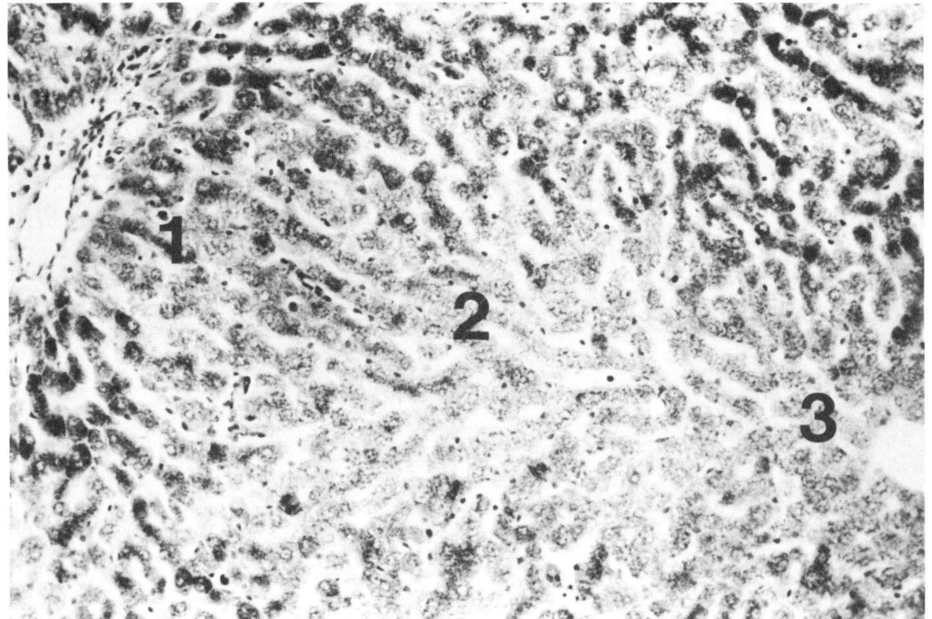

Figure 1. Section of liver from a copper-loaded rat showing periportal distribution of positive cytoplasmic granules. Hepatic zones 1–3 are indicated. Rubeanic acid. X128.

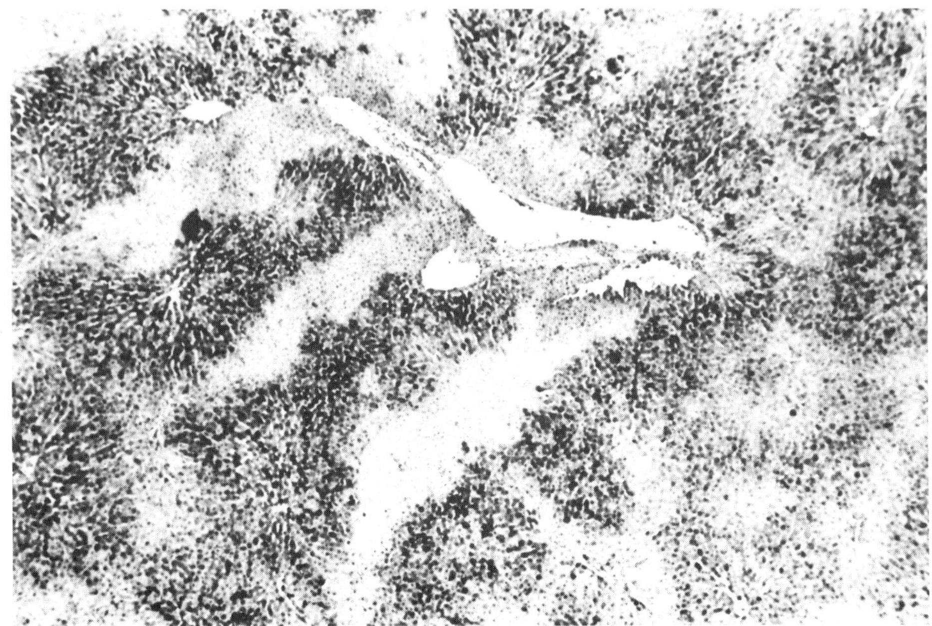

Figure 2. Section of copper-loaded rat liver showing variable immunostaining of hepatocytes. Haematoxylin counterstain. X 32.

In copper-loaded rats rubeanic acid staining was prominent in cytoplasm of zone 1 hepatocytes (Fig. 1) and renal PCT. Small intestine was negative for rubeanic acid. Immunoreactivity appeared variable between liver zones (Fig. 2, zones 1, 2 and 3), and between individual cells (Fig. 2). Hepatocytes in zones 1 and 2 stained most intensely, with both nuclear and cytoplasmic staining observed. Heavy MT immunoreactivity was seen in both nuclei and cytoplasm of renal cortical PCT (Fig. 3). Staining had a prominent segmental distribution extending from the capsular surface in rays towards the medulla (Fig. 4). Occasional distal convoluted tubule (DCT) epithelial cell cytoplasm and nuclei also stained moderately. In the small intestine, cytoplasm and nuclei of villi columnar epithelium and of Paneth cells displayed moderate to intense MT staining (Fig. 5).

4. DISCUSSION

The immunoperoxidase staining identified MT in both Cu-loaded and control rat liver, kidney and small intestine, whereas rubeanic acid only identified Cu in Cu-loaded liver and kidney. In liver, rats with the most intense immunostaining seemed to have the highest Cu concentrations as determined by AAS; this relationship was less clear with rubeanic acid staining. This finding corresponds to observations in Cu-associated diseases in humans (9), and Cu-loaded rats (3). It has been previously demonstrated that rubeanic acid detects cupric (CuII) forms of Cu, whereas the injurious cuprous (CuI) form is undetected (10). Copper is bound to MT in a CuI form (3,10). Therefore, distribution and intensity of MT immunostaining may better reflect potential Cu toxicity than rubeanic acid staining.

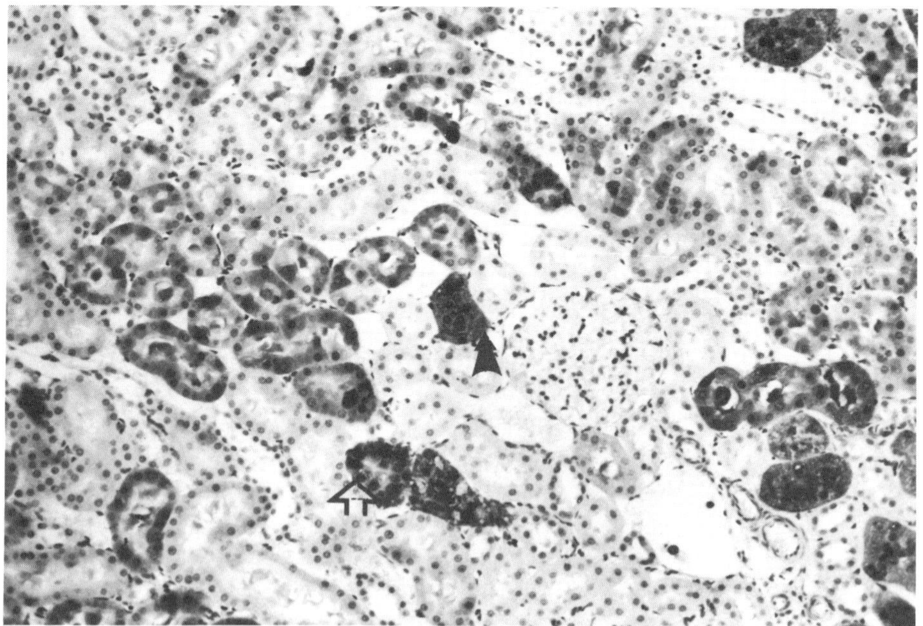

Figure 3. Section of copper-loaded rat kidney showing immunostaining of proximal convoluted tubule nuclei (open arrow) and cytoplasm (closed arrow). Haematoxylin counterstain. X 128.

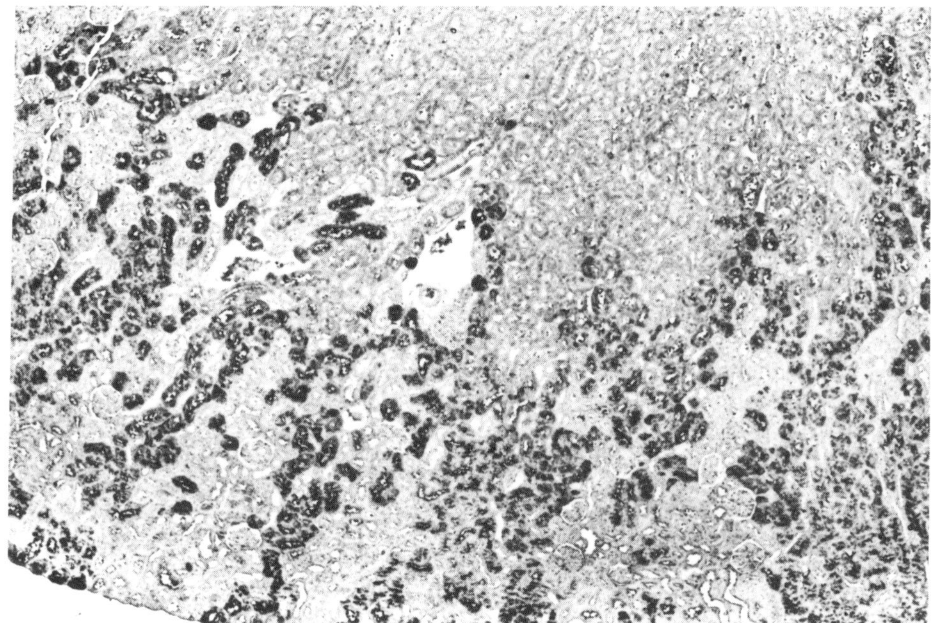

Figure 4. Section of copper-loaded rat kidney showing segmental distribution of metallothionein immunostaining. Haematoxylin counterstain. X 32.

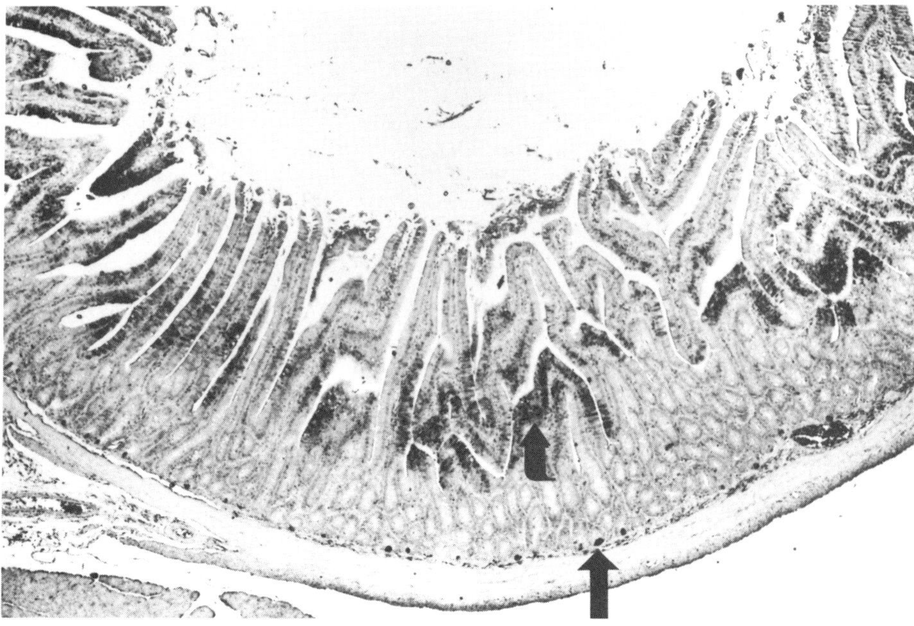

Figure 5. Section of small intestine of a copper-loaded rat with immunoreactivity in villous columnar epithelial cells (curved arrow) and Paneth cells (closed arrow). Haematoxylin counterstain. X 32.

The MT immunostain and rubeanic acid stained periportal areas (zone 1) of the liver most intensely, as reported by other authors (3,8). Both cytoplasm and nuclei stained in a periportal distribution. However, staining was variable between cells and within hepatocytes. This is in agreement with other studies (2,3) and reflects differences in the zonal distribution of hepatic enzymes and subcellular structures (11).

The kidney of Cu-loaded rats had marked immunoreactivity for MT. This was especially pronounced in outer cortex, and in segmental extensions into the medulla. Predominance of MT immunoreactivity in the outer cortex has been observed previously (3), as has the segmental distribution (12). Evering et al. (3) suggested that this pattern may represent a zonal distribution of cells which have prior committment to MT synthesis, such as exists in the liver.

Metallothionein staining was diffuse and amorphous within cytoplasm and nuclei of hepatocytes, whereas rubeanic acid stained discrete cytoplasmic granules. In Cu-fed rats, most of the Cu in hepatocytes is within lysosomes (13), which is detected by rubeanic acid (14). Results in the present study are similar to previous studies in which MAb 1E9 did not stain lysosomal MT, but only soluble cytoplasmic and nuclear MT in human fetuses (15), and in Cu-loaded rats (3).

In conclusion, MT immunostaining was found to be a sensitive indirect means of determining relative abundance of metal within tissues of Cu-loaded rats. It is not specific for Cu, since MT can bind to other metals such as Cd and Zn. However, immunostaining appears to have a good relationship with the amount of Cu within rat livers as determined by AAS. Immunostaining also appears more sensitive than histochemical staining for Cu since it was positive, and rubeanic acid staining was negative, in control rat livers despite the presence of low concentrations of Cu determined by AAS.

5. REFERENCES

1. S. Goldfischer, H. Popper and I. Sternlieb, *Am. J. Pathol.* **99,** 715–730 (1980).
2. L.M. Williams, H. Cunningham, A. Ghaffar, G.I. Riddoch, I. Bremner, *Toxicology* **55**, 307–316 (1989).
3. W.E. Evering, S. Haywood, M.E. Elmes, B. Jasani and J. Trafford, *J. Pathol.* **160**, 305–312 (1990).
4. D.H. Hamer, *Ann. Rev. Biochem.* **55**, 913–951 (1986).
5. P.E. Olsson and C. Hogstrand, *J. Chromatogr.* **402**, 293–299 (1987).
6. J.S. Garvey, R.J. Vander Mallie and C.C. Chang, *Methods Enzymol.* **84**, 121–138 (1982).
7. I.I. Uzman, *Lab. Invest.* 5, 229–305 (1956).
8. I.C. Fuentealba, S. Haywood and J. Trafford, *J. Comp. Pathol.* **100**, 1–11 (1989).
9. M.E. Elmes, J.P. Clarkson, N.J. Mahy and B. Jasani, *J. Pathol.* **158**, 131–137 (1989).
10. A.G.E. Pearse, in *Histochemistry, Theoretical and Applied*, vol. 2, A.G.E. Pearse, ed., Churchill Livingstone, Edinburgh, U.K., pp. 991–995 (1985).
11. R.G. Thurman, F.C. Kauffman and K. Jugermann, *Regulation of hepatic metabolism: Intra and intercellular compartmentation*, Plenum Press, New York (1986).
12. W.E. Evering, *PhD thesis*, University of Liverpool, UK (1989).
13. S. Haywood, *J. Pathol.* **145**, 149–158 (1985).
14. S.R. Gooneratne, J. M.C.C. Howell and E. Aughey, *J. Comp. Pathol.* **96**, 593–612 (1986).
15. C.D. Fuller, M.E. Elmes and B. Jasani, *J. Pathol.* **161**, 167–172 (1990).

47

TRACE ELEMENTS AND BONE METABOLISM

Anne Peretz

Rheumatology Clinic
Internal Medicine Department
Hôpital Universitaire Brugmann
place Van Gehuchten 4, B-1020 Brussels, Belgium

1. INTRODUCTION

The role of trace elements (TE) in bone has been mostly investigated for diagnostic procedures (isotopic scans, absorption test, fossil analysis) because they are bone-seekers that concentrate at the surface of bone. The presence of essential TE at physiological levels represents a reservoir (zinc, copper) while the presence of other elements which are environmental contaminants may be toxic for bone. Analysis of human skeletons from prehistoric to present time (1) revealed that TE concentrations vary with time and geographic situation. For example, bone lead concentration increased dramatically in Europe from prehistoric to middle-age, followed by a decrease until present period which seems in contradiction with the industry burst observed in our countries. The analysis of TE concentrations in bone for clinical or research purposes is a complicated task because reference data are lacking for "normal trace elements concentrations" in human bone biopsy. Very few authors (2) analysed human iliac crest bone for their bulk and trace elements in healthy and osteoporotic subjects.

2. BONE METABOLISM

Bone is a complex tissue composed of two phases: the mineralised matrix with 90% hydroxyapatite (Ca/P) and other minerals including trace elements laying on a protein framework of collagen type I and a small number of non collagenous proteins. Bone is in a dynamic state, being continually broken (resorption) and reformed (formation) by the co-ordinate actions of osteoclasts and osteoblasts (figure 1). The succession of these events is tightly regulated by a coupling process. Bone turnover is regulated by systemic hormones such as parathyroid hormone (PTH), 1,25-dihydroxy vitamin D3, thyroxine and also by factors generated in the bone microenvironment. Osteoclast activation is the initial step probably induced by the release of a soluble substance liberated by resting osteoblasts (collagenase?). PTH, $1,25\text{-}OH_2D_3$, interleukins 1 and 6 stimulate the differentiation of pre-osteoclasts to mature osteoclasts. During the resorption process, local factors (TGFb, IGF-1)

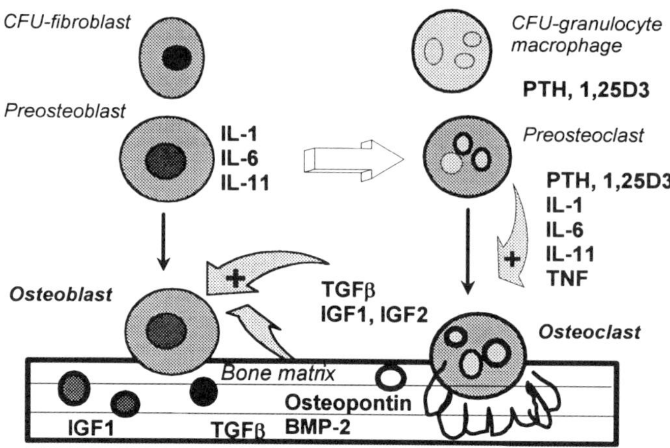

Figure 1. Activation of osteoblasts and osteoclasts.

buried in the matrix are liberated. These factors attract the precursors of osteoblasts to the site of bone resorption and initiated their differentiation and activation with the help of other growth factors. Any disturbance in the bone remodelling process in favour of resorption or formation could lead to bone disorders (3). In addition to calcium regulating-hormones and local factors, TE are known to be essential for the normal development and the maintenance of the skeleton, at least by virtue of their role in the activity of enzymes involved in collagen and proteoglycans and in collagenous and non-collagenous bone proteins synthesis. Recent studies (4) have shown that TE may act directly or indirectly on osteoclasts and osteoblasts (table 1).

3. ZINC

It is well established that zinc is important for growth in humans and animals. Bone growth retardation is reported in most conditions associated with zinc deficiency. Several studies reported skeletal abnormalities in zinc deficient animals, including osteopenia, disproportionate bone growth and cartilage dystrophy (5). Zinc is intimately involved in nucleic

Table 1. Effects of TE on bone formation and bone resorption

Trace Element	Bone formation	Bone resorption
Zinc	↑	↓
Copper	(↑)	↓
Fluoride	↑	
Strontium	↑	↓
Zeolite (Si)	↑	↓
Selenium	↑	?
Gallium	?	↓
Cadmium	?	↑
Lead	?	↑
Boron	?	↓

acid and protein synthesis and may therefore participate in many metabolic pathways in bone.

3.1. Bone Formation

Zinc has a stimulatory effect on bone formation in cell culture of osteoblastic lineages and in tissue culture of calvaria from weaning rats, demonstrated by the increase in DNA and protein content (6). In parallel, zinc increases the bone alkaline phosphatase activity, an enzyme involved in the mineralisation process and reflecting osteoblasts differentiation (7). The addition of cycloheximide, an inhibitor of protein synthesis at the translation level, completely abolishes these effects. The stimulatory effect of zinc is dose-dependent since zinc sulfate at 10^{-8} M has no effect contrary to the range 10^{-7} M -10^{-5} M (7). Yamaguchi et al (7) developed a new zinc compound β-alanyl-*L*-histidinato zinc (AHZ) in which zinc is chelated to β-alanyl-*L*-histidine. This compound seems more potent than zinc sulfate, particularly on osteoblastic proliferation suggesting that the mode of action of AHZ differs from that of zinc sulfate (7). The authors showed that zinc sulfate and AHZ stimulate the activity of osteoblastic aminoacyl-tRNA synthetase, a translational, rate-limiting enzyme which synthesises aminoacyl-tRNA , the initiator of protein synthesis (8). The same authors demonstrated that zinc sulfate markedly enhanced the anabolic effect of insulin-like growth factor 1 (IGF-1) on alkaline phosphatase activity, protein concentration, DNA content and cell number in osteoblastic cells (9). Such an effect was not observed for AHZ, reinforcing the hypothesis that zinc salts and zinc-chelating peptides may act differently. Zinc salts may be involved in the IGF-1 cellular signals, the tyrosine kinase and the protein phosphatase pathways. Additionally, zinc can directly stimulate the production of IGF-1 in osteoblastic cell culture.

3.2. Bone Resorption

The effect of zinc on bone resorption has received little attention until it was observed that zinc inhibits bone metabolism in organ culture of femoral bone and in ovariectomised rats. Yamaguchi et al (10) undertook in vitro studies using the model of osteoclasts-like cells in mouse marrow cultures. They reported that zinc at 10^{-6} - 10^{-5} M and AZH at 10^{-7} M - 10^{-8} M inhibited the proliferation of osteoclasts-like cells induced by bone-resorbing agents. By contrast, Ni, Cu, Mn or Co at similar concentrations had no effect on PTH-induced osteoclast-like cell proliferation (10). Using another model, Moonga & Dempster (11) also found that zinc is a potent inhibitor of osteoclasts in vitro, at low concentrations (10^{-14} M). Such property was not observed for Ba, Sr, Ga, and Mn (11). The authors ascertained that the effect of zinc could not be due to cell toxicity or to an interaction of zinc with the surface of the bone.

3.3. *In Vivo* and Clinical Studies

The administration of zinc salts or zinc compounds to rats was able to prevent the bone loss associated with several conditions. Ovariectomy in rats induced a rapid and important bone loss which could be completely prevented by the administration of AHZ at 10, 30, and 100 mg/Kg (12). Positive results obtained after surgical ovariectomy, which is a good model to study osteoporosis, strongly suggests a therapeutic role for zinc. However, the administration of zinc to prevent osteoporosis has not been studied in humans. The administration of AHZ in corticosteroid-induced osteopenia in rats was also able to prevent the associated bone loss (13). The evoked mechanisms are both the inhibition of bone resorption induced by secondary hyperparathyroidism and the stimulation of bone formation. Zinc concentration in human bone was found normal (2), or decreased (14) in osteoporotic patients. Elderly Japanese patients had normal bone zinc unless a chronic disease is present. Patients with cancer presented an elevated rib zinc content, attributed to a zinc redistribution state as found in inflammatory states (15). Increased urinary zinc excretion was reported in spinal-cord-injured patients, a condition associated with rapid bone demineralisation (16), in osteoporotic women and in elderly women (17). Hyperzincuria was correlated with biochemical indices of bone turnover. This suggests that zinc is released from the skeleton together with collagen fragments and other minerals in diseases with high bone turnover (17). Therapeutic intervention with zinc alone has not yet been reported. The administration of 1000 mg calcium together with TE (zinc 15 mg, copper 2,5 mg, manganese 5 mg) in a small group of postmenopausal osteoporotic women was reported to abolish the bone loss (+0.2%, bone mineral density) compared to calcium alone (-1.3%), TE alone (-2%) and placebo-treated patients (-3.1%) (4). Therefore, zinc appears as one of the most promising agent in metabolic bone disorders but possible clinical applications need to be confirmed by further studies.

4. COPPER AND MANGANESE

Copper is the cofactor of lysyloxidase, an enzyme involved in the cross-linking of collagen and elastin to form stable fibrils. Copper deficiency is associated with bone abnormalities in several animal species: osteoporosis, thinned corticles, wide marrow space, absence of trabeculae, spontaneous fractures. Skeletal defects have been reported in Menkes disease and in infants consuming inadequate copper amounts, notably in neonates and during parenteral nutrition. The defects were osteopenia, cupping of metaphyses, and spontaneous fractures in trabecular bone (ribs, vertebrae) (18). Manganese deficient animals have low galactosyltransferase activity, an enzyme associated with chondroitin sulfate synthesis (4). In rats, moderate copper (and manganese) diet restriction decreased the femur calcium content without modification of phosphorus, suggesting that modifications in crystal composition (hydroxyapatite Ca/P) could be an early manifestation of copper deficiency. In the serum, indices of bone turnover were increased, similarly to osteoporosis. To study cell bone metabolism, two models of subcutaneous implants with either mineral-containing bone particles or demineralised bone powder were used in rats. They allowed to document osteoinduction, the recruitment and differentiation of osteolcast-like cells at the site of the implant. Copper (and manganese) deficiency reduced both bone induction and bone resorption (4). Decreased serum manganese has been reported in postmenopausal osteoporosis without modification in bone content as compared to age-matched controls (19). In postmenopausal women, serum copper levels were in the normal range, unless the women is on oestrogen replacement therapy which causes an important increase in copper levels. Whether such increase could play a role in the prevention of bone loss by oestrogens remains speculative.

5. SELENIUM

Selenium has attracted much attention as the constituent of glutathione peroxidase which is involved in the host-defence against oxidative damage. Nevertheless, selenium deficiency has been associated with Kashin-Beck disease in which cartilage and bone defects are characteristic. Experimental selenium deficiency for several months in rats induced a decrease in bone mineral density of the femur associated with a decrease in selenium serum concentrations and in alkaline phosphatase activity without changes in serum calcium and phosphorus concentrations. Chondrocytes were also affected, showing in deep layer a ballooning of endoplasmic reticulum and nuclei fragmentation. The activity of the sulfotransferase activity, an enzyme involved in glycosaminoglycan synthesis was decreased (20). In mouse, selenium deficiency increased bone fragility. Collagen type I (bone) and type II (tendon) showed overhydroxylation of lysine residues which impairs collagen resistance and the process of fibril formation (21).

5. SILICON

Silicon was reported as a TE important in bone formation and mineralisation. It seems particularly abundant in sites of active calcification. Animals fed a silicon-deficient diet presented impaired collagen synthesis and skeletal defects. Recently, a new class of silicon-containing compounds, the zeolites, were shown to possess osteogenic properties (22). They are composed of hydrated alkali-aluminium silicates which form cage-like structures in crystalline form. The rationale for using zeolite A in bone disorders originates from the observation that the addition of zeolite A to the diet of chicken increased their eggshell thickness and, in predisposed chickens, decreased the expression of chondrodysplasia. In vitro studies demonstrated that Zeolite A increased the proliferation and differentiation of human osteoblasts-like cells. The effect is explained by an increase of the mRNA levels of TGFβ, a mitogen for osteoblasts (23). In addition to a positive effect on bone formation, zeolite was reported to inhibit osteoclasts-mediated bone resorption (24) . Silicate or aluminate salts alone were not able to reproduce the effect. Therefore, the activity of zeolite A could depend on the crystalline structure or to a substructure of the intact compound.

6. OTHER TRACE ELEMENTS

Fluoride is currently used to treat osteoporosis although the optimal dose and the most adequate chemical form are not clearly established. Strontium, a potent inhibitor of bone resorption, has recently gained much interest in the treatment of post menopausal osteoporosis. Both elements are extensively considered elsewhere. Gallium nitrate is used in USA to treat accelerated bone turnover states (Paget's disease) and bone metastases (25). Boron was shown to decrease osteoarthritis symptoms (26) and to decrease parameters of accelerated bone turnover in parallel with an increase in bone mineral density (27).

7. CONCLUSION

TE are constituents of the mineralised matrix of bone. They are active partners in bone metabolism and could interfere with bone cells proliferation and differentiation. In vitro studies have demonstrated that some TE may have dual properties depending on the type of the cells in culture: stimulation of osteoblast-like cells and inhibition of osteoclasts-like cells. In accelerated bone turnover disorders, an ideal therapeutic agent should possess such properties. However, it is rarely observed because bone resorption and bone formation are tightly coupled in vivo. The next important step in the field of TE in bone metabolism would be to further document clinical applications.

8. REFERENCES

1. A. Chausmer, S. Wallach, in *Trace Metals and Fluoride in Bones and Teeth*, N.D. Priest, F.L. Van De Vyver, eds, CRC Press, Boca Raton, Floride, pp 253–270 (1990).
2. M. F. Baslé, Y.Mauras, M. Audran, P. Clochon, A. Rebel, P. Allain, *J. Bone Miner. Res.* 1, 41–47 (1990).
3. G.R. Mundy, *Bone remodeling and its disorders*, Martin Dunitz Ltd, London (1995).
4. L. Strause, P. Saltman, K. Smith, M. Andon, in *Nutritional Aspects of Osteoporosis*, vol. 85, P. Burckhardt, R.P. Heaney, eds, Raven Press, New York, pp. 223–233 (1991).
5. R. E. Burch, H.K.J. Hahn, J.F. Sullivan, *Clin. Chim.* 21, 501–520 (1975).
6. M. Yamaguchi, R. Yamaguchi, *Biochem. Pharmacol.* 35, 773–777 (1986).
7. M. Yamaguchi, J. Ohtaki, *Pharmacology* 43, 225–232 (1991).
8. M. Yamaguchi, S. Kishi, M. Hashisume, osteoblastic MC3T3-E 1 cells: activation *Peptides* 15, 1367–1371 (1994).
9. T. Matsui, M. Yamaguchi, *Peptides* 16, 1063–1068 (1995).
10. S. Kishi, M. Yamaguchi, *Biochem. Pharmacol.* 48, 1225–1230 (1994).
11. B. Moonga, D.W. Dempster, *J. Bone Miner. Res.*10, 453–457 (1995).
12. Y. Segawa, N. Tsuzuike, E. Tagashira, M. Yamaguchi, *Biol. Pharm. Bull.* 16, 486–489 (1991).
13. Y. Segawa, N. Tsuzuike, Y. Itokazu, E. Tagashira, M. Yamaguchi, *Res. Exp. Med.* 192, 317–322 (1992).
14. E.M. Alhava, H. Olkkonen, J. Puittinen, V.M. Nokko-Koivisto, *Acta Orthop. Scand.* 48, 1–4 (1977).
15. J. Yoshinaga, T. Suzuki, M. Morita, M. Hayakawa, *Sci. Total Environ.* 162, 239–252 (1995).
16. A. Ohry,Y. Shemesh, R. Zak, M. Herzberg, *Paraplegia* 18, 174–180 (1980).
17. M. Herzberg, J. Foldes, R. Steinberg, J. Menczel, *J. Bone Miner. Res.* 5, 251 -257 (1990).
18. J.J. Strain, *Med. Hypotheses* 27, 332–338 (1988).
19. J.Y. Reginster, L.G. Strause, P.Saltman, P. Franchimont, *Med. Sci. Res.* 16, 337–338 (1988).
20. S. Sasaki, H. IWATA, N. Ishiguro, O. Habuchi, T. Miura, bone, and articular *Nutrition* 10, 538–543 (1994).
21. C. Yang, C. Niu, M. Bodo, E. Gabriel, H. Notbohm, E. Wolf, P.K. Müller, acid *Biochem. J.* 289, 829–835 (1993).
22. M.J. Keenan, M. Hegsted, P.J. Wozniak, J. D. Bott, A. Reisenauer, colonies. *J. Vit. Nutr. Res.* 62, 228–232 (1992).
23. P.E. Keeting, M.J. Oursler, K.E. Wiegand, S.K. Bonde, T.C. Spelsberg, B.L. Riggs, *J. Bone Miner. Res.* 7, 1281–1289 (1992).
24. N. Schütze, M.J. Oursler, J. Nolan, B.L. Riggs, Spelsberg, *J. Cell. Biochem.* 58, 39–46 (1995).
25. P.T. Guidon, J.R. Roberto Salvatori, R. Bockman, *J. Bone Miner. Res.* 8, 103–112 (1993).
26. R.E. Newnham, *Environ. Health Perspect.* 102 (suppl7), 83–85 (1994).
27. S.L. Meacham, L.J. Taper, S. L. Volpe, *Environ. Health Perspect.* 102 (suppl 7), 79–82 (1994).

48

EFFECTS OF STRONTIUM ON BONE TISSUE AND BONE CELLS

P. J. Marie

INSERM Unit 349
Cell and Molecular Biology of Bone and Cartilage
Lariboisière Hospital
6 rue Guy-Patin, 75475, Paris Cedex 10, France

1. INTRODUCTION

Beside hormones and growth factors, minerals and trace elements are known to affect bone formation and resorption through direct and indirect effects on bone cells. In addition, some trace elements, such as strontium, have been shown to induce pharmacological effects on cells when administered at levels higher than those required for normal cell physiology. This brief review summarizes the known effects of strontium on bone tissue and bone cells, and discusses the potential interest of strontium for the treatment of osteoporosis.

2. STRONTIUM IN SOFT TISSUES

Strontium was first discovered in 1809 near the Scottish village of Strontian. This member of the alkaline earth elements has an atomic mass of about twice that of calcium. Strontium is present in soil and water at variable concentrations (0.001–39 mg/l) and a normal diet contains about 2–4 mg/day strontium (1). Most organs are capable of discriminating among calcium and strontium. In the gut, the intestinal absorption of calcium is higher than that of strontium. Renal, placental and cardiac cells are also able to discriminate between calcium and strontium (2, 3). Following oral administration of strontium, plasma and intracellular strontium concentrations rise dose-dependently, without affecting intracellular calcium concentrations (4). In contrast, intravenous administration of high doses of strontium induces a fall in plasma calcium resulting from an increase in renal excretion of calcium (5).

The effects of strontium on soft tissue and cells are dependent on the dose used. At doses lower than 1% in the diet, no toxic effect at short or long term is observed in animals. In chicken given a calcium-deficient diet, high doses of strontium (> 1% in the diet) were reported to reduce the 25(OH)vitamin D_3–1α-hydroxylase activity (6). In this model, however, hyperparathyroidism secondary to calcium deficiency may increase the 25(OH)vitamin

D_3–1α-hydroxylase activity, which can be reduced by either strontium or calcium. Indeed, although high concentrations of strontium may reduce the production of $1,25(OH)_2$vitamin D_3 (7), calcium induces similar effects (8). Thus, both cations act identically in vivo to decrease the $25(OH)$vitamin D_3-hydroxylase activity (8).

In animals fed a low calcium diet, high doses of strontium may also induce skeletal abnormalities. In these animals, the lower than normal calcium absorption associated with the low calcium diet induce bone abnormalities such as rickets (9,10), defective bone mineralization (11) and alteration of the mineral profile (12,13). In contrast, as discussed below, low doses of strontium have no effect on bone mineralization in any species.

3. STRONTIUM IN CALCIFIED TISSUES

Strontium content in bone declines through Evolution, which probably reflects the transition from the aquatic life to the terrestrial life for most vertebrates. The high disponibility of calcium in the soil may have led to the selection of calcium rather than strontium as the main mineral component of the skeleton. Strontium is a bone-seeking trace element, and biochemical studies using tracers (85Sr, 47Ca) have shown that strontium and calcium incorporate at the same rate into the skeleton (14). This led to use strontium as a marker of bone metabolism and to the development of clinical tests measuring mineral absorption and fixation in the skeleton using stable strontium (4). Strontium incorporates into the skeleton in a dose-dependent manner and bone strontium content is directly proportional to plasma levels. At the mineral level, strontium is known to substitute for calcium at random in hydroxyapatite cristals. Small hydroxyapatite cristals are rapidly formed and incorporate much more strontium than bigger cristals that are formed more slowly (15). However, the size of cristals containing strontium is higher than that of hydroxyapatite cristals, which leads to a more regular form of cristals containing strontium and to stabilisation of these cristals. Hence, substitution of calcium by strontium and its incorporation in dental enamel induces stability of the apatitic structure and high resistance to degradation by bacterial acids (15). In addition, the combination of strontium and fluoride is more effective on cristal resistance to acids than each element alone (15). Epidemiological studies have indeed described a very low incidence of dental caries in Ohio and Wisconsin (USA) as a result of high strontium content (5–10 mg/l) and fluoride content (1 mg) in drinking water (16).

The first studies on the effects of strontium element on bone tissue were reported by Lenherdt in 1910, who showed that stable strontium stimulates bone formation and inhibits bone resorption in animals given a normal calcium diet (17). Later, Shorr and Carter (18), and Janes and McCaslin (19) reported clinical studies in osteoporotic subjects treated with strontium salts. These authors found a clinical and radiological improvement after treatment with strontium lactate in women with postmenopausal osteoporosis. Other authors reported an increased mineral uptake after administration of strontium lactate (20). More recently, Skoryna reported an increased bone mineral density in lytic lesions associated with metastatic cancer in patients receiving stable strontium, suggesting preferential incorporation of strontium in sites of increased bone remodeling (5). These preliminary studies suggested that low doses of stable strontium may influence bone formation and resorption directly, which generated new experimental studies on the effects of strontium on bone tissue and bone cells.

4. EFFECTS OF STRONTIUM ON BONE FORMATION AND RESORPTION

Since 1984, we have developed a number of studies aimed at determining the effects of strontium salts on bone and bone cells. We initially showed that oral administration of strontium salts at low dosage levels (0.19–0.40 %) for 2 months in rats fed a normal calcium diet, induced a rise in plasma and bone strontium levels, stimulated bone formation and increased the trabecular bone volume evaluated by bone histomorphometry (21, 22). This effect was not mediated by changes in parathyroid hormone or 1,25-dihydroxyvitamin D_3 and did not affect the mineral profile or the mineralization process in these animals (21–23). We also reported that stable strontium supplementation for one month in normal mice dose-dependently increased bone formation and reduced bone resorption, as evaluated by bone histomorphometry (24). Other authors showed that the local administration of strontium leads to increased bone formation in alveolar bone in rats (25). In a preliminary clinical study, we also reported that the administration of strontium carbonate at low dosage levels (600 mg/day) for six months in six osteoporotic patients induced a rise in bone strontium, and stimulated bone formation, as evaluated by bone histomorphometry performed on iliac crest bone biopsies (26). In this study, strontium supplementation induced an increased amount of matrix and number of osteoblasts, whereas bone mineralization was not affected. These in vivo data suggested that strontium administration at low doses can stimulate the endosteal bone formation in normal animals and humans. On the other hand, it was found that strontium inibits dose dependently bone resorption in organ culture of rat bone (27). Altogether, these in vitro and in vivo data suggest that strontium has unique properties since this element is able to stimulate bone formation and to inhibit bone resorption.

5. STRONTIUM AND OSTEOPOROSIS

The experimental studies described above showed that strontium inhibits bone resorption and stimulates bone formation in vivo, suggesting that strontium salts may have a potential interest in the treatment of osteoporosis. Based on these data, a divalent strontium salt (S12911) was developed by a French pharmaceutical company (Institut de Recherches Internationales Servier, IRIS, France). This compound containing two strontium atoms allows a good bioavailability for strontium. This strontium derivative has therefore been tested in a number of in vitro and in vivo experimental studies.

In vivo experiments showed that a short term (8 weeks) oral administration of the strontium derivative S12911 in normal growing rats stimulated bone formation in the tibia metaphysis, and increased the femoral bone mineral content which was associated with an increased number of trabeculae (28). Long-term administration (2 years) in normal rats improved the mechanical properties of the midshaft humerus associated with an increase in the shaft diameter, and increased the mechanical properties of the lumbar vertebrae, leading to a better bone resistance (29).

These results led to test this compound as a possible effective treatment in two models of osteopenia. Estrogen deficiency in rats resulting from ovariectomy induces osteopenia due to an accelerated rate of bone resorption which predominates over bone formation. Therefore, the ovariectomized rat has been widely used as a model of postmenopausal osteoporosis. Using this model, we have shown that treatment with the divalent strontium salt S12911 inhibited bone resorption and prevented the trabecular bone loss induced by oestrogen deficiency in ovariectomized rats, as evaluated by bone ash, bone mineral content and trabecular bone volume (30). In contrast to other compounds (bisphosphonates, calcitonin) capable of inhibing bone resorption in this model, S12911 decreased bone resorption whereas bone formation was maintained, suggesting that this strontium derivative may act as an uncoupling agent (30). In addition, the administration of this compound was shown to partially restore the bone mineral content in rats treated three months after ovariectomy (31). This strontium derivative, therefore, has potential preventive and curative effects on ovariectomy-induced bone loss in rats.

The effect of S12911 was also tested in another model of osteopenia induced by immobilization. Hind-limb immobilization in rats induces trabecular osteopenia resulting from increased bone resorption and decreased bone formation. We found that treatment of immobilized rats with the strontium derivative S12911 partially prevented the trabecular bone loss in the immobilized limb, as evaluated by bone ash, bone mineral content and trabecular bone volume. This effect was associated with a reduction of biochemical and histomorphometrical indices of bone resorption, whereas bone formation remained unaffected (32). No mineralization defect was observed in all these experimental experiments, including in non-rodents species, where strontium was found to be incorporated more in trabecular bone than in cortical bone after repeated oral administration of S12911 for 3 months in Cynomolgus monkeys (33).These in vivo results confirm that this strontium derivative has uncoupling activity and suggest that strontium salts may have potential beneficial interest in the treatment of osteopenic disorders.

Several experiments were conducted in vitro to evaluate the cellular mechanisms of action of the strontium derivative S12911 on bone cells. It was found that S12911 inhibited the bone resorbing activity of mouse osteoclasts in organ culture (34). Neither S12911 calcium or sodium salts showed any significant inhibitory effect in this culture system. It was also found that pre-incubation of bone slices with S12911 strontium salt inhibited the resorption induced by isolated rat osteoclasts in vitro without affecting the cellular attachment and viability (35). In addition, S12911 inhibited the osteoclast differentiation induced by 1,25-dihydroxyvitamin D_3 in chicken bone marrow cultures (35). These in vitro data demonstrate that strontium salts may inhibit directly and indirectly the osteoclast function.

The cellular effects of S12911 strontium salt on bone formation have also been investigated in vitro in a model evaluating the osteoblastic cell recruitment and function. Using a quantitative evaluation of cell replication of periosteal cells, pre-osteoblasts and osteoblasts in rat calvaria in vitro, it was found that the strontium derivative S12911 enhanced the replication of pre-osteoblastic cells and subsequently stimulated collagen and non-collagen protein synthesis (36). The stimulation of pre-osteoblastic cell replication and collagen synthesis by this divalent strontium salt was confirmed in rat calvaria cell cultures (36). The available data on the effects of strontium salts, summarized in Table 1, show that strontium has direct effects on bone cells, inducing the proliferation of osteoblast precursor cells and inhibiting osteoclast function.

The precise cellular and molecular mechanisms of action of strontium salts on bone cells remain to be clarified. Strontium may directly affect the osteoclast activity, or induce changes in cristal conformation, leading to stabilisation of the cristal, and inhibition of osteoclastic resorption. On the other hand, part of the strontium effect on bone cells may be due to a calcium-like effect, since high doses of calcium were shown to inhibit osteoclast activity in vitro. In osteoblasts, strontium may function as calcium within the cell, leading to stimulate cell growth and subsequently inducing an increase in bone matrix synthesis and bone formation. Strontium may also be involved as a co-factor stimulating specific enzymatic activities

Table 1. Effects of strontium salts on bone tissue and bone cells

	Bone formation/ osteoblasts	Bone resorption/ osteoclasts	References
In vivo			
Rats	+	−	21-23, 28-32
Mice	+	−	24
Humans	+	−	26
In vitro			
Organ culture	+	−	25, 27, 36
Osteoblastic cells	+		36
Osteoclast-like cells		−	34, 35

+: Stimulatory effect; −: Inhibitory effect.

involved in bone cell metabolism. Further studies are therefore required to determine more precisely the cellular and molecular mechanisms of action of strontium on bone cells.

6. CONCLUSIONS

Several experimental studies showed that strontium salts increase bone formation and reduce bone resorption in different models and animals. The available information indicates that low doses of stable strontium have multiple effects on bone metabolism. In vitro studies showed that strontium salts may directly or indirectly affect bone cells recruitment and activity in vivo and in vitro, and have inhibitory effects on bone resorption and stimulatory effects on bone formation (Table 1). On the other hand, a divalent strontium derivative (S12911) was shown to inhibit bone resorption and to stimulate bone formation in vitro, and to have beneficial effects on bone formation and resorption and trabecular bone volume in different models of osteopenic disorders. If these unique effects are also observed in humans, strontium may be of potential therapeutic value in the prevention or treatment of osteopenic disorders, such as in postmenopausal osteoporosis. Further experimental and clinical studies have to be developed to evaluate these promising possibilities.

7. REFERENCES

1. S.C. Skoryna, *Can. Med. Assoc. J.* **125**, 703–712 (1981).
2. A.M. Shane, in *The Transfer of Calcium and Strontium Across Biological Membranes*, R.H. Wasserman ed., Acadamic, New York (1963).
3. S.G. Kshirsagar, *J. Biosci.* **9**, 129–135 (1985).
4. I.R. Reid, J. Pybus, T.M.T. Lim, S. Hannon, H.K. Ibbertson, *Calcif. Tissue. Int.* **38**: 303–305 (1986).
5. S.C. Skoryna, M.Fuskova, in *Hanbook of Stable Srontium*, S.C. Skoryna ed., Plenum Press, New York, pp 593–617 (1985).
6. J.L. Omdhal, H.F. DeLuca, *J. Biol. Chem.* **247**, 5520–5525 (1972).
7. R.A. Corradino, R.H.Wassermann, *Proc. Soc. Exp. Biol. Med.* **133**, 960–963 (1970).
8. J.L. Omdhal, P. Evan, *Arch. Biochem. Biophys.* **184**, 179–188 (1977).
9. P.G. Shipley, E.A. Park, E.V. McCollum, N. Simmonds, E.M. Kihney, *John Hopkins Hosp. Bull.* **33**, 216–220 (1922).
10. E. Storey, J. *Bone Jt Surg.* **443**, 194–208 (1962).
11. E. Storey, V.C. West, *Calcif. Tissue Res.* **6**, 290–300 (1971).
12. A.E. Sobel, Y. Cohen, B. Kramer, *Biochem. J.* **29**, 2640–2645 (1935).
13. E.B. Neufeld, A.L. Boskey, *Bone,* **15**, 425–430 (1993).
14. A.R. Johnson, W.D. Armstrong, L.Singer, *Calcif. Tissue Res.* **2**, 242–252 (1968).

15. J.D.B. Featherstone, C.P. Shields, B.Khademazad, M.D.Oldersham, *J. Dent. Res.* **62**, 1049–1053 (1983).

16. M.E.J. Curzon P.C. Spector, *Arch. Oral Biol.* **23**, 647–653 (1978).

17. F. Lenherdt, *Beitr. Path. Anat. Allg.* **47**, 215–245 (1910).

18. E. Shorr, Carter, A.C. *Bull. Hosp. Jt Dis. Orthop. Inst.* **13**, 59–66 (1952).

19. J.M. Janes, F. McCaslin, *Proc. Mayo Clin.* **34**, 329–334 (1959).

20. J.M. Warren, H. Spencer, *Clin. Othop.* **117**, 307–320 (1976).

21. P.J. Marie, M.T. Garba, M. Hott, L. Miravet, *Min. Elect. Me*tab., 11, 5–13 (1984).

22. M.D. Grympas, P.J.Marie, *Bone* **11**: 313–319 (1990).

23. M.D. Grynpas, E. Hamilton , R. Cheung, Y. Tsouderos, P. Deloffre, M. Hott, P.J. Marie, *Bone* (in press).

24. Marie PJ, M. Hott, *Metabolism*, **35**, 547–551 (1986).

25. E.F. Ferraro, R. Carr, K.A. Zimmerman, *Calcif. Tissue Int.* **35**: 258–260 (1983).

26. P.J. Marie, G.Chabot, F.H. Glorieux, P.J. Piron, S.C. Skoryna, G. Donnay, In *Trace Subst. Env. Halth*" ed. DD. Hemphill, University of Missouri Press, Missouri, **19**, 193–208 (1985).

27. A. Matsumoto, *Arch. Toxicol. 62.* 240–241 (1988).

28. M.E. Arlot, J.P. Roux, G. Boivin, B. Perrat, Y. Tsouderos, P. Deloffre, P.J. Meunier, *J. Bone Min. Res.* **10**, S1, abstract M415 (1995).

29. P. Ammann, R. Rizzoli, P. Deloffre, Y. Tsouderos, J.M. Meyer, J.P. Bonjour, *J. Bone Min. Res.* **10**, S1, abstract M422 (1995).

30. P.J. Marie, M. Hott, D. Modrowski, C. de Pollak, J. Guillemin, P. Deloffre, Y. Tsouderos. J. Bone Min. Res. **8**, 607–617 (1993).

31. M.E. Arlot, P. Braillon, J.P. Roux, P. Deloffre, Y. Tsouderos, P.J. Meunier, *Bone Morphometry Meeting.* Lexington, 4 (Abstract).

32. P.J. Marie, M. Hott, D. Modrowski, P. Chappuis, J. Guillemain, P. Deloffre, Y. Tsouderos, *J. Bone Min. Res.*, **10**, S1, abstract 105 (1995).

33. G. Boivin, P. Deloffre, B. Perrat, G. Panczer, M. Boudeulle, Y. Mauras, Y. Tsouderos. P.J. Meunier, *J. Bone Min. Res.* (in press).

34. Y. Su, Bonnet J., P. Deloffre, Y. Tsouderos, R. Baron, *Bone Mineral*, **17**, S1, abstract 449 (1992).

35. Y. Su, Bonnet J., P. Deloffre, Y. Tsouderos, R. Baron, *J. Bone Min. Res.* **7**, S1, abstract 853 (1992).

36. E. Canalis, M. Hott, P. Deloffre, Y. Tsouderos, P.J. Marie, *Bone* (in press).

49

FLUORIDE AND BONE

Toxic Effects and Therapeutic Role

Georges Boivin and Pierre J. Meunier

INSERM Unité 403
Faculté A. Carrel
69372 Lyon Cedex 08, France

1. INTRODUCTION

Trace elements have an essential effect in minute doses, in contrast with other elements such as calcium and phosphorus that are necessary to the organism in large amounts. However, trace elements may be very toxic when present in large amounts. Thus, the effects of trace elements may be considered as physiological manifestations caused by very small amounts of substances which are toxic at higher concentrations. The wide distribution of fluoride (F) in food and water, plants and animals, as well as in human tissues, suggests that fluoride must have a definite physiological effect. Fluoride is present in the hard and soft tissues of the body in extremely small amounts of the same order of magnitude as other essential trace elements (1). The metabolism and toxicity of fluoride have been reviewed by Whitford (2) and Jenkins (3). Fluoride is necessary for the human organism and plays a role in controlling certain enzymatic reactions in various organs. Rapidly excreted from the body, the main retention sites for fluoride are calcified tissues. This retention depends on the quantity of fluoride administered or ingested, on the bioavailability of the fluoride salt used, as well as on the duration of exposure to fluoride.

At small doses (up to 1.5 mg F^-/day), fluoride has a prophylactic action against the development of dental caries. At low doses (about 15 to 25 mg F^-/day), given either as sodium fluoride or disodium monofluorophosphate, this element is currently used as a therapeutic agent, essentially for the treatment of established osteoporosis with vertebral crush fractures. At high doses, fluoride appears toxic but rarely lethal and the most well-known expression of its chronic toxicity is the development of skeletal fluorosis (4–6). It is a bone disease resulting in radiologically demonstrable abnormal bone densification. When it is measured, bone fluoride content is very high. Histological observations of bone tissue show osteosclerosis, mottled bone, formation defects and often hyperosteoidosis reflecting a calcification defect (4, 5, 7–10). Skeletal fluorosis is characterized by an unbalanced coupling in favor of bone formation, and a great number of osteoblasts with a high proportion of flat and poorly active

Therapeutic Uses of Trace Elements, edited by Nève et al.
Plenum Press, New York, 1996

283

osteoblasts. This support the view that fluoride may have a dual effect on osteoblasts: an increased birthrate at the tissue-level due to a mitogenic effect of fluoride on precursors of osteoblasts, and a toxic effect at the individual cell-level. The addition of these effects represents, however, a marked increase of bone formation at the organ level.

Vertebral osteoporosis is characterized by a predominant atrophy of trabecular bone inducing non-traumatic vertebral fractures. Because bone mass decrease is the main determinant of fracture risk, the ideal treatment should be capable of substantially increasing bone mass beyond the fracture threshold. Thus, the rate of new vertebral fracture would decrease. This ideal treatment should also be capable of restoring the normal architecture of cancellous bone tissue in order to maintain all mechanical properties of bone and a normal quality of bone tissue, in particular a normal mineralization of osteoid. Treatments that decrease bone resorption are able to maintain or slightly increase bone mass and to protect bone architecture. In contrast, there are very few substances that stimulate bone formation and thereby increase significantly cancellous bone mass other than fluoride salts. Sodium fluoride, proposed as a curative treatment for vertebral osteoporosis for the first time by Rich and Ensinck (11), has been extensively evaluated in many studies which have shown its ability to linearly increase cancellous bone mass. This was first shown by radiological techniques, then by histomorphometric analysis of iliac crest biopsies and more recently by measurements of bone mineral density of the axial skeleton with dual photon and dual energy X absorptiometry, quantitative computed tomography and neutron activation analysis. All these investigative procedures have proven the ability of sodium fluoride combined with adequate calcium supplementation to have an anabolic effect on bone mass. The effects of this increase in cancellous bone mass on the vertebral fracture rate seem to be controversial as long the cumulative dose of fluoride reaching bone is not taken into account. High doses given as high-potency preparations for more than 3 years have been shown to be without significant influence on the vertebral fracture risk because of the poor quality of bone, with calcification defects. In contrast, lower doses given as low-potency preparations for 2 years have been shown to significantly reduce vertebral fracture rate (12–14).

2. METABOLISM OF FLUORIDE IN THE HUMAN BODY

The normal dietary intake of fluoride is 1–3 mg F$^-$ daily, although there is a wide range mainly due to differences in fluoride content in water and food (15, 16). Fluoride is absorbed in the gastrointestinal tract, mainly at the duodenum level (2). In man, absorbed fluoride enters the blood stream and is distributed between two compartments, one with a short half-life of a few hours, probably comprising blood and soft tissues, and one with a much longer half-life (up to several years) corresponding to the hard tissues (teeth and skeleton essentially). Normal human blood fluoride level measured in 17 controls was found 0.032 ± 0.012 mg/l in our laboratory (Chapuy unpublished results) in agreement with values reported by Singer and Ophaug (17).

As previously mentioned, bone constitutes the most important site of fluoride retention in the human body. However, the fluoride is not homogeneously distributed in bone but is mainly localised in regions undergoing active calcification during fluoride exposure (4, 8, 18–20). Fluoride incorporation in bone occurs at the bone mineral level and results in various modifications of the characteristics of apatite crystals (21–23). Fluoride deposited in the skeleton is progressively removed by osteoclastic resorption processes during remodeling activity. A part of this recycled fluoride is later fixed again in newly forming bone regions, while another part is excreted in the urine. Thus, if the fluoride exposure has ceased, the fluoride content in bone decreases progressively but slowly with a half-life in bone of about 20 years (24, 25). Bone fluoride clearance takes over four times longer than does fluoride uptake (25). The fluoride content in bone samples constitutes a good reflection of the fluoride exposure history of a patient. This biochemical measurement allows confirmation of fluoride intoxication and permits fluoride therapy of osteoporosis to be followed. In subjects not previously exposed to excess fluoride (controls or patients never treated with fluoride salts), the mean bone fluoride content (calcined bone samples; specific ion electrode) is less than or equal to 0.10% in bone ash (24). These values (Table 1) are of the same order of magnitude as those reported in other studies (26–28).

The kidneys play a major role in fluoride excretion but fluoride clearance is complex because it is both secreted and reabsorbed by the renal tubule. Thus, moderate impairment of renal function could lead to increased retention of fluoride. The fluoride content of urine is, like blood fluoride, a good reflection of the amount of fluoride ingested. Urinary fluoride excretion measured in 17 human controls in our laboratory, is 0.25 ± 0.16 mg/24h (Chapuy, unpublished results). This focuses the attention on the relationship between fluoride retention and kidney function. Disturbed renal function predisposes to excessive retention of fluoride (30, 62) and increases the risk of skeletal fluorosis when patients with an impaired renal function are exposed to fluoride. Furthermore, fluoride should only be used as a therapeutic agent with great caution in the presence of impaired renal function (29, 30), because of the absence of information on the bioavailability of fluoride in different degrees of renal failure.

Table 1. Bone fluoride content in human iliac crest biopsies, modified and actualized from Boivin et al. (24)

Diagnosis		Biopsies (N)	Fluoride content (% of bone ash) *
Controls (Switzerland)		76	0.08 ± 0.05
Osteoporosis before treatment			
	Lyon	185	0.06 ± 0.03
	Geneva	102	0.08 ± 0.05
Osteoporosis after NaF treatment **			
4 months	(Lyon)	31	0.24 ± 0.12
1 year	(Geneva)	23	0.26 ± 0.16
2 years	(Lyon)	60	0.40 ± 0.16
2 years	(Geneva)	21	0.48 ± 0.15
3-4 years	(Geneva)	13	0.45 ± 0.25
5-6 years	(Geneva)	18	0.67 ± 0.18
Osteoporosis after NaF treatment ***			
1 year	(Lyon)	12	0.41 ± 0.12
2 years	(Lyon)	13	0.56 ± 0.16
Skeletal fluorosis	(Lyon)	67	0.82 ± 0.36

* Mean ± standard deviation for the number (N) of biopsies.
** 50-66 mg NaF/day administered with calcium (1g/day) and vitamin D (800 IU/day); *** 200 mg MFP/day administered continuously.

3. EFFECTS OF FLUORIDE ON BONE CELLS AND BONE MINERAL

The main histological findings in skeletal fluorosis and in osteoporotic patients treated with fluoride salts strongly suggest that fluoride acts at two different levels, the osteoblasts (direct or indirect action on birthrate and proliferation, possible toxic effect at high doses) and bone mineral substance. All the features of fluoride metabolism have to be taken into account to evaluate the cumulative dose of fluoride reaching bone.

3.1. Fluoride and Bone Cells

The direct effect of fluoride on osteoclasts is still unknown, but it can be indirectly assessed by measuring the extension of osteoclastic resorption surfaces in bone tissue. In skeletal fluorosis, this resorption is moderately increased (4, 10). In osteoporosis, after two years of combined therapy (sodium fluoride + calcium + vitamin D), the extent of resorption surfaces were generally unchanged, except in primary osteoporosis with osteoblastic depression where this parameter significantly increased (35–37).

It is now well known that fluoride is a potent stimulator of bone formation, resulting in a net increase in bone mass both in skeletal fluorosis (4) and in sodium fluoride-treated osteoporosis (35). Fluoride appears to have a dual and antagonistic action on osteoblasts: an increase of the osteoblast population due to an increased birth rate at the tissue level and a putative toxic effect at the individual cell level. Indeed, fluoride augments the number of osteoblasts, as shown by an extension of cancellous surfaces covered with these bone cells. However, the augmentation is due to a proportionnally greater accrual of flat osteoblasts (elongated cells showing a moderate secretory activity) as compared to that of plump osteoblasts (large and active cells) (4). How fluoride stimulates bone formation activity is still poorly understood. The main studies (38, 39), using in vitro approaches (cultures of calvaria cells and whole chicken bones) have been reviewed by Gruber and Baylink (40). It has recently been demonstrated (41) that the function of osteoblast was not modified after in vitro exposure to fluoride. By contrast, given in vivo, fluoride has a mitogenic effect on osteoblasts and stimulates their activity. These data emphasize the hypothesis that fluoride may act either on osteoprogenitor cells or through an indirect mechanism mediated by a cofactor as for example a phosphotyrosyl protein phosphatase (42) or the transforming growth factor ß (43). Recent *in vitro* studies (44) suggest that fluoride, in combination with traces of aluminum, modulates a growth factor-dependent tyrosine phosphorylation pathway enhancing mitogen-activated protein kinase and osteoblastic proliferation which could explain the increased bone mass observed with fluoride *in vivo* .

Finally, the action of fluoride on osteoclasts and osteoblasts, is characterized during remodeling activity by an unbalanced coupling in favour of bone formation, i.e. formation activity is more strongly stimulated than resorption activity (4, 45) with an increase in the mean wall thickness of cancellous packets which represent the end-product of the osteoblastic activity.

3.2. Fluoride and Bone Mineral Substance

Fluoride incorporation into bone mineral substance is considered as a detoxication mechanism inducing several modifications in the mineral substance. In bone tissue from humans having ingested fluoride, the fluoride ion is captured by apatite crystals forming bone mineral substance. This uptake may occur by ionic exchange or by the adsorption of fluoride ions onto the surface of the mature crystals, although, this mechanism is of secondary importance in the uptake of fluoride (31). The main mechanism of fluoride capture consists of ionic substitutions of hydroxyl ions by fluoride ions in the crystal lattice of apatite (23, 32). These substitutions take place inside a triangle of calcium ions forming a column in the structure of the crystal. Modifications in the crystal lattice contribute to changes at the mineral level (20, 32) which cause : a) an increase in crystallinity and crystal size ; b) a reduction in unit cell size (parameter a) ; c) a reduction in the crystal strain due to the increase in the stability of the crystal lattice ; d) a reduction in the solubility or dissolution rate (33) which provokes a decrease in bone resorption often observed at the tissue level ; e) a decrease in the specific surface area of the crystals with an increase in bone packing. Finally, because of its action at the bone mineral level, fluoride accumulates progressively in bone. The cumulative amount of fluoride taken up into bone mineral may attain, under certain conditions, a toxic level provoking diffuse and/or focal calcification defects which could compromise the biomechanical properties of bone, despite an increase in global bone mass. The main defects are : an increase in osteoid volume and thickness (4, 6), the presence of numerous linear calcification defects (4, 6, 34), the existence of mottled periosteocytic lacunae (4, 6, 34). In addition, poor healing of microfractures, due to a calcification defect of the microcallus which focally take up a large amount of fluoride, is responsible for the lower extremity pain syndrome (18). Calcification defects, which are readily observed in bone from patients with skeletal fluorosis (4), are rarely seen in bone biopsies of osteoporotic patients after long-term fluoride treatment. However, they appear to occur more frequently when the cumulative bone fluoride content was two-fold higher than after a normal fluoride treatment (34).

4. TOXIC EFFECTS OF FLUORIDE ON BONE

Skeletal fluorosis results from the effects of a prolonged intoxication (often over several years or sometimes decades) by high amounts of fluoride. The total amount of fluoride ingested is only rarely known with precision and when it is, this amount varies from 20 to 300 g F⁻ (7). This toxicity affects principally the skeleton but chronic toxicity affecting other organs (kidney, liver, thyroid, blood and reproduction) has been suggested (2, 3). There is no convincing evidence supporting the hypothesis of a carcinogenic or a mutagenic action of fluoride (46, 47).

4.1. CAUSES OF SKELETAL FLUOROSIS

Endemic hydric fluorosis is due to prolonged ingestion of water rich in fluoride in geographic regions (mainly India) where the soil is rich in fluoride (9, 10, 48, 49). In the most severe cases, the effects of malnutrition with calcium and/or vitamin D and/or protein deficiency may contribute to enhance fluorosis (6). Sporadic hydric fluorosis results from the prolonged consumption of F-rich bottled mineral waters, mainly to our knowledge the French "Saint Yorre Royale" water containing 8.5 mg F/l (4, 7). Telluric fluorosis is due to the inhalation of dusts containing high quantities of fluoride (51). Industrial fluorosis mainly results from the inhalation of fluoridated gases in fluoride processing or manufacturing industries (4, 6, 8, 51–53). Iatrogenic fluorosis is due to prolonged administration of F-rich drugs such as non-steroid anti-inflammatory agents (54). Skeletal fluorosis may also have other origins but cases are uncommon and the bone pathology has not yet been well documented. Wine fluorosis has been described in Spain and was due to large amounts of fluoride added to stop the fermentation process (6). Criminal fluorosis may result from chronic poisoning with very high doses of fluoride. Fluorosis of drug addicts occurs because of the practice of some drug dealers who mix sodium fluoride with the drug. Skeletal fluorosis have also been described after illicit used of an organofluoride anesthetic agent for its sedative or psychoactive effects (55). Fluorosis sometimes occurs in women in whom the source of fluoride was traced to soil that they ate because of habitual geophagia (56). Finally, fluorosis is a potential future hazard in osteoporotic patients who have poorly monitored treatment or are self-treated with fluoride, especially if renal insufficiency is also present.

4.2. Clinical and Biological Data

In adults, the main clinical features observed in skeletal fluorosis are as follows : generalised bone and joints pains, stiffness and rigidity, restricted movements of the spine and joints, deformities of spine and limbs, knock knees, and a crippled bed-ridden state may eventually result. These clinical findings correlate approximately with the severity of skeletal fluorosis (10, 52). Furthermore, nausea, loss of appetite, vomiting and constipation have been reported as acute symptoms of fluoride exposure (52). Respiratory disorders have also been described and include asthma, dyspnea, emphysema and lung fibrosis (52). Skeletal fluorosis is generally associated with a major osteosclerosis predominant in the spine, pelvis and thorax, which can be easily quantified by non-invasive methods of measurement of bone mass. Radiological findings have also been described (9, 51) : irregular periosteal bone formation, ectopic calcifications (ligaments, capsules, interosseous membrane, tendons), irregular osteophytosis and exostoses. Atypical radiological manifestations (distal or diffuse osteopenia) have also been reported in endemic fluorosis (India) and a relationship was found with a poor nutritional intake of calcium and proteins (57). Recently, radiographic features have been reviewed in endemic fluorosis of the skeleton and showed a wide variety of radiographic appearances, including calcification of the attachments, osteosclerosis, osteopenia, and osteomalacic zones (58).

The fluoride content in urine, blood and bone were always increased (8, 24, 48). Bone fluoride content (Table 1) was always more than five times the normal value, and was almost always upper than 0.5% and often close to or upper than 1% in skeletal fluorosis.

4.3. Bone Histomorphometry and Histological Aspects

The most striking histological modifications of bone tissue, frequently associated with skeletal fluorosis (Figs. 1, 2), include an increased cortical porosity, hypervascularisation of bone tissue, marked remodeling activity with an unbalanced coupling in favour of bone formation, the existence of either large or linear formation defects, the presence of numerous mottled periosteocytic lacunae and, occasionnally, the presence of newly formed and hypermineralized periosteal bone (4, 7–9). Large or linear formation defects corresponded to areas of more or less well organised osteoid tissue, totally embedded in calcified bone tissue. Under polarised light, calcified bone tissue - mottled or not - and osteoid tissue generally had a lamellar texture. To date, very few studies have reported histomorphometric data associated with the measurement of bone fluoride content in skeletal fluorosis (4, 7, 8, 10, 54, 59), while histological dynamic observations after tetracycline double labeling have been mentioned only by Teotia et al. (10) and Boivin et al. (4). Fluorotic patients showed significant increases in cortical thickness and porosity but not reduction of compact bone mass. The radiologically evident osteosclerosis observed in patients with skeletal fluorosis was confirmed

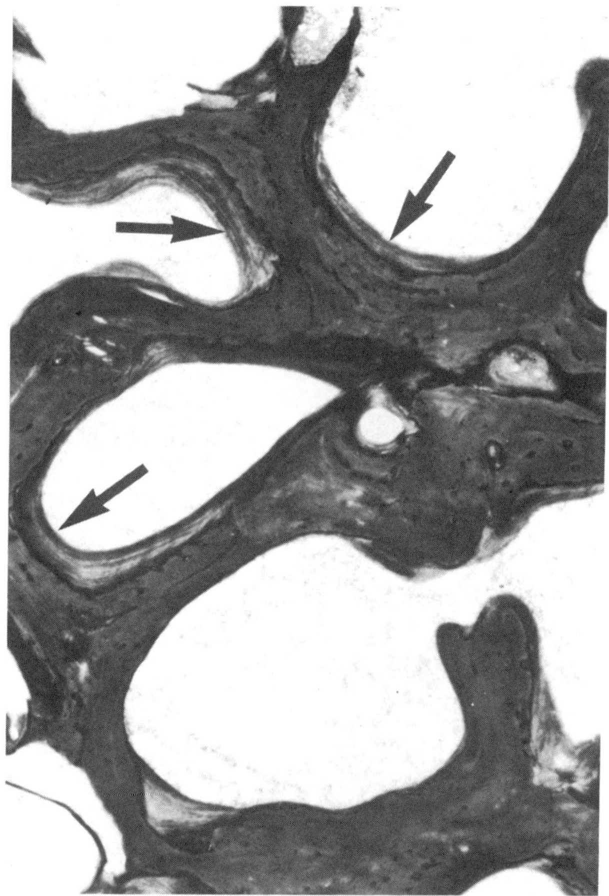

Figure 1. Undecalcified transiliac bone biopsy from a patient suffering from toxic skeletal fluorosis (bone fluoride content=0.66% of bone ash). Histological section stained by Goldner's trichrome illustrating the cancellous bone densification and the extension of osteoid seams (arrows) along bone trabeculae.

by a significant increase of cancellous bone volume. Since the cancellous calcified bone volume was augmented, the increase of cancellous bone volume was not only due to an augmentation of cancellous osteoid volume. The parameters reflecting bone formation (cancellous osteoid volume, cancellous osteoid surfaces, thickness of osteoid seams) were markedly elevated in fluorosis (Fig. 1). Furthermore, the increase in osteoid surfaces was almost three-fold that noted in cancellous resorption surfaces. Concerning the osteoblastic surfaces, the fluorotic patients had a greater number of osteoblasts than controls and this appeared to be due to the presence of a very high proportion of flat osteoblasts as compared to plump osteoblasts. In fluorotic patients doubly labeled with tetracycline, cancellous singly labeled surfaces were much increased, while double labeled surfaces exhibited the opposite trend. These changes were associated with a significant reduction in the calcification rate and a great extension of the mineralization lag time. Bone formation rate, calculated at the basic multicellular unit level, apparently decreased in fluorotic patients. Skeletal fluorosis was thus characterised by an unbalanced coupling in favour of bone formation, and a great number of osteoblasts with a high proportion of flat osteoblasts.

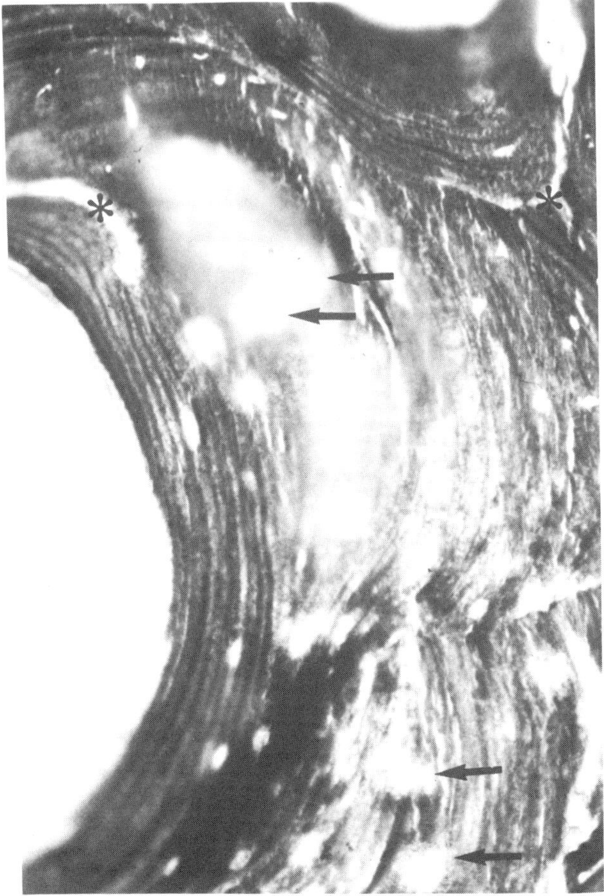

Figure 2. Undecalcified transiliac bone biopsy from a patient suffering from toxic skeletal fluorosis (bone fluoride content = 0.66% of bone ash). Histological section stained by solochrome cyanin R and showing the presence of focal calcification defects such as mottled periosteocytic lacunae (arrows) and linear defects (asterisk).

5. THERAPEUTIC ROLE OF FLUORIDE

It is now well established that fluoride salts prevent the development of dental caries, one of the most widespread diseases in man (60, Magloire et al. in the same meeting). It is important to emphasize that a daily intake of fluoride salts, derived from a multiplicity of sources and now generally considered safe, may in fact be potentially harmful over long periods of time. The therapeutic role of fluoride is essentially the treatment of postmenopausal and corticosteroid-induced osteoporosis.

5.1. Clinical Aspects

Osteoporosis is a human bone disease characterized by an absolute decrease in the amount of bone to an extent such that fragility fractures may occur after minimal trauma or relative lifetime risk of fractures is unacceptable. In osteoporosis, during the bone remodeling, there is a negative unbalanced coupling between bone resorption and bone formation, resulting in bone loss (62). Thus, the ideal therapeutic agent for osteoporosis should not only prevent bone loss but also increase bone mass to a level that exceeds the fracture threshold. A state of positive unbalanced coupling (bone remodeling in favor of bone formation) should thus be achieved with a formation period greater than the resorption period. Unfortunately, the most currently available therapeutic agents (estrogen, calcitonin, bisphosphonates, calcium, vitamin D) act directly or indirectly by reducing bone resorption rather than by increasing bone formation and are only capable of preventing bone loss and not of substantially and linearly increasing bone mass. In contrast, fluoride has been shown to increase bone formation more than bone resorption and is for this reason considered as the best curative agent for established osteoporosis with vertebral fractures (30, 62). However, the efficacy and safety of long term fluoride treatment in osteoporosis remain a matter of debate (62, 63). Unfortunately, many of the studies published were uncontrolled and the doses used ranged from 10 to 200 mg F⁻ daily. Furthermore, various formulations of sodium fluoride have been used and the regimens sometimes included calcium supplements and vitamin D preparations is varying doses. Trials have been conducted for variable times, using widely differing techniques for assessment and often on relatively small numbers of patients. Four factors influence the response to treatment: the daily dose of fluoride salts, the intestinal absorption of the compound used, the duration of treatment and the administration of calcium supplements with fluoride. The following treatment has been shown adequate for patients suffering from type I osteoporosis with vertebral crush fractures: 50 mg of sodium fluoride daily (i.e. 22.6 mg F⁻), given continuously for at least two years, in association with 1 g calcium daily and with a prevention of putative vitamin D deficiency (12). It has also been suggested that, in association with calcium, sodium monofluorophosphate may be useful in the treatment of type I osteoporosis (64). New double-blind placebo-controlled studies are now ongoing for confirming the positive effects of fluoride therapy on the vertebral fracture rate.

Side-effects of fluoride therapy are common (gastric irritation, pain in the lower extremities), but their incidence is clearly dose-dependent, and it is important to take into account not only the dosage but also the bioavailability of the compounds used (i.e. NaF or MFP, capsules, enteric-coated or effervescent tablets...), i.e. the overall amount of fluoride taken up in bone (63, 65, 66). The importance of a cumulative dose of fluoride on histological bone changes has been recently documented (67). For NaF, highly soluble preparations such as non-enteric-coated capsules (13) are responsible for frequent gastrointestinal symptoms due to irritation of the gastric mucosa. Slow release tablets (68) minimize gastrointestinal side-effects and have a low fluoride absorption. Enteric-coated tablets are used in France (12), and the concomitant administration of calcium does not significantly reduce fluoride bioavailability. MFP, the only available form of this highly soluble salt which forms, with calcium carbonate, a soluble salt, is rapidly absorbed in the duodenum and the small intestine. After administration of an almost identical amount of fluoride-ion given, either as NaF or as MFP, the bioavailability of fluoride was found to be greater for MFP than for NaF (63). However, it has been shown that, in postmenopausal women, two fluoride compounds (NaF and MFP-Ca) had a comparable bioavailability (66). Commercially available MFP effervescent tablets also contain calcium carbonate which guarantees a better compliance. For treatment with NaF tablets, calcium has to be taken separately from fluoride.

A decrease in the incidence of vertebral crush fractures has been claimed in patients with sodium fluoride (28, 35). Vertebral crush fractures appear to be significantly less frequent in the second year of the treatment than in the first (35). Furthermore, treatment with fluoride seems neither to decrease nor to increase substantially the incidence of hip fractures (69). The effects of NaF given at high dose (a mean of 75 mg/day, i.e. 60 mg one day and 90 mg the other day) for 4 years do not appear to be favourable (14). But results were totally different in a French prospective randomized study (12) on 466 patients with vertebral osteoporosis, defined as at least one non-traumatic vertebral crush fracture. The fracture rate per year was found to be significantly lower in the NaF group than in the non-fluoride group. The mean number of new vertebral crush fracture was also lower in the NaF group. It is very likely that the reduction in the fracture rate depends on the cumulative dose of fluoride-ion reaching the bone. This cumulative dose depends on the bioavailability of the fluoride salt used, on the dose of fluoride administered, and on the duration of the treatment.

Stress fractures have been reported in the appendicular skeleton (mainly legs) from fluoride-treated osteoporotic patients. They could represent the addition of numerous microfractures in the same skeletal localisation and correspond to an increased uptake of bone tracers on scintigraphs. However, it should be emphasized that these fractures appeared more frequently when high doses of fluoride were used. During healing of a microfracture, a callus is formed all around the cancellous fracture line. Boivin et al. (18) have demonstrated that fluoride is preferentially concentrated in calluses in fluoride-treated osteoporotic patients when compared to untreated patients. It is likely but remains to be proven that this local increase in fluoride uptake compromises the healing of microfractures. Osteomalacia induced by fluoride treatment is rare if fluoride is given at mild doses with a calcium supplement. Renal failure contraindicates fluoride treatment.

5.2. Effects of Treatment on Bone Mass and Biochemical Findings

Many studies have reported increased bone density of the axial skeleton. Changes included coarsening and increased prominence of cancellous markings and thickening of the vertebral end plates (30). Radiographic improvement has been noted in 30–65% of patients receiving long term fluoride treatment (28, 63). Fluoride is capable of producing an increase in bone mass. Histomorphometric analysis of iliac crest biopsies in patients treated with a daily dose of 50 mg NaF, 1 g calcium-element and vitamin D during 2 years has been performed in our laboratory, and the histomorphometric findings demonstrate the potent anabolic effect of fluoride on cancellous bone, with a significant increase in cancellous bone volume of about 70% in 2 years and of about 100% of osteoid surfaces (35). In spite of this, 20 to 30% of the patients have no significant increase in cancellous bone volume and are considered to be non-responders to fluoride treatment.

In general, there were few or no changes in plasma or urine calcium and phosphate levels during long term treatment (35, 63). However, some transitory changes could be observed depending on the presence and quantity of supplements (calcium and/or vitamin D). Plasma alkaline phosphatase has been shown to be unchanged when low doses of NaF (50 mg/day) are used (28, 35, 70), or sometimes slightly increased after a treatment with higher doses. Urinary excretion of hydroxyproline, which reflects the degradation of collagen and is thus considered to be related to resorption activity, was not changed during fluoride therapy (28, 35, 70). No changes in vitamin D metabolite levels have been reported with fluoride treatment alone but 25-hydroxyvitamin D was increased when treatment was supplemented with vitamin D. No long term changes have been noted in parathormone secretion. Plasma, urinary and bone fluoride contents increased after fluoride therapy (24, 27, 28, 35, 63, 71), and urinary fluoride is a very convenient tool for controlling the compliance of the fluoride-treated patients. In fluoride-treated osteoporotic patients, the mean bone fluoride content depended partly on the duration of therapy and increased with it (Table 1). The cumulative amount of fluoride absorbed by bone has also been measured in patients treated with either 200 mg/day of MFP or 50 mg/day of NaF. Results showed that bone fluoride contents were significantly higher in MFP-treated patients than in NaF-treated patients (Table 1).

5.3. Histological Bone Changes after Treatment

The histological defects (linear formation defects, mottled bone) currently observed in bone from patients with skeletal fluorosis, were only rarely seen in bone biopsies of osteoporotic patients after long term fluoride treatment. These defects seemed more frequent when high fluoride doses were used (34, 71). Furthermore, their presence depended on the type of treatment, i.e. on the administration or absence of supplements.

The main histomorphometric findings clearly demonstrated an increase in cancellous bone volume with cancellous thickening and an increase in osteoid volume. The augmentation in osteoid was due in large part to an increase in osteoid surface and, to a lesser extent, to an increase in osteoid thickness (35, 63, 72, 73). Trabecular calcified bone volume was unchanged (28), as was cortical thickness (35). The bone resorption surfaces were not significantly modified after treatment (35, 72). Direct measurements of calcification rate (double tetracycline labeling) revealed that it was stable in most patients (35) but in a significant minority a decrease was noted indicating delayed mineralization even in patients given calcium and vitamin D (35). A striking increase in osteoid surfaces without augmentation in resorption surfaces is the explanation for the anabolic effects of fluoride on cancellous bone mass. The mean wall thickness of cancellous packets was increased by the treatment demonstrating an increased formation period in the absence of significant changes in the osteoblastic appositional rate. After two years of therapy, a large percentage of osteoid surfaces was covered by flat osteoblasts and rarely by plump osteoblasts.

Recently, attention has been focused on patients under fluoride therapy for osteoporosis in whom no beneficial effect was seen (74, 75). About 25% of the patients were non-responders to fluoride therapy. The causes of the individual variability in response and, thus, the mechanisms of unresponse are still unknown. The main histological findings in osteoporotic patients treated with fluoride strongly confirm that fluoride may have a dual effect on bone formation : an increased birthrate of osteoblasts at the tissue level and, after long term exposure or at high doses, a depressive effect on osteoblasts at the individual cell level. The first effect seems very predominant in the treatment of osteoporosis with fluoride and accounts for the positive cancellous bone balance.

6. REFERENCES

1. W. Mertz, *Science* **213**, 1332–1338 (1981).
2. G.M. Whitford, *The metabolism and toxicity of fluoride*, Karger, Basel (1989).
3. G.N. Jenkins, in *Trace Metals and Fluoride in Bones and Teeth*, N.D. Priest, F.L. van de Vyver eds., CRC Press, Boca Raton, pp. 141–173 (1990).
4. G. Boivin, P. Chavassieux, M.C. Chapuy, C.A. Baud, P.J. Meunier, *Bone*, **10**, 89–99 (1989).
5. G. Boivin, P. Chavassieux, M.C. Chapuy, C.A. Baud, P.J. Meunier, *J. Bone Miner. Res.* **5** suppl 1, S185-S189 (1990).
6. G. Boivin, P.J. Meunier, in *The Metabolic and Molecular Basis of Acquired Disease*, R.D. Cohen, B. Lewis, K.G.M.M. Alberti, A.M. Denman eds., Baillière Tindall, London, pp. 1803–1823 (1990).
7. G. Boivin, P. Chavassieux, M.C. Chapuy, C.A. Baud, P.J. Meunier, *Pathol. Biol.* **34**, 33–39 (1986).
8. C.A. Baud, R. Lagier, G. Boivin, M.A. Boillat, *Virchows Arch. Pathol. Anat. Histol.* **380**, 283–297 (1978).
9. S.P.S. Teotia, M. Teotia, N.P.S. Teotia, *Fluoride* **9**, 91–98 (1976).
10. S.P.S. Teotia, M. Teotia, and D.P. Singh, in *Fluoride Research 1985, Studies in Environmental Science*, H. Tsunoda, M.H. Yu eds., Elsevier, Amsterdam, pp. 347–355 (1986).
11. C. Rich, F. Ensinck, *Nature* **191**, 184–185 (1961).
12. N. Mamelle, P.J. Meunier, R. Dusan, M. Guillaume, J.L. Martin, A. Gaucher, A. Prost, G. Zeigler, P. Netter, *Lancet* **ii**, 361–364 (1988).

13. B.L. Riggs, S.F. Hodgson, W.M. O'Fallon, E.Y.S. Chao, H.W. Wahner, J.M. Muhs, S.L. Cedel, L.J. Melton III, *N. Engl. J. Med.* **322**, 802–809 (1990).

14. B.L. Riggs, W.M. O'Fallon, S.F. Hodgson, E.Y.S. Chao, H.W. Wahner, J.M. Muhs, L.J. Melton III, *Bone Mineral* **17** suppl. 1, 74 (1992).

15. G.S. Rao, *Annu. Rev. Nutr.* **4**, 115–136 (1984).

16. G.E. Smith, *Xenobiotica* **15**, 177–186 (1985).

17. L. Singer, R. Ophaug, *C.R.C. Crit. Rev. Clin. Lab. Sci.* **18**, 111–140 (1982).

18. G. Boivin, B. Grousson, P.J. Meunier, *J. Bone Miner. Res.* **6**, 1183–1190 (1991).

19. S. Bang, C.A. Baud, *J. Bone Miner. Res.* **5** suppl. 1, S87-S89 (1990).

20. C.A. Baud, S. Bang, in *Fluoride in Medicine*, T.L. Vischer ed., Huber, Bern, pp. 27–36 (1970).

21. C.A. Baud, S. Bang, J.M. Very, *J. Biol. Buccale* **5**, 195–202 (1977).

22. G. Boivin, P.J. Meunier, *Res. Clin. Forums* **15**, 13–19 (1993).

23. M.D. Grynpas, *J. Bone Miner. Res.* **5** suppl. 1, S169-S175 (1990).

24. G. Boivin, M.C. Chapuy, C.A. Baud, P.J. Meunier, *J. Bone Miner. Res.* **3**, 497–502 (1988).

25. C.H. Turner, G. Boivin, P.J. Meunier, *Calcif. Tissue Int.* **52**, 130–138 (1993).

26. Y. Suzuki, *Tohoku J. Exp. Med.* **129**, 327–336 (1979).

27. R.D. van Kesteren, S.A. Duursma, W.J. Visser, J. van der Sluys Veer, O. Backer Dirks, *Metab. Bone Dis. Rel. Res.* **4**, 31–37 (1982).

28. B. Courvoisier, C.A. Baud, J.M. Very, A. Assimacopoulos, H.J. Tochon-Danguy, G. Boivin, A. Donath, J. Garcia, A. Gasser, J. Fischer, S. Korol, P. Burckhardt, *Schweiz. Med. Wochenschr.* **115**, 922–931 (1985).

29. J.C. Gerster, S.A. Charhon, P. Jaeger, G. Boivin, D. Briançon, A. Rostan, C.A. Baud, P.J. Meunier, *Br. Med. J.* **287**, 723–725 (1983).

30. J.A. Kanis, P.J. Meunier, *Q. J. Med.* **210**, 145–164 (1984).

31. E.D. Eanes, A.H. Reddi, *Metab. Bone Dis. Rel. Res.* **2**, 3–10 (1979).

32. N.C. Blumenthal, in *Trace Metals and Fluoride in Bones and Teeth*, N.D. Priest, F.L. van De Vyver eds., CRC Press, Boca Raton, pp. 307–313 (1990).

33. M.D. Grynpas, P.T. Cheng, *Bone Mineral* **5**, 1–9 (1988).

34. G. Boivin, J. Duriez, M.C. Chapuy, B. Flautre, P. Hardouin, P.J. Meunier, *Osteoporosis Int.* **3**, 204–208 (1993).

35. D. Briançon, P.J. Meunier, *Orthop. Clin. North Am.* **12**, 629–648 (1981).

36. M.W. Lundy, D.J. Baylink, S.F. Hodgson, B.L. Riggs, *J. Bone Miner. Res.* **2** suppl. 1, 379 (1987).

37. R.R. Recker, D.B. Kimmel, J.C. Gallagher, A. Vaswani, J. Aloia, *J. Bone Miner. Res.* **2** suppl. 1, 381 (1987).

38. J.R. Farley, J.E. Wergedal, D.J. Baylink, *Science* **22**, 330–332 (1983).

39. M.W. Lundy, J.R. Farley, D.J. Baylink, *Bone* **7**, 289–293 (1986).

40. H.E. Gruber, D.J. Baylink, *Clin. Orthop.* **267**, 264–277 (1991).

41. P. Chavassieux, G. Boivin, C.M. Serre, P.J. Meunier, *Bone* **14**, 721–725 (1993).

42. W.K.H. Lau, J.R. Farley, T.K. Freeman, D.J. Baylink, *Metabolism* **38**, 858–868 (1989).

43. B.Y. Reed, J.E. Zerwekh, P.P. Antich, C.Y.C. Pak, *J. Bone Miner. Res.* **8**, 19–25 (1993).

44. J. Caverzasio, T. Imaï, P. Ammann, D. Burgener, J.P. Bonjour, *J. Bone Miner. Res.* **11**, 46–55 (1996).

45. D.J. Baylink, P.B. Duane, S.M. Farley, J.R. Farley, *Caries Res.* **17** suppl 1, 56–76 (1983).

46. E. Zeiger, M.D. Shelby, K.L. Witt, *Environ. Mol. Mutagen.* **21**, 309–318 (1993).

47. P. Grandjean, J.H. Olsen, O.M. Jensen, K. Juel, *J. Natl. Cancer Inst.* **84**, 1903–1909 (1992).

48. J.M. Faccini, S.P.S. Teotia, *Calcif. Tissue Res.* **16**, 45–57 (1974).

49. F. Manji, S. Kapila, *Odontostomatol. Trop.* **9**, 15–20 and 71–75 (1986).

50. Y. Haikel, J.C. Voegel, R.M. Frank, *Arch. Oral Biol.* **31**, 279–286 (1986).

51. M.A. Boillat, C.A. Baud, R. Lagier, J. Garcia, P. Rey, S. Bang, G. Boivin, C. Demeurisse, M. Gössi, H.J. Tochon-Danguy, J.M. Very, P. Burckhardt, B. Voinier, A. Donath, B. Courvoisier, *Schweiz. Med. Wochenschr.* **109** suppl 8, 1–28 (1979).

52. P. Grandjean, *Am. J. Ind. Med.* **3**, 227–236 (1982).

53. P. Grandjean, G. Thomsen, *Br. J. Ind. Med.* **40**, 456–461 (1983).

54. P.J. Meunier, P. Courpron, J.S. Smoller, D. Briançon, *Clin. Orthop.* **148**, 304–309 (1980).

55. P.J. Klemmer, N.M. Hadler, *Ann. Int. Med.* **89**, 607–611 (1978).

56. J.R. Fisher, M.L. Sievers, R.T. Takeshita, H. Caldwell, *Arizona Med.* **38**, 833–835 (1981).

57. A. Mithal, N. Trivedi, S.K. Gupta, S. Kumar, R.K. Gupta, *Skeletal Radiol.* **22**, 257–261 (1993).

58. Y. Wang, Y. Yin, L.A. Gilula, A.J. Wilson, *Amer. J. Roentgenol.* **162**, 93–98 (1994).

59. I. Arnala, E.M. Alhava, P. Kauranen, *Acta Orthop. Scand.* **56**, 161–166 (1985).

60. Y. Ericsson, *J. Dental Res.* **59** DII, 2131–2136 (1980).

61. G.K. Stookey, P.F. DePaola, J.D.B. Featherstone, O. Fejerskov, I.J. Möller, S. Rotberg, K.W. Stephen, J.S. Wefel, *Caries Res.* **27**, 337–360 (1993).

62. B.L. Riggs, in *Bone and Mineral Research*, Annual 2, W.A. Peck ed., Elsevier, Amsterdam, pp. 366–393 (1983).
63. G. Boivin, J. Dupuis, P.J. Meunier, in *Nutrition and Osteoporosis*, A. Simopoulos, C. Galli eds., World Rev. Nutr. Diet. vol. 73, Karger, Basel, pp. 80–103 (1993).
64. Y. Ericsson, *Caries Res.* **17** suppl. 1, 46–55 (1983).
65. P. Arnold, M. Wermeille, M.C. Chapuy, J. Biollaz, E.M. Grandjean, J.L. Schelling, P.J. Meunier, *Bone* **10**, 401–407 (1989).
66. F. Lioté, C. Bardin, A. Liou, A. Brouard, J.L. Terrier, D. Kuntz, *Calcif. Tissue Int.* **50**, 209–213 (1992).
67. M. Kleerekoper, R. Balena, J. Foldes , M.S. Shih, D. Rao, A.M. Parfitt, *J. Bone Miner. Res.* **5** suppl. 2, S140 (1990).
68. C.Y.C. Pak, K. Sakhaee, C. Gallagher, C. Parcel, R. Petersen, J.E. Zerwekh, M. Lemke, F. Britton, M.C. Hsu, B. Adams, *J. Bone Miner. Res.* **1**, 563–571 (1986).
69. B.L. Riggs, D.J. Baylink, M. Kleerekoper, J.M. Lane, L.J. Melton III, P.J. Meunier, *J. Bone Miner. Res.* **2**, 123–126 (1987).
70. P. Charles, L. Mosekilde, F. Taagehoj Jensen, *Bone* **6**, 201–206 (1985).
71. M.W. Lundy, M. Stauffer, J.E. Wergedal, D.J. Baylink, J.D.B. Featherstone, S.F. Hodgson, B.L. Riggs, *Osteoporosis Int.* **5**, 115–129 (1995).
72. M.P. Whyte, M.A. Bergfeld, W.A. Murphy, L.V. Avioli, S.L. Teitelbaum, *Am. J. Med.* **72**, 193–202 (1982).
73. V.J. Vigorita, M.K. Suda, *Clin. Orthop.* **177**, 274–282 (1983).
74. S.A. Duursma, J.H. Glerum, A. van Dijk, R. Bosch, H. Kerkhoff, J. van Putten, J.A. Raymakers, *Bone* **8**, 131–136 (1987).
75. P.J. Marie, M.C. de Vernejoul, A. Lomri, *J. Bone Miner. Res.* **7**, 103–113 (1992).

50

MEASUREMENT OF TRACE ELEMENTS IN BONE BY ICP-MS

J-E. Beck Jensen, M. M. Larsen, B. Kringsholm, G. Pritzl, and
O. H. Sørensen

[1] Osteoporosis Research Centre
Copenhagen Municipal Hospital, Denmark
[2] Department of Environmental Chemistry
National Environmental Research Institute, Roskilde, Denmark
[3] Medico-legal Institute
Copenhagen, Denmark

1. INTRODUCTION

Osteoporosis constitutes a growing problem first of all because of the increase in the elderly population (1). In addition, several studies have demonstrated significant increases in age-corrected incidence of osteoporotic fractures during the last decades which makes declining quality of bone an important factor (2). The classical osteoporotic fractures are located at the distal forearm, the spine and the hip, sites that are all dominated by trabecular bone (3). Women are primarily affected but a rising number of men suffer from osteoporotic fractures. Many factors such as sex-hormones, exercise, vitamin D and calcium intake are known to affect the bone turnover and the risk of developing osteoporosis. Specially interesting are marked unexplained differences in the prevalence of osteoporosis in different parts of the world (4) and even within the same geographical areas with lower incidence of osteoporosis in the rural districts compared to the cities (5). A number of trace elements have influence on bone. Some elements are toxic others beneficial and some essential for a normal bone metabolism. Intoxications with aluminum in patients with chronic renal diseases (6), treatment with fluoride in osteoporosis (7), and the Itai-Itai disease caused by cadmium in Japan (8) are important examples. Furthermore, many enzymes need trace elements as cofactors for optimal function. Important enzymes for bone such as alkaline phosphatase and lysyl oxidase are thus dependent on zinc and copper, respectively (9,10).

Therapeutic Uses of Trace Elements, edited by Nève et al.
Plenum Press, New York, 1996

Many different methods have been used to analyse multiple trace elements in bone (11). In the present study Inductively Coupled Plasma Mass Spectrometry (ICP-MS) (12) was used due to its multielement capacity and low limits of detection.

2. MATERIAL AND METHODS

Bone biopsies from three females and three males representing young, middle, and old age were studied. All persons were healthy and had died acutely. Cylinders with a length of 35–40 mm and a diameter of 7 mm were drilled out horizontally from the vertebral body. From each individual eight samples of trabecular bone, 4 from the second (L2) and 4 from the fourth (L4) lumbar vertebral body were analysed (N=48). Only 10 mm from the middle part of the cylinder constituted the sample. The equipment used for sampling was made of pure titanium and teflon and specially rinsed by washing in half concentrated "Suprapur" nitric acid and millipore water (> 18 MΩ cm) . The samples were freeze dried for 24 hours and opened in micro vials teflon bombs containing 2.5 ml "Suprapur" nitric acid 1:1 with millipore water in a micro wave owen under pressure control. Fifty parts per billion rhodium were used as digestion standard. After dilution to 20 ml with millipore water the samples were analysed by a PE-sciex ELAN 5000 ICP-MS instrument with platinum cones running in the peak hop quantitative mode with 3 sweeps and 3 replicates using 50 parts per billion iridium as internal standard for instrument drift. Argon gas flow were: plasma flow 15 l/min, nebulizer flow 0.9 l/min and auxiliary flow 0.85 l/min. Sample uptake was 1.0 ml/min, RF power 1045 watt and lens settings as instrumental defaults. Calibration was done as standard addition calibration. A three point calibration curve was obtained for every five measurements in a human bone matrix with approximately the same concentrations of matrix proteins, sodium, chloride and calcium as in the samples. The added concentrations in the calibration matrix were 5, 10 and 100 ppm except for strontium and zinc for which 10, 110 and 1100 ppm were used due to the higher concentrations of these elements. For every 5 samples a blank was taken through all the steps and measured in order to assess contamination. Results were calculated from the original counts corrected for acid blank, sample blank and instrument drift in accordance with the calibration curves. Calcium had a tendency to precipitate at the ICP-MS cones during the measurements and thus lower the sensitivity. Although this was satisfactory corrected for by the internal Ir-standard, the overall drop in sensitivity forced to clean the cones If the pressure fell below 10^{-5} Atm.

All results are given per gram dry bone weight. The following elements were measured: 27Al, 55 Mn, 63Cu, 64Zn, 69Ga, 88Sr, 98Mo, 120Sn, 138Ba and 208 Pb. Each measurements was based on the mean of 3 replicate values. From these 3 values the analytical coefficients of variation (CV_A) were calculated for each analysed element. The total variance of a quantity can be separated into analytical (V_A), the within subject (V_I) and the between subject (V_G) variance or expressed as coefficient of variation (CV). The total variation is expressed as $CV_{TOT}^2 = CV_A^2 + CV_G^2 + CV_I^2$. The intraindividual coefficient of variation ($CV_{TI} = (CV_I^2 + CV_A^2)^{1/2}$) was calculated as the mean of the individual values. From the CV_{TI}, the CV_A and the total variation the intraindividual (CV_I) and interindividual (CV_G) coefficient of variation were calculated . The index of individuality (I_{Index}) was calculated as CV_I divided by CV_G. Furthermore, maximum, minimum, and median were calculated for each quantity using the pooled data from all 48 measurements. For statistical evaluation of the trace elements from L2 and L4 in each person the Mann-Whitney U test was used. ($P < 0.05$ was considered significant).

Table 1. The concentration of ten elements from trabecular bone

Quantity	Minimum	Maximum	Median
88 Sr *	21.8	175	57.0
64 Zn *	28.2	239	48.9
138 Ba *	0.89	7.94	2.69
27 Al *	0.18	7.13	1.60
208 Pb *	0.46	4.35	1.04
63 Cu #	320	7600	590
120 Sn #	28	42015	254
55 Mn #	4	803	81
69 Ga #	6	242	33
98 Mo #	5	5716	16

* Results expressed as μg/g dry weight.

\# Results expressed as ng/g dry weight.

The figure beside the element refers to the mass measured.

3. RESULTS

For each element measured there were no differences between the concentrations in L2 and L4 in the individual person. Data from each person were therefore pooled. Results from elements with various concentrations are shown in Table 1. At the microgram per gram dry weight level strontium and zinc had the highest concentrations followed by barium, aluminum and lead. Furthermore lead showed a pattern of increasing concentration with age. The range of concentrations were in the order of a factor 10 except for aluminum in which case the maximum concentration was 40 times higher than the minimum one. At the nanogram per gram dry weight level copper was followed by tin, manganese, gallium and molybdenum. The concentration range was much wider for these quantities up to a factor 1500 for tin. Table 2 illustrates the calculated values for CV_A, CV_I, CV_G and I_{Index}. The analytical coefficients of variation were relatively small and below 10 percent except for gallium, molybdenum and aluminum where the coefficients were between 27 and 37 percent. The CV_I

Table 2. Analytical and biological coefficients of variation as well as index of individuality for ten elements in trabecular bone

Quantity	CV_A %	CV_G %	CV_I %	I_{Index}
88 Sr	3.23	40.0	41.6	1.04
64 Zn	3.46	56.0	53.0	0.95
138 Ba	2.79	39.6	44.8	1.13
27 Al	37.5	45.3	58.6	1.29
208 Pb	1.90	60.7	46.8	0.77
63 Cu	5.84	102	69.7	0.68
120 Sn	8.21	407	83.9	0.21
55 Mn	9.08	93.3	63.5	0.68
69 Ga	27.0	89.7	64.6	0.72
98 Mo	29.7	279	64.7	0.23

CV_A % : analytical coefficient of variation . CV_G % : between subject coefficient of variation. CV_I %: within subject coefficient of variation. I_{Index} : index of individuality (CV_I/CV_G). The figure besides the element refers to the mass measured.

values ranged from 42 to 84 percent and the CV_G values from 40 and up to 407 percent for tin. The highest I_{Index} was 1.29 for aluminum and the lowest 0.21 for tin.

4. DISCUSSION

The analysed elements are all important in relation to bone (13). Trabecular bone with its high content of organic elements as well as high concentrations of calcium, phosphorous, chloride and sodium constitutes a complicated matrix to analyse by ICP-MS (14) and it is of major importance to use addition calibration in order to control matrix interference. The small analytical coefficient of variation in spite of the analysis of several elements at different concentration ranges makes the method attractive. In order to get an impression og the biological variations the within and between subject variation were calculated after pooling the samples from both vertebrae. Both the within and between subject biological variation were considerable and could be explained by variable antropogen exposition, different intestinal absorption, variable uptake in the bone tissue due to different bone turnover or simply by different compositions of the samples i.e. one sample could contain more bone tissue and less blood or fat than another sample. This last variability could partly be eliminated by washing and extracting the samples but such a procedure would however increase the risk of contamination and wash out of elements. Biological variation can be expressed by means of index of individuality which gives information about the usefulness of reference intervals (15). If the index is low (<0,6) reference intervals are of no use in contrast to a high I_{Index} (>1,4) which is able to separate abnormal and normal values in relation to a reference intervals. Analytical goal setting based on biological variation is widely accepted as the analytical coefficient of variation being less or equal to half the within subject coefficient of variation (16). Otherwise the analytical imprecision will increase the variation to an unacceptable value . This was fulfilled for all quantities in the present study except for aluminum. Lead is known to accumulate in bone with increasing age (17) which was confirmed in the present

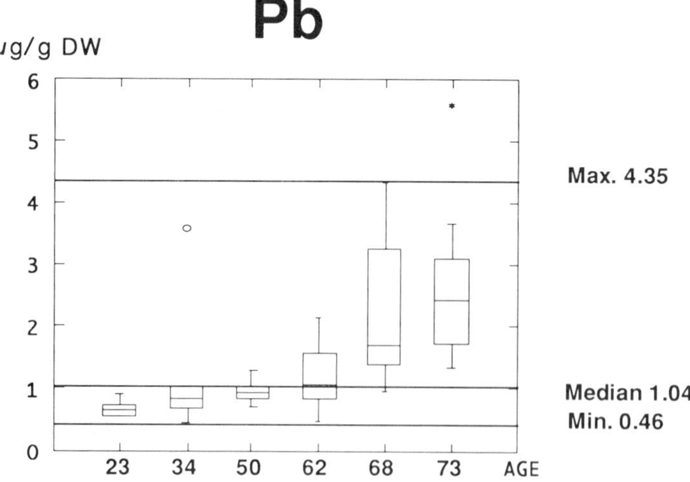

Figure 1. Box and whiskers plots of lead concentrations for each of the six persons studied. Maximum, median and minimum refers to the pooled data (N=48).

investigation (Figure 1). The number of persons studied was, however, small, and larger studies are under way to investigate possible sex and age related differences.

5. CONCLUSION

ICP-MS is a reliable method for determination of 27Al, 55 Mn, 63Cu, 64Zn, 69Ga, 88Sr, 98Mo, 120Sn, 138Ba and 208 Pb in trabecular bone. There were no differences between the concentrations of any element in L2 and L4 in the individual person. The within and between subject biological variations as well a the analytical variations were established and analytical goals in relation to biological variation were fulfilled. The method seems to have a great potential.

6. REFERENCES

1. L.J. Melton, *Bone* **14**, S1–8 (1993).
2. K.J. Obrant, U. Bengnér, O. Johnell, B.E. Nilsson, I. Sernbo, *Calcif Tissue Int* . **44**,157–67 (1989).
3. B.L. Riggs, L.J. Melton, *N Engl J Med* . **327**, 620–26 (1992).
4. L.J. Melton, in *New dimensions in Osteoporosis in the 1990s.*, Hong Kong, Excerpta Medica Asia, pp. 13–18(1991).
5. P. Gärdsell, O. Johnell, B.E. Nilsson, I. Sernbo, *J Bone Miner Res* . **6**, 67–75 (1991).
6. H. Malluche, M-C. Fauger, *Kidney International* . **38**, 193–211(1990).
7. R. Lindsay, *N Engl J* . **322**, 845–46 (1990).
8. K. Nogawa, in *Cadmium in the Environment,* Wiley, New York, pp. 1–38 (1981).
9. J. Savory, M.R. Wills, *Clin. Chem.* **38**, 1565–1573 (1992).
10. N.W. Solomons, *J. Am. Coll. Nutr.* **4**, 83–105 (1985).
11. H. Zwanziger, *Biol. Trace Elem. Res.* **19**, 195–232 (1989).
12. R.S. Houk, *Anal Chem.* **58**, 97–107 (1981)
13. N.D. Priest, F.L. Van de Vyver, T*race Metals and Fluoride in Bone and Teeth.*, CRC Press, Boca Raton, Florida (1990).
14. U. Vollkopf, K. Barnes, *At Spec.* **16**, 19–22 (1995)
15. E.K. Harris, *Clin Chem.* **20**,1535–42 (1974).
16. C.G. Fraser, E.K. Harris, *Critical Reviews in Clinical Laboratory Sciences.* **27**, 409–37 (1989).
17. L.E. Wittmers, J. Wallgren, A. Alich , A.C. Aufderheide, G. Rapp, *Arch Environ Health.* **43**, 381–391(1988).

51

AGE AT LEAD EXPOSURE INFLUENCES LEAD RETENTION IN BONE

Shenggao Han, Xianwen Qiao, Francis W. Kemp, and John D. Bogden

Department of Preventive Medicine and Community Health
UMDNJ-New Jersey Medical School
185 South Orange Avenue, Newark, New Jersey 07103-02714

1. INTRODUCTION

More than 90% of the body lead burden in rats and people is in the skeleton (1–3). The half-life of bone lead is long, 5–20 years or more, with lead in cortical bone having a longer biological residence time than lead in trabecular bone (1–3). The long half-life suggests that environmental exposure of children to lead can result in bone lead accumulation that may persist for many years. However, it has been hypothesized that the high rate of bone remodeling during childhood and the related high turnover of calcium and lead results in a substantial reduction in bone lead stores, so that much of the lead incorporated into bone during childhood does not persist into adulthood (4). We tested the alternative hypothesis that younger age at lead exposure results in a greater accumulation of lead stored in the skeleton as well as in soft tissues. The objective of this study was to determine the effect of age during lead exposure on blood and organ lead concentrations one month later. Organ concentrations of the essential divalent metals calcium, copper, iron, magnesium, and zinc were also measured for comparison to lead.

2. MATERIALS AND METHODS

Weanling female Sprague-Dawley rats (Taconic Farms, Germantown, NY) (n = 24, 4 weeks old) were housed individually in plastic cages. After 1 week of acclimation, rats were fed a modified AIN-76 diet; the composition of this diet has been previously described (5). The rats were then randomly assigned to one of three treatment groups. The first group was given drinking water containing 250 mg/L of lead as the acetate beginning at 5 weeks of age. Lead exposure of the other 2 groups began at 10 (mid-adolescence) and 15 (young adulthood) weeks of age. Glacial acetic acid was added to the drinking water solution at a concentration of 12.5 µL/L to prevent the precipitation of lead carbonate. Drinking water consumption was monitored daily.

Therapeutic Uses of Trace Elements, edited by Nève et al.
Plenum Press, New York, 1996

Administration of lead-containing drinking water to each group for 5 weeks was followed by a 4 week period without lead exposure to ensure that most of the body lead burden was in the skeleton (4,6). Following this, blood was withdrawn by cardiac puncture from rats anesthetized with 25 mg/kg sodium pentobarbital. Rats were killed by decapitation while under anesthesia and their organs harvested. Whole blood lead and organ metal concentrations were determined by previously described methods (5,7). Calculations of concentrations were based on wet tissue weight.

ANOVA was used to evaluate the data. If ANOVA indicated that there were statistically significant ($p < 0.05$) differences among the three treatment groups for a specific measurement, then pair-wise comparisons were made by Duncan's multiple range test at $\propto = 0.05$.

3. RESULTS

Daily drinking water consumption during the 5 week period of lead exposure was 16.4 $\pm$ 2.3, 19.6 $\pm$ 3.0, 15.1 $\pm$ 0.1 mL/d for the 5, 10, 15 week old rats, respectively. These values did not differ significantly (ANOVA, $p < 0.05$), indicating that lead exposure via the drinking water was comparable for the three treatment groups.

Table 1 contains the organ divalent metal concentrations and Table 2 the organ total metal content. Lead concentrations and total content in brain, kidney, liver, femur, and spinal column bone were highest in the rats exposed to lead beginning at 5 weeks of age and lowest in the group exposed at 15 weeks of age. Rats exposed beginning at 10 weeks of age had intermediate values. Blood lead concentrations followed a similar pattern, with values of 1.39 $\pm$ 0.05, 1.18 $\pm$ 0.12 and 0.82 $\pm$ 0.05 μmol/L in the groups exposed to lead beginning at 5, 10, and 15 weeks of age, respectively. The ratios of the lead concentrations of the group exposed beginning at 5 weeks to that at 15 weeks was 2.46 for brain, 6.83 for kidney, 1.88 for liver, 4.31 for femur, 3.02 for spinal column bone, and 1.20 for blood. The oldest rats had significantly higher kidney and spinal column calcium, brain copper, liver and femur iron, and femur and spinal column zinc concentrations than the youngest rats. Kidney copper and liver calcium were highest in the youngest rats. Brain copper and liver and femur iron content increased with age. Kidney copper content decreased with age. Other organ concentrations and contents did not differ significantly among the 3 groups of rats or displayed only small differences.

4. DISCUSSION

The results for the blood and organ lead concentrations do not support the hypothesis of more rapid depletion of bone lead stores in younger versus older animals, but rather suggest that younger age at lead exposure is associated with greater lead retention and toxicity subsequent to cessation of lead exposure. In fact, the differences in organ lead concentrations between the youngest and oldest rats are substantial, as reflected in the ratios calculated.

It is well known that the short-term toxic effects of lead, especially on the central nervous system are greater for young children than for adults (9). The present study suggests that retention of lead in the skeleton by children exposed at a young age may also pose long-term health problems. A particular concern is the release of lead from the skeleton in women during pregnancy and lactation, which can then be transferred to the fetus or neonate (8,9).

The cause of the consistently higher blood and organ lead concentrations in the youngest rats was not determined in this investigation, but is not due to differences in lead inges-

Table 1. Metal concentrations in organs of rats in three treatment groups[1]

Organ	Age[2]	Lead	Calcium	Copper	Iron	Magnesium	Zinc
	weeks	nmol/g	μmol/g	nmol/g	μmol/g	μmol/g	nmol/g
Brain	5	2.51 ± 0.18[A]	1.05 ± 0.10	39.1 ± 0.56[C]	0.29 ± 0.01[B]	6.48 ± 0.07[AB]	195 ± 2.2[A]
	10	1.40 ± 0.20[B]	2.08 ± 0.89	42.4 ± 0.96[B]	0.31 ± 0.01[AB]	6.27 ± 0.09[B]	186 ± 3.3[B]
	15	1.02 ± 0.06[B]	0.91 ± 0.03	47.2 ± 0.87[A]	0.32 ± 0.01[A]	6.58 ± 0.13[A]	188 ± 2.2[AB]
Kidney	5	148.2 ± 57.3[A]	27 ± 14[B]	339 ± 64[A]	1.59 ± 0.10	10.5 ± 0.5[B]	457 ± 18
	10	74.4 ± 18.6[AB]	197 ± 49[A]	159 ± 11[B]	1.88 ± 0.04	14.4 ± 1.6[A]	447 ± 25
	15	21.7 ± 3.7[B]	142 ± 42[A]	154 ± 12[B]	3.90 ± 1.70	12.6 ± 1.3[AB]	438 ± 24
Liver	5	4.68 ± 0.54[A]	0.78 ± 0.03[A]	112 ± 9	5.20 ± 0.43[B]	9.92 ± 0.24	455 ± 9.4
	10	2.60 ± 0.29[B]	0.70 ± 0.02[B]	113 ± 6	6.44 ± 0.34[B]	9.57 ± 0.20	432 ± 9.7
	15	2.49 ± 0.25[B]	0.67 ± 0.02[B]	100 ± 5	8.28 ± 0.75[A]	9.80 ± 0.21	442 ±12.1
Femur	5	988 ± 108[A]	4503 ± 230	50.5 ± 2.0	1.14 ± 0.04[B]	135 ± 6.4	2439 ± 105[B]
	10	374 ± 52[B]	4469 ± 132	49.0 ± 1.9	1.31 ± 0.09[B]	131 ± 3.6	2551 ± 91[B]
	15	229 ± 21[B]	4841 ± 83	50.4 ± 0.8	1.54 ± 0.08[A]	139 ± 4.2	2866 ± 62[A]
Spinal	5	915 ± 54[A]	3420 ± 76[B]	50.0 ± 2.7	1.15 ± 0.04	110 ± 3	2269 ± 86[B]
Column	10	469 ± 60[B]	3670 ± 43[A]	47.1 ± 0.8	1.31 ± 0.05	113 ± 1	2515 ± 53[A]
Bone	15	303 ± 28[C]	3667 ± 78[A]	47.7 ± 0.9	1.34 ± 0.09	110 ± 2	2659 ± 57[A]

[1] Data are Mean ± SE; n=8 for all values. Means with different letter superscripts differ signifcantly (Duncan's test, P<0.05)
[2] Age = age in weeks at start of 5 weeks of lead exposure

Table 2. Total organ metal content of rats in three treatment groups[1]

Organ	Age[2]	Lead	Calcium	Copper	Iron	Magnesium	Zinc	Organ Weight
	weeks	nmol/g	μmol/g	nmol/g	μmol/g	μmol/g	nmol/g	grams
Brain	5	4.42 ± 0.35[A]	1.85 ± 0.20	68.6 ± 1.7[C]	0.50 ± 0.01	11.4 ± 0.2	342 ± 6	1.76 ± 0.04
	10	2.49 ± 0.34[B]	3.71 ± 1.56	76.0 ± 1.7[B]	0.56 ± 0.03	11.3 ± 0.3	333 ± 6	1.80 ± 0.03
	15	1.77 ± 0.11[B]	1.60 ± 0.08	82.4 ± 2.1[A]	0.56 ± 0.02	11.5 ± 0.3	329 ± 9	1.75 ± 0.04
Kidney[3]	5	124 ± 50[A]	22 ± 11[B]	270 ± 49[A]	1.30 ± 0.11	8.5 ± 0.4[B]	371 ± 17	0.81 ± 0.02[B]
	10	69 ± 16[AB]	197 ± 53[A]	151 ± 12[B]	1.79 ± 0.11	13.9 ± 1.9[A]	428 ± 39	0.95 ± 0.05[A]
	15	19 ± 3[B]	130 ± 39[AB]	134 ± 11[B]	3.44 ± 1.59	11.1 ± 1.4[AB]	382 ± 31	0.87 ± 0.03[AB]
Liver	5	37.8 ± 6.2[A]	6.18 ± 0.41	901 ± 115	41.0 ± 4.0[B]	78.1 ± 4.2	3580 ± 172	7.90 ± 0.44
	10	22.5 ± 2.5[B]	6.31 ± 0.67	992 ± 60	56.6 ± 3.7[A]	85.4 ± 6.6	3837 ± 250	8.90 ± 0.63
	15	19.2 ± 1.3[B]	5.38 ± 0.42	801 ± 61	64.7 ± 5.0[A]	77.9 ± 4.4	3493 ± 153	7.95 ± 0.43
Femur	5	603 ± 66[A]	2743 ± 180	30.7 ± 1.8[AB]	0.71 ± 0.06[B]	82.1 ± 5.0	1492 ± 102[B]	0.62 ± 0.06
	10	267 ± 39[B]	3200 ± 131	34.9 ± 1.2[A]	0.94 ± 0.08[A]	94.1 ± 4.0	1834 ± 106[A]	0.72 ± 0.04
	15	133 ± 11[C]	2865 ± 144	29.8 ± 1.2[B]	0.91 ± 0.05[A]	82.0 ± 3.9	1696 ± 85[AB]	0.59 ± 0.03
Spinal	5	839 ± 71[A]	3115 ± 149	45.7 ± 3.6	1.05 ± 0.05	100 ± 5	2066 ± 114	0.91 ± 0.04
Column	10	431 ± 55[B]	3479 ± 261	44.5 ± 3.1	1.24 ± 0.09	107 ± 8	2401 ± 211	0.95 ± 0.07
Bone	15	278 ± 25[B]	3375 ± 131	44.0 ± 2.0	1.23 ± 0.07	101 ± 4	2457 ± 124	0.92 ± 0.04

[1] Data are Mean ± SE; n=8 for all values. Means with different letter superscripts differ signifcantly (Duncan's test, P<0.05).
[2] Age = age in weeks at start of 5 weeks of lead exposure.
[3] Metal content for one kidney.

tion, which were comparable in the 3 treatment groups. One explanation may be the greater gastrointestinal absorption of lead in young versus older rats (6), though other differences in lead metabolism and in excretion may also contribute. The consistent pattern of decreasing organ lead concentrations with increasing rat age at exposure differed from that of the other five metals studied. This may reflect the cessation of lead exposure 4 weeks prior to blood and organ harvesting; in contrast, ingestion of the essential divalent metals from food occurred throughout the study. However, differences in the metabolism of lead and the other metals may also play a role. If the results of this study are applicable to humans, excessive lead exposure or lead poisoning as a child may result in the retention of more lead in the skeleton than exposure at a later age.

5. ACKNOWLEDGMENT

Supported by a grant from the American Heart Association - New Jersey Affiliate.

6. REFERENCES

1. G.R. Nordberg, K.R. Mahaffey, and B.A. Fowler, *Environ. Health Perspect.* **91**, 3–7 (1991).
2. P.J. Landrigan, *Environ. Health Perspect.* **91**, 81–86 (1991).
3. P.S.I. Barry, *Brit. J. Industr. Med.* **32**, 119–139 (1975).
4. E.J. O'Flaherty, *Toxicol. Appl. Pharmacol.* **118**, 16 (1993).
5. J.D. Bogden, S.B. Gertner, S. Christakos, F.W. Kemp, Z. Yang, S.R. Katz, and C.Chu, *J. Nutr.* **122**, 1351 (1992).
6. E.J. O'Flaherty, *Toxicol. Appl. Pharmacol.* **111**, 313 (1991).
7. Y. Naveh, P. Weis, H.R. Chung, and J.D. Bogden, *J. Nutr.* **117**, 1576 (1987).
8. J.D. Bogden, F.W. Kemp, S. Han, M. Murphy, M. Fraiman, D. Czerniach, C.J. Flynn, M.L. Banua, A. Scimone, L. Castrovilly, and S. Gertner, *J. Nutr.* **125**, 990 (1995).
9. P. Mushak, J.M. Davis, A.F. Crocetti, and L.D. Grant, *Environ. Res.* **50**, 11 (1989).

52

RELATIONSHIPS BETWEEN BONE MINERAL DENSITY, GROWTH, AND ALUMINUM IN HEALTHY FORMER PREMATURE INFANT

D. Bouglé, F. Bureau, B. Guillois, R. Morello, J. F. Duhamel, and
J. P. Sabatier

Laboratoire de Physiologie Digestive et Nutritionnelle
Laboratoire de Biochimie A, Service de Néonatologie
Laboratoire d'Informatique Médicale, Service de Pédiatrie A
Laboratoire des Radioisotopes
Centre Hospitalo Universitaire, F 14033 Caen Cedex, France

1. INTRODUCTION

The healthy adult is protected from aluminum (Al) accumulation by a low (< 1 %) digestive absorption of ingested intake and by the renal excretion of the absorbed fraction (1). The main target organs for (Al) are bone, brain, liver and the hematopoietic system (1). In the premature infant (PT), tissue Al loading has been described at necropsy of critically ill subjects, due to IV feeding and renal failure (2–5). However, increased Al plasma levels have been described in healthy PT or fullterm infants (6–8). The functional relationship between such variations of serum Al and biological functions remains to be defined in the infant inasmuch as it has been claimed that Al blood levels might be of poor interest in adult subjects (1).

A large number of PT suffer from osteopenia (9). Al plasma levels have been found to be higher in PT with than without fractures or rickets (10). It was therefore assumed that, in the PT, Al side effects could explain part of the variations of their bone mineralisation pattern. This study looked at blood Al and Ca related parameters which could be predictive of bone mineralisation in the healthy premature infants.

2. SUBJECTS AND METHODS

Infants studied were well growing, formula fed babies whose gestational age range at birth was 28 - 36 weeks and chronological age at study 4 - 426 weeks; none took drug known to contain Al. Their parents have given their informed consent to the study which has been approved by the Ethic Committee of the County of Basse-Normandie. A blood sample was obtained for the dosage of serum calcium (Ca), phosphorus (P), alkaline phosphatase activity, 25-OH Vitamin D and Al. Tubes were previously checked to be free of Al contamination (6). Al was measured by Atomic Absorption Spectrometry on a Perkin Elmer 3030, as previously described (6). The day of sampling, a bone densitometry of the lumbar spine was performed by dual energy X-ray absorptiometry (DEXA) using an ODX 240 DEXA, ORIS CEA, Saclay, France. DEXA uses two X rays of different energy levels in order to differentiate bone from surroundings tissues; its precision in measuring mineralised bone expressed in grams of hydroxyapatite is 2 - 3 % in new-borns and infants. The relationships between bone mineral density (BMD) and clinical (height, gestational age, weight) and biological (plasma Al, Vitamin D, serum Ca and P, alkaline phosphatase activity) data were assessed by simple and multiple stepwise regression analysis using StatView SE+ (Graphics TM,Abacus Concept Inc.).

3. RESULTS

A total of 26 measurements was made in 19 PT (7 infants could be studied twice). Mean (± 1 SD) BMD was 263 ± 102 mg/cm^2. Mean plasma Al was 0.92 ± 0.80 μmol/l (adult norms < 1.10 μmol/l). Other parameters were within the laboratory's reference values. Simple regressions displayed a significant correlation ($p < 0.01$) between BMD and chronological age, weight, height, serum P and serum Al. No correlation was found between BMD and gestational age at birth, serum Ca, 25-OH Vitamin D and alkaline phosphatase activity, neither between gestational age at birth and BMD or Al. The significant results displayed by multiple stepwise regressions between BMD as dependent variable and related parameters are given in the Table 1. Age was added to biological independent variables in multiple regressions because of its effect on their evolution.

4. DISCUSSION

Almost all of premature infants show some degree of hypomineralisation or frank osteopenia during their evolution (9). The aetiology of this metabolic bone disease is multifactorial. The main suggested factors are low Ca and/or P supply and bioavailability, and Vitamin D deficiency (9).

In our study the simple regression analysis confirmed the relationships between bone mineralisation and age, height, and weight (11–14). Multiple regression, however, showed that linear growth is the only determining factor, to which age and weight are related. Concerning biological data, bone mineralisation was not found to be related to blood levels of Ca, nor with 25-OH Vitamin D or phosphatase alkaline activity. These results confirm most of the previous studies (11–14) and the low predictive value of usual markers of Ca metabo-

Table 1. Multiple stepwise regression of parameters contributing to bone mineral density in healthy premature infants

	F	r	r^2
Height	150.2	0.91	0.83
Age	24.7	0.79	0.62
Plasma aluminum	19.7	0.86	0.74

lism (11). Simple regression showed a strong relationship between serum P and BMD emphasising its role as limiting factor in bone mineralisation of the low-birth-weight-infant (11,13,14). However this relationship was no longer observed in multiple regression analysis which showed a significant link only between bone mineralisation as dependent variable and age and serum Al. Al effects on bone are both direct by inhibiting its mineralisation and indirect through interactions with Ca and P metabolism. The direct inhibiting effects of Al on bone mineralisation are not fully explained (15) but Al is known to deposit at the calcification front, then preventing further mineralisation of osteoids (1), and to reduce the total quantity of mineralisable osteoids. These effects are enhanced in case of high bone turnover (16) that could be of importance in the high turnover osteopenia of the PT. Interactions between Al and Ca occur at every step of their metabolism (17). These interactions are enhanced by the higher Al supply of PT compared to fullterm infants. Indeed, due to neonatal illnesses, PT are often IV fed with Al rich solutes (18) and the low-birth-weight formulas are more Al contaminated than standard formulas (19, 20).

Finally, it is well known that the excretion of the absorbed or perfused Al is impaired even by a moderate lowering of renal functions (1). The immature renal functions of the PT explain why the more gestationally immature of them display the higher serum Al concentrations (6) and the reduced decrease in Al/creatinine ratio in urines (10). Therefore high intakes and/or absorption and low excretion of Al could increase the exposure duration of tissues to Al, that makes the healthy PT liable to store it.

In conclusion, Al serum levels and age appear to be better correlated to bone mineralisation in preterm infants than usual parameters of Ca metabolism.

5. REFERENCES

1. A.C. Alfrey, in *Trace elements in human and animal nutrition*, 5th ed., vol.1, W. Mertz, ed., Academic Press Inc., Orlando, pp. 399–413 (1986).
2. A.B. Sedman, G.L. Klein, R.J. Merritt, M.L. Miller, K.O. Weber, W.L. Gill, H.A. Anand. and C. Alfrey, *N. Engl. J. Med.* **312**, 1337–43 (1985).
3. M. Freundlich, G. Zilleruelo, C. Abitbol, J. Strauss, M.C. Faugère and H.H. Malluche, *Lancet* **2**, 527–9 (1985).
4. W.W.K. Koo, L.A. Kaplan, R. Bendon, P. Succop, R.C. Tsang, J. Horn and J.J Steichen, *J. Pediatr.* **109**, 877–83 (1986).
5. M.E.A. Bozynski, A.B. Sedman, R.A. Naglie and E.J. Wright, *J.P.E.N.* **13**, 428–31 (1989).
6. D. Bouglé, F. Bureau , J. Voirin , D. Neuville and J.F. Duhamel, *J. P. E. N.* **16**, 167–9 (1992).
7. R. Litov, V.S. Sickles, G. Chan, M.A. Springer and A.C. Cordano, *Pediatrics* **84**, 1105–7 (1989).
8. N.M. Hawkins, S. Coffey, M.S. Lawson and H.T. Delves, *J.Pediatr.Gastroenterol. Nutr.* **19**, 377–81 (1994).
9. W.W.K. Koo, S.K. Krug-Wispe, P. Succop, R. Bendon and L.A. Kaplan, *Pediatrics* **89**, 877–81 (1992).
10. F.R. Greer, *Annu. Rev. Nutr.* **14**, 169–85 (1994).
11. S.A. Abrams, R.J. Schanler and C. Garza, *J. Pediatr.* **112**, 956–60 (1988).
12. G.M. Chan, *J. Pediatr.* **123**, 439–43 (1993).
13. S.W. Ryan, J. Truscott, M. Simpson and J. James, *Acta Paediatr.* **82**, 518–21 (1993).
14. R. Namgung, R.C. Tsang, B.L. Specker, R.I. Sierra and M.L. Ho, *J. Pediatr.* **122**, 269–75 (1993).
15. G.L. Klein and J.W. Coburn, *Annu. Rev. Nutr.* **14**, 135–67 (1994).
16. L.D. Quarles, *Am. J. Physiol.* **258**, E576–81 (1990).
17. J. Moon, A. Davison and B. Bandy, *Can. Med. Assoc. J.* **147**, 1308–13 (1992).
18. P. Chappuis, J. Poupon, J. Arnaud, M.C. Jaudon and R. Zawislak, *Am. J. Clin. Nutr.* **54**, 951–2 (1991).
19. W.W.K. Koo, L.A. Kaplan and S.K. Krug-Wispe, *J.P.E.N.* **12**, 170–3 (1988).
20. D. Bouglé, P. Foucault, J. Voirin and J.F. Duhamel, *Arch. Fr. Pédiatr.* **46**, 768 (1989).

53

INFLUENCE OF CHEWING GUM WITH SODIUM FLUORIDE ON THE HUMAN TOOTH ENAMEL FLUORIDE CONTENT IN 13-YEAR-OLD CHILDREN AFFECTED BY CARIES

T. Ogonski,[1] J. Radlinska,[1] D. Samujlo,[2] and I. Nocen[1]

[1] Biochemistry and Chemistry Department
[2] General Dentistry Department, Pomeranian Medical Academy
Aleja Powstanc ów Wielkopolskich 72; 70–111 Szczecin, Poland

1. INTRODUCTION

The carious lesion is initiated by the production of organic acid, products of the carbohydrate metabolism by plaque bacteria and the subsequent dissolution of the enamel. Such a process is a dynamic one, and a number of factors must be considered. These factors vary from the type and relative amounts of microorganisms colonizing tooth surfaces to the chemical and ultrastructural properties of the enamel, what ultimately leads to the unique phenomenon of subsurface demineralization (1). Numerous authors have proved that hard dental tissues are permeable to certain sizes of molecules and ions. One of them can be fluoride. There is general agreement that availability of fluoride reduces caries. The beneficial influence of fluoride on caries has been confirmed in numerous studies (2). In spite of a vast amount of researches, the mode of action of fluoride as a caries preventive agent is still discussed. The most important of the current views seems to be that fluoride in the fluids surrounding the enamel crystals enhances the rate of the remineralization after the acid attack (3,4). The plaque fluid contains approximately 0.85 mmol/dm^3 of calcium ions (5), and it has been suggested that calcium fluoride may be precipitated when plaque is exposed to fluoride present in mouthrinse solutions, toothpastes, or other fluoride-containing preparations used for caries prevention. The fluoride level in the oral fluids necessary for the saturation with respect to calcium fluoride is estimated to be approximately 0.30 mmol/dm^3, assuming an ionized concentration of 0.50 mmol/dm^3, similar to that found in the parotid saliva (6). This fluoride concentration may be compared to those found in the whole saliva ranging from 1.4 to 38 mmol/dm^3 during the first minutes after administration of fluoride tablets, fluoride-containing chewing gums or dentifrices, or 1.2 % fluoride solutions (7). The aim of the present study was to investigate the influence of a sodium-fluoride-calcium-gluconate-containing

Therapeutic Uses of Trace Elements, edited by Nève et al.
Plenum Press, New York, 1996

chewing gum on the human dental enamel fluoride content among children affected by advanced caries.

2. MATERIALS AND METHODS

The study comprised a total of 300 enamel biopsies from teeth of 30 schoolchildren, 13 years old, indicated by a discriminant DFT = 8,9 ± 3,0 (Decay + Filling + Teeth) chosen from the high caries risk group of 297 children. Each tablet of chewing-gum (P.W.C. Odra Ltd.; Brzeg, Poland) contained 25 % xylitol, 0.02 % sodium fluoride and 10 % calcium gluconate. Children, with parent's approval, were put through a five series of examinations. The first one was performed before the experiment had started. Next three examinations were after 1, 2, 3 months of the chewing-gum application (3 times a day after main meals) and the last one was 1 month later, after discontinuance of chewing.

An *in-vivo* acid-etch biopsy technique was developed to produce artificial caries lesions and to determine the fluoride content in the superficial layers of the dental enamel (8,9). Biopsy samples were obtained from clinically sound labial surfaces of upper permanent central incisor teeth. The area to be sampled was 2,5 mm diameter circular cut-out in a plastic adhesive tape. Using the forceps, the tape with the hole delimiting the selected biopsy area was placed over the central vertical line of the tooth surface. Two superimposed biopsies were taken from each site. The surface sample was taken by placing 5 μl of 0.125 M HNO_3 on the sampling site for 30 seconds and recovering it quantitatively into 1 ml of TI-SAB II buffer (Orion Cat. No. 940909) diluted 1:2 with dd-H_2O. The subsurface sample was taken exactly 30 seconds later. The procedure was identical but 0.500 M HNO_3 was used. The fluoride content of the collected samples was determined on microaliquots (0.2 ml) in duplicates using a combination F-electrode (Orion). Calcium analyses were carried out in duplicates by atomic absorption spectroscopy (PU 9100X). Absolute biopsy depths were calculated from the sample area and the amount of calcium present in samples, assuming that the shape of the biopsies was cylindrical and that sound enamel contains 37 % (w/w) Ca and its density is 2.95 (10):

$$\text{Absolute depth of biopsy } (\mu m) = \frac{\text{Weight of enamel in sample } (\mu g)}{\text{Biopsy area } (mm^2) \times 2.95 \ (g/cm^3)}$$

As the fluoride concentration of enamel decreases with distance from the surface within each biopsy layer it is more closely related to the central depth (half-layer thickness) than to the absolute depth of the layers. The depth values were derived as follows: the central depth of the surface biopsy is half of its absolute depth, whereas that of the subsurface biopsy is its half depth added to the absolute depth of the surface biopsy. The total cumulative depth is the sum of the absolute depths of the two superimposed biopsies.

3. RESULTS

The fluoride concentrations (Mean ± SD) in enamel areas treated with diluted HNO_3 are shown on table 1. The considerable variation in the fluoride concentrations among various groups was found. The mean values of the fluoride concentrations in both central layers of enamel increase with the increasing time of the experiment. The standard deviation values

Table 1. Fluoride concentration of enamel by biopsy layer and
according to fluoride exposure (mmol/kg)

Month	0	1	2	3	4
Layer 1	50 ± 16	57 ± 14	98 ± 38	91 ± 23	103 ± 61
Layer 2	22 ± 18	23 ± 13	36 ± 23	41 ± 17	48 ± 25

are comparable except values evaluated for the last group (month 4). Table 2 shows the mean values and standard deviations of the central and total cumulative depths in respect of biopsy layers, by fluoride exposure during the time of the experiment. Both biopsy depths decrease with very good correlation. The intergroup variation was examined by means of Student's t-test at the level of significance $P < 0.05$

4. DISCUSSION

One of the important functions of fluoride is to contribute to remineralisation. There is evidence that during the demineralisation, the more soluble apatite crystals are removed from the enamel and, in the presence of fluoride, are replaced with more acid-resistant fluoridated crystals (2). Thus a poorly mineralised surface can be progressively improved. As the teeth develop the presence of fluoride improves the crystallinity and stability of the enamel, reducing its solubility. When fluoride is used the tooth can withstand a higher frequency of acid. At the begining of our experiment mean fluoride concentrations in enamel biopsies from upper permanent central incisor teeth among 13-year-old children belonging to a high caries risk group were found to be 50 and 22 mmol/kg in two succesive layers with a mean thickness of 1.68 and 8.34 µm respectively. These fluoride concentrations are about 50 % lower in comparison with values measured among children of the same age from the communities with 0.5 mg fluoride/dm^3 in the supplied water (8). Taking a biopsy depth (distance from the enamel surface of artificial lesion) as a reflection of the enamel dissolution, our results may be interpreted as a support for *in vitro* findings that fluoride reduces the solubility of hydroxyapatite (11). The significantly greater mean biopsy depth obtained in the F-deficient-teeth group is in good agreement with a finding that in populations exposed to a very low fluoride concentrations in the water the amount of enamel removed by biopsy procedure and fluoride concentration of enamel are inversely associated (11,12). When fluoride is used teeth can withstand a higher frequency of acidic agents. The mentioned process of the fluoride incorporation into F-deficient tooth enamel takes about two months and is effective for the next two months even when the fluoride administration is lower or totally suspended.

Table 2. Biopsy depth (distance from enamel surface) according to
fluoride exposure (µm)

Month	0	1	2	3	4
Layer 1	1.7 ± 0.5	1.4 ± 0.7	0.8 ± 0.2	1.0 ± 0.3	1.1 ± 0.3
Layer 2	8.3 ± 2.7	7.3 ± 2.8	5.4 ± 2.0	4.8 ± 1.1	4.6 ± 0.9
Layer 1+2	13.3 ± 5.1	11.6 ± 4.8	9.3 ± 3.9	7.6 ± 1.8	7.1 ± 1.5

5. CONCLUSIONS

Fluoride ions contained in the chewing-gum could be incorporated into children, belonging to a high caries risk group, teeth enamel by the time of minimum two months when systematically used. In this case the teeth enamel resistance against the acidic environment, checked by the examination of biopsy depths, grows which can be considered as a caries resistance factor. The positive effect of the chewing gum on the teeth enamel remained one month after finishing the trial which supports the idea that a fluoride containing chewing gum can be proposed as a good and easy in use caries protective agent.

6. REFERENCES

1. L.M. Silverstone, *Oral Sci. Rev.* **3**, 100–160 (1973).
2. R.P. Shellis and R.M. Duckworth, *Int. Dent. J.* **44** (3), 263–273 (1994).
3. F.M.C. Driessens, *Mineral aspects of dentistry*. Basel: Karger, 129–142 (1982).
4. R.S. Levine, *Dent.* Update (April), 105–110 (1991).
5. C.M. Carey, A. Tatevossian and G.L. Vogel, *Caries Res.* **21**(A1), 158 (1987).
6. F. Lagerlöf, *Clin. Chim. Acta* **102**, 127–35 (1980).
7. C. Bruun, D. Lambrou, M.J. Larsen, O. Fejerskov and A. Thylstrup, *Community Dent. Oral Epidemiol.* **10**, 124–129 (1982).
8. C. Bruun, E.C. Munksgaard and K. Stolze, *Community Dent. Oral Epidemiol.* **3**, 217–222 (1975).
9. D.H. Retief, J.M. Navia and H. Lopez, *Arch. Oral Biol.* **22**, 207–213 (1977).
10. E. Kirkegaard, *Caries Res.* **11**, 16–23 (1977).
11. R. Soremark and K. Samsahl, *Arch. Oral Biol.* **6**, 275–283 (1961)
12. E. Brown, T.M. Gregory and L.C. Chow, *Caries. Res.* **11**(Suppl.1), 118–141 (1977).

54

INFLUENCE OF CHEWING GUM WITH SODIUM FLUORIDE ON THE ORAL HYGIENE, GINGIVAL STATUS, SUSCEPTIBILITY OF AN ENAMEL, SALIVARY LEVEL OF *STREPTOCOCCUS MUTANS* AND *LACTOBACILLUS* IN 13-YEAR-OLD CHILDREN AFFECTED BY CARIES

J. Radlinska[1] and T. Ogonski[2]

[1] General Dentistry Department
[2] Biochemistry and Chemistry Department, Pomeranian Medical Academy
Aleja Powstancόw Wielkopolskich 72
70–111 Szczecin
Poland

1. INTRODUCTION

Dental caries are multi bacterial diseases depending on the interaction between enamel, saliva, dental plaque and consumed carbohydrates. In the process of enamel demineralisation, *Streptococcus mutans* and *Lactobacillus* bacteria were shown to play a mayor role (1). This process can take place only when the destruction of the structure of minerals occurred (2, 3).

Fluoride improves the process of the enamel remineralisation by increasing the size of crystals. A simultaneous saturation of the environment with calcium phosphates is necessary. Fluorides can also influence the ecology of dental plaque by decreasing the numerical force of acid-forming bacteria such as *Streptococcus mutans* (4, 5). According to Söderling et al. (6) xylitol can limit the adherence of the dental plaque, and according to Smits et al. (7) it can decrease demineralisation of the enamel.

The aim of this paper was the evaluation of the influence of chewing a gum containing sodium fluoride, calcium gluconate and xylitol on the hygiene of the oral cavity, the condition of gums, the susceptibility of the enamel to acids, the level of fluorides in the enamel and the saliva as well as the number of caries-forming bacteria in the saliva of children of high caries risk group.

Therapeutic Uses of Trace Elements, edited by Nève et al.
Plenum Press, New York, 1996

2. MATERIALS AND METHODS

Thirty school children of the age of 13, selected among 297 ones and residing in one of Szczecin districts were the subject of the studies. The criteria for the children to be selected to the group of a high risk of caries was the average of DFT figures (Decay + Filling + Teeth). Children of DFT figure > 4 (average 8.9 ± 3.0) were classified to the experimental group. The following was evaluated: the hygiene of the oral cavity by the application of Oral Hygiene Index (OHI-S: evaluated within the scale from 0 to 3), the condition of gums by using Gingival Index (GI: scale from 0 to 3), and the susceptibility of the enamel to acids by the Colour Reaction Time test (CRT test: measured in seconds). The CRT test was carried out on the 11 (central upper right incisor) and the 31 (central lower left incisor) tooth. Moreover, samples of the resting saliva and the enamel - with the acidic biopsy method - were taken for biochemical analysis in order to determine the level of fluorides. The saliva for tests was taken about 2 hours after breakfast. The salivary level of *Streptococcus mutans* and *Lactobacillus sp.* was determined with the help of Dentocult SM and Dentocult LB tests that are available on the market (Orion Diagnostica - Finland). After the clinical examination and sampling for the analysis, the children would get a chewing gum containing sodium fluoride (1 mg of NaF per 4.8 g of one chewing gum tablet), calcium gluconate (480 mg per tablet) and xylitol (1.2 g per tablet) (P.W.C. Odra Ltd., Brzeg, Poland) with the recommendation to chew it 3 times a day for 30 minutes after every main meal. The experiment lasted for 3 months. The clinical tests and sampling for the analysis were carried out by 1 person.

3. RESULTS

Average values of parameters measured at subsequent stages of the experiment, are shown in Table 1. The index of the hygiene of the oral cavity and the condition of gums were not considerably different during the course of the experiment. In case of children, where a high OHI-S index was found out during preliminary examinations, it remained as such for the following months. In all CRT tests, the micro demineralisation time was shorter on the 31 tooth than on the 11 one. The increase of the enamel microdemineralisation time occurred both on the 11 tooth as well as on the 31 one, but statistically, a more clear difference concerned the 31 tooth. The average level of fluoride in the saliva was showing the rising inclination up to the second month of chewing the gum. Then, it remained at that level during the month that followed after the gum chewing treatment had stopped. In the same time the average level of fluoride in the enamel showed a constant rising inclination, inclusive of the month that followed after the gum chewing treatment had been stopped. During preliminary

Table 1. Average values of measured parameters of the examined children at subsequent stages of examinations

Month	0	1	2	3	4
OHI-S	1.67	1.48	1.65	1.95	1.96
GI	0.37	0.36	0.45	0.41	0.44
CRT (tooth 11) sec.	40	-	42	44	48
CRT (tooth 31) sec.	30	-	35	39	45
Saliva fluoride µM/L	2.02 ± 0.34	2.68 ± 0.58	3.48 ± 0.58	1.99 ± 0.70	1.68 ± 0.29
Enamel fluoride µg/g	475 ± 208	545 ± 260	1296 ± 392	1860 ± 399	2210 ± 402

examinations, it was found out that 62% of the examined children had a high titre of *Streptococcus mutans* ($\geq 10^6$ CFU/ml) and Lactobacillus ($\geq 10^5$ CFU/ml) After 3 months of the experiment, 41% of the examined children showed a decrease of the numerical force of Lactobacillus, and only 17% of them indicated a decrease of Streptococcus mutans, however, the decrease of Lactobacillus in the saliva occurred in case of as much as half of the examined children with a high level of Lactobacillus.

4. DISCUSSION

The results of the experiment pointed out that chewing the gum with an addition of sodium fluoride, calcium gluconate and xylitol did not improve the hygiene of the oral cavity. Such a result is not surprising, because the children were not informed in details how to maintain the hygiene of the oral cavity, and in the conditions of experiments. The activity of chewing the gum, that also involves soft tissues of the oral cavity - cheeks and the tongue, does not have influence on the effective removal of the dental plaque from the surface of smooth teeth, where the plaque was evaluated. On one side, the presence of the plaque on teeth can be regarded as an caries-forming factor, but on the other side, it can be a long lasting reserve of fluoride that take part in processes of the enamel repair [8, 9, 10]. Poulsen et al. [11] pointed out that a professional teeth cleaning can prevent dental caries and gingival inflammation, however, Horowitz et al. [5] as well as Frans et al. [12] indicated a lack of good correlation between the hygiene of the oral cavity and the development of caries. In our experiment, the result of preliminary studies proved that a high level of Streptococcus mutans and Lactobacillus occurs in the saliva of 62 of the examined children. Literature data [1, 13] point out that the growth of colonies of Streptococcus mutans $>10^6$ CFU/ml and/or Lactobacillus $>10^5$ CFU/ml in the saliva increases the risk to develop dental caries. Loesche [14] is of the opinion, that Lactobacillus has a limited influence on the initiation of dental caries and it is rather connected with already existing caries, so it only contributes to the development of that disease. After having used the chewing gum with additions, which can decrease the number of cariogenic bacteria in the saliva, such a reaction was observed only with reference to Lactobacillus. A drop in the number of this kind of bacteria was obtained in case of 41 of all the examined individuals, out of which, 50 of children with a high number that was confirmed during the preliminary examination.

The long-lasting influence of fluoride on the flora of the plaque is difficult to determine due to the possibility of a phenotypic and genetic adaptation of the bacteria to fluorides[15]. So far, distinct differences in the composition of the plaque of individuals exposed to high and low densities of fluorides have not been proved and comparing the acid forming abilities of the plaque of individuals residing in areas with fluoridated water and water with no fluoridation did not give uniform results [16]. Nelson et al. and Wong et al. suggest that the content of fluorine in concentrations that usually occur in the enamel (approximately 1000 mg/g) does not effectively secure it from caries. Only higher concentrations, close to those occurring in fluoroapatite, can significantly decrease its degree of solubility in acids (17, 18). Other experiments show that already a concentration of 230 mg/g decreases significantly the degree of solubility of the enamel hydroxyapatite [18, 19]. The results of our experiments point out that sodium fluoride used in the chewing gum incorporates with enamel structure and brings about a decrease of its susceptibility to acids. The decrease of susceptibility, that was found out, can be the result of a joint action of sodium fluoride, calcium gluconate and xylitol [6, 7].

5. REFERENCES

1. J. Suhonen, Schweiz. Monatsschr. Zahnmed. 102, 286–291 (1992)
2. J. M. ten Cate and P. P. E. Duijsters, Caries Res. 17, 193–199 (1983)
3. J. M. ten Cate and P. P. E. Duijsters, Caries Res. 17, 513–519 (1983)
4. J. R. Hamilton, Caries Res. 11 (Suppl.1)., 262–278 (1977)
5. A. M. Horowitz, J. D. Suomi, J. K. Peterson and B. A. Lyman, J. Public. Health Dent. 37, 180–188 (1977)
6. E. Söderling, L. Alaräisänen, A. Scheinin and K. K. Mäkinen, Caries Res. 21, 109–116 (1987)
7. M. T. Smits and J. Arends, Caries Res. 22, 160–165 (1988)
8. R. M. Duckworth and S. N. Morgan, Caries Res. 25, 123–129 (1991)
9. R. M. Duckworth, S. N. Morgan, G. S. Ingram and D. J. Page, in Clinical and Biological Aspects of Dentifrices, G. Embery and G. Rolla, ed., Oxford University Press, Oxford , pp. 91–104 (1992)
10. D. A. M. Gedds and S. G. Mc Nee, Arch. Oral Biol. 27, 765–769 (1982)
11. S. Poulsen, N. Agrebeak, B. Melson, L. Glavind and G. Rolla, Community Dent. Oral Epidemiol. 4, 159–199 (1976)
12. F. E. Frans and L. J. Baume, SSO, 93, 1183–1188 (1983)
13. K. Kristofferson, H. G. Gröndhl and D. Brathall, J. Dent. Res. 64, 58–61 (1985)
14. W. J. Loesche, Microbiol. Rev. 4, 353–380 (1986)
15. G. H. W. Bowden, J. Dent. Res. 69, 653–659 (1990)
16. W. M. Edgar, in The Environment of the Teeth. Frontiers of oral physiology, vol 3, D. B. Ferguson, ed., Basel, Karger pp. 19–37 (1981)
17. D. G. A. Nelson, J. D. B. Featherstone, J. F. Duncan and T. N. Cutress, Caries Res. 17, 200–211 (1983)
18. L. Wong, T. W. Cutress and J. F. Duncan, J. Dent. Res. 66, 1735–1741 (1987)
19. D. J. Crommelin, W. J. Higuchi, J. L. Fox, P. J. Spooner and A. V. Katdare, Caries Res. 17, 289–296 (1983)

55

DEGREE, CLINICAL CONSEQUENCES, AND ERADICATION OF IODINE DEFICIENCY IN EUROPE

F. Delange

International Council for Control of Iodine Deficiency Disorders (ICCIDD)
153, avenue de la Fauconnerie, B-1170, Brussels, Belgium

1. INTRODUCTION

Iodine is a trace element whose only confirmed function is to constitute an essential substrate for the synthesis of thyroid hormones, tetraiodothyronine (thyroxine or T_4) and tri-iodothyronine (T_3) (1). The daily requirements of iodine for humans are 150 µg in adults, 200 µg in pregnant and lactating women and 90–120 µg in infants and children (2). When the physiological requirements of iodine are not met in a given population, a series of functional and developmental abnormalities occur, including hypothyroidism, endemic goiter, cretinism and mental retardation, decreased fertility, increased perinatal death and infant mortality. These complications, which constitute a hindrance to the development of the affected populations, are grouped under the general heading of Iodine Deficiency Disorders, IDD (3). The data presently available indicate that approximately 1.5 billion people are at risk of IDD, i.e. 28.9 % of the earth population, including 140 million in Europe (4). Iodine deficiency therefore constitutes one of the most common preventable causes of mental deficiency. The objective of this paper is to review the present situation of IDD and its control in Europe.

2. EPIDEMIOLOGY

Endemic goiter, occasionally complicated by endemic cretinism, has been reported in Europe up to the turn of the 20th century, especially from remote, isolated, mountainous areas in central parts of the continent including Switzerland, Austria, Northern Italy, the former Czechoslovakia, Bulgaria and Poland (5). The problem of IDD has been entirely eradicated in Switzerland thanks to the implementation and monitoring of a program of salt iodization (6). Probably because of the impact on the medical world of this remarkable program, IDD seem to have been considered no longer as significant public health problems in Europe during the last five decades.

Therapeutic Uses of Trace Elements, edited by Nève et al.
Plenum Press, New York, 1996

However, reevalaution of the problem in the late 1980's clearly indicated that, with the exception of most of the Scandinavian countries, Austria and Switzerland, most of the European countries or at least certain areas of these countries were still affected, especially in the Southern part of the continent (7, 8). These surveys also revealed a lack of information on IDD in countries of the Eastern part of the continent. The status of iodine nutrition was reevaluated in all European countries, including in the Eastern part of the continent, at an international workshop entitled "Iodine Deficiency in Europe. A continuing concern" held in Brussels in April 1992 (9).

The outcome of this workshop, summarized in Figure 1, indicated that, in 1992, iodine deficiency was under control in only five countries, namely Austria, Finland, Norway, Sweden and Switzerland. Iodine deficiency was marginal or present mainly in "microfoci" (pockets of goiter) in Belgium, the Czech and Slovak Republics, Denmark, France, Hungary, Ireland, Portugal and the United Kingdom. IDD have recurred after transitory resolution in Croatia, the Netherlands and possibly some Eastern Europe countries. Finally, iodine deficiency persisted and varied from moderate to severe in all the other European countries,

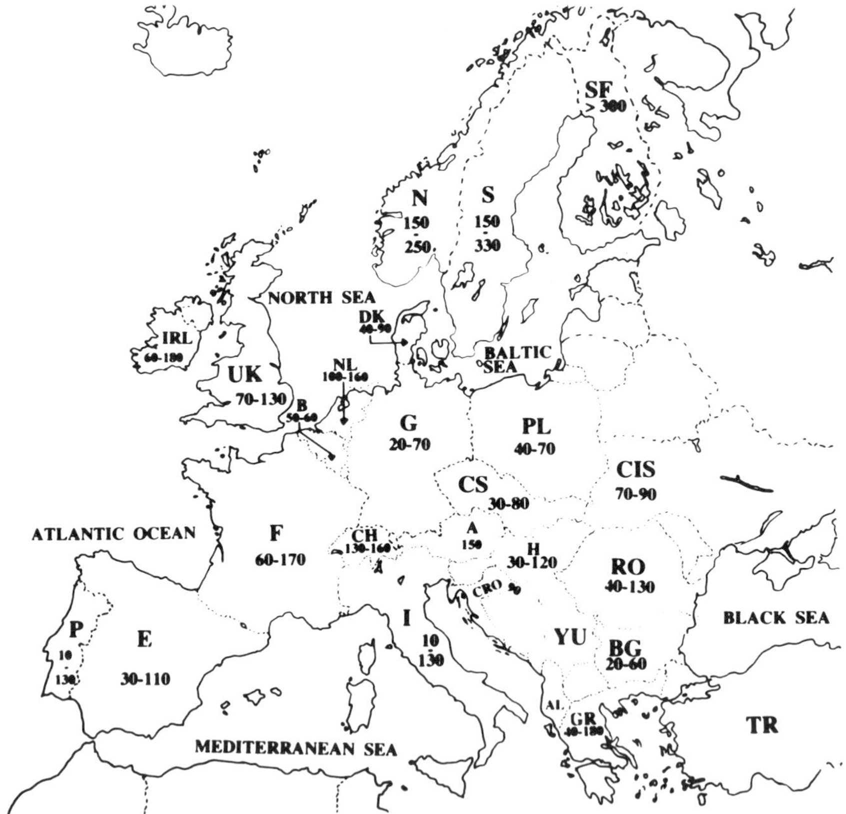

Figure 1. Evaluation of iodine intake in Europe as at 1992 (mg/day). Range of the values observed during regional or national surveys. N: Norway, S: Sweeden, SF: Finland, DK: Denmark, IRL: Ireland, UK: United Kingdom, B: Belgium, NL: The Netherlands, G: Germany, PL: Poland, CS: Former Czechoslvakia, CIS: The Commonwealth of Independent States, F: France, CH: Switzerland, A: Austria, H: Hungary, Ro: Romania, P: Portugal, E: Spain, I: Italy, CRO: Croatia, Y: Yugoslavia, BG: Bulgaria, GR: Greece, AL: Albania, TR: Turkey. From Delange 1994 (16). With permission.

namely Bulgaria, the Commonwealth of Independent States (CIS), Germany, Greece, Italy, Poland, Romania, Spain and also in Turkey. In some of these countries, such as Bulgaria and Romania, the prevalence of goiter in schoolchildren varied from 16 to 81 % and the median urinary iodine concentrations could be only 2 µg/dL.

From 1992 onwards, and especially following the endorsement by WHO Euro of the recommendations resulting from the Brussels meeting (9) to appoint national iodine committees in all countries and to generalize the use of iodized salt, quite significant progress has been achieved in many European countries.

Presently ongoing reevaluation of the status of iodine nutrition in 12 European countries by means of standardized measurements of thyroid volume by ultrasonography and urinary iodine in schoolchildren (ThyroMobil project, unpublished) indicates that substantial progress towards the control of iodine deficiency occurred in some European countries or regions such as Southern Germany, France and the Czech and Slovak Republics. In others, however, such as Belgium, the situation of borderline iodine deficiency did not improve yet.

3. PUBLIC HEALTH CONSEQUENCES

The state of mild to severe iodine deficiency persisting in many European countries or regions has important consequences from a public health point of view, including on the intellectual development of infants and children.

3.1. In Adults

The frequency of simple goiter is elevated in many countries and the cost of therapy of thyroid problems resulting from iodine deficiency in the adult population is enormous. For example, the cost for the diagnosis and treatment of goiter due to iodine deficiency in Germany for the year 1986 was estimated at 900 million DM (approximately 700 million US$) (10) while prevention by iodized salt would cost only 2–8 US cents per person and per year (11). Thyroidal uptake of radioiodine varies markedly from one European country to another and is inversely related to the iodine intake . Elevated thyroidal uptake due to iodine deficiency aggravates the risk of thyroid irradiation and development of thyroid cancer in case of a nuclear accident (12).

Thyroid function is usually normal in adults in Europe. In contrast, it is frequently altered in pregnant women. During pregnancy, the gland undergoes stimulatory events due to the synergic effects of three mechanisms : direct stimulation by HCG, stimulation through the usual feedback mechanism via the increase in TBG and the lowering of free hormone concentrations, and finally, the overall enhancing role of limited iodine availability (13). It has been shown that, at least in conditions of borderline iodine intake as seen in Belgium (50–70 µg iodine/day), pregnancy is accompanied by a progressive decrease of serum free T_4 and, consequently, by an increase of serum TSH. This state of chronic TSH hyperstimulation results in the development of goiter in about 10 % of the pregnant women and in a progressive increase in the serum concentration of thyroglobulin (Tg). Goiter can persist after pregnancy in an important number of women. Pregnancy, especially in conditions of borderline iodine intake, at least partly explains the higher frequency of thyroid problems in women than in men.

The consequences of marginal iodine deficiency during pregnancy in Belgium on the thyroid function of the neonate include even more elevated serum levels of TSH and Tg on cord blood than in the mothers and a slight enlargement of the thyroid gland. The role played by iodine deficiency in these changes is demonstrated by the fact that they are prevented by iodine supplementation of the mothers during pregnancy (14) and that they do not occur in iodine replete areas in Europe such as some parts of The Netherlands (15).

3.2. In Adolescents and Children

Euthyroid pubertal goiter is especially frequent in adolescents and occasionally requires substitutive therapy by T_4 or iodide. A very important issue is the demonstration that even in Europe today, clinically euthyroid schoolchildren born and living in an iodine deficient environment exhibit subtle or even overt neuropsychointellectual deficits as compared to controls living in the same ethnic, demographic, nutritional and socio-economic system, except that they are not submitted to iodine deficiency (16). These deficits are of the same nature, although less marked, than those found in schoolchildren in areas with severe iodine deficiency and endemic mental retardation (17).

3.3. In Neonates

The most important and frequent alterations of thyroid function due to iodine deficiency in Europe occur in neonates and young infants.

The frequency of transient primary hypothyroidism is almost 8 times higher in Europe than in North America (18). This syndrome is characterized by postnatally acquired severe primary hypothyroidism lasting for a few weeks and requiring substitutive therapy (19). The risk of transient hypothyroidism in the neonates increases with the degree of prematurity (20). The specific role played by iodine deficiency in the etiology of this type of hypothyroidism is demonstrated by the disappearance of the syndrome following systematic iodine supplementation of preterm infants.

There is an inverse relationship between the urinary iodine concentration in newborn populations in Europe used as an index of their status of iodine nutrition and the frequency of serum TSH above 50 µU/mL at day 5, at the time of screening for congenital hypothyroidism, i.e. the recall rate under suspicion of congenital hypothyroidism (16). Consequently, neonatal thyroid screening appears as a particularly sensitive index of the presence and action of goitrogenic substances in the environment (21) and can be used as a monitoring tool in the evaluation of the effects of iodine prophylaxis at a population level.

The reason for the particular sensitivity of the newborn, especially of the preterm infant, to the effects of iodine deficiency appears from the data summarized in Table 1. In Toronto, where the iodine intake is elevated, the iodine content of the thyroid in full term infants is 300 µg. In Brussels, with a borderline iodine intake, the iodine content of the thyroid is 82 µg and in Leipzig, which used to be severely iodine deficient, the content is only 43 µg. The table shows that the turnover rate of intrathyroidal iodine is markedly accelerated in iodine deficient neonates. Therefore, thyroid failure is more likely to occur.

4. PREVENTION AND THERAPY OF IDD IN EUROPE

It is hard to understand and impossible to admit that iodine deficiency, the most common preventable cause of mental deficiency in the world to-day, is still so prevalent in Europe. The most probable cause of the phenomenon has been the insufficient awareness until recently of the problem of IDD by the health authorities, including the medical and paramedical world, and by the public.

Table 1. Relationship between the iodine content of urines in adults and neonates used as an index of iodine supply and thyroid weight, iodine contentand estimated turnover rate of thyroidal iodine in neonates in three areas with markedly different iodine intake. Results given as mean ± SEM. The number of patients is shown in parentheses. Adapted from Delange et al. 1993 (25)

| | | Neonates | | | | |
| | | Urine Iodine | | Thyroids | | |
Cities	Adults Urine iodine(µg/day)	Median (µg/dl)	Values < 5 µg/dl (%)	Weight (g)	Iodine Content (µg)	Estimated turnover rate[1] (%/day)
Toronto(Canada)	600-800	14.8 (81)	11.9	1.0±0.1 (13)	292±47	17
Brussels(Belgium)	51	4.8 (196)***	53.2	0.8 ± 0.3 (4)	81 ± 9**	62
Leipzig(Germany)	16	1.6 (70)***	97.2	3.3 ± 0.4(10)**	43±6**	125

[1] Based on a requirement of IT4 of 50 µg/day).

Levels of significance as compared to Toronto ** p < 0.01, ***p < 0.001.

During the past 10 years, a series of major decision making meetings took place, including the World Health Assembly in 1990 (Geneva), the World Summit for Children (New York 1990), the Policy Conference on Ending Hidden Hunger (Montreal 1991) and the International Conference on Nutrition (Rome 1992). As an outcome of these meetings, WHO and UNICEF committed themselves at the virtual elimination of iodine deficiency disorders in the world, including in Europe, by the year 2000.

The strategy for the prevention and therapy of iodine deficiency in Europe has to start with information and health education not only of the public but also of the health professionals who are often unsufficiently aware of the problem. This includes the dissemination through appropriate channels of the presently available information on IDD and on the national status of iodine nutrition in each country. Undue concern about the possible side effects of iodine supplementation as well as the uncontrolled abuse of iodine, such as self medication with Lugol solutions have to be carefully avoided. Appropriate food habits including the regular consumption of adequately prepared seafoods have to be encouraged.

The major measure, however, for the prevention of iodine deficiency in Europe, is the systematic introduction and monitoring of iodine supplementaiton through programs of Universal Salt Iodization (USI), i.e. the fortification of all salt for human and animal consumption.

In 1992, iodized salt was available in most European countries, usually nationwide, with the exception of Denmark where it was prohibited (22). It was compulsary in 8 countries. The price was most usually barely higher than non iodized salt; it was even lower in France. In spite of this apparently satisfying situation and with the remarkable exceptions of Switzerland, Austria, Sweden and Finland, national programs of salt iodization were not sufficiently operational and efficient in Europe (9). This partial failure could result from the fact that many of the iodized salt samples were grossly inadequate with respect to their iodine content, at least when available at the level of the consumer (23). It also could result from a gross overestimation of household salt consumption : the actual ingestion of that salt, measured by the lithium marker technique, was only 15 % of the total salt intake (23). Therefore, iodized salt should be made available not only for household but also for industrial food production, including cheese and bread, as well as for animal consumption.

Of course, in the context of USI programs, particular attention must be paid to the usual recommendation to limit as much as possible the daily intake of salt for medical reasons. The level of iodization of salt has to be adapted in order to meet permanently the appropriate daily requirements of iodine. For example, in Switzerland, because of the progressive decline in salt intake, the iodine intake slightly decreased in the early eighties. The level of salt iodization was therefore increased from 7.5 to 15 ppm. This resulted in an increase of iodine intake from a borderline value of 90 µg/day to a perfectly adequate value of 150 µg/day. Interestingly enough, this change was followed after one year by a steadily decline in the incidence of both toxic nodular goiter and Graves' disease (24).

As long as USI is not systematically implemented in Europe, special attention has to be devoted to the protection of the two main target groups, i.e. pregnant and lactating mothers and young infants. If iodine deficient, these age groups should be supplemented with physiological quantities of iodine, for example by including iodine to the polyvitamins administered to these two age groups. Moreover, the iodine content of formula milk should be increased in Europe above the classical recommendation of 5 µg/dL. The present recommendation, endorsed by WHO Euro and by ICCIDD, is 10 µg/dL milk for fullterms and 20 µg/dL for preterms (2).

In addition, in some European areas affected in the past by overt endemic cretinism, the iodine deficiency was and remains extremely severe, with impairment of neonatal thyroid function (25) and, consequently, with potentially harmful consequences for brain development. In such situations, emergency and transient more drastic measures can be justified from a public health point of view, such as the oral administration of iodized oil.

It is now demonstrated that in many industrialized countries such as the United States, Great Britain and Northern European countries, the main source of dietary iodine is neither salt nor seafoods but dairy milk (26). This results either from the use of iodophors in the industrial processing of milk or from iodine supplemented diets for the animals, or from both. European rules for monitoring the iodine content of milk and a precise evaluation of the possible role of milk as a substrate for iodine supplementation in Europe are desirable.

In summary, iodine deficiency still affected some 140 million people in Europe in 1992 and 97 million had a goiter. Substantial but insufficient progress has been achieved during the last few years. More precise evaluation of some national situations, dissemination of the information, health education, universal salt iodization and adequate monitoring are the priorities in order to reach the goal of elimination of IDD in Europe by the year 2000.

5. REFERENCES

1. A. Taurog, in *The Thyroid.* S.C. Werner and S.H. Ingbar eds., Harper and Row publ. New York, pp. 31–61 (1986).
2. F. Delange, in *Iodine deficiency in Europe. A continuing concern.* F. Delange, J.T. Dunn and D. Glinoer eds., Plenum Press publ., New York, pp. 5–16 (1993).
3. B.S. Hetzel, *Lancet* **ii** , 1126–1129, (1983).
4. Micronutrient Deficiency Inforamtion System. Global prevalence of iodine deficiency disorders. MDIS working paper N° 1 WHO-Nutrition Unit , pp. 1–80 (1993).
5. D.A.Koutras, in *Endemic goiter and endemic cretinism.* J.B. Stanbury and B.S. Hetzel eds., John Wiley publ., New York, pp. 79–100 (1980).
6. H. Bürgi, Z. Supersaxo and B. Selz, *Acta Endocrinol. (Kbh)* **123**, 577–590 (1990).
7. R. Gutekunst and P.C. Scriba, *J. Endocrinol. Invest* **12**, 209–220 (1989).
8. F. Delange, P. Heidemann, P. Bourdoux, A. Larsson, R. Vigneri, M. Klett, C. Beckers and P. Stubbe, *Biol. Neonate* **49**, 322–330 (1986).
9. F. Delange, J.T. Dunn and D. Glinoer, *Iodine deficiency in Europe. A continuing concern.* Plenum Press publ., New York, pp. 1–491 (1993).
10. P. Pfannenstiel, *IDD Newsletter*, **5**, 7–8 (1989).
11. V.M.G. Mannar, in *S.O.S. fo a billion. The conquest of Iodine Deficiency Disorders.* B.S. Hetzel and C.S. Pandav eds., Oxford University Press publ., Dehli, pp. 89–107 (1994).

12. F. Delange, in *Iodine prophylaxis following nuclear accidents*. E. Rubery and E.Smales Eds. Pergamon Press publ., pp. 45–53 (1990).

13. D. Glinoer, P. De Nayer, P. Bourdoux, M. Lemone, C. Robyn, A. Van Steirteghem, J. Kinthaert and D. Lejeune, *J. Clin. Endocrinol. Metab.* **71**, 276–287 (1990).

14. D. Glinoer, P. De Nayer, F. Delange, V. Toppet, M. Spehl, J.P. Grun , J. Kinthaert and B. Lejeune, *J. Clin. Endocrinol. Metab.* In Press.

15. A. Berghout, E. Endert, A. Ross, H.V. Hogerzell, N.J. Smits and W.M. Wiersinga, *Clin. Endocrinol.* **41**, 375–379 (1994).

16. F. Delange, *Thyroid* **4**, 107–128 (1994).

17. F. Delange, in *The Thyroid. A fundamental and clinical text*. L.E. Braverman and R.D. Utiger eds., J.B. Lippincott publ., Philadelphia, pp. 942–955 (1991).

18. G.N. Burrow and J.H. Dussault, *Neonatal Thyroid Screening*. Raven press publ., New York, pp. 1–322 (1980).

19. F. Delange, J. Dodion, R. Wolter, P. Bourdoux, A. Dalhem, D. Glinoer and A.M. Ermans, *J. Pediatr.* **92**, 974–976 (1978).

20. F. Delange, A. Dalhem, P. Bourdoux, R. Lagasse, D. Glinoer, D.A. Fisher, P.G. Walfish and A.M. Ermans, *J. Pediatr.* **105**, 462–469 (1984).

21. F. Delange, P. Bourdoux and A.M. Ermans, in *Thyroid disorders associated with iodine deficiency and excess*. R. Hall and J. Köbberling eds., Raven Press publ., New York pp. 273–282 (1985).

22. H. Bürgi, in *Iodine deficiency in Europe. A continuing concern*. F. Delange, J.T. Dunn and D. Glinoer eds., Plenum Press publ., New York , pp. 261–268 (1993).

23. F. Delange and H. Bürgi, *Bull. WHO* **67**, 317–326 (1989).

24. B.L. Baltisberger, C.E. Minder and H. Bürgi, *Eur. J. Endocrinol.* In Press.

25. F. Delange, P. Bourdoux, M. Laurence, L. Peneva, P. Walfish, H. Willgerodt, in *Iodine deficiency in Europe. A continuing concern*. F. Delange, J.T. Dunn and D. Glinoer eds., Plenum Press publ., New York , pp. 199–209 (1993).

26. J.T. Dunn, in *Iodine deficiency in Europe. A continuing concern*. F. Delange, J.T. Dunn and D. Glinoer eds., Plenum Press publ., New York, pp. 17–24 (1993).

56

IODINE DEFICIENCY IN CZECH REPUBLIC

V. Zamrazil, J. Čeřovská, M. Dvořáková, A. Šimečková, J. Vrbíková,
R. Bílek, F. Tomíška, and J. Kvíčala

Institute of Endocrinology
Prague, Czech Republic

1. INTRODUCTION

Iodine deficiency was present in nearly all Czech Republic and various forms of iodine deficiency disorders (IDD) were common in past, especially euthyroid endemic goitre. After iodine supplementation of salt, started at the beginning of the fifties, a marked improvement was recorded (6). During last years some clinical findings focused our interest to this problem (8). Therefore, we recently reevaluated iodine intake and thyroid status in our country.

2. SUBJECTS AND METHODS

Five groups of randomly selected persons from five different regions of CR were evaluated. Regions were selected according to original severity of iodopenia and IDD before iodination of salt and according to actual degree of pollution. Investigated groups involved children aged 6, 10 and 13 years respectively and adults aged 18–35, 36–50 and 52–65 years respectively. Number of investigated persons in subgroups according to age and gender varied from 35 to 55. Total number was 2 600 persons.

All persons were evaluated by the same medical staff as follows: clinical examination, estimation of thyroid volume (using ultrasonographic method) calculated according to Gutekunst (4), and assessment of iodine intake by determination of iodine concentration in a morning urine sample according to ICCIDD recommendation (3), using Sandell-Kollthoff colorimetric method after mineralisation. For estimation of TSH (thyrotropin) and total T4 (thyroxine), immunofluorescent analysis was performed using ABBOTT analyser.

A complex evaluation of questionary data, selected anthropometric values, palpatory findings of the thyroid and some further laboratory findings were performed, but results are not given in this report. For statistical evaluation, Statgraphic Plus program was used.

Therapeutic Uses of Trace Elements, edited by Nève et al.
Plenum Press, New York, 1996

Table 1. Regional differences in ioduria (mg/L)

Region	1.00	2.00	3.00	4.00	5.00
Children					
n	260.00	230.00	224.00	250.00	226.00
x	98.00	94.00	106.00	152.00	151.00
S.D.	51.00	31.00	112.00	119.00	82
Adults					
n	277.00	213.00	266.00	344.00	213.00
x	78.00	80.00	90.00	113.00	134.00
S.D.	36.00	37.00	66.00	101.00	113.00

n = number of investigations; x = arithm. mean; S.D. = standard deviation.
Region 1 = Jindøichùv Hradec; region 2 = Ústi n/L; region 3 = Praha;
region 4 = Vsetín; region 5 = Znojmo.

3. RESULTS

Table 1 shows regional differences of ioduria. Significant differences among regions were found both in children and adults. It might be of interest that lower values were recorded in regions of severe pollution (Prague, and Ústí nad Labem in North Bohemia). Distribution of categories of ioduria is given in Table 2. Ioduria below 50 mg/L was recorded in 7.2 % of boys, in 10.4 % of girls, in 12.2 % of men and even in 21.1 % of women. Satisfactory ioduria according to ICCIDD (7) was found in less than 1/3 of women. Analysis of thyroid volume (Table 3) omitted such important factors as influence of age, but still proved significant differences among investigated regions in children and adults. Relations to ioduria were only weak. Total thyroxine levels (Table 4) were evaluated in 4 regions only. Values obtained were within normal range for our population in all regions, with significant regional differences both in children and adults. Thyrotropine (TSH) levels (Table 4) were also in normal range and regional differences were also present.

4. DISCUSSION

Salt supplementation with KI to final concentration 25 mg of iodine per Kg of salt nearly compensated a genuine iodine deficiency and ameliorated prevalence and severity of IDD in the Czech Republic. As evident from presented data, the problem of iodopenia has not been completely solved. In some regions, the average ioduria was below limits of IC-CIDD and 7.8 - 21.2 % of evaluated inhibitants did not reach 50 mg/L, although did not drop below 20 mg/L (the limit of severe iodopenia connected with occurence of endemic cretin-

Table 2. Distribution of ioduria in children and adults (in per cent.)

Ioduria	Boys	Girls	Men	Women
< 50 mg/L	7.20	10.40	12.20	21.10
50 - 99 mg/L	35.80	40.10	49.10	45.00
100 - 299 mg/L	51.20	46.30	35.80	30.40
> 300 mg/L	5.80	3.20	2.90	3.50

Table 3. Regional differences of thyroid volume (ml)

Region	1.00	2.00	3.00	4.00	5.00
Children					
n	263.00	231.00	217.00	249.00	224.00
x	4.60	6.20	5.40	6.30	6.40
S.D.	2.90	3.30	2.60	3.90	3.30
Adults					
n	284.00	225.00	261.00	347.00	263.00
x	17.70	16.80	19.40	17.50	16.90
S.D.	8.60	7.60	9.50	9.30	9.40

For explanations, see Table 1.

ism). It seems to be proved that only iodine supply resulting in ioduria at least over 100 mg/L is able to ensure complete eradication of IDD (1). This situation is true for only 1/3 - 1/2 of investigated persons in CR. According to previous reports (5), very low ioduria was found in pregnant and lactating women and in newborns.

Although the problem of iodopenia and IDD in CR as well as in Europe is complex and not yet fully elucidated (2), informations obtained is sufficient for some activity for improvement of iodopenia. An interdisciplinary approach is necessary for achieving the following projects: treatment of pregnant and lactating women with KI (100 mg of iodine per day) according to recommendations of Czech Endocrine and Czech Pediatric Society; improvement of salt iodinization: kalium iodate has been used instead of kalium iodide since 1995 at a dose of 35 mg of iodine per Kg of salt; fortification of infant formulas with KI is now in preparation and new iodinised products will be probably available during 1996.

In the future, some other ways to improve iodine intake in Czech population, will be considered emphasizing groups at risk. Nevertheless it is necessary to pay attention to other factors important for optimal thyroid function - for example to interactions of iodine and selenium and to the role of natural and anthropogenic goitrogens in CR.

ACKNOWLEDGMENT

This work was partly supported by grant of GA ÈR No. 311/93 0982 and IGA MZ ÈR No. 1436–3.

Table 4. Regional differences of TSH (ml/L) and T4 (nmol/L)

| Region | Children | | | | | Adults | | | | |
	1.00	2.00	3.00	4.00	5.00	1.00	2.00	3.00	4.00	5.00
TSH										
n	249.00	214.00	216.00	242.00	203.00	276.00	218.00	266.00	247.00	255.00
x	1.20	1.80	1.90	2.40	2.00	1.00	1.60	1.70	2.00	1.70
S.D.	0.60	0.90	1.00	1.20	1.00	0.80	1.00	1.50	1.40	1.60
T4										
n	—	218.00	217.00	242.00	203.00	—	217.00	266.00	347.00	255.00
x	—	111.00	131.00	131.00	123.00	—	113.00	120.00	125.00	111.00
S.D.	—	17.00	18.00	24.00	19.00	—	24.00	20.00	24.00	21.00

For explanations, see Table 1.

5. REFERENCES

1. B.M. Baltisberger, C.E. Minder, H. Burgi, *Eur. J. Endocrinol*. **132**, 546–549,(1995).
2. F. Delange, K.T. Dunn, D.Glinoer, *Iodine deficiency in Europe*, Plenum, NewYork, London (1993).
3. J.T. Dunn, H.E. Crutchfield, R. Gutekunst, D. Dunn, *Methods for measuring iodine in urine*, ICCIDD, Netherlands (1993).
4. R. Gutekunst, H. Martin-Teichert in *Iodine deficiency in Europe,* F. Delange, J.T. Dunn, D.Glinoer, eds. Plenum Press, New York, London, pp. 83–89 (1993).
5. O. Hníková et al., *Paediat. Paedol.* **29,** 140 (1994).
6. K. Šilink, K. Èerný, *Endemic goiter et allied diseases*, Veda, Bratislava (1966).
7. *WHO/NUT Indicators for assessing iodine deficiency disorders and their control though salt iodization,* WHO/NUT, Geneva (1994).
8. V. Zamrazil et al., *Europ. J. Endocrinol.* **130**, Suppl 2, 133 (1994).

57

EVALUATION OF THE THYROID FUNCTION IN NEWBORNS BY MEANS OF FACTOR ANALYSIS

P. Zagrodzki,[1,2] R. Ratajczak,[3] M. Rybakowa,[3] and R. Podsiadły[4]

[1] Department of Food Chemistry and Nutrition
Collegium Medicum, Jagiellonian University
Podchorazych 1, 30–084 Kraków
Poland
[2] Institute of Nuclear Physics
Radzikowskiego 152, 31–342 Kraków
Poland
[3] Department of Pediatric Endocrinology
Polish-American Children's Hospital
Collegium Medicum, Jagiellonian University
Wielicka 265, 30–663 Kraków
Poland
[4] Chemical Physics Laboratory
Institute of Chemistry, Jagiellonian University
Ingardena 3, 30–060 Kraków
Poland

1. INTRODUCTION

In neonates the thyroid gland is very sensitive to iodine deficit. This hypersensitivity is due to a less efficient mechanism of adaptation to iodine deficiency (1). The main purpose of the present study was to evaluate the thyroid function in newborns for whom four different thyroid Iodine Deficiency Disorders (IDD) were diagnosed: hypothyroxinemia, transient hypothyroidism, hyperthyrotropinemia without goitre, and hyperthyrotropinemia with goitre. The study was carried out shortly after the iodine prophylaxis (salt iodisation) in Poland had been resumed after nearly 10 years of break. So, it was interesting to compare the evolution of particular disorders in subsequent years and the evolution of correlations between parameters used to diagnose cases. Some relevant insights were done by help of factor analysis, which aimed to explain the observed relations among parameters and to find basic casual influences on them. As a research instrument, factor analysis and other pattern recognition methods have been used in numerous studies dedicated to clinical practice and epidemiology (2–4).

Therapeutic Uses of Trace Elements, edited by Nève et al.
Plenum Press, New York, 1996

2. MATERIALS AND METHODS

The population studied lived in the Southern Poland region. The exact iodine supply is unknown in neonatal population, but the frequency of elevated neonatal TSH, Total Goitre Rate (TGR, defined as a fraction of population with any grade of goitre by palpation) and urinary iodine excretion in school children, which were estimated in previous studies (5–7), suggest that the population studied lived on the area which can be recognised as iodine deficient region. Cases were selected on the basis of neonatal TSH screening tests. Newborns, whose TSH blood spot values measured between 3-rd and 6-th day of life (in the frame of initial screening test) were higher than 20 mUI/L (cut-off for the congenital hypothyroid screening program), next visited the Outpatient Clinic of Department of Endocrinology at the Polish-American Children's Hospital within first three or four weeks after delivery. The neonates with confirmed congenital hypothyroidism as well preterm and small for age were excluded from our study. The cohort encompassed a total 260 infants and their mothers. Since some parameters were not recorded in all cases, smaller subsets of data were examined in subsequent analyses. The cohort of neonates was divided into 4 groups depending on severity of disorders. More iodine deficient groups 1 and 2 included hypothyroxinemic and transient hypothyroid neonates, less iodine deficient groups 3 and 4 included hyperthyrotropinemic neonates with or without goitre, respectively. The duration of study covered the period of time 1991–1994, with exception of the year 1992, when screening program was suspended for several months, because of lack of financial support. The following parameters were taken into account: triiodothyronine (T3-RIA), thyroxine (T4-RIA), thyroid stimulating hormone (TSH-RIA), free thyroxine (fT4-RIA), molar T3/T4 ratio, size of goitre (by palpation, according to simplified WHO classification) (8), maternal T3M (letter M means mothers' parameter). The molar T3/T4 ratio was calculated. Results were expressed as means ± SD, although for the means coming from nongaussian populations, data were transformed in logarithms and retransformed after calculations. Statistical analyses were carried out using commercially available statistical package STATGRAPHICS.

3. RESULTS

Almost all mothers (97 %) had TSH values below 5 mUI/L, but they could not be classified as any homogenous group for few reasons: Total Goitre Rate was 50.3 %, T3M was higher than upper limit of normal range (2.00 ng/mL) in 13.0 % and T4M was higher than upper limit of normal range (12.5 μg/dL) in 8.1 % of women's population. In the whole studied population of newborns, T3 levels in 0.8 % of newborns were below normal range (1.0 - 2.4 ng/mL), while in 34.9 % over that range, with mean value 2.33 ng/mL. However, the TGR was extremely high and yielded 55 %. In the group of newborns with hypothyroxinemia or with transient hypothyroidism the mean values for T4, T3, TSH, T3/T4 were: 9.3 ± 4.0 μg/dL, 2.71 ± 0.87 ng/mL, 14.1 (9.4 - 28.7) mUI/L, 0.034 (0.016 - 0.030); the total T4, T3, TSH and T3/T4 levels in newborns with hyperthyrotropinemia with or without goitre were: 15.1 ± 3.6 μg/dL, 2.30 ± 0.60 ng/mL, 2.92 (1.58 - 3.44) mUI/L, 0.018 ± 0.006, respectively. During the subsequent years of prophylaxis the most severe disorders (groups 1 and 2) became very rare or disappeared. The parameters which differed most frequently between distinct groups were TSH and T3/T4 ratio. The two-dimensional (TSH and T3/T4) representation of the whole population studied in year 1991 was shown on Figure 1. Consideration of these two parameters alone gave less than 85 % of results correctly classified. On the

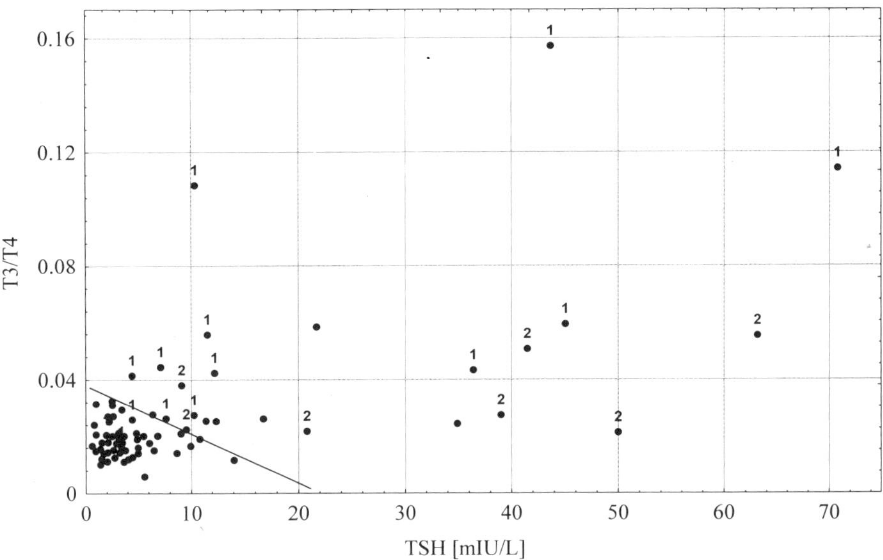

Figure 1. Scatterplot of cases from year 1991. Diagnoses 1–2 and the border line between groups 1+2 and 3+4 are depicted.

other hand factor analysis led up to 96 % of cases correctly classified. In order to check if our approach works properly (in terms of discrimination) not only for the particular year 1991 but also in a wider range, it was applied to the whole studied period (1991–1994). The results of factor analysis were summarised in Tables 1 and 2 and in Figures 2 and 3. Only 3 first factors are shown in the Table 1 because the further ones explain not more than 3% of variance. Karhunen-Loeve projection used to get insight into the scattering of cases was shown on Figure 4 (concerning the population from the year 1991) and Figure 5 (period 1991–1994). The χ^2 test was applied to check the location of the groups of points representing factor loadings on the first two latent factors with respect to the best fit of two sets of factor loadings corresponding to the 1991 and 1991–1994 observations. The arbitrary choice of the fitting function was the simplest one, linear. The χ^2 value for the 1991 factor loadings was 1.13 and for the 1991–1994 factor loadings 0.69. Such result is the evidence that the two groups are differently located in the parameters space, i.e. that they represent different relations between the parameters studied. The difference is better pronounced after transformation to the logarithmic scale and consideraton of the agreement with the exponential fit. The corresponding χ^2 values become then 1.55 (1991) and 0.51 (1991–1994).

Table 1. Principal components in the factor analysis for year 1991 and for the period 1991–1994

	Eigenvalue		Variance (%)		Cumulative variance (%)	
Factor	1991	1991-1994	1991	1991-1994	1991	1991-1994
1	2.603	2.279	64.7	60.4	64.7	60.4
2	1.071	1.220	26.6	32.3	91.4	92.7
3	0.308	0.197	7.7	5.2	99.0	97.9

Table 2. Estimated communalities for the variables in the factor analysis for year
1991 and for the period 1991–1994

Period	T4	T3	TSH	fT4	Size of goiter	T3M	T3/T4
1991	0.847	0.819	0.451	0.356	0.046	0.198	0.956
1991-1994	0.880	0.835	0.270	0.387	0.039	0.157	0.931

4. DISCUSSION

At the moment of the visit in the Department of Endocrinology of the Polish-American
Children's Hospital, a majority of neonates, especially from the groups 1 and 2 exhibited
preferencial T3 secretion, relative hypothyroxinemia, high molar T3/T4 ratio and elevated
TSH. This phenomenon is characteristic for iodine deficiency (9). The discrimination between groups of diagnoses obtained in the coordinate system (TSH, T3/T4) is not satisfactory. Much better results were achieved when pattern recognition method was employed.
Figures 2 and 3 demonstrate that the subgroups 1,2 and 3,4 exhibited some internal similari-
ties. Because cases 1,2 did not occur at all and cases 3,4 were extremely rare in the last two
years, as a suggested result of reinstituted iodine prophylaxy, the correlations between thyroid function parameters changed (different factor loadings), but the discrimination obtained
by means of Karhunen-Loeve projection was generally preserved. Cumulative variance of
variables explained by first two common factors in both cases (1991 and 1991–1994) was
greater than 90 %. The highest factor loadings were for T3/T4, T3, T4 (1991) and again
T3/T4, T3, T4 (1991–1994). All of those parameters are useful for assessment of hyperstimulation of thyroid gland due to iodine deficiency (5). As shown on Figures 2 and 3 in year
1991 and similarly in years 1991–1994 there are strong dependencies between neonatal T4

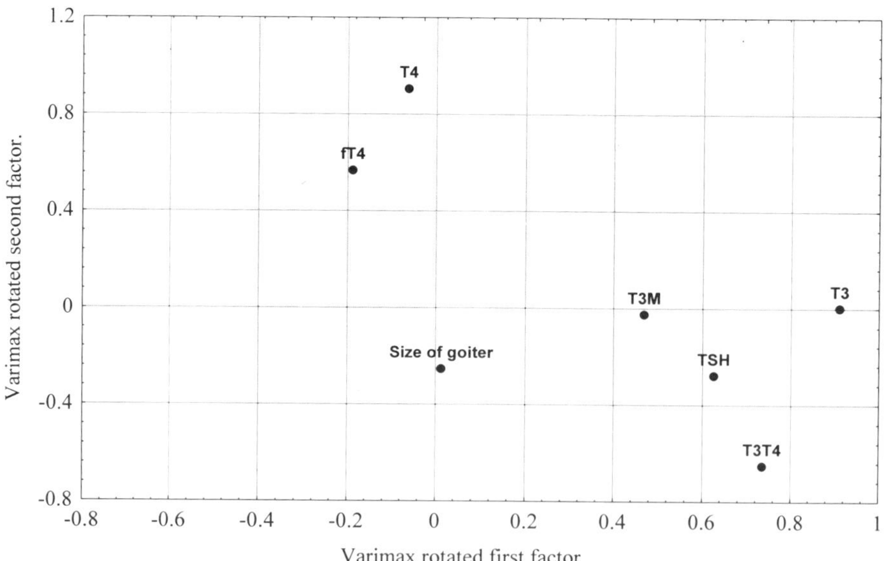

Figure 2. Plot of first two factor loadings for year 1991. Factors are Varimax rotated.

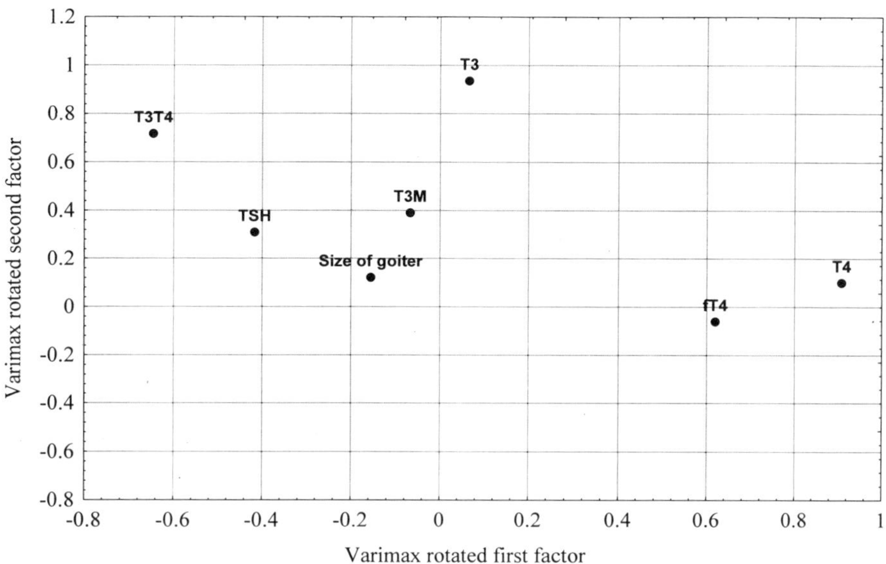

Figure 3. Plot of first two factor loadings for years 1991–1994. Factors are Varimax rotated.

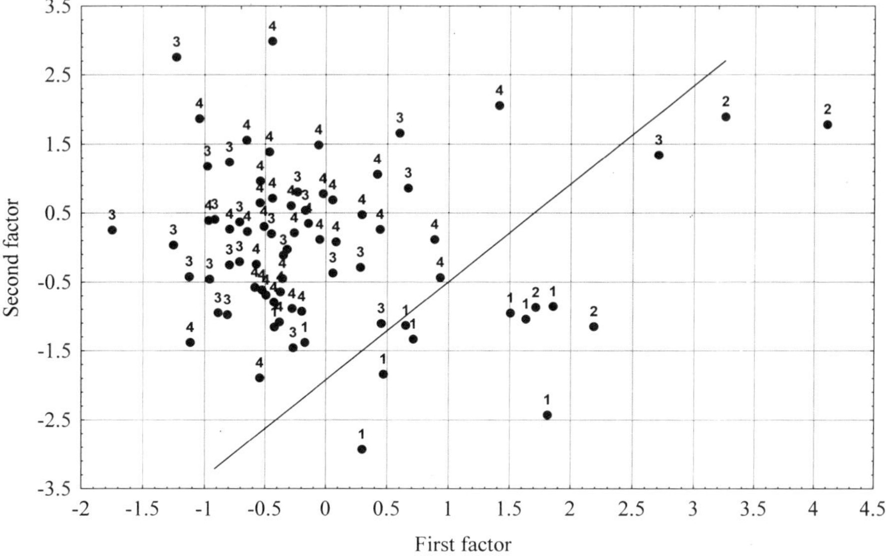

Figure 4. Karhunem-Loeve projection of the population studied in year 1991. Diagnoses 1–4 and the border line between groups 1+2 and 3+4 are depicted.

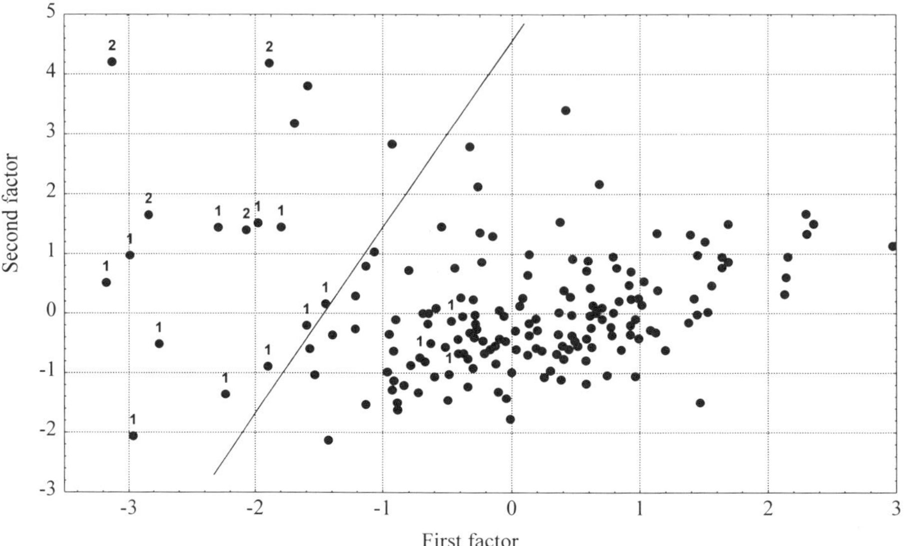

Figure 5. Karhunem-Loeve projection of the population studied in year 1991–1994. Diagnoses 1–2 and the border line between groups 1+2 and 3+4 are depicted.

and fT4, between neonatal T3/T4 and TSH and between maternal T3 and neonatal TSH. The first one relation is obvious. The last two seem to be characteristic for iodine deficiency.

5. CONCLUSION

Introduction of iodine prophylaxy in 1991 (and perhaps other factors) caused the marked decrease in severe IDD occurrence. During the period 1991–1994, there were some changes in correlations between parameters relevant to neonatal and maternal thyroid functions. Factor analysis allowed us to carry out more objective and complete analysis of cases. Finally, in neonatal population, estimations of both thyroid hormones T3, T4 (or fT4) and TSH bring best information about thyroid status in iodine deficiency.

ACKNOWLEDGMENTS

The authors express their deep gratitude to Dr. Barbara Petelenz for her help in manuscript preparation and valuable critical suggestions.

6. REFERENCES

1. F. Delange, P.Bourdoux, M.Laurence, L.Paneva, P.Walfish, H.Willgerodt, in *Iodine Deficiency in Europe: A Continuing Concern,* F. Delange, J.T.Dunn, D.Glinoer, ed., Plenum Press, New York., pp. 199–207 (1993).
2. D.L. Duewer, B.R. Kowalski, K.J. Clayson, R.J. Roby, *Computers and Biomed.Res.***11**, 567–580 (1978).
3. D. Coomans, M. Jonckheer, D.L. Massart, I. Broeckaert, P. Blockx, *Anal.Chim.Acta* **103**, 409–415 (1978).
4. R.T.P. Jansen, F.W550. Pijpers, G.A.J.M.de Valk, *Anal.Chim.Acta* **133**, 1–18, (1981).

5. D.F. Nordenberg, R. Ratajczak, M. Rybakowa, D. Tylek, G.F. Maberly, in: *The damaged brain of iodine deficiency*, J.B. Stanbury, ed., Cognizant Communication Corporation, New York, pp. 79–283 (1994).
6. Z. Szybiński et al., *Endokrynol. Pol.* **44**, 235–248 (1993).
7. M. Rybakowa et al., *Endokrynol. Pol.* **44**, 249–258 (1993).
8. Document WHO/Nut/94.6, *Indicators for assessing Iodine Deficiency Disorders and their control through salt iodisation,* WHO, UNICEF, ICCIDD, Geneva, p.16 (1994).
9. D. Glinoer, F. Delange, I. Laboureur, P. De Nayer, B. Lejeune, J. Kinthaert, P. Bourdoux, *J.Clin. Endocrinol. Metab.* **75**, 800–805, (1992).

58

THE RELATIONSHIP OF DIETARY SELENIUM TO CARCINOGENESIS

P. D. Whanger

Oregon State University
Corvallis, Oregon

Selenium has come full circle in two respects--from toxicity to an essential element and from a carcinogenic agent to an anticarcinogenic element. Until 1957 the only significance of selenium was thought to be its toxicity. Since that time research has been conducted to confirm that it is an essential element, and several seleno-enzymes and selenoproteins are now known (1). In 1943 a paper was published indicating that selenium was a carcinogenic element (2). This created a very damaging image for selenium in that it was thought to be a very toxic element which caused cancer. Even with all the information to indicate otherwise, this image still lingers. For example, in 1975 a IARC review on selenium and selenium-containing compounds concluded that "the available data provide no suggestion that selenium is carcinogenic to man" (3). Nevertheless, even today a number of selenium compounds are marketed with a warning label that indicates the substance may be a carcinogen. There are numerous reviews on the relationship of selenium to cancer and a few of them are listed here (1, 4–6).

In the late 60's and early 70's several epidemiological studies suggested an inverse relationship between selenium intakes and certain cancers in humans (7,8). This caused researchers to reevaluate the topic of selenium and cancer and studies with laboratory animals have provided mostly positive data on this element and carcino-genesis. Experimental studies with rodents indicate that selenium supplementation at levels well above dietary requirements is capable of protecting against tumorigenesis induced by chemical carcinogens and viruses (1, 4–6). However, the type of diet can have a marked influence on the effects of selenium against tumorigenesis (4,5 and 9). Until the last few years, selenocompounds that were examined were available from commercial sources such as selenite and selenomethionine (Semet). The evidence indicate that both selenite and Semet must be metabolized in order to exert their anticarcinogenic action. Diets deficient in methionine reduce the protective effect of Semet even though the tissue selenium was actually higher in these animals as compared to those with sufficient amounts of methionine (10). This is thought to be due to greater incorporation of Semet into body proteins in place of methionine, and thus is diverted from the pathway of selenium utilization and detoxification. Secondly, the chemopreventive action of selenite was shown to be almost completely abolished by coadministration of

Therapeutic Uses of Trace Elements, edited by Nève et al.
Plenum Press, New York, 1996

arsenite (11), which is known to interfere with the formation of methylated selenium metabolites. This information along with findings that a continuous intake of supplemental selenium is necessary to achieve maximal protection against cancer (12) suggest that some active species of selenium with antitumorigenic potential is generated only when the supply of selenium is maintained at a certain level.

With high intake of selenium, the levels of intermediate excretory metabolites are expected to rise, particularly the methylated derivatives including methylselenol (in urine), dimethylselenide (in breath) and trimethylselenonium (TMSe) ion (in urine). Selenite and Semet are converted to a common intermediate, selenide, which is then methylated when excess amounts are present. Several synthetic compounds were used which would enter the metabolic pathway at different points (4,13). Seleno-betaine and Se-methylselenocysteine were used because they are converted to methylselenol, and selenobetaine methyl ester and dimethylselenoxide were used because they are converted to dimethylselenide in the animal. The efficacy of these compounds were compared to selenite and Semet. By correlating the chemical form and metabolism of different selenium compounds with their chemopreventive efficacy, the following conclusions were reached. a) Good anticarcinogenic activity of selenium was obtained when selenium was introduced into the pathway beyond the hydrogen selenide step but before the TMSe level. b) Certain selenium compounds such as selenobetaine and Se-methylselenocysteine are more efficacious than selenite and have good anticarcinogenic activity. These compounds are able to produce a steady stream of methylated metabolites, particularly the monomethylated species. c) The anticarcinogenic activity is lower for selenoamino acids that can be incorporated into proteins. This is the reason Semet is thought to be less effective than some other selenocompounds. d) There is a correlated reduction in the anticarcinogenic potency as the selenocompounds become more methylated. Forms of selenium such as dimethylselenoxide and selenobetaine methyl ester which are metabolized rapidly and quantitatively to dimethylselenide and TMSe are poor choices for use in chemoprevention. Thus, it appears that methylselenol is likely the active species in cancer chemoprevention by either selenite or Semet.

The term chemopreventive index was introduced to compare the efficacy of different agents (14). This index is calculated as the ratio of the maximum tolerable dose to the effective dose which produces a 50% inhibition in total tumor yield. The chemopreventive index is lowest for Semet and highest for two novel seleno-compounds called, 4-phenylenebis(methylene)selenocyanate (pXSC) and triphenylselenonium ion. Even though both of these selenocompounds are excellent chemopreventive agents, the triphenylselenonium ion is superior to pXSC with regards to the chemopreventive index (15). Neither of these selenocompounds, however, have high nutritional value since their bioavailabilities for regeneration of the activities of selenoenzymes are very low (14, 15). Since these selenocompounds are not found in plants, their greatest value appear to be as therapeutic agents.

The ultimate goal is to find selenocompounds with low toxicity, but high efficacy and nutritional values. One way to achieve this is to find ways to deliver selenium efficiently in a food product. As shown in table 1, many selenium enriched vegetables have been shown to reduce the incidence of mammary tumors in rats (16). Of these, selenium enriched garlic appears to be the most effective. Selenium enriched garlic is superior to regular garlic, selenite or Semet in mammary cancer prevention in the rat DMBA model (17).

It is believed that selenium enriched vegetables provide an ideal system for delivering selenium in cancer prevention (16, 18). There are several reasons for this conclusion. First, vegetables provide a wide range of anticancer activity which can easily be adapted for human consumption. Second, this avenue could provide a wide distribution in a cost effective manner, a prerequisite to the implementation of cancer prevention for the general population.

Table 1. Mammary cancer prevention in rats fed different sources of selenium

Dietary treatment	Dietary selenium (ppm)	Tumor incidence	Total number of tumors	Final body weight (g)
Control[*]	0.1	25/30 (83.3%)	81	331 ± 7
Selenite[*]	3.0	15/30 (50.0%)	42[a]	328 ± 8
Selenomethionine[*]	3.0	18/30 (60.0%)	54[a]	326 ± 8
High-selenium garlic[*]	3.0	10/30 (33.3%)	26[a]	334 ± 7
High-selenium onions[*]	3.0	14/30 (46.7%)	44[a]	335 ± 8
Brazil nuts[*]	3.0	14/30 (46.7%)	38[a]	330 ± 8
High-selenium broccoli[**]	3.0	12/25 (48.0%)	28[a]	325 ± 7

[a]$P < 0.05$, compared to the corresponding control valve
[*]Taken from Ip and Lisk, *Carcinogenesis* **15**, 573 (1994)
[**]Personal communication (C. Ip, Roswell Park Cancer Institute, Buffalo, NY)

Third, this would provide selenium as an essential nutrient in maintaining full activity of functional selenoenzymes. Lastly, the ingestion of high levels of selenium enriched vegetables would not result in excessive accumulation of tissue selenium which is a concern associated with Semet consumption (19). Semet is the most common form of selenium in selenium enriched yeast (20), which is the major commercial product available for human consumption.

The free selenocompounds in selenium enriched garlic, onions and broccoli have recently been examined (21). Interestingly, the major selenocompound in these vegetables is Se-methylselenocysteine. Thus, the work indicating this seleno-compound to be a very effective anticarcinogenic compound appears to be confirmed with these vegetables. Presumably, this selenocompound is one of the major ones contributing to protection against tumors when selenium enriched vegetables are fed.

Although most of the work has been done in laboratory animals, there are sufficient reasons to believe that supernutritional levels of selenium will reduce the incidence of certain cancers in humans. After five years of supplementation with selenium as selenium enriched yeast (and vitamins A and E) the incidence of stomach and throat cancers was significantly reduced in a Chinese population (22). This was in an area where these types of cancers are very prevalent. There is currently a trial in the United States where the effects of selenium on colon cancer in humans is under investigation. This data is soon to be published in the *New England Journal of Medicine*, and the information indicates a statistically significant reduction in cancer of the colon in humans taking selenium as selenium enriched yeast. Therefore, these two human trials provide sufficient data to indicate that selenium will reduce certain cancers in people.

REFERENCES

1. J.A. Milner, CRC Press, Boca Raton, Chap. 8 (1995).
2. A.A. Nelson, O. G. Fitzhugh and H. O. Calvery, *Cancer Res.* **3**, 230 (1943)
3. International Agency for Research on Cancer (IARC) **9**, 245 (1993).
4. C. Ip, and H. E. Ganther, in *Cancer Chemoprevention*, L. Wattenberg, et al., eds, CRC press, Boca Raton, pp. 479–488 (1992).
5. G.F. Combs, and S. B. Combs. Academic press, New York, Chap. 10 (1986).
6. P.D. Whanger, *J. Trace Elem. Elect. Health Dis.* **6**, 209 (1992).
7. R.J. Schamberger, and C. E. Willis, *CRC Crit. Rev. Lab. Sci.* **2**, 211 (1971).
8. G.N. Schrauzer, D. A. White and C. J. Schneider, *Bioinorgan. Chem.* **7**, 23 (1977).

9. P.D. Whanger, J. A. Schmitz and J. H. Exon. *Nutr. Cancer* **3**, 240 (1982).
10. C. Ip, *J. Natl. Cancer Inst.* **80**, 258 (1988).
11. C. Ip, and H. E. Ganther, *Carcinogenesis* **9**, 1481 (1988).
12. C. Ip, *Cancer Res.* **41**,4386 (1981).
13. C. Ip, C. Hayes, R. M. Budnik and H. E. Ganther, *Cancer Res.* **51**, 595 (1991).
14. C. Ip et al., *Carcinogenesis* **15**, 187 (1994).
15. C. Ip, H. Thompson and H. E. Ganther, *Carcinogenesesis,* **15**, 2879 (1994).
16. C. Ip, and D. J. Lisk, *Carcinogenesis* **15**, 573 (1994).
17. C. Ip, D. J. Lisk and G. S. Stoewsand, *Nutr. Cancer* **17**, 279 (1992).
18. C. Ip and D. J. Lisk, *Carcinogenesis* **15**, 1881 (1994).
19. P.D. Whanger, and J. A. Butler, *J. Nutr.* **118**, 846 (1988).
20. M.A. Beilstein and P. D. Whanger, *J. Nutr.* **116**, 1701 (1986).
21. X.J. Cai, et al., *J. Agric. Food Chem.* **43**, 1754 (1995).
22. J.Y. Li et al., *J. Natl. Cancer Institute* **85**, 1483 (1995).

59

DEFICIENCY OF SELENIUM IN INHABITANTS OF HIGHLY POLLUTED AREA OF NORTH-WEST BOHEMIA

J. Kvíčala,[1] V. Zamrazil,[1] and B. Tlučhoř[2]

[1] Institute of Endocrinology
Národní 8, 116 94 Praha 1, Czech Republic
[2] Immunotech a.s.
Radiová 1, 102 27 Praha 10, Czech Republic

1. INTRODUCTION

Essential role of selenium (Se) for animals and man has been well established during the past three decades. Se is genetically incorporated into active centre of some enzymes, i.e. four glutathione peroxidases which are parts of antioxidative defence system (1), and iodothyronine 5′deiodinase I, which plays a substantial role in thyroid hormone regulation (2).

Optimal Se concentration, 100–140µg/l serum (3), and Se intake 55–70µg/day for adults and 20–40 ug/day for children (4) have been deduced from supplementation studies. Fifty to sixty per cent of Se intake are excreted by urine in the low and medium intake groups (3). It fits well with the daily excretion of about 30µg Se or less in adults of most European countries (5), Se status of which is known to be marginally deficient. Other Se indices used in various studies for assessment of Se status are Se concentrations of blood, hair and nails (3).

Se status and intake of whole population (both sexes in the age from 6 to 65 y) of very industrially affected region Ústí nad Labem in North-West Bohemia were estimated in this study by the use of three Se indices (serum, urine and hair Se concentrations).

2. MATERIALS AND METHODS

Chemicals and biological reference standards together with instruments were used as stated elsewhere (6). Moreover, CRM human hair GBW 07601 (IGGE, Langfang, China) was used as biological reference standard for hair analysis.

Blood serum and hair were analyzed by neutron activation analysis (NAA) as described elsewhere (7,8). Human serum of the second generation, IAEA standard reference material H-4 and IGGE CRM human hair were simultaneously analysed for quality assurance of the Se detection: results were 1.03 - 1.07µg/g dry serum (declared value 1.05µg/l), 0.27µg/g dry muscle (certified 0.28µg/g), and 0.59–0.63µg/g hair (certified 0.60µg/g). Within batch precision 8.5% (n=10) and between batch precision 8.7% (n=19) were obtained for human serum. Urine selenium was measured by fluorimetry as complex with DAN after one-tube wet-ashing as stated elsewhere (6). Precision of urine selenium determination was checked by Lyphocheck urine analysis. Our values were 53 - 59µg/l; declared values were 50µg/l with acceptable values 40–60 mg/l. Within batch precision of analyses of Lyphocheck urine Se was 6.4% (n=6). Between batch precision of analyses in apparently healthy adults was 6.5% (n=9, mean 11.2µg Se/l urine). Analytical results were evaluated by the usual statistical methods.

3. RESULTS AND DISCUSSION

Randomly selected inhabitants of the highly industrially polluted region Ústí nad Labem in North-West Bohemia were searched for selenium status. Serum Se level is the most often analysed and compared parameter (3,9,10). An overall arithmetic mean of 49µg Se/l was obtained for 237 inhabitants 6 to 65 y old (Table 1). Difference between men and women was not significant contrary to differences between groups of children and adults (p < 0.05). Data of individual categories according age and gender are depicted in Figure 1. Significant differences were found only between young men and 10 and 13 y old boys, and between young women and 13 y old girls, probably due to small number of cases (16–23 per category). Gender differences between the same age groups were not significant.

Low serum Se concentrations reflect serious Se deficiency of the inhabitants in Ústí n.L. This conclusion is supported by frequency distribution of serum Se concentrations in Figure 2. Pronounced deficit levels below 20µg Se/l (3,9–11) did not occur, but serum Se concentrations of 70% of our population belong to the values of severe Se deficiency (20–55µg Se/l) and only 3% of the population have this parameter in the levels of optimal values (more than 70µg Se/l). Overestimation of the deficiency limits for individual and its diagnosis may lead to errors (3), but these statistical results for whole population of the region are alarming because of the role of selenium in the organism and of risks connected with selenium deficiency (1,2).

Analysis of hair as long-term parameter was used mainly in various regions in China where Se-deficient populations simultaneously exist with those with selenosis. Concentration of 0,36µg Se/g hair was found as optimal during these studies (9). Eighty seven% of hair samples from our men (in the age 36–49 y) below this limit show again deficiency of Se in our population.

The third parameter - selenium concentration in urine - shows essentially the same picture. An overall mean of 12.6µg Se/l was obtained for 361 inhabitants 6 to 65 y old (Table

Table 1. Se concentrations of groups according to age and gender

	Serum Se(µg /l)	Urine Se(µg /l)	Urine Se/creat(µg /g creat)
Boys	46 ± 9 (60)	15.3 ± 7.7 (97)	20 ± 13 (90)
Men	51 ± 14 (58)	9.5 ± 3.1 (82)	12 ± 9 (81)
Girls	46 ± 9 (60)	13.0 ± 5.5 (96)	16 ± 10 (94)
Women	54 ± 10 (59)	12.1 ± 5.4 (86)	18 ± 11 (85)
All	49 ± 11 (237)	12.6 ± 6.1 (361)	18 ± 11 (350)

Results are expressed as mean ± S.D. (n)

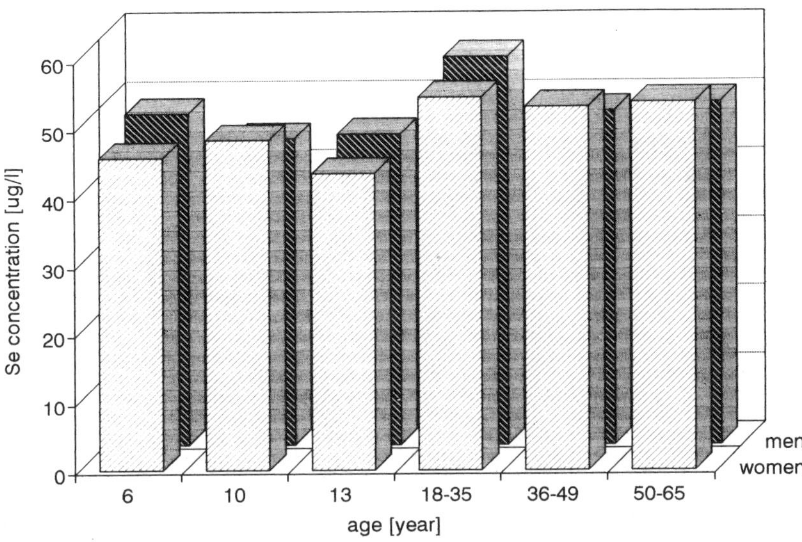

Figure 1. Serum selenium concentrations by age and sex.

1). The value for men (9.5µg Se/l) is rather low in comparison with other countries, close to the critical values of Keshan disease districts in China (5,11,12). The difference between men and women is highly significant (p < 0.001). Specific spectrum of urine Se values was obtained both for males and females of various age in this industrially extremely polluted region (Figure 3) in comparison with other Czech and Moravian regions (13,14). Age-dependent significant differences were found for both sexes, higher between groups of men and

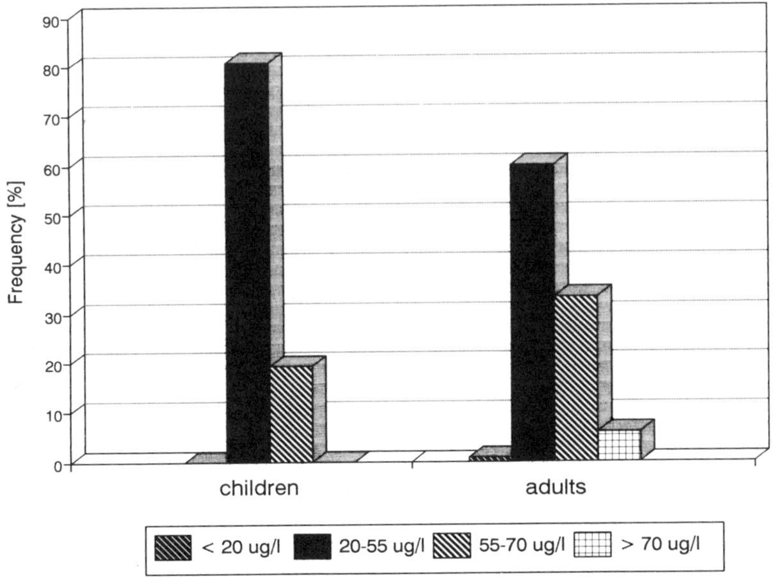

Figure 2. Frequency of serum selenium concentrations.

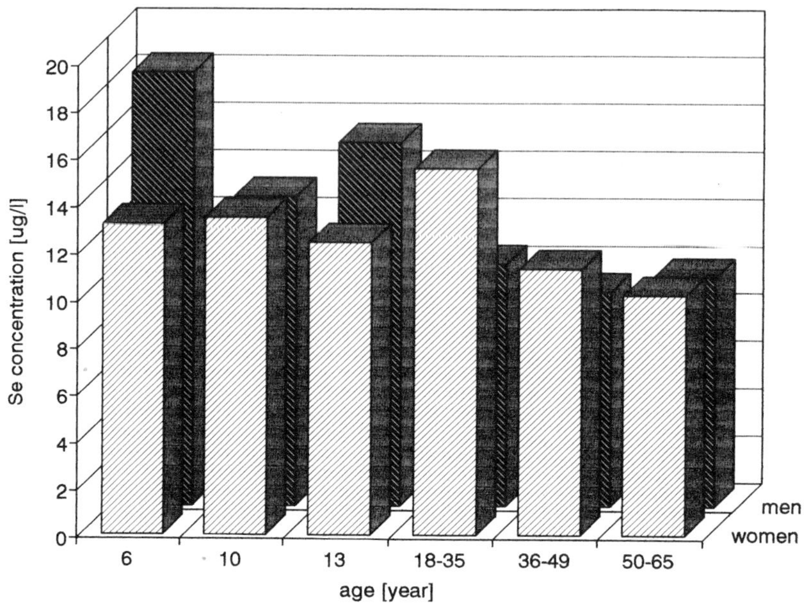

Figure 3. Urine selenium concentrations by age and sex.

boys (Table 1) as well as within individual male categories (Figure 3). Sex related significant differences in urine Se concentrations were also found (p < 0.05 for children, p < 0.01 for young adults - Figure 3). Frequency distributions of urine Se in Figure 4 show relatively higher Se intake by children (40% with more than 24µg Se/l) probably because of higher ratio of Se rich food (meat and eggs) in comparison with adults (only 17% of samples above 24µg Se/l).

Se in urine was also recalculated to urine creatinine (Table 1) for some authors observed urine selenium and creatinine excretion (15) and preferred their relation to Se concentration values. But creatinine levels differ for different groups (age, sex, locality differences, or various diseases connected with the change of protein metabolism) and the comparison of Se status might be influenced by various conditions not connected with selenium. Moreover, serious Se deficiency might also change creatinine production through alteration of thyroid hormones that control catabolic-anabolic steady state of organism. Therefore we do not consider the relation Se/creatinine in urine good for a parameter of Se status.

Ústí n.L. is the only region from five up to now studied in the Czech Republic with higher Se/creatinine than Se/urine values. Also variability of this parameter is higher compared both with other regions (13,14) and Se in serum and urine (Table 1). Urine selenium concentrations are slightly higher than those in Prague (13) or Znojemsko (14), but the main difference of the Se/creatinine value is due to changes in creatinine excretion. It seems that industrial pollution results in changes of protein metabolism and slightly higher Se excretion is the consequence of these changes.

Significant correlations were not found between serum and urine concentration values, which is in agreement with literature (3,12). Slight correlation (r=0.31, p < 0.05) was found between serum Se and Se/creatinine in urine of women. Expected correlations between concentration of Se in urine and its relation to creatinine were found in all

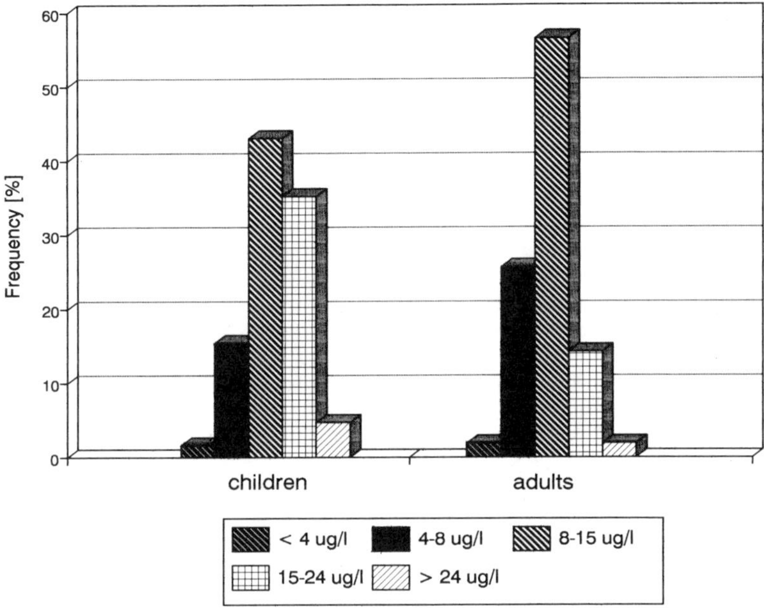

Figure 4. Frequency of urine selenium concentrations.

groups, but with correlation coefficients lower than expected - between 0.36 for girls and 0.45 for women. These numbers are lower than correlations of Oster and Prellwitz (12). Significant correlation was detected between 23 hair samples and urines of middle aged men (r=0.45, p < 0.05).

Excretion of selenium by urine is the main regulatory mechanism of selenium metabolism (3). About 60 per cent of Se intake is excreted by urine (3,10,11). Therefore it is possible to assess Se intake from urine Se (10). Comparison of assessed values (children between 6 and 10 y 25–35µg Se/day, older inhabitants 17–30µg Se/day with negative correlation between age and intake) with WHO recommendations shows very low intakes of Se for inhabitants older then 10 y. This low intake seems to be the main reason of low Se status, proved by all three measured Se parameters.

Ioduria was measured in the same sample of population as urine selenium, and marginal to mild deficiency has been found for the population of Ústí n.L. with the means of 94µg I/l for children and 80µg I/l for adults. Slighter but highly significant correlations between selenium and iodide in urine (r=0.23, p < 0.0001) may signify the possibility of serious harms from alteration of thyroid hormone metabolism at least for a part of the population with both the selenium and iodine deficiency.

Low intake of selenium by the nutrition seems to be the main reason for a very low selenium status in region Ústí n.L. The differences in serum selenium within categories by the age are small in this region with the maximum for young adults. Serious industrial pollution probably affects protein metabolism and the changes in Se/creatinine parameter are the consequence. Some type of selenium supplementation should be applied either to all or at least to the groups of inhabitants with the greatest risk of selenium deficiency (pregnant and lactating women, old people). The correlations between selenium and iodine also necessitate concurrent supplementation of iodine during the selenium supplementation.

ACKNOWLEDGMENTS

This work was supported partly by grant IGA MZ ÈR No. 3417-3, and grant GA ÈR No. 311\93\0986. Prof. Versieck (Rijksuniversiteit Ghent, Belgium) is gratefully acknowledged for the kind gift of the reference material of the second generation human serum.

4. REFERENCES

1. Trace Elements and Free Radicals in Oxidative Diseases, Proc. 4th Int. Congress on Trace Elements in Medicine and Biology, A.E. Favier, J. Neve, P. Faure, eds., AOCS Press, Champaign, Illinois, 1994
2. G.J. Beckett, F. Nicol, P.W.H. Rae, S. Beech, Y. Guo, J.R. Arthur, *Am. J. Clin. Nutr. Suppl.* 57, 240S-243S (1993)
3. J. Neve, *J. Trace Elem. Electr. Hlth Dis.* 5, 1–17 (1991)
4. US National Academy of Sciences, Recommended Dietary Allowances, 10th ed., Natl.Acad.Press, Washington,D.C.,1989
5. H.J. Robberecht, H.A. Deelstra, *Clin. Chim. Acta* 136, 107–120 (1984)
6. J. Kvíèala, V. Zamrazil, M. Soutorová, F. Tomíška, *Analyst* 120, 959–965 (1995)
7. J. Kvíèala, J. Havelka, *J. Radioanal. Nucl. Chem.* 121, 261–270 (1987)
8. J. Kvíèala, J. Havelka, *J. Radioanal. Nucl. Chem.* 121, 271–277 (1987)
9. O.A. Levander, *Federation Proc.* 44, 2576–2583 (1985)
10. P. Van Dael, H. Deelstra, *Flair Concerted Action No 10 Status Papers* 312–316 (1993)
11. Y. Xia, X. Zhao, L. Zhu, P.D. Whanger, *J. Nutr. Biochem.* 3, 211–216 (1992)
12. O. Oster, W. Prellwitz, *Biol..Trace Elem. Res.* 24, 119–146 (1990)
13. J. Kvíèala, J. Havelka, V. Zamrazil, J. Èeøovská, S. Èermák, in *Trace Elements in Man and Animals - TEMA 8*, Proc. 8th Int. Symp. on Trace Elements in Man and Animals, M. Anke, D. Meissner, C.F. Mills, eds., Verlag Media Touristik, Gersdorf, Germany, 1993, pp. 233–234
14. J. Kvíèala, V. Zamrazil, V. Jiránek, in *Natural Antioxidants & Food Quality in Athero- sclerosis & Cancer Prevention*, QSFNE & ALPAC Conference Proceedings, J. Kumpulainen, ed., The Royal Society of Chemistry, Cambridge, UK, 1996, in press.
15. Y. Hojo, *Bull. Environm. Contam. Toxicol.* 29, 37–42 (1982)

60

COMPARISON OF SERUM SELENIUM LEVELS IN INHABITANTS FROM DIFFERENT PORTUGUESE REGIONS

A. M. Viegas-Crespo,[1] M. L. Pavão,[2] M. L. Mira,[3] I. Torres,[4] M. J. Halpern,[5] and J. Nève[6]

[1] Department Zoology and Anthropology
University of Lisbon, Portugal
[2] Department Technological Sciences
University Azores, P. Delgada, Portugal
[3] Department Chemistry
University of Lisbon, Portugal
[4] University Madeira, Funchal, Portugal
[5] Department Biochemistry
New University Lisbon, Portugal
[6] Department Pharmaceutical Chemistry
Free University Brussels, Belgium

1. INTRODUCTION

Selenium is an essential trace element in human nutrition, which is an integral part of the peroxide destroying enzyme, glutathione peroxidase (1). It also takes part in the constitution of other selenoproteins (2) and seems to play other biological functions, such as an important role in the peripheric deiodination of thyroid hormones (3). Normal selenium status varies widely from one part of the world to another. This situation, which is the consequence of differences in selenium intake, depends on the selenium content in soils and hence in food, as well as its bioavailability (4). Moreover, different physiological conditions and also the occurrence of pathological situations are liable to influence the status of the element in human beings.

In order to assess the selenium status of Portuguese inhabitants in association with other parameters involved in the antioxidant function, and their relationship with atherosclerotic risk factors, we carried out an extensive epidemiological study concerning populations with different socio-economic profiles and feeding habits. In the present study, serum selenium levels were determined in inhabitants from five populations (two urban, two rural and one piscatorial) located in three different Portuguese regions - Mainland, Madeira

Therapeutic Uses of Trace Elements, edited by Nève et al.
Plenum Press. New York. 1996

Island and Azores Archipelago. The relationship of serum selenium concentration with age and sex was also considered.

2. SUBJECTS AND METHODS

1 The studied groups consisted of 101 (39 women and 62 men), 98 (50 women and 48 men), 35 (19 women and 16 men), 17 (men) and 65 (men) volunteer Portuguese subjects, aged 20 to 60 y and living respectively in Lisbon-Mainland (urban region), *Ponta Delgada* - Azores Archipelago (urban region), *Salvaterra de Magos* - Mainland (rural region), *Curral das Freiras* - Madeira Island (rural region), and *Câmara de Lobos* - Madeira Island (piscatorial region). The donors were non-alcoholic persons who as well did not take drugs. Age, gender and the date of sampling were registered. The existence of chronic diseases and a history of any cardiovascular condition or stroke were also considered. The subjects were asked to begin to fast 12 h before blood sampling, which occurred in the morning and was carried out from April to July, 1994.

Blood was collected by venipuncture in polyethylene tubes Serum was removed after centrifugation without addition of anticoagulants and an aliquot kept frozen at -20 °C until analysed for selenium. Serum selenium was quantified by a direct electrothermal atomic absorption spectrometric procedure with Zeeman background correction (5). Accuracy of the procedures was checked with standard reference material (Seronorm, Nycomed).

Normality of the distribution was evaluated by the Kolmogorov-Smirnov test. Distribution was studied by drawing frequency polygons after division into class intervals. Individual mean comparisons were tested for significance by the Student's t-test or by the Mann-Whitney test. The correlation of serum selenium with age was analysed by linear regression or correlation coefficient.

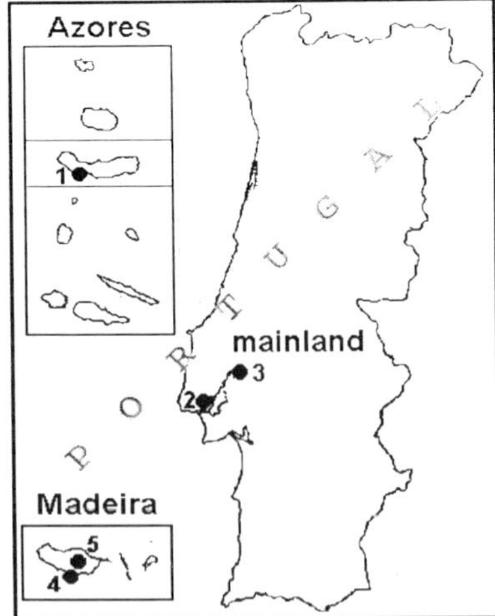

Figure 1. Localization of the five studied populations within the three Portuguese regions: 1 - *Ponta Delgada;* 2 - *Lisbon;* 3 - *Salvaterra de Magos;* 4 - *Câmara de Lobos;* 5 - *Curral das Freiras.*

Table 1. Serum selenium concentration of populations from *Lisbon-mainland, Ponta Delgada - Azores archipelago, Salvaterra de Magos-mainland, Curral das Freiras and Câmara de Lobos - Madeira island*

Age (year)	Gender	n	Selenium (µg/L)	Population
41 ± 15	W	39	94 ± 17	Lisbon (urban)
41 ± 11	M	62	99 ± 21	
41 ± 12	W	50	88 ± 15	*P. Delgada* (urban)
41 ± 10	M	48	98 ± 16	
42 ± 11	W	19	84 ± 14	*Salvaterra* (rural)
41 ± 11	M	16	85 ± 17	
45 ± 13	M	17	60 ± 17	*C. das Freiras* (rural)
40 ± 12	M	65	104 ± 21	*C. de Lobos* (piscatorial)

Values represent mean ± SD; W = women; M = men.

3. RESULTS

The localisation of the studied populations within the three regions is shown in the Figure 1. A significant difference (p < 0.01) of average serum selenium concentrations was found between the two genders in the Azores population. A less important variation (p < 0.05) was also observed in the population from Lisbon (Table 1). Serum selenium levels of men from *Câmara de Lobos* were similar to those observed in urban populations (Table 1). Women and men of the rural regions (*Salvaterra de Magos* and *Curral das Freiras*) exhibited significant differences (p < 0.01) in the average serum selenium concentration when compared with both the urban (Lisbon and Azores) and piscatorial (*Câmara dos Lobos*) populations. An exception was found, since similar values of this parameter were observed in women from the Azores and from *Salvaterra de Magos* (Table 1). When considering men and women together, results of the Azorean population revealed an increase (p < 0.02) of serum selenium levels with age. For the other populations no correlation between age and selenium was observed.

4. DISCUSSION

The serum selenium levels for most of the studied Portuguese subjects, reported not only in the present study, but also in a previous one concerning inhabitants from the area of Lisbon (Women: 89 ± 21 µg/L, n = 60; Men: 97 ± 15 µg/L, n = 66) (6), are in the same range than values obtained in other European countries (7,8). Nevertheless, they seem to be higher than those obtained in the southern European countries, according to data observed in most populations from Greece (8) Yugoslavia (9) and Spain - Catalunya (10), but they are similar to those found in some populations from Italy (8).

The increase of serum selenium levels with age in the Azorean population is in accordance with results obtained by other authors for the same range of age (11,12). The tendency observed in the studied Portuguese populations concerning the sex difference in serum selenium levels, agrees with data reported by some authors, but was not found in majority of studies (11,12). Such as suggested by Robberecht *et al* (11), the race and hormonal status may jeopardise the conclusions.

Data on the selenium status and its relationship with atherosclerotic risk factors (the serum lipid profile as well as life habits) in these populations (except for the subjects from Madeira Island) were reported elsewhere (13). The most striking result of this comparative study is the significantly lower (p < 0.01) mean serum selenium concentration observed in subjects of the rural regions from *Salvaterra de Magos* and *Curral das Freiras* (particularly the latter), when compared with the values of the urban and piscatorial populations. This fact suggests that the selenium status is related to the feeding habits, being lower in the rural regions, where the consumption of animal proteins is still lower than that from urban and fishing regions. A study on the diet and the respective nutrient composition of populations from *Madeira* island is being carried out, as well as the extension of the present work to women. Besides the evaluation of the cited parameters, further investigation on selenium intake in all these Portuguese regions will be also an important goal to elucidate this matter.

ACKNOWLEDGMENTS

We thank to Dr. Paulo Sá de Sousa for his help drawing the map.

5. REFERENCES

1. K. Schwartz and C. M. Foltz, *J. Am. Chem. Soc.* **79**, 3292–3293 (1957).
2. R.F. Burk , *J. Nutr.*, **119**, 1051–1054 (1989).
3. D. Behne, A. Kyuakofoulos, H. Meinhold & J. Kohle, *Biochem. Biophys. Res. Commun*, **173**, 1143–1149 (1990).
4. J. Néve, *Experientia*, **47**, 187 - 193 (1991).
5. J. Nève, S. Chamart and L. Molle, *Trace Element Analytical Chemistry in Medicine and Biology*, Vol. 4, Walter de Gruyter, Berlin-N. Y., 349 - 358 (1987).
6. A. M. Viegas-Crespo, J. Nève, M.L. Monteiro, M. F. Amorim, O.S. Paulo and M.J. Halpern *J.Trace Elem. Electrolytes Health Dis.*, **8**, 119–122 (1994).
7. J. Nève, *J.Trace Elem. Electrolytes Health Dis.*, **5**, 1 - 17 (1991).
8. R. Van Cauwenbergh, H. Robberecht, H. Deelstra, D. Picramenos and A. Kostakopoulos, *J. Trace Elem. Electrolytes Health Dis.*, **8**, 99 - 109 (1994).
9. Z. Maksimovic, I. Drujic, V. Jovic, M. Rsumovic, *Biol. Trace Elem. Res.*, **3**, 187 - 196 (1992).
10. F. Fernandez-Banares, C. Dolz, M.D. Mingorance, E. Cabré, M. Lachica, A. Abad-Lacruz, A. Gil, M. Esteve, J.J. Giné and M.A. Gassul, *Eur. J. Clin Nutr.*, **4**, 225 - 229 (1990).
11. H. Robberecht and H. Delstra, *J. Trace Elem. Electrolytes Health Dis.*, **8**, 129 - 143 (1994).
12. J. Versieck and R. Cornelis, *Trace Elements in Plasma or Serum*, C.R.C. Press, Boca Raton, Florida, U. S. (1989).
13. A. M. Viegas-Crespo, M.L. Pavão, V. Santos, M.L. Cruz, J. Leal, N. Sarmento, M.J. Halpern and J. Nève *In: Natural Antioxidants & Food Quality in Atherosclerosis & Cancer Prevention, Royal Soc.of Chemistry, in press,* 1995.

61

SELENIUM AND KASHIN-BECK DISEASE

Xu Jinpeng[*]

Health Department of Shaanxi Province
34 Lian Hu Ave., Xi'an 710003, China

1. INTRODUCTION

Kashin-Beck Disease (KBD) which hardly attracts attention in the west has been mainly found in remote and poor areas in China. The characteristics of the disease are short stature and deformed joints due to a focalised destruction of the chondrocytes of the articular cartilage and the growth zone (1). Although some factors have been found to be related to the prevalence of KBD, the etiology of the disease still is not clear since its distinct recognition around 1850. Shaanxi province with a population of 34 million is the most prevalent in P.R.China or in the world (2). 315252 cases were registered officially and over 4 million people were in the risk zone for KBD in Shaanxi province (3), where first reports and many intensive studies of selenium with KBD were carried out.

2. EPIDEMIOLOGICAL STUDIES

The first report of finding low Se micro-environment in endemic village appeared in 1970s. Li Jiyun et al. (4) found that Se contents in grains and drinking water were lower than those in non-endemic areas. The further investigations showed that around 90% of grains from endemic villages had the Se content of less than 20 mg/kg and 95% of drinking water less than 0.2 mg/kg. The total Se in soil from 80% of endemic areas were below 150 mg/kg and that from 70% non-endemic areas were above 150 mg/kg (5).

People in endemic areas were found in a Se deficient status. A border of 200 mg/kg in hair Se was indicated between the people in endemic and non-endemic areas, especially between children. A recent investigation on 298 new KBD cases showed that 111 mg/kg was the border between case children and the normal (6). Residents from endemic and non-endemic areas shared different blood Se of below 35 µg/L and 67 ± 17 µg/L respectively. Recently, some case-control investigations were carried out in Yaoxian County, Shaanxi province. 28 parameters related to KBD in past studies were used to determine the epidemic characteristics and aetiology of the disease. Results showed that odds ratios of Se contents in grain from endemic areas were, compared with two control groups (related non-endemic and

---

* *Present address:* 153/D2 Laarbeeklaan 107, B-1090 Brussels, Belgium

Therapeutic Uses of Trace Elements, edited by Nève et al.
Plenum Press, New York, 1996

non-endemic) respectively, as high as 6 and 4.2. The other odds ratios of acid fuchsin staining of wheat blastema were 6.1 and 3.1 respectively (7).

3. EXPERIMENTAL STUDIES

Oral sodium selenite tablets have been introduced to endemic villages since 1970s. Different results were reported. Most studies showed positive effects and some found it no distinct or no effect (1) in the prevention and treatment of the disease. A comparative study found that vitamin C (600 mg/day) and sodium selenite (2 mg/week) shared similar prevention and therapeutic effect. The explanation was that Se contents in blood, urine and hair, as well as enzymological activity of glutathione peroxidase among the two groups were increased significantly after the administration, which indicated that excessive dose of Vitamin C may increase the absorption of Se (8). Anyway, Se-enriched salt was observed to be effective in the prevention. Some experiments showed that even Se-fertiliser could play a positive role in preventing KBD occurrence (1).

An interesting study was done in two endemic villages which shared very similar geology and landform, living condition, staple food as well as the prevalence, except for drinking water. The test village (Se in water: 1.38 mg/L) instead of the previous rain water (0.09 mg/L) and the control used rain water (0.08 mg/L) constantly. A significant prevention and therapeutic effect was observed from 1977 to 1987. A sharp hair Se increase (from 41 ± 9 mg/kg to 301 ± 9 mg/kg) in test village accompanied by no change in controls (from 40 ± 13 mg/kg to 49 ± 5 mg/kg) was also observed (9).

4. DISCUSSION

The above epidemiological surveys and experiments on the spot indicated very strong relationship between Se and KBD. Some scholars even suggested the formula that cause of KBD = A + Σ B, where A is the basic causing factor and B accessory factors. Se deficiency was speculated to cause KBD through lowering in glutathione peroxidase activity (1). The positive role of Se in enhancing antigen stimulated proliferation of lymphocytes or the maintenance of an optimal immune response in humans (10,11) may partially explain the Se supplementation function in prevention and treatment of KBD, since fungal toxicity was also strongly suggested as the causing factor of the disease (1). Se deficiency should be an important or dominant factor in causing KBD.

5. REFERENCES

1. Office for Controlling Endemic Dis., *The Collections of Scientific Investigation on KBD in Yonshou county of Shaanxi Province*, People's Health Press, Beijing (1983).
2. WHO workshop meeting on KBD, draft report, May 28–29 (1992).
3. Li Jiyun, et al., *Newsl. Endemic. Dis.* **2**, 22–28 (1979).
4. Shaanxi Provincial Health Dpt,. *Ann. Health Sta., Shaan xi Province* (1984).
5. Li Jiyun, et al, *J. Environ. Sci.* **2**, 91 (1982).
6. Wang Zhilun, et al., *J. Xi'an Med. Univ.* **14**, 31–32 (1993).
7. Wang Zhilun, et al, in *Research on Regional Environment and Development*, Wang H, ed., China Environmental Sciences Press, (1993).
8. Wang Zhilun, et al., *J. Xi'an Med. Univ.* **14**, 7–9 (1993).
9. Wang Zhilun, et al., *J. Xi'an Med. Univ.* **14**, 22–27 (1993).

10. Peretz A, in *Selenium in Medicine and Biology*, Nève J and Favier A., ed. Walter de Gruyter, New York, pp 234–235 (1988).
11. Peretz A. and Nève J. et al., *Nutrition.* **7**, 215–221 (1991).

62

SELENIUM DEFICIENCY TRIGGERING INTRACTABLE SEIZURES

A Case Study

M. R. Calomme,[1] V. Th. Ramaekers,[2] W. Makropoulos,[3] and D. A. Vanden Berghe[1]

[1] Department of Pharmaceutical Sciences
University of Antwerp (UIA)
Belgium
[2] Department of Paediatric Neurology
[3] Department of Occupational Health
University of Aachen
Germany

1. INTRODUCTION

Diseases related to selenium deficiency are rare in humans. In some endemic areas of China, selenium deficiency secondary to low soil content contributes to the pathogenesis of Keshan cardiomyopathy (1). In the Western world, selenium deficiency has been reported after prolonged parenteral nutrition (2), and after dietary treatment for phenylketonuria, maple syrup urine disease and propionic acidaemia (3). The selenoenzyme glutathione peroxidase (GSH-Px) act as part of the protective mechanisms (4) against the ubiquitous formation of damaging oxygen radicals and catalyses the reduction of hydrogen peroxide and organic peroxides. Phospholipid hydroperoxide GSH-Px (PHGSH-Px) is membrane associated and reduces different phospholipid-hydroperoxides (4). Since the brain is deficient in catalase (5), the central nervous system can only be protected against hydrogen peroxide and lipid peroxidation by virtue of GSH-Px or PHGSH-Px.

In the present study a child with poorly controlled seizures and elevated liver function tests was found to have a low selenium status, resulting in impaired protection against peroxidation in the brain. Selenium-enriched lactobacilli were administered to correct the decreased selenium status. This recently characterized selenium supplement (6,7) contains selenoproteins in which selenium is present as selenocysteine. We selected a non-seleno-methionine-containing form since selenomethionine can substitute methionine in proteins (8), which should be avoided in our view when supplementing young children.

Therapeutic Uses of Trace Elements, edited by Nève et al.
Plenum Press, New York, 1996

2. PATIENT AND METHODS

The patient came from two unrelated families without history of epilepsy or neurological disease. Known metabolic disorders, genetic or neurodegenerative conditions were ruled out by extensive investigations including a nerve, muscle and skin biopsy. Normal values were obtained for urinary organic and amino acids, galactose, as well as normal plasma values for amino acids, pyruvate, lactate, electrolytes, calcium, phosphate, zinc, copper, manganese, ceruloplasmin, superoxide dismutases, biotinidase, protein and lipid electrophoresis, vitamin E, fatty acids and phytanic acid, and lysosomal enzymes.

The selenium concentration in plasma was measured with electrothermal atomic absorption spectrometry equipped with deuterium background correction (ETAAS, Perkin Elmer) (9,10). Reference material (batch 10017, Nycomed and co, Oslo, Norway) was purchased to validate the method. The selenium concentration in urine, and in selenium supplements was determined with flow injection hydride generation atomic absorption spectrometry (HGAAS, Perkin Elmer) after wet, acid digestion (10). The accuracy of this method was checked against bovine liver (BCR 185, Community Bureau of Reference, Brussels, Belgium). Reference values for selenium in plasma were obtained from children with a febrile illness in whom a diagnostic lumbar puncture was necessary to rule out meningo-encephalitis. GSH-Px activity was measured according to the method of Paglia and Valentine modified by Beutler (11) with hydrogen peroxide as the substrate. The activity in erythrocytes is reported per g Hb. Glucose-6-phosphate dehydrogenase (G-6-PDH) was measured according to Beutler (11).

Clinical and biochemical features for selenium depletion were specially sought for, i.e.: hair depigmentation and whitening of nail beds, cardiac arrhythmias, myopathy and elevated creatinine kinase (CK), disturbed liver function tests, anaemia, lowered triiodothyronine (T_3) compared to thyroxine (T_4) level, and radiological features for osteopenia and Kashin-Beck disease.

After detection of a lowered selenium status and obtaining informed parental consent, daily oral supplements with 3 μg Se kg^{-1} bodyweight were administered in the form of either selenium-enriched Lactobacillus (Se-Lact.) (6,7) which contains selenocysteine or, in a later supplementation period, sodium selenite.

3. RESULTS

The patient who was born at term after an uneventful pregnancy and delivery, became cyanosed with heart failure on the fourth day of life. A ventricular septal defect with a large left-right shunt was found and treated with fluid restriction, diuretics and digoxin. The patient's dietary intake has been normally balanced since birth. Subsequent neurodevelopment progress was slow. CT-brainscan showed moderate frontal atrophy. From the age of 11 months the onset of multifocal myoclonic seizures failed to respond to pyridoxine or various combinations of anticonvulsants. Thereafter, petit mal status epilepticus occured for which treatment with ACTH and later dexamethasone provided temporary control. However, as soon as steroids were phased out, myoclonic seizures recurred. At the age of 16 months his hair became depigmented, dry and straw-like (12). From the age of one year and ten months liver transaminases became moderately increased to 2–3 times the normal reference values. Serum CK levels remained normal. From the age of 2 years knee joints enlarged. At the age of three years and six months non-accidental injury with spontaneous fractures were present in the lower lumbovertebral bodies, left hip, arm and shoulder. The skeleton showed severe

Table 1. Follow-up measurements of plasma selenium concentration and glutathione peroxidase (GSH-Px) activity in plasma and red blood cells (RBC) of a patient with severe neurodevelopment retardation and intractable seizures

	Age[1]	Plasma Se (μmol l^{-1})	Plasma GSH-Px (U l^{-1})	RBC GSH-Px (U g^{-1}Hb)	Comment
Patient	3 9/12	0.44	191	9.4	Petit mal status
	3 10/12	0.62	266	ND	Seizure free, after first trial with Se-Lact.
	3 11/12	0.53	180	15.2	Relapse of seizures without selenium.
	4	0.78	491	14.1	Seizure free, after second trial with Se-Lact.
		0.82	439	ND	After continuous selenite treatment; complete cessation of seizures.
		0.83	452	ND	
	4 1/12	0.98	445	17.2	
Controls[2]	1-6	0.67-1.32	268-472	9.2-17.4	
	(n = 16)	(n = 7)	(n = 11)	(n = 11)	

[1] Age indicated in years and months.
[2] Control values for selenium in plasma were obtained from children with a febrile illness. ND: not determined.

osteopenia and extraosseous calcium deposits (12). At the age of three years and nine months petit mal status recurred and became uncontrollable despite benzodiazeprine drips or high doses dexamethasone. Plasma selenium concentration and GSH-Px activity in both erythrocytes and plasma were low (Table 1), while the urinary selenium concentration of 6.42 μg/24 h was well below the reported normal range of 37–184 μg/24 h (13). At the time of selenium deficiency the activity of glucose-6-phosphate dehydrogenase (G-6-PDH) in erythrocytes was lowered with 88 U g^{-1} Hb (normal range: 118–144 U g^{-1} Hb). After oral supplementation with Se-Lact., plasma selenium concentration and GSH-Px activity became nearly normal while petit mal status and myoclonic seizures together with generalized spike-wave activity in the EEG stopped completely after two weeks (Figures 1a, b). After selenium substitution the G-6-PDH activity in erythrocytes and the liver transaminases normalized.

After discontinuation of the selenium treatment, plasma selenium concentration and GSH-Px activity dropped again followed by a relapse of seizures. Thereafter the patient was put on a daily regimen of Se-Lact. and later selenite supplements (3 μg Se kg^{-1} bodyweight), which resulted in marked clinical improvement with virtually complete cessation of myoclonic seizures and petit mal. The EEG showed still some bilateral parieto-occipital discharges, but remained without generalized discharges and did not relapse into petit mal status. Following treatment the depigmented hair regained normal pigmentation. The enlarged knee joints and radiological findings were considered to be due to long-standing selenium depletion, i.e. consistent with Kashin-Beck disease.

4. DISCUSSION

Apart from low dietary intake, selenium deficiency has been described in short bowel syndrome (14) and Crohn disease (15). In fact, the small intestine is known to be the site of selenium absorption. Because no signs of intestinal disease or malnutrition have been found, the basic selenium requirement in the present case is suspected to be higher than normal. This is substantiated by the fall in both selenium concentration and glutathione peroxidase activity in plasma following cessation of selenium supplementation. Possibly, the selenium deficiency is caused by a disturbed transport across the small intestinal border.

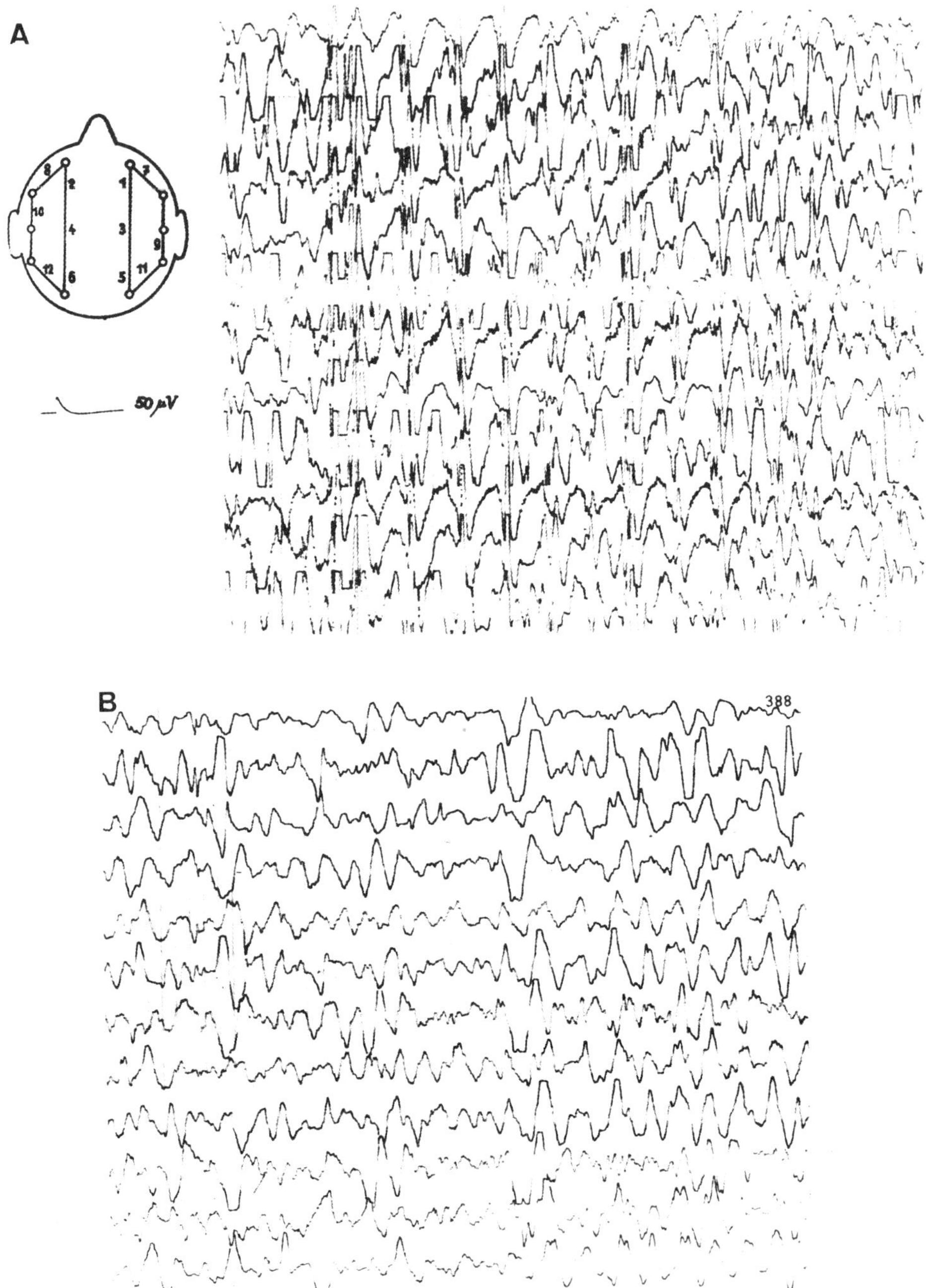

Figure 1. The EEG recordings of a patient with severe neurodevelopment retardation: a) at the time of selenium depletion, showing continuous generalised spike-wave activity associated with petit mal status; b) after two weeks Se-Lact. supplementation (daily 3 µg Se kg-1 bodyweight) arrest of petit mal status and virtually complete cessation of epileptic discharges.

Selenocysteine exerts several functions as an essential constituent of different enzymes, i.e. type I 5'-deiodinase and glutathione peroxidases (4). Together with other defence mechanisms the selenium-dependent glutathione peroxidases are essential in the protection against peroxidative stress. Catalase is another ubiquitously present enzyme which is able to transform H_2O_2 into H_2O and O_2. However, since the central nervous system is deficient in catalase, selenium deficiency will render neurons more vulnerable to peroxidative damage. During seizures, particularly those neurons with increased neurometabolic activity produce oxygen radicals with consecutive peroxidative damage which affects the electro-physiological integrity and stability of neuronal membranes. This could be substantiated by in vitro studies using iron-induced epileptogenic foci in the rain brain (16). After intracortical injection of ferric chloride ($FeCl_3$) in the rat cerebral cortex, epileptic foci and epileptiform discharges were identified by electrocorticography together with large increases in lipid peroxidation. Measurements of local enzyme activities at the iron epileptic focus showed a relative deficiency of catalase and GSH-Px. In the presence of selenium depletion the consequent further lowering of GSH-Px and PHGSH-Px activity would cause a higher susceptibility to lipid peroxidation with resulting neuronal cell and membrane damage. This might well explain the failure of anti-epileptic drugs acting on glutamate receptor-mediated ion channels and GABA-ergic receptors with the occurrence of status epilepticus. In the present case, postnatal hypoxic ischemic brain injury might have been involved in the origin of epilepsy and frontal brain atrophy. The later onset of selenium depletion was believed to constitute an important triggering event predisposing to the occurrence of a vicious cycle with petit mal status unresponsive to steroids and anticonvulsants. The failure of protection due to selenium depletion will increase the amount of oxidative stress to firing neurons. In a recent review (17) large circumstantial evidence was presented that glutamate receptor-mediated processes will promote oxidative stress. In this view, excessive activation of glutamate receptors and oxidative stress represent sequential as well as interacting processes providing a final common pathway for cell vulnerability and cell death in the brain (17). We speculate that correction of selenium depletion in the present case could stop the vicious cycle by restoring the defence mechanisms against the chain reaction of harmful oxygen radicals and lipid peroxidation.

In conclusion, we suggest that the selenium status of children with epilepsy who developed intractable seizures should be screened since selenium deficiency could be a trigger of neuronal membrane damage and instability. Inborn errors of selenium uptake or metabolism could be involved in the pathogenesis of intractable epilepsy, or progressive neuronal degeneration of childhood.

5. REFERENCES

1. Y. Giangqui, G. Keyou, C. Junshi, C. Xiaoshu, World Rev. Nutr. Diet . 55, 98–152 (1988).
2. C.L. Kien, H.E. Ganther, Am.J.Clin. Nutr. 37, 319–328 (1983).
3. I. Lombeck, K.H. Ebert, K. Kasperek, L.E. Feinendegen, H.J. Bremer, Eur. J. Pediatr. 143, 99–102 (1984).
4. B.A. Zachara, J. Trace Elem. Electrolytes Health Dis. 6, 137–151 (1992).
5. B. Halliwell, J.M.C. Gutteridge, TINS 8, 22–26 (1985).
6. M. Calomme, J. Hu, K. Van Den Branden, D. Vanden Berghe, Biol. Trace Elem. Res. 47, 379–383 (1995).
7. M.R. Calomme, K. Van Den Branden, D.A. Vanden Berghe J.Appl. Bacteriology 79, 331–340 (1995).
8. M.A. Beilstein, P.D. Whanger, J. Nutr. 116, 1711–1719 (1986).
9. R. Van Cauwenbergh, H. Robberecht, P. Van Dael, H. Deelstra, J. Trace Elem. Electrolytes Health Dis. 4, 127–131 (1990).
10. P. Van Dael, R. Van Cauwenbergh, H. Robberecht, H. Deelstra, M. Calomme, Atom. Spectroscopy 16, 251–255 (1995).

11. E. Beutler, in Red cell metabolism, ed. Grune & Stratton, pp. 66–68 (1971).

12. V. Th. Ramaekers, M. Calomme, D. Vanden Berghe, W. Makropoulos, Neuropediatrics 25, 217–223 (1994).

13. R. Cornelis, A. Speecke, J. Hoste, Anal.Chim. Acta 78, 317–327 (1975).

14. J. Nève, J. Trace Elem. Electrolytes Health Dis. 5, 1–17 (1991).

15. K. Kawakubo, M. Lida, T. Matsumoto, Y. Mochizuki, K. Doi, K. Aoyagi, M. Fujiishima, Postgrad. Med. J. 70, 215–219 (1994).

16. R. Singh, D.N. Pathak, Epilepsia 31, 15–26 (1990).

17. J.T. Coyle, P. Puttfarcken, Science 262 , 689–695 (1993).

63

TRACE ELEMENT LEVELS IN CHILDREN AND ADOLESCENTS FROM SELECTED REGIONS OF SLOVAKIA

A. Béderová, A. Brtková, T. Magálová, and K. Babinská

Research Institute of Nutrition
Limbova 14, 833 37 Bratislava, Slovak Republic

1. INTRODUCTION

Essential trace elements, zinc, copper and selenium, are important for optimal function of specific biochemical reactions. They influence the immune system and as a component of enzymes, they are involved in redox reactions. They may be important in development of cardio-vascular diseases and cancer. This hypothesis is supported by results of epidemiological and ecological studies. Increased levels of zinc and copper were observed in women with breast cancer (1,2), in lung cancer and in and stomach cancer (3). Increased levels of copper increase the risk of cardio-vascular diseases (CVD) (4,5). Similarly, decreased values of selenium under 0.57 μmol/L represent a risk for development of cancer and or CVD (6–9). Aim of this study was to obtain serum levels of zinc, copper and selenium in our children and adolescent population, and to assess the possible risk.

2. SUBJECTS AND METHODS

Estimation of trace elements was part of large epidemiological study focused on nutritional status and health of children (11–14 y) and adolescents (15–18 y). Blood samples were collected in the morning after overnight fasting. Serum zinc and copper were determined by flame atomic absorption spectrophotometry (Varian SpectrAA 30, Australia) with deuterium background correction. Exa-test (Lachema, Brno, Cz) was used as the control serum. Serum selenium levels were determined according to Jacobson and Lockitch (10) by flameless method using the same equipment. "Second generation" biological references material (Freeze-dried human serum, Versieck, Belgium) was used as control serum for accuracy of estimation of selenium. Statistical analysis was performed using STATGRAPHICS program.

Therapeutic Uses of Trace Elements, edited by Nève et al.
Plenum Press, New York, 1996

Table 1. Mean serum zinc, copper and selenium levels in children (aged 11–14 yr)
from selected regions of Slovakia

Parameter (µmol/L)	Modra - West part of Slovakia			Cadca - North part of Slovakia		
	n	Mean ± SD	(range)	n	Mean ± SD	(range)
Girls						
Zinc	54	15.3 ± 2.2	(9.5 - 25.6)	52	15.1 ± 1.6	(10.5 - 19.0)
Copper	56	17.0 ± 2.4	(11.7 - 23.0)	51	17.6 ± 1.8	(14.9 - 22.2)
Selenium	28	0.59 ± 0.12	(0.33 - 0.93)	26	0.68 ± 0.14	(0.49 - 1.05)
Boys						
Zinc	43	15.4 ± 1.9	(12.5 - 21.2)	67	15.9 ± 2.0	(11.4 - 19.7)
Copper	47	17.8 ± 2.7	(10.3 - 25.6)	67	18.9 ± 2.6	(14.0 - 26.8)
Selenium	25	0.65 ± 0.15	(0.37 - 0.93)	23	0.68 ± 0.2	(0.39 - 1.12)

3. RESULTS

Mean serum levels and ranges of estimated trace elements are shown Tables 1 and 2. Values out of reference range were calculated according to Iyengar and Woittiez (7), Šimek (11), and Malvy (12). For zinc, there were no sex or regional differences in children. We observed decreased serum levels below 10.7 µmol/L only in group of girls (1.9%), and increased values above 18.35 µmol/L in girls (2.8%) and boys (9.5%). In the group of adolescents sex and regional differences (p < 0.001) were found: increased values were observed in 8.6% of girls and 16% of boys. For copper, there were significant interregional differences in boys 11–14 y (p < 0.05). Sex differences (p < 0.001) were observed in this age group. Copper values of our children were in the middle of reference range (11). We found decreased values below 12.6 µmol/L in 0.9% of children. In the group of adolescents, decreased values were observed in 1.4% girls and in 4% boys and increased values only in 0.7% girls. For selenium in the 11–14 y girls, we found mean serum levels on the bottom of reference range (7). Cut- off point (4) did not reach 25.9% of girls and 31.2% of boys. In the group of adolescents, we observed decreased values in 18.5% of girls and 8.8% of boys.

Table 2. Mean serum zinc, copper and selenium levels in adolescents (aged 15–18 y)
from selected regions of Slovakia

Parameter (µmol/L)	Modra - West part of Slovakia			Cadca - North part of Slovakia		
	n	Mean ± SD	(range)	n	Mean ± SD	(range)
Girls						
Zinc	106	16.0 ± 2.2	(11.4 - 24.1)	45	14.8 ± 1.5	(10.9 - 17.4)
Copper	106	17.3 ± 2.9	(11.8 - 30.8)	39	16.5 ± 2.1	(12.9 - 20.6)
Selenium	41	0.71 ± 0.15	(0.49 - 1.24)	24	0.76 ± 0.23	(0.36 - 1.14)
Boys						
Zinc	70	16.7 ± 2.1	(12.4 - 21.2)	36	16.0 ± 1.7	(12.7 - 20.7)
Copper	69	16.1 ± 2.1	(12.7 - 25.6)	32	15.8 ± 2.8	(12.2 - 24.5)
Selenium	28	0.72 ± 0.11	(0.48 - 1.13)	29	0.74 ± 0.21	(0.41 - 1.31)

4. DISCUSSION

Diet is the main source of trace elements. Content of trace elements in foods differ in different countries, especially that of selenium. We estimated mean serum levels of zinc, copper and selenium in children and adolescents from two economic and ecological different regions of Slovakia. Mean serum zinc levels were in our groups comparable with other studies (12,13,14). Serum copper levels were similar to data from Finland (4) and slightly lower than in other studies from Netherlands, England and Italy (1,2,15). There were low occurrence of the decreased or increased values of zinc and copper out of reference range (11,12). There are indications about the role of zinc in the aetiology of CVD and cancer (1,2,15). Some studies suggested a correlation between serum zinc and copper levels and risk of breast cancer in women. Increased serum copper levels were found in persons who died from CVD or breast cancer (5,9). Relative risk is increased when levels are out of recommended range. Serum selenium concentrations were lower when compared with other European countries (12,16,17). Occurrence of values below cut- off point was high, markedly in 11–14 y children. There may be relationship between low levels of selenium and their lower dietary intake, low content in soil and low bioavailability.

5. REFERENCES

1. F. Cavallo, M. Gerber, E. Marubini et al., *Cancer,* 67, 783–745, (1991).
2. J. Niskanen, J. Marniemi, O. Piironen, *Acta Pharmacol. Toxicol.* 59,340–343, (1986).
3. P. Knekt, A. Aromaa, J. Maatela et al., *Br.J.Cancer*, 65, 292–296, (1992).
4. J. T. Salonen, G. Alfthan, J. K. Huttenen, P. Puska, *Amer. J. Epidem,* 120, 342–349, (1984).
5. K. Overvad, D .Wang, J. Olsen et al., *Amer. J. Epidem.* 137, 409–414, (1993).
6. G. W. Comstock, T. L. Bush, K. Helzlsouer, *Am. J. Epidem.* 135, 115–121, (1992).
7. V. Iyengar, J. Woitiez, *Clin. Chem.* 34, 474–481, (1988).
8. A. Reunanen, P. Knekt, R. K. Aoran, *Amer. J. Epidem.* 136,1082–1090, (1992).
9. J. T. Salonen, R. Salonen, H. Korpela et al., *Am.J.Epidem.* 134, 268–276, (1991).
10. E .B. Jacobson, G. Lockitch, *Clin. Chem.* 34, 709–714, (1988).
11. J. Šimek, *Physiological values in human being*, 2nd ed., Avicenum,Praha (1981).
12. D. J. M. Malvy, J. Arnaud, B. Burtschy et al., *Eur. J. Epidemiol.* 3, 155–161, (1993).
13. T. Magalová, A. Brtková, I. Kajaba et al., *Čs. Gastroent. Vý•.* 44, 315–321, (1990).
14. R. Laitinen, E. Vuori, S. Dahlstrom et al., *Pediatric Res.* 25, 323–326, (1989)
15. F. J. Kok, C. M. Van Duijn, A. Hofman et al., *Am. J. Epidem,* 128, 352–359, (1988).
16. G.Marano, A.Spagnolo, G.Morisi et all., *J.TraceElem.Elec.Health Dis.* 5, 59–61, (1991).
17. M. G. Medhin, U. Ewald, T. Tuvemo, *Upsala J. Med.Sci.* 93, 57–62, (1988).

64

SERUM SELENIUM LEVELS AND ERYTHROCYTE GLUTATHIONE PEROXIDASE ACTIVITY IN WOMEN WITH BREAST CANCER

A. Brtková,[1] V. Bella,[2] T. Magálová,[1] M. Kudlácková,[1] K. Babinská,[1] and A. Béderová[1]

[1] Research Institute of Nutrition
Limbova 14, 833 37 Bratislava, Slovak Republic
[2] National Institute of Oncology
Bratislava, Slovak Republic

1. INTRODUCTION

The most common type of cancer in women is breast cancer. Its incidence has increasing trend in Slovakia, especially in young women. Animal studies indicated a possible protective effect of selenium on virally or chemically induced mammary tumor (1). Ecological studies showed a negative association between selenium and breast cancer mortality (2,3). Case-control and prospective studies provided inconsistent results about the protective effect of selenium in breast cancer. Several case-control studies showed an inverse association (4–8), while the other studies showed no significant association between blood selenium levels and breast cancer risk (9–13). In addition, no significant association between blood and nail selenium levels and breast cancer risk was reported (11,12,14,15). The present case-control study was undertaken to study serum selenium concentration and erythrocyte glutathione peroxidase activity in women with breast cancer, benign breast disease as well as in women who underwent surgery, and in healthy controls. This case-control study was carried out in cooperation with the National Institute of Oncology in Bratislava.

2. MATERIAL AND METHODS

2.1. Subjects

A total of 28 women with breast cancer (aged 37–78 y) and 24 women with benign breast disease (dysplasia, fibrocystic disease and fibroadenoma tumors, aged 36–65 y) as well as 14 women that underwent surgery (partial or total mastectomy, aged 39–78 y) participated in this study. Breast cancer and benign breast disease were firstly diagnosed in these subjects when they underwent a preventive medical examination in mammary out-patients´department. The diagnosis was based on radiological (mammography and sonography) and histological examinations. The control group (aged 39–64 y) consisted of 92 apparently healthy women.

2.2. Methods

Blood samples were collected between 8:00–10:00 AM after overnight fasting. Serum selenium levels were determined by direct EAAS method (16) on Varian SpectrAA-30 (Varian, Australia) equipped with deuterium background corrector. The Super Lamp system (Starna Cells, Inc., Atascadero, CA) was used to achieve better sensitivity. The accuracy of determination was verified by analyzing an external quolity control serum "Second Generation" Biological Reference Material (Freeze Dried Human Serum) for Trace Element Determination (17). Erythrocyte GSH-Px activity was assayed according to the method of Paglia and Valentine (18) using cumene-hydroperoxide as a substrate. The GSHPx activity was expressed in international units per ml of whole blood. Statistical analyses were perfomed using linear regression analyses and Student´s t-test.

3. RESULTS

Table 1 shows the basic characteristics of women with malignant and benign breast diseases as well as of women who underwent surgery and of healthy controls. The mean (± SD), the 95% confidence interval, and the range of serum selenium levels are shown in Table 2. The mean selenium levels in women with malignant breast disease and benign breast disease as well as women that underwent partial or total mastectomy were not statisticallly significant in comparison with the control group. The mean (± SD), the 95% confidence interval, and the range of erythrocyte GSHPx activity are presented in Table 3. No significant differences between GSHPx activity in women with malignant breast cancer and women with benign breast disease were observed as well as women who underwent partial or total mastectomy when compared to the control group. There was a positive correlation between serum selenium concentration and erythrocyte GSHPx activity in control group (r = 0.236, p < 0.05). No significant correlation was noticed between serum selenium levels and erythrocyte GSHPx activity in the patient groups.

Table 1. Basic characteristics of the study groups

Group	n	Age (years)	Height (cm)	Body weight (kg)
Malignant disease	28	59 ± 12	160 ± 6	71 ± 11
Benign disease	24	44 ± 12	164 ± 6	64 ± 14
After surgery	14	56 ± 13	162 ± 5	68 ± 13
Control group	92	48 ± 15	163 ± 4	69 ± 14

Results are expressed as mean ± SD

Table 2. Serum selenium levels (μmol/L) in women with malignant and benign breast diseases, after surgery and in control group

Group	n	Mean ± SD	95% CI	Range
Malignant disease	28	0.90 ± 0.22	0.82 - 0.98	0.46 - 1.41
Benign disease	24	0.93 ± 0.21	0.85 - 1.05	0.55 - 1.36
After surgery	14	0.88 ± 0.18	0.78 - 0.98	0.52 - 1.09
Control group	92	0.88 ± 0.19	0.84 - 0.97	0.47 - 1.49

4. DISCUSSION

As already mentioned, there is no agreement between results of case-control and prospective studies. Several case-control studies on serum or plasma selenium levels showed decreased selenium levels in patients with breast cancer (4–7). Plasma and erythrocyte selenium levels were lowered in breast cancer patients with large tumor (11) and selenium was inversely associated with breast cancer stage (4). In our case-control study, we did not find a significant association between serum selenium level and breast cancer risk. Serum selenium in women with breast cancer and benign breast disease was not statistically significant in comparison with healthy controls. No association with breast cancer risk was found in other case-control studies when selenium was measured in whole blood, erythrocytes or plasma (10,14). Most of the prospective studies indicated no significant association between selenium and human mammary carcinogenesis (9,10,13). Only one study from Finland (19) showed an increased breast cancer risk for the lowest category of serum selenium. Studies conducted with nail selenium levels that may reflect long-term exposure to selenium showed no significant association between selenium and breast cancer risk as well (11,12,14,15).

In our case-control study, there were no significant differences between glutathione peroxidase activity in the patient groups and controls although the mean GSHPx activity was higher in women with malignant breast disease (about 20%) than in controls. On the contrary, GSHPx was lower (about 21%) in women with benign breast disease and about 23% lower in women that underwent surgery. Previous studies (8) showed that erythrocyte GSHPx activity does not correlate well with serum selenium and is not a marker for breast cancer risk. Our study does not provide evidence of a protective effect of selenium in breast cancer and benign breast disease.

Table 3. Glutathione peroxidase activity in erythrocytes (U/ml whole blood) in women with malignant and benign breast diseases, after surgery and in control group

Group	n	Mean ± SD	95% CI	Range
Malignant disease	28	4.1 ± 2.4	3.3 - 5.1	0.5 - 9.3
Benign disease	24	2.7 ± 1.8	2.1 - 3.5	0.5 - 6.8
After surgery	14	2.6 ± 2.4	1.4 - 3.9	0.1 - 9.9
Control group	92	3.4 ± 2.4	3.0 - 4.0	0.7 -10.0

5. REFERENCES

1. C. Ip, *J. Am. Coll. Toxicol.* **5**, 7–20 (1986).
2. G.N. Schrauzer, D.A. White and C.J. Schneider, *Bioinorg. Chem.* **7**, 23–34 (1977).
3. L.C. Clark, K.P. Cantor and W.H. Allaway, *Arch. Environ. Health* **46**, 37–42 (1991).
4. S. Chaitchik, C. Shenberg, Y. Nir-El and M. Mantel, *Biol. Trace Elem. Res.* **15**, 205–212 (1988).
5. M.S. Bratakos, T.P. Vouterakos,and P.V. Ioannou, *Sci. Total Environ.* **92**, 207–222 (1990).
6. H. Kršnjavi and D. Beker, *Breast Cancer Res. Treatment*, **16**, 57–61 (1990).
7. Z. Pawlowicz, B.A. Zachara, V. Trafikowska, A. Maciag, E. Marchaluk and A. Nowicki, *J. Trace Elem. Electrolytes Health Dis.* **5**, 275–277 (1991).
8. L. Hardel, M. Danell, C.A. Anquist, S.L. Marklund, M. Fredriksson, A.L. Zakari and A.Kjellgren, *Biol. Trace Elem. Res.* **36**, 99–108 (1993).
9. W. C. Willet, B.F. Polk, J.S. Morris, M.J. Stampfer, S. Pressel, J. Taylor, K. Schneider and C.G. Hames, *Lancet* **2**, 130–134 (1983).
10. I. Peleg, S. Morris and C.G. Hames, *Med. Oncol. Tumor. Pharmacother.* **2**, 157-163 (1985).
11. P. van´t Veer, R.P.J. Van der Wielen, F.J. Kok, R.J.J. Hermus and F. Sturmans, *Am. J. Epidemiol.* **131**, 987–994 (1990).
12. M. Gerber, S. Richardson, R. Salkeld and P. Chappuis, *Cancer Invest.* **9**, 421–428 (1991).
13. K. Overvad, D.Y. Wang, J. Olsen, D.S. Allen, E.B. Thorling, R.D. Bulbrook and J.L. Hayward, *Eur. J. Cancer* **7**, 900–902 (1991).
14. D.J. Hunter and C.W. Willet, *Ann. Rev. Nutr.* **14**, 393–418 (1994).
15. P.A. van den Brandt, R.A. Goldbohm, P. van´t Veer, P. Bode, E. Dorant, R.J.J. Hermus and F. Sturmans, *Am. J. Epidem.* **140**, 20–26 (1994).
16. B.E. Jacobson and G. Lockitch, *Clin. Chem.* **34**, 709–714 (1988).
17. J. Versieck, L. Vanballenberghe and A. De Kesel, *Anal. Chim. Acta* **204**, 63–75 (1988).
18. D.E. Paglia and W.N. Valentine, *J. Lab. Clin. Med.* **70**, 158–169 (1967).
19. P. Knekt, A. Aromaa, J. Maatela, G. Alfthan, R.K. Aaran, M. Hakama, T. Hakulinen, R. Peto and L. Teppo, *J. Natl. Cancer Inst.* **82**, 864–868 (1990).

65

ZINC AND COPPER IN BREAST CANCER

T. Magálová,[1] V. Bella,[2] K. Babinská,[1] A. Brtková,[1] M. Kudláčková,[1] and
A. Béderová[1]

[1] Research Institute of Nutrition
Bratislava, Limbová 14, Slovak Repulic
[2] National Institute of Oncology
Bratislava, Klenová 1, Slovak Republic

1. INTRODUCTION

Present incidence of breast cancer as well as its ratio to all cancer ranks Slovakia to European countries with high and still increasing incidence of breast cancer. Mortality rates became stabilised in the recent time but evidently increasing incidence of breast cancer is observed especially in lower age groups. Zinc and copper are essential trace elements that play important roles in different biochemical reactions. Relevance of zinc to cellular growth is well known. In recent years, their disbalances were linked to chronic disease etiology due to antioxidant properties. Moreover, zinc affects immunological properties and acts together with copper as cofactor of Cu,Zn-superoxide dismutase (SOD). This enzyme protects cells and their important components against free radical damage.

This study is a part of a case-control study concerning breast cancer and nutrition. Diet is a main source of essential trace elements, copper and zinc. Nutritional deficiency and/or excess of these elements may affect their status. Trace elements as micronutriens may be involved in etiology of some types of cancer.

2. SUBJECTS AND METHODS

2.1. Selection of Cases and Controls

Study subjects were enrolled from the breast clinic at National Institute of Oncology. Women participated at screening examination for breast cancer and their diseases were firstly diagnosed (palpation, sonography, mammography and histology). The study included 23 women with benign breast disease (moderate and severe dysplasia, fibroadenoma and cystic disease), 29 breast cancer and 96 apparently healthy controls randomly selected among those with no diagnosis of breast disease. Mean age of control group was 49.3 y (38.4

Therapeutic Uses of Trace Elements, edited by Nève et al.
Plenum Press, New York, 1996

- 75.8), mean age of group of benign breast diseases was 44.5y (28.7 - 64 .7) and mean age of breast cancer group was 57.9 y (36 .4 - 78.2).

2.2. Methods

Blood samples were collected in the morning after overnight fasting. The acid-washed test tube collected from each person was used for the trace element analysis. For zinc and copper determination the samples were stored at -18 °C after clot separation. Serum copper and zinc were determined after 1:4 dilution with deionised water by flame atomic absorption spectrophotometry, with deuterium background correction using a Varian SpectrAA 30. Exa-test (Lachema Brno, Czech Republic) was used as the control serum. SOD (E.C.1.15.1.1.) was measured in erythrocytes by RANDOX test. Statistical analyses were done by commercial software package STATGRAPHICS.

3. RESULTS

Women in groups were divided according to menopausal status, except women in the group of benign diseases, in which there were only two postmenopausal women. Mean serum levels of copper and zinc are shown in Table 1. Mean serum copper in the whole breast cancer group was significantly higher than in the whole control group. Increased serum copper levels were also found in postmenopausal women. There were no significant differences of serum copper levels in premenopausal women. Serum zinc levels were slightly but significantly decreased in the group with benign disease and in the whole breast cancer group. Menopausal status did not influence serum zinc levels in women with breast cancer. There were no difference in activity of SOD (Table 1). In control group, we observed a positive correlation between copper and zinc (r = 0.37, n = 96, p < 0.001), while in groups of cases this relationship was not observed. A weak negative correlation between SOD activity and serum copper levels in premenopausal women with breast cancer was calculated (r = 0.53, n = 6, p < 0.02). There was a weak nonsignificant positive correlation between SOD activity and zinc serum level in women with benign diseases (r = 0.24, n = 23, p < 0.21).

Table 1. Serum levels of copper and zinc and erythrocyte activity of Cu,Zn-superoxide dismutase.

	Copper (µmol/L)		Zinc (µmol/L)		Cu,Zn-SOD(U/mL)	
	n	x ± SD	n	x ± SD	n	x ± SD
Benign diseases	23	17.7 ± 2.5	23	13.9 ± 2.1*	25	149 ± 42
Breast cancer						
All	29	18.9 ± 4.6*	29	14.0 ± 1.7*	27	149 ± 39
Premenopausal	7	17.5 ± 3.8	7	13.5 ± 0.8	6	142 ± 30
Postmenopausal	22	19.4 ± 4.8[a]	22	14.2 ± 1.9	21	151 ± 41
Control group						
All	96	17.3 ± 2.6	96	15.0 ± 2.3	94	148 + 34
Premenopausal	55	17.2 ± 2.5	55	14.9 ± 2.4	53	146 ± 29
Postmenopausal	41	17.4 ± 2.7	41	15.0 ± 2.3	41	149 ± 40

*p < 0.05 vs control group
[a] p < 0.05 vs control group postmenopausal

4. DISCUSSION

In present time great attention is devoted to the antioxidant defence system which may affect initiation and progression of neoplastic diseases, including breast cancer. From this point of view, it is very important to assess essential trace elements and other parameters of the antioxidant defence system. In this study serum zinc and copper levels and activity of erythrocyte SOD were determined. The relation of copper and zinc in women with breast cancer was investigated only in a few studies and their results are contradictory (1,2,3,4). We observed increased serum copper levels in women with breast cancer. In order to exclude the effect of age, results were evaluated according to menopausal status. While in premenopausal women, the levels of copper in cases were comparable than in controls. There were significant differences in postmenopausal women. These findings are in agreement with Yücel (2). In another study (5), authors also found increased values of copper levels but they did not exclude an influence of the age. The same authors found significant differences between women with benign and malignant breast lesions. In present study, we observed a slight nonsignificant decrease in serum copper levels when compared benign and malignant group.

With regard to zinc, mean serum levels of women with benign lesions and malignant cases were lower in comparison with controls. We did not observe an age dependence. These results are in agreement with other studies (2,5), but Cavallo (1) found increased values of zinc in breast cancer. An unexpected positive and significant correlation was noted between zinc and copper in the control group. In the group of cases, this relationship is not significant, contrary to findings of Cavallo et all (1). Mean serum zinc levels tended be higher in Slovak population (6) when compared to other European countries. Activity of SOD was similar in all groups and not dependent on age such as alredy reported (7). Other authors (8) observed in urban Spanish population the highest activity of SOD in young adults. The negative correlation between SOD activity and serum copper levels in premenopausal women with breast cancer was surprising. It is well known, that restricting dietary copper quickly impairs of SOD activity (9). However, cases are rather few and need further studies.

5. REFERENCES

1. F.Cavallo, M.G.Gerber, E.Marubini et al., *Cancer*, **67**, 738–745 (1991).
2. I.Yücel, F.Arpaci, A.Özet et al., *Biol. Trace Elem. Res.* **40**, 31–38 (1994).
3. G.D.Kanias, E.Kouri, H. Arvaniti et al., *Biol.Trace Elem. Res.* **43–45**, 363–370 (1991).
4. M.Faber,C.Coudray,H.Hida et al., *Biol.Trace Elem.Res.* **47**, 117–123 (1995).
5. J.Garofalo,H.Ashikari,M.L.Budrick et al., *Cancer*, **15**, 793–794 (1980).
6. T.Magálová,A.Brtková,A.Béderová et al., *Biol.Trace Elem.Res.* **40**, 225–235 (1994).
7. Y.Niwa,O.Iizawa,K.Ishimoto et al., *Am.J.Pathol.* **143**, 312–320 (1993).
8. M.R.De la Torre,A.Casado and M.E.López-Fernández, *Experientia*, **90**, 854–856 (1990).
9. E.D.Harris, *J.Nutr.* **122**, 636–640 (1992).

66

ANTIOXIDANT MICRONUTRIENT STATUS DURING ONCOLOGICAL TREATMENT IN CHILDREN WITH CANCER

D. J-M. Malvy,[1, 2] J. Arnaud,[3] B. Burtschy,[4] D. Sommelet,[5] G. Leverger,[6] L. Dostalova,[7] and O. Amédée-Manesme[1]

[1] INSERM U056, Hospital Center of Bicêtre
F-94275 Bicêtre cedex, France
[2] Laboratory of Public Health, University of Medicine
F-37032 Tours cedex, France
[3] Laboratory of Biochemistry C, University Hospital Center
F-38043 Grenoble cedex, France
[4] Stat. Unit. Télécom Paris
F-75634 Paris cedex 13, France
[5] Department of Pediatrics"II", Children's Hospital
F-54511 Vandœuvre cedex, France
[6] Department of Pediatric Hematology, Hospital St. Louis
F-75010 Paris, France
[7] Department of Clinical Nutrition, F. Hoffmann-La Roche & Co.
CH-4002 Basel, Switzerland

1. INTRODUCTION

Cancer is frequently associated with a combination of metabolic abnormalities leading to a complex, abnormal biochemical state in the tumor-bearing host, including alterations in vitamin and mineral concentrations. Moreover, the tumoricidal action of several anti-cancer drugs is known to be mediated by a free radical dependent mechanism (1). It has been suggested that lipid peroxidation is one of the main causes of irradiation damage. Conditioning regimens for cancer treatments in children often consist of high-dose chemotherapy, possibly combined with surgery or irradiation. These regimens may approach tolerance limit for several tissues. Vitamins and other micronutrients with antioxidant properties (2) have not been clearly assessed during childhood malignancies. We therefore investigated whether abnormal breakdown of antioxidants such as beta-carotene, alpha-tocopherol, zinc and selenium occurs at the time of diagnosis and might follow the conditioning therapy in certain groups of children with cancer. The study was conducted in France as a satellite investigation of a large multicenter case-control survey designed to document the relationship between serum micronutrient values and childhood malignancy (3). We therefore measured retinol (vitamin A), beta-carotene, alpha-tocopherol (vitamin E), cholesterol, zinc, selenium, and related proteins in serum collected from 1986 to 1989 from 170 children aged 1–16 years with newly diagnosed cancer, and from 632 healthy controls who were cancer-free. In the patient group, sample processing was performed twice, once at diagnosis and before treatment (month-0), and once 6 months after initiation of treatment (month-6).

2. PATIENTS AND METHODS

2.1. Patients

The study group consisted of 170 eligible patients, with 82 males (48.2 percent), and 88 females (51.8 percent). Patients were distributed in five age-groups: from one to three months (n = 6), from four months to 11 months (n = 15), from one to two years (n = 20), from three to five years (n = 34), from six to eight years (n = 25), from nine to 11 years (n = 30), from 12 to 13 years (n = 20) and from 14 to 16 years (n = 20). Diagnosis was confirmed histologically in 90 percent of cases. The pilot study was conducted in France, in the cities and surrounding countryside of Paris (n = 30), Lyon (n = 30), Bordeaux (n = 20), Tours (n = 20) and Nancy (n = 70). The study group consisted of children admitted to ten Departments of Pediatric Oncology for diagnosis and cancer treatment between November 1, 1986, and November 31, 1989. Pathology Departments notified the study team of all new reports of malignancies. The classification of tumors corresponds to the pathology and histology patterns of the malignant growth, and refers to the site of the primary tumor (*International Classification of Diseases for Oncology*, ICD-0) (4) (Table 1).

A preoperative blood sample was obtained from the 170 patients before surgery, radiation, or chemotherapy were started. A second sample was collected from the same children 6 months after starting treatment. Concurrently, 632 healthy control subjects 0–16 years of age, 317 males (50.2 percent) and 315 females (49.8 percent) were recruited. The sampling procedure used has been described elsewhere, and was conducted to provide a reference for age

Table 1. Classification of tumours in 170 children registered in the "Cancer in Children and Antioxidant Micronutrients" study, France, 1986–1989, and followed up 6 months after initiation of cancer therapy, according to the criteria of the International Classification Scheme for Childhood Cancer

Diagnostic group (ICD-0*)		n
I.	Leukemias	62
II.	Lymphoma and other reticuloendothelial neoplasms	24
III.	Central nervous system and miscellaneous intracranial and intraspinal neoplasms	20
IV.	Sympathetic nervous system tumours	10
V.	Retinoblastoma	1
VI.	Renal tumours (Wilms' tumour)	4
VII.	Malignant bone tumours	24
VIII.	Soft tissue sarcoma	16
IX.	Hepatic tumours (hepatoblastoma)	2
XII.	Other unspecified malignant neoplasms	7
	All cancers	170

International Classification of Diseases for Oncology

and sex-related serum micronutrient concentrations for healthy French children (5–7). Serum specimens from cancer patients and controls were grouped into sets to ensure that they would be analyzed together during the same period to avoid technical confounding factors between patients and controls. Serum specimens were identified by a number only to ensure that evaluators were blind to disease status.

Blood was collected from fasting, seated individuals, between 8 and 10 a.m. Blood samples were taken from a forearm vein into polystyrene free zinc tubes, using steel needles. Serum was rapidly separated by centrifugation (1300 x g, 10 min). Separated 0.5 ml serum aliquots were removed and stored frozen at -70 °C for up to 6 months until analysis.

2.2. Methods

A separate high-performance liquid chromatography assay comprising an isocratic system using silica gel (adsorption) as the stationary phase was used for retinol, alpha-tocopherol and beta-carotene, as described previously (3, 5). Results for alpha-tocopherol were also expressed as ratio of serum vitamin E to cholesterol. The molar relation between retinol and retinol-binding protein (RBP) in serum was calculated to assess the vitamin A status. Retinol-binding protein and other serum protein values - immunoglobulins: IgG, IgA, IgM, two visceral proteins, namely albumin and transthyretin (TTR), and two acute-phase reactants: alpha$_1$-acid glycoprotein (orosomucoid) and C-reactive protein (CRP) - were quantified by nephelometry (8). Four of these parameters were aggregated in the following formula, allowing determination of a prognostic inflammatory and nutritional index (PINI) (9):

$$PINI = \frac{alpha_1\text{-}acid\ glycoprotein\ (mg\,/\,l) \times C\text{-}reactive\ protein\ (mg\,/\,l)}{albumin\ (g\,/\,l) \times transthyretin\ (mg\,/\,l)}$$

This tool should be of value in inflammatory and nutritional disorders, regardless of sex and age. Finally, a flame atomic absorption technique was used for plasma zinc determination (7, 10), and selenium was assayed with a modification of the method described by Nève *et al* (11).

2.3. Statistical Analysis

Since nutrient blood levels were generally skewed toward higher values, we used log-transformed variables which provided good approximations to the normal distribution. Means and standard deviations of chemistry variables were calculated for patients at month-0, before treatment began, patients at month-6 after diagnosis and for controls. Data analysis was conducted with regression residual analysis after adjustment for age and sex on the basis of the control group data (3, 5, 6). Statistical comparisons between groups were performed by means of t tests and analysis of variance. First the average residual serum micronutrient and chemistry values for patients at month-0 were compared with the average residual value for the controls using the two sample t test. Data for patients at month-0 were subsequently compared with those reported for patients at month-6. Comparison was carried out separately for the different cancer sites. The statistical package SAS using the general linear modeling procedure (GLM) was used for the analysis (12).

3. RESULTS

Mean serum concentrations of retinol, beta-carotene, zinc and alpha-tocopherol were lower in cancer patients at month-0 than in controls (Table 2). Since the levels of cholesterol tended to be very significantly lower in patients than in controls and were positively associated with vitamin E values in controls (p < 0.001), the ratio of vitamin E to cholesterol tended to diminish and suppress the significant difference. There was no appreciable difference in serum selenium concentrations. It should be noted that variability, as indicated by the standard deviation relative to the mean, was high for CRP and the PINI determined by the above-mentioned formula, particularly in patients. The PINI score was below 1.5 in healthy controls and its mean value was around 20 in patients at diagnosis (9). There were no appreciable differences between patients and controls for Ig and retinol-binding protein, revealing a wider scatter range than that of albumin. Absolute means for albumin and transthyretin levels were lower for overall cancers than in controls, and higher for CRP and alpha$_1$-acid glycoprotein, recognized to be a sensitive marker of the phlogistic phenomenon. The changes in mean values of the micronutrient status parameters for all childhood cancers and for types of malignancy for each cancer site are given in Tables 2, 3 and 4. It is obvious from Table 2 that children in the "all cancers" group showed an increase in serum retinol and serum visceral proteins, albumin, transthyretin and retinol-binding protein, and a decrease in levels of acute phase reactants after 6 months of treatment, whereas beta-carotene and alpha-tocopherol values remained unchanged and low in comparison with the initial value. Some significant differences between diagnosis and month-6 were also found according to the cancer site. Patients with leukemia tended to have higher retinol, molar retinol to retinol-binding protein ratio, cholesterol and selenium levels after 6 months of treatment. Patients with malignant lymphoma, who were treated during the whole observation period with combined cytostatic

Table 2. Mean (± standard deviation) or median (range) serum micronutrients and biochemical parameters in 170 children with malignancies, before and after cancer therapy, "Cancer in Children and Antioxidant Micronutrients" Study, France, 1986–1989

Serum micronutrients	Controls(n = 632)	Pre-treatment	Post-treatment	p value
Retinol (µg/dl)	36 ± 12*	32 ± 16	39 ± 16	0.003
Retinol / RBP	0.83 ± 0.30*	0.71 ± 0.30	0.80 ± 0.32	NS
β-carotene (µg/l)	579 (206-812)*	390 (151-620)	397 (160-580)	NS
α-tocopherol (mg/dl)	8.8 ± 2.8*	7.3 ± 4.6	7.0 ± 3.0	NS
Cholesterol (g/l)	1.69 ± 0.46*	1.51 ± 0.47	1.58 ± 0.47	0.02
α-tocopherol/cholesterol	5.2 ± 1.3	4.9 ± 2.9	4.5 ± 1.7	NS
Zinc (µmol/l)	14.6 ± 3.1*	13.4 ± 4.0	11.5 ± 3.1	0.004
Selenium (µmol/l)	0.81 ± 0.34	0.77 ± 0.30	0.77 ± 0.26	NS
IgG (g/l)	14.0 ± 5.3	12.3 ± 6.5	11.0 ± 5.5	0.002
IgA (g/l)	1.8 ± 1.2	2.2 ± 1.6	1.9 ± 1.7	< 0.001
IgM (g/l)	2.4 ± 1.5	3.0 ± 1.9	1.5 ± 1.4	< 0.001
Albumin (g/l)	53 ± 11*	38 ± 10	43 ± 10	< 0.001
Transthyretin (mg/dl)	27 ± 10*	20 ± 11	26 ± 12	< 0.001
α$_1$-acid glycoprotein (mg/dl)	122 ± 51*	170 ± 103	105 ± 60	< 0.001
Retinol-binding protein (mg/dl)	3.5 ± 1.2	3.3 ± 1.7	3.9 ± 1.9	< 0.001
C-reactive protein (mg/dl)	0.83 (0.30-1.35)*	2.53 (1.98-5.56)	1.64 (1.18-4.65)	0.003
PINI	0.9 (0.1-1.8)	20.0 (4.2-40.1)	4.1 (2.2-18.9)	0.004

* P < 0.001 for significant t test between age and sex adjusted healthy controls and pretreatment cases
RBP, retinol-binding protein; Retinol/RBP, molar retinol/retinol-binding protein ratio; CRP, C-reactive protein; PINI, Prognostic Inflammatory and Nutritional Index ; NS, not significant

Table 3. Absolute mean (± standard deviation) or median (range) serum levels among children with cancer, France, 1986–1989

Nutrient	Leukemia (n = 62)		Lymphoma (N = 24)	
	Pre-treatment	Post-treatment	Pre-treatment	Post-treatment
Retinol (µg/dl)	28 ± 15	45 ± 19*	29 ± 11	32 ± 10
Retinol/RBP	0.67 ± 0.26	0.79 ± 0.36	0.68 ± 0.29	0.76 ± 0.22
β-carotene (µg/l)	414 (120-650)	308 (110-550)	395 (95-570)	361 (85-550)
α-tocopherol (mg/l)	8.2 ± 6.4	6.6 ± 2.9	7.1 ± 2.8	7.1 ± 3.3
Cholesterol (g/l)	1.36 ± 0.43	1.52 ± 0.52*	1.47 ± 0.39	1.56 ± 0.43
α-tocopherol/Cholesterol	5.9 ± 3.9	4.4 ± 1.8	4.9 ± 2.3	4.6 ± 1.8
Selenium (µmol/l)	0.68 ± 0.27	0.81 ± 0.28*	0.68 ± 0.28	0.66 ± 0.19
Zinc (µmol/l)	12.5 ± 3.9	11.3 ± 3.7	13.3 ± 3.2	12.1 ± 2.3
IgG (g/l)	11.7 ± 5.3	9.8 ± 5.1	14.2 ± 10.4	10.0 ± 4.3
IgA (g/l)	2.0 ± 1.4	1.4 ± 1.0	2.4 ± 1.8	1.6 ± 1.1*
IgM (g/l)	2.6 ± 1.7	1.0 ± 1.1*	3.2 ± 2.2	1.4 ± 1.0
Albumin (g/l)	37 ± 9	42 ± 10*	39 ± 11	40 ± 9
Transthyretin (mg/dl)	16 ± 8	27 ± 12*	18 ± 10	22 ± 8
RBP (mg/dl)	3.0 ± 1.6	4.4 ± 2.0*	3.2 ± 1.6	3.4 ± 1.5
α$_1$-acid glycoprotein (mg/dl)	182 ± 104	103 ± 54*	169 ± 102	107 ± 54
CRP (mg/dl)	2.90 (2.30-5.55)	1.78 (1.20-3.55)	2.57 (2.10-5.26)	1.23 (1.26-1.40)
PINI	24 (15-40)	4 (2-12)	21 (16-50)	2 (1-5)

*$P < 0.001$ for significant t test against pretreatment controls.
RBP, retinol-binding protein; Retinol/RBP, molar retinol/retinol-binding protein ratio; CRP, C-reactive protein; PINI, Prognostic Inflammatory and Nutritional Index; NS, not significant.

therapy, tended to have increased retinol levels than at diagnosis. No significant decreases in mean values were observed. Patients with brain tumors had lower zinc levels, with a slightly significant decrease in inflammation parameters in comparison with the initial values. For patients with bone tumors, a decrease in serum alpha-tocopherol to cholesterol ratio and in zinc was found during the period of treatment.

4. DISCUSSION

Recognition of the fact that nutritional requirements may be altered in a complex way by the development of malignant disease is of practical significance. Radiotherapy and aggressive oncological cytostatic chemotherapy frequently affect the nutritional status of tumor patients. The effects of the cytostatic agents on the tumor-bearing host seem to overcome the physiological protein-sparing mechanisms (13, 14). There is little evidence to point out that specific deficiencies in vitamins and minerals exist in cancer. Nevertheless, deficiencies are part of a total picture of malnutrition.

In general, the vitamin indicators measured in this report were in a lower range than the range of the control group. The mean values of beta-carotene, retinol, alpha-tocopherol, zinc and the related nutritional proteins were lower at the beginning of the observation period in comparison with controls.

No significant decreases in mean values were observed at 6 months, except for the alpha-tocopherol to cholesterol ratio in patients with bone tumors and serum zinc in bone tumors and central nervous system malignancies. Indeed an increase in mean values for some parameters was found during the period of treatment, namely for retinol and selenium in leu-

Table 4. Absolute mean (± standard deviation) or median (range) serum levels among children with cancer, France, 1986–1989

Nutrient	Central nervous system (n = 20)		Malignant bone tumours (n = 24)	
	Pre-treatment	Post-treatment	Pre-treatment	Post-treatment
Retinol (µg/dl)	46 ± 17	47 ± 17	34 ± 11	33 ± 10
Retinol/RBP	0.80 ± 0.21	0.87 ± 0.31	0.92 ± 0.47	0.89 ± 0.39
β-carotene (µg/l)	407 (115-676)	486 (140-654)	316 (81-505)	361 (110-578)
α-tocopherol (mg/l)	6.9 ± 3.2	7.6 ± 3.4	7.5 ± 3.4	7.1 ± 3.2
Cholesterol (g/l)	1.79 ± 0.42	1.89 ± 0.45	1.56 ± 0.38	1.61 ± 0.35
α-tocopherol/Cholesterol	3.8 ± 1.8	4.1 ± 1.5	4.9 ± 1.8	4.3 ± 1.8*
Selenium (µmol/l)	0.90 ± 0.31	0.86 ± 0.28	0.83 ± 0.22	0.75 ± 0.29
Zinc (µmol/l)	14.1 ± 3.0	11.2 ± 1.9*	14.7 ± 3.8	11.8 ± 1.9*
IgG (g/l)	9.9 ± 3.4	8.8 ± 4.5	16.9 ± 6.4	14.6 ± 4.9
IgA (g/l)	1.8 ± 1.1	1.5 ± 1.0	3.5 ± 2.4	3.3 ± 2.6
IgM (g/l)	3.2 ± 1.8	2.0 ± 1.1*	4.1 ± 2.8	2.0 ± 0.9
Albumin (g/l)	38 ± 8	44 ± 12	44 ± 9	46 ± 12
Transthyretin (mg/dl)	30 ± 11	33 ± 14	25 ± 10	28 ± 13
RBP (mg/dl)	4.3 ± 1.7	3.8 ± 1.2	3.1 ± 1.5	3.3 ± 2.1
α_1-acid glycoprotein (mg/dl)	137 ± 65	83 ± 48*	203 ± 144	107 ± 37*
CRP (mg/dl)	1.60 (1.20-2.45)	1.26 (0.41-2.55)	3.02 (1.28-5.56)	1.48 (0.55-2.93)
PINI	7 (3-15)	2 (1-5)	20 (16-40)	2 (1-4)

* $P < 0.001$ for significant t test against pretreatment controls
RBP, retinol-binding protein; Retinol/RBP, molar retinol/retinol-binding protein ratio; CRP, C-reactive protein; PINI, Prognostic Inflammatory and Nutritional Index; NS, not significant.

kemia patients. The status for other nutrients in these patients was not affected. Beta-carotene was maintained at the initial concentrations determined prior to therapy.

Obviously, if the anti-oxidant properties of beta-carotene, zinc and vitamin E are taken into account, the disturbances described in central nervous system and bone tumors may suggest a loss of antioxidants in patients undergoing therapy. There may be multifactorial causes for loss of antioxidants in patients undergoing cancer treatment. The tumoricidal action of several anti-cancer drugs and radiation is believed to be mediated by an oxygen free radical dependent mechanism (1). Activated granulocytes appearing during hematopoietic reconstitution may serve as a source of oxygen radicals, in combination with impairment of iron storage and release during chemotherapy (15, 16). Moreover, in view of the decrease in zinc and vitamin E parameters observed in some cancer groups and the maintenance of low levels of beta-carotene in the "all cancer group", the results suggest that intake of nutrients was insufficient. In such situations, the present daily recommended dietary allowance for alpha-tocopherol (RDA: 8 mg, or 10 mg tocopherol equivalents) usually provided by the amounts of vitamin E administered in nutritional therapy procedures, may be too low (17).

Nevertheless, the vitamin and nutritional status of the groups of children with cancer investigated appeared in general to be adequate, even during intensive treatment or diseases imposing prolonged chemotherapy. Moreover, the conditioning treatment was related to improvement in the balance between the occurrence of malnutrition and inflammation, with increase in serum albumin and transthyretin levels, two substances used as markers of nutritional status, in view of the link between hepatic synthesis and substrate availability related to nutritional status (13). The improvement in serum retinol concentration, independent of beta-carotene, suggests characteristics of the vitamin separate from carotene. However, special attention should be paid to other micronutrients, as other fat-soluble vitamins (18), vitamin B12 and folic acid (19) have been found to be decreased during treatment of cancer patients.

In conclusion, supplementation with antioxidants, even at a level higher than that administered in enteral or parenteral (for zinc) nutrition, could be required in patients undergoing highly toxic cancer treatment. Ongoing intervention with nutrients including beta-carotene, alpha-tocopherol and trace minerals, *eg*, zinc, should lead to more effective maintenance of optimal nutrition and retention in children with cancer, according to different patterns of tumor excision, radiation therapy and chemotherapy.

5. REFERENCES

1. P. Sangeetha, U.N. Das, R. Koratkar and P. Suryaprabha, *Free Rad. Biol. Med.* **8**, 15–19 (1990).
2. E. Seifter, J. Mendecki, S. Holtman, J.D. Kanofsky, E. Friedenthal, L. Davis and J. Weinzweig, *Pharmac. Ther.* **39**, 357–365 (1988).
3. D.J.M. Malvy, B. Burtschy, J. Arnaud, D. Sommelet, G. Leverger, L. Dostalova, J. Drucker and O. Amédée-Manesme, *Int. J. Epidemiol.* **22**, 761–771 (1993).
4. WHO, International Classification of Diseases for Oncology (ICD-O), WHO, Geneva, (1980).
5. D.J.M. Malvy, B. Burtschy, L. Dostalova and O. Amédée-Manesme, *Int. J. Epidemiol.* **22**, 137–246 (1993).
6. D.J.M. Malvy, J.D. Povéda, M. Debruyne, B. Montagnon, C. Herbert and O. Amédée-Manesme, *Clin. Chem.* **38**, 394–399 (1992).
7. D.J.M. Malvy, J. Arnaud, B. Burtschy, M.J. Richard, A. Favier, D. Houot and O. Amédée-Manesme, *Int. J. Epidemiol.* **9**, 155–161 (1993).
8. D.J.M. Malvy, J.D. Povéda, M. Debruyne, B. Burtschy, L. Dostalova and O. Amédée-Manesme, *Eur. J. clin. Chem. Clin. Biochem.* **31**, 47–48 (1993).
9. Y. Ingenbleek and Y.A. Carpentier, *Internat. J. Vit. Nutr. Res.* **55**, 91–101 (1985).
10. J. Arnaud, J. Bellanger, F. Bienvenu, P. Chappuis and A. Favier. *Ann. Biol. Clin.* **44**, 77–87 (1986).
11. J. Nève, S. Chamart and L. Molle, in *Trace Element Analytical Chemistry in Medicine and Biology*, P. Bratter and P. Schramel, eds., Walter de Gruyter, Berlin; **4**, pp. 1–10 (1987).
12. SAS user's guide; statistics, Version 7ed. Cary, NC: SAS Institute, Inc., (1990).
13. G. Ollenschlaeger, K. Konkol, P.D. Wickramanayake, M.S. Schrappe-Baecher and J.M. Mueller, *Am. J. Clin. Nutr.* **50**, 454–459 (1989).
14. N. Vaisman, V.A. Stallings, H. Chan, Sh. S. Weitzman, R. Clarke and P.B. Pencharz, *Am. J. Clin. Nutr.* **57**, 679–684 (1993).
15. J.Y. Follézou and M. Bizon, *Neoplasma* **33**, 225–231 (1986).
16. D.V. Godin and S.A. Wohaieb, *Free Rad. Biol. Med.* **5**, 165–176 (1988).
17. A.T. Diplock, *Free Rad. Biol. Med.* **3**, 199–201 (1987).
18. J. Conly, J. Suttie, J. Loftson, K. Ramotar and T. Louie, *Am. J. Clin. Nutr.* **50**, 109–113 (1989).
19. W.H.P. Schreurs, J. Odink, R.J. Egger, M. Wedel and P.F. Bruning, *Internat. J. Vit. Nutr. Res.* **55**, 425–432 (1985).

67

RELATIONSHIPS BETWEEN SERUM COPPER CONCENTRATION AND CARDIOVASCULAR RISK FACTORS IN NORMAL SUBJECTS

Domenico Girelli,[1] Oliviero Olivieri,[1] Antonella Bassi,[2] Margherita Azzini,[1] Simonetta Friso,[1] Carla Russo,[1] Sara Lombardi,[1] and Roberto Corrocher

[1] Institute of Medical Pathology, Chair of Internal Medicine
[2] Clinical Chemistry
University of Verona, Policlinico Borgo Roma, 37134 Verona, Italy

1. INTRODUCTION

High serum copper (s-Cu) has been reported as an independent risk factor for cardio-vascular disease (CVD) in both case-control (1) and large prospective population studies (2, 3). The mechanisms underlying these associations are largely unclear. *In vitro* Cu is highly efficient in promoting the oxidation of low-density-lipoprotein (4), which is considered an important step in atherogenesis. A synergistic effect between the pro-oxidant action of Cu and low status of selenium (an antioxidant), leading to atherogenesis via an imbalance of defence against free radicals has been suggested *in vivo* (5). Also, experimental animal studies suggest a role of Cu in lipid metabolism, particularly in cholesterol (6) and fatty acid metabolism (7). Within the framework of a cross-sectional survey aimed to study the status of several trace elements in a sample of healthy adults (8), we examined the relationships between s-Cu and several, well-known, risk factors for CVD, including plasma lipoproteins and fatty acids, and coagulation factors.

2. SUBJECTS AND METHODS

2.1. Subjects

The study was performed in adult healthy volunteers living in Nove, a village near Vicenza (northern Italy), recruited as previously described in detail (8). Briefly, an initial sex-balanced selection of 500 subjects was obtained by applying the tables of random numbers to the population of Nove (4950 inhabitants) on the electoral register. A further careful selection was performed with the collaboration of the three general practitioners operating in the area. To avoid confounding effects by any underlying pathological process on s-Cu and on the other parameters studied, strict criteria were adopted to define the "healthy" subjects. Subjects recognized to be affected by any chronic diseases (clinically overt atherosclerosis; hyperlipidaemia; hypertension; diabetes; coagulative disorders; liver, neoplastic, renal, endocrinological or immunological diseases) or acute intercurrent illness were therefore excluded. Pregnant women and subjects taking any medication including the contraceptive pill or vitamin preparations were also excluded. 90 subjects (45 M and 45 F) were finally admitted to the study. Smoking was recorded and quantified on an arbitrary scale (0 = non-smokers; 1 = smokers). 75 subjects were non-smokers, and 15 smoked moderately (less than 15 cigarettes/day).

2.2. Biochemical Analysis

Blood samples were collected after overnight fasting, transported to Verona within 1 hour, and processed immediately. For serum copper, selenium and zinc assay, blood was collected in vacutainer tubes for metal and metalloid determinations (Becton Dickinson, Rutherford, NJ), containing no additives. Analysis of plasma FAs, plasma α-tocopherol and retinol and serum selenium were performed as previously described (8). Serum concentrations of Cu and Zn were determined by flame atomic absorption spectrometry using a Perkin Elmer 372 double-beam spectrophotometer equipped with an air-acetylene flame burner and hollow cathode lamp (operated at 15 mA for Zn and at 20 mA for Cu). Atomic absorption was measured at 324.8 nm for Cu and at 213.9 nm for Zn. The spectral bandwidth was 0.7 nm (9). Standard solutions containing 30 µmol/l of Cu or Zn were prepared in double-distilled water. The coefficient of variation for both Cu and Zn was < 5%.

2.3. Statistics

The comparisons between s-Cu values in the different groups (i.e. smokers/non-smokers) were performed using Student's t-test. Pearson's coefficient was used to estimate the strengths of the correlations between s-Cu and the other parameters. The independence of these associations was then assayed by a stepwise multiple-linear-regression analysis. To avoid by-chance significance, only variables at the $p < 0.015$ significance level were included in the final model. The tolerance limit between covariates was 0.01.

3. RESULTS

The main clinical and biochemical data of the subjects are reported in Table 1. S-Cu did not substantially differ between non-smokers and moderate smokers (17.7 ± 3.1 versus 16.5 ± 1.9, respectively; p = NS). S-Cu was significantly higher in women than in man (18.3 ± 3.2 versus 16.8 ± 2.5, respectively; $p < 0.02$). Because of this difference, univariate and multiple regression analyses were performed not only in the whole population but also in sex-specific groups. Univariate analysis showed significant ($p < 0.05$) associations between s-Cu and several parameters (Table 2). The stepwise multiple-linear-regression analysis in the whole population revealed age and fibrinogen as significant and independent predictors of a consistent proportion of s-Cu variability (Adjusted R^2 = 0.5, $p < 0.001$; ß-standardized coefficients: for age = 0.33, $p < 0.001$; for fibrinogen = 0.5, $p < 0.001$). The results of the sex-specific stepwise multiple-linear-regression analysis substantially confirmed the strong association between s-Cu and fibrinogen, whereas age entered the final equation only in men (data not shown).

Table 1. Summary of the variables recorded in all subjects (means ± SD)

Serum copper (μmol/l)	17.5 ± 3.0
Age (years)	56 ± 18
Body mass index (kg/m2)	24.6 ± 3.0
Systolic blood pressure (mmHg)	135 ± 15
Diastolic blood pressure (mmHg)	81.8 ± 6.4
Total serum cholesterol (mmol/l)	5.5 ± 1.0
LDL cholesterol (mmol/l)	3.5 ± 0.9
HDL cholesterol (mmol/l)	1.5 ± 0.4
Serum triglycerides (mmol/l)	1.3 ± 0.5
Fibrinogen (mg/dl)	341 ± 81
Factor VIIc (% of standard)	110 ± 35
Plasma vitamin A (μmol/l)	3.1 ± 0.9
Plasma vitamin E (μmol/l)	40.6 ± 9.0
Serum selenium (μmol/l)	1.1 ± 0.2
Serum zinc (μmol/l)	14.4 ± 2.0
Major plasma Fatty Acids (% by wt)	
C 16:0 (palmitic)	23.8 ± 2.0
C 18:0 (stearic)	8.7 ± 1.2
Σ SFA	35.3 ± 2.5
C 18:1 (oleic)	23.7 ± 3.8
Σ MUFA	23.9 ± 3.8
C 18:2 (linoleic)	27.2 ± 4.4
C 20:4 (arachidonic)	9.6 ± 1.7
Σ ω-6 PUFA	37.1± 4.7
C 22:6 (docosahexaenoic)	2.1 ± 0.6
Σ ω-3 PUFA	3.4 ± 0.7
Σ total PUFA	40.5 ± 4.9

4. DISCUSSION

This study investigated the relationship between s-Cu and several biochemical parameters thought to be important in determining the cardiovascular risk. Most of the latters, i.e. fatty acids (10), antioxidant vitamins (11) and selenium (5), have never been previously evaluated for their association with s-Cu in a stepwise multiple regression analysis, whereas a relationship with Cu was conceivable on the basis of experimental data. For example, Cu is involved in fatty acid metabolism, being the essential cofactor of the Δ-9 desaturase which converts stearic acid to oleic acid (7). Also, Cu is able to chemically inactivate selenium (12), and has been reported to have a synergistic effect with low serum selenium in promoting atherogenesis by favouring the oxidation of low-density lipoproteins (5). Whereas the simple correlation analysis revealed associations of potential interest between s-Cu and some of the above mentioned parameters (e.g. the inverse relationships with selenium and vitamin A), they disappeared at the multivariate analysis, which clearly allowed the identification of two covariates strongly and independently associated with s-Cu: age and plasma fibrinogen. The direct relationship between age and both s-Cu is consistent with earlier reports (13). The effect of age was greater in men than in women, as confirmed by the results of the sex-specific multivariate analysis. The sex-related difference in s-Cu is also a well-known phenomenon due to the effect of female hormones on Cu metabolism (13, 14). S-Cu tend to increase during the use of oral contraceptives and during pregnancy (14). The relatively high number

Table 2. Univariate analysis: correlations between serum
copper and some of the variables studied

		p
Age	0.56	.001
Systolic blood pressure	0.37	.01
Diastolic blood pressure	0.27	.001
Total serum cholesterol	0.35	.001
LDL cholesterol	0.27	.05
HDL cholesterol	0.24	.05
Fibrinogen	0.65	.001
Factor VIIc	0.33	.005
Plasma vitamin E	0.25	.05
Serum selenium	- 0.25	.05
Plasma vitamin A	- 0.34	.001

of post-menopausal women (25/45) in our study probably accounts for the attenuation of the effect of age on s-Cu in the female group.

To the best of our knowledge, the strong relationship between s-Cu and fibrinogen in healthy subjects, which was evident in both sexes, is a new observation. In recent years, fibrinogen level has emerged as a major independent risk factor for CVD, with a predictive power even stronger than that of cholesterol (15, 16). It is worthy of note that in all the prospective studies claiming s-Cu as an *independent* risk factor for CVD, the effect of s-Cu was adjusted for several well-known risk factors (plasma lipids, smoking, etc.), but none of these studies included the evaluation of fibrinogen and/or other haemostatic factors (e.g. factor VII) as potentially confounding covariates. Our descriptive data do not permit to draw conclusion about the mechanism underlying the strong association between s-Cu and fibrinogen in the population examined. A pathophysiological link between the two parameters may lie on their common relationship with the inflammation process. Fibrinogen acts as an acute-phase protein, being its hepatic biosynthesis stimulated from monocyte-derived cytokines (15). Since atherosclerosis is increasingly thought to be a chronic inflammatory disease (17), it is debated whether fibrinogen is a true causal risk *factor* by altering haemorheologic and haemostatic balance, or it represent a risk *marker* for early atherosclerosis (16). S-Cu is also increased in inflammatory conditions, mainly because caeruloplasmin, which binds the vast majority of S-Cu, is also an acute-phase protein (18). We cannot exclude that the presence of a subclinical inflammatory status, such as early atherosclerosis, could explain the association between serum copper and fibrinogen. However, the restrictive criteria adopted in this study for the selection of "healthy" subjects tend to argue against this hypothesis.

On the other hand, there is increasing evidence of a link between Cu and some factors involved in the coagulation cascade. It has been reported that the coagulation factors V and VIII share a large sequence homology with ceruloplasmin, indicating a common ancestral precursor gene (19) and that factor V is capable of tightly binding Cu (20); Cu nutritional status has been shown to influence the activity of both factor V and factor VIII (21) and of the fibrinolytic system (22). Moreover, in vitro experiments have suggested that Cu may catalyze the formation of clot-like material from fibrinogen (23) and is able to inactivate plasmin (24), both actions probably being mediated by a free-radical mechanism. Finally, in our study we found a direct relationship (although only in the univariate analysis) between s-Cu and factor VII. Differently from fibrinogen, this key-factor of the coagulation cascade does not act as an acute-phase reactant. Despite these suggestion, mostly from in vitro and/or animal studies, the role of Cu in the coagulation system is at present only putative. Whatever is the mechanism involved in the association between s-Cu and fibrinogen, our data suggest that the role of s-Cu as an *independent* risk factor for CVD should be reconsidered. Further studies addressed to clarify the relationship between Cu and coagulation are suggested.

5. REFERENCES

1. M.M. Singh, R. Singh, A. Khare, et al., *Angiology* **36**, 504–510 (1985).
2. F.J. Kok, C.M. Van Duijn, A. Hofman, et al., *Am. J. Epidemiol.* **128**, 352–359 (1988).
3. J.T. Salonen, R. Salonen, H. Korpela, S. Suntioinen and J. Tuomiletho, Am. J. Epidemiol. **134**, 268–276 (1991).
4. H. Esterbauer, J. Gebicki, H. Puhl and J. Jürgens, *Free Rad. Biol. Med.* **13**, 341–390 (1992).
5. J.T. Salonen, R. Salonen, K. Seppänen, M. Kantola, S. Suntioinen and H. Korpela, *Br. Med. J.* **302**, 756–760 (1991).
6. N.Y. Yount, D.J. McNamara, A.A. Al-Othman and K.Y. Lei, *J. Nutr. Biochem.* **1**, 27–33 (1990).
7. S.C. Cunnane, *Prog. Lipid. Res.* **21**, 73–90 (1982).
8. O. Olivieri, A.M. Stanzial, D. Girelli, et al., *Am. J. Clin. Nutr.* **60**, 510–517 (1994).
9. B. Welz, *Atomic absorption spectrometry*, 2nd ed., VCH Publishers, Weinheim, (1985).
10. D.F. Horrobin, *Semin. Thromb. Haemostasis*, **19**, 129–137 (1993).
11. D. Steinberg, *N. Engl. J. Med.* **328**, 1487–1489 (1993).
12. R.J. Shamberger, in Biochemistry of the essential ultratrace elements, E. Frieden, ed., Plenum Press, New York pp. 201–237 (1984).
13. P.E. Johnson and D.B. Milne, *Am. J. Clin. Nutr.* **56**, 917–925 (1992).
14. N.W. Solomons, *Am. J. Clin. Nutr.* **32**, 856–871 (1979).
15. N.S. Cook and D. Ubben, *Trends Pharmacol. Sci.* **11**, 444–451 (1990).
16. E. Ernst and K.L. Resch, *Ann. Intern. Med.* **118**, 956–963 (1993).
17. R. Rosse, *Nature* **362**, 801–809 (1993).
18. R.A. Di Silvestro, *Nutr. Res.* **10**, 355–358 (1990).
19. W.R. Church, R.L. Jernigan, J. Toole, et al., *Proc. Natl. Acad. Sci. USA* **81**, 6934–6937 (1984).
20. K.G. Mann, C.M. Lawler, G.A. Vehar, W.R. Church, *J. Biol. Chem.* **21**, 12949–12951 (1984).
21. S.M. Lynch, L.M. Klevay, *J. Nutr. Biochem.* **3**, 387–391 (1992).
22. S.M. Lynch, L.M. Klevay, *Nutr. Res.* **13**, 913–922 (1993).
23. G. Marx, M. Chevion, *Thromb. Res.* **40**, 11–18 (1985).
24. S.E. Lind, J.R. McDonagh and C.J. Smith, *Blood* **82**, 1522–1531 (1993).
25. T.W. Meade, S. Mellows, M. Brozovic, et al., *Lancet* **ii**, 533–537 (1986).

68

TRACE ELEMENT STATUS IN CUBA

Relationships with Epidemic Neuropathy through SECUBA Protocol

T. Verdura,[1] J. Arnaud,[2] R. Perez Cristia,[3] J. C. Tressol,[4] P. Fleites,[3] M. Chassagne,[5] M. J. Richard,[2] J. C. Renversez,[6] A. Favier,[2] and J. Barnouin[5]

[1] Instituto Finlay
Ave 27, n°19805, La Lisa, La Habana, Cuba
[2] Laboratoire de Biochimie C, CHUG
BP 217, 38043 Grenoble Cedex 9, France
[3] Centro Nacional de Toxicologia
Ave 31 y calle 114, La Habana, Cuba
[4] Unité Maladies Métaboliques et Micronutriments, INRA
63122 Saint Genès Champanelle, France
[5] Laboratoire d'Ecopathologie, INRA
63122 Saint Genès Champanelle, France
[6] Laboratoire de Biochimie A, CHUG
BP 217, 38043 Grenoble Cedex 9, France

1. INTRODUCTION

From 1991 to 1993, 50,862 cases (461.4 per 100,000 inhabitants) of optic and peripheral neuropathy (NEE) of unknown etiology were reported in Cuba (1–4). First cases and highest incidence were observed in Pinar del Rio province. Incidence rates peaked in March and April 1993 when there were 3000 to 4000 new cases per week. There were no fatal cases.

Many potential causal factors have been investigated by 55 international institutions. They were related to sudden changes in food and fuel shortages and subsequent increased physical exercice. Most of the studies were inconclusive but several hypothesis have been proposed. The evidence first suggested that vitamins were centrally important. Most patients (99.9%) improved significantly with B-group, A and E vitamin treatment. The number of new cases decreased after vitamin supplementation to the entire Cuban population (2.5 mg thiamin, 1.6 mg riboflavin, 20 mg niacin, 2 mg vitamin B6, 6 µg B12, 250 µg folate and 2500 IU retinol) initiated in March - April 1993 (3). Vitamin (B1, niacin, folate, B12, A) levels were found to be low for both patients and controls. However, vitamin status was similar in patients and controls (1). Although no clinical and biological evidence of protein-calorie malnutrition (3), the dramatic fall of protein, fat and micronutrient intakes combined with excessive sugar and alcohol consumptions appeared to be the primary risk factor for NEE. The number of cases was reduced in subjects who received additionnal food [children < 15 y old, pregnant women and elderly > 65 y old]. Nutritional supplementation programs for these populations have been maintained despite severe food rationing (3). Therefore, the greatest incidence of NEE was observed in unprotected adults (1,3,4). Moreover, serum lycopene levels were found to be dramaticaly decreased in patients *vs* controls (4). Blood selenium, alpha and beta carotenes were also decreased in patients but to a lesser extent (4). These results suggested a contribution of the impairment of protective antioxydant pathways in NEE. However, high levels of TBARS were reported both in patients and controls (2). The increase in physical exercice associated with hot weather and limitations in food availability were also involved in large reported weight losses among the population (3). Larger weight losses were observed in patients *vs* controls (2). Finally, there were no evidence of infectious diseases (3), or genetic defects (1). No particular chemical toxic to nerve was identified (1,3). However, tobacco use was identified as the main risk factor in all studies (1,3,4).

The presented results are part of a large multidisciplinary prospective study: Seguridad Alimentaria y Buena Alimentacion in Cuba (SECUBA). The aims of this study were to evaluate incidence of previous hypothetic toxic, nutritional and seasonal risk factors for health status (ecopathological methodology) in 200 middle-aged HIV-seronegative Cuban men. Informations were obtained about social and demographic characteristics, occupation, toxic exposure, tobacco use and physical exercice. Participants were also asked to report their weekly consumption of foods and beverages, four times a year (April 1995, July 1995, October 1995 and February 1996). Clinical examination, anthropometric measurements, blood and urine collections were performed the week of dietary questionnaire. Indices of trace elements, vitamins, lipids, proteins, enzymes and amino acids status were determined. The number of new NEE cases had increased during this study and 51 of them were enrolled in October 1995.

The present paper is focussed on the influence of tobacco use on trace element status in healthy subjects and on trace element differences between controls and NEE patients studied within the same period (October 1995).

2. MATERIAL AND METHODS

One hundred and sixty three healthy middle-aged men (30–50 y) living in Havana were enrolled. Eighty three were smokers (S) and 80 were non smokers (NS). Among the smokers, 35 smoked from 1 to 19 cigarettes/d, 27 smoked 20 cigarettes/d and 16 smoked more than 20 cigarettes/d. Few of them received vitamin supplement. Fifty one new cases of optic neuropathy, living in Pinar del Rio province, were also enrolled. The patients were matched to the controls for sex and age. All subjects were HIV-seronegative. The protocol was approuved by the Cuban Ministry of Public Health and all subjects gave written informed consent.

Serum trace elements (Cu, Zn, Se) were measured by atomic absorption spectrometry using Seronorm[R] Trace Element (Nycomed) as internal quality control. Serum iron was determined by colorimetry using Ferrozine[R] as complexant and Biotrol sera as internal quality control. Serum transferrin (TRF) was determined by immunoprecipation using Behring nephelometer and Ciba-Corning serum as internal quality control. Serum ferritin was measured by enzyme linked immunosorbent assay using IFCC liver ferritin and Ramco and Probioqual sera as internal quality control. Erythrocyte Se-glutathione peroxidase (GPX) was determined by a modified method of Gunzler (5) using pool of human erythrocytes as internal quality control. Erythrocyte Cu-Zn superoxide dismutase (SOD) was measured by the method of Marklund and Marklund (6) using pool of human erythrocytes and bovine SOD solution as internal quality controls. Serum thiobarbituric acid reactants (TBARS) analysis was performed by fluorometry (7). Differences between groups were assessed using Kruskal-Wallis or Anova tests after normalization by logarithmic transformations.

3. RESULTS AND DISCUSSION

Main results are presented in Table 1. In smokers, the increase of serum copper did not vary with the amount smoked. Ferritin was negatively correlated to the number of cigarettes smoked per day (r = -0.252, p < 0.005). The increase of copper in smokers is in agreement with previous papers (8–9). The high levels of plasma copper in smokers have been related to the delivery of copper to the body by tobacco and cigarette paper (9). Our results did not corroborate this statement. The increase of copper could also be related to hormonal changes induced by smoking (9). Decrease of ferritin was significant in subjects who smoked 20 or more cigarettes per day (p < 0.05). These results are not in agreement with those of Strain et al. (10) who reported an increase in serum ferritin and iron in English male smokers. In Western countries, the decrease of selenium and zinc status (11–14) in smokers has been explained by Cd-Se or Cd-Zn interactions (11,15) and the lower dietary intakes and density of smokers (13,16). Nevertheless, differences between smokers and non smokers in trace element intakes strongly depend on social classes (16). Therefore, absence of tobacco influence on zinc and selenium status in Cuba could be a consequence of low differences in income among social classes, limited choice of food items and general food shortage. In pregnant women, Kuhnert et al. (15) reported similar plasma zinc levels in smokers and non smokers with similar zinc intakes. In Bangladesh, copper was found decreased and zinc increased in smokers *vs* nonsmokers (17). Finally, tobacco had no influence on selenium status in Slovak republic (18).

In contrast to severe carotenoid deficiencies (results not shown), there was no evidence of trace element deficiencies in the studied Cuban population and serum nutritional protein were in the reference range (result not shown). Therefore, the available foods could probably

Table 1. Trace element status, expressed as mean ± SD or median (range), in healthy controls and NEE patients (optic form)

		Controls		
	All	(non-smokers)	Smokers	Patients (all)
n	163	80	83	51
Cu, µmol/l[1,2]	16.5 ± 2.8	16.0 ± 2.5	17.0 ± 3.0	17.8 ± 3.4
SOD, U/mg Hb	1.22 ± 0.11	1.21 ± 0.09	1.23 ± 0.12	1.19 ± 0.10
Zn, µmol/l	13.8 ± 2.4	13.9 ± 2.6	13.7 ± 2.2	13.7 ± 1.8
Se, µmol/l[3]	1.20 ± 0.15	1.21 ± 0.14	1.19 ± 0.15	1.16 ± 0.22
GPX, U/g Hb[4]	41.8 ± 9.3	41.4 ± 8.3	41.9 ± 9.9	38.2 ± 9.5
Fe, mg/l	0.96 ± 0.40	0.97 ± 0.33	0.95 ± 0.45	0.92 ± 0.40
TRF, g/l	2.97 ± 0.52	2.99 ± 0.49	2.98 ± 0.55	3.04 ± 0.56
Ferritin, µg/l[5]	52 (5-730)	56 (5-730)	50 (5-480)	80 (12-340)
TBARS, µmol/l	3.29 ± 0.47	3.30 ± 0.46	3.29 ± 0.48	3.23 ± 0.48

[1] Serum copper increases in control smokers *vs* control non smokers, p < 0.05, [2] Serum copper increases in patients *vs* controls, p < 0.005, [3] Serum selenium decreases in patients *vs* controls, p < 0.05, [4] Erythrocyte GPX decreases in patients *vs* controls, p < 0.05, [5] Serum ferritin increases in patients *vs* controls, p < 0.01.

have high trace element densities. Although high TBARS levels in serum, trace element status remained in the reference ranges. Serum cholesterol levels were also in the reference range (1.70 ± 0.37 g/l in controls and 1.60 ± 0.32 g/l in patients). High TBARS levels in Cubans are not explained by smoking habits in contrast to previous works (9,14,19,20). Dramatic decrease of carotenoids in serum (results not shown) could explain the increase of lipid peroxidation (21,22).

Compared to 1993, ranges of selenium, ferritin and zinc values are similar, whereas TBARS levels are higher (2,4). As in 1993, serum selenium decreases in patients compared to controls and there is no difference in serum zinc and TBARS levels between patients and controls. In the present study, additional parameters were measured. GPX was significantly lower and serum copper and ferritin significantly higher in patients *vs* controls. The present results, as well as those of carotenoids confirm previous observations of the impairment of protective antioxidant pathways in NEE development in Cuba (4).

ACKNOWLEDGMENTS

We thank Nestec-Nestlé, Trace Element Institute for UNESCO, Merck-Biotrol Diagnostics and Pharmaciens Sans Frontières for their help.

4. REFERENCES

1. K. Tucker and R.R. Hedges, *Nutr.Rev.* **51**, 349–357 (1993)
2. R. Perez-Cristia and P. Fleites-Mestre, in *Taller Internacional sobre la Neuropatia Epidemica ocurrida en Cuba,* La Habana (12–15 Julio 1994*).*
3. G.C. Roman, *J.Neurol. Sci.* **127**, 11–28 (1994).
4. The Cuba Neuropathy Field Investigation Team, *N. Engl. J. Med.* **333**, 1176–1182 (1995).
5. W.A. Gunzler, H. Kremers and L. Flohe, *Z. Klin. Chem. Klin. Biochem.* **12**, 444–448 (1974).
6. S. Marklund and G. Marklund, *Eur. J. Biochem* **47**, 469–474 (1974).
7. M.J. Richard, C. Portal, J. Meo, C. Coudray, A. Hadjian and A. Favier, *Clin Chem.* **38**, 704–709 (1992).
8. G.N. Davidoff, M.L. Votaw, W.W. Coon, D.E. Hultquist, B.J. Filter and S.A. Wexler, *Am. J. Clin. Pathol.* **70**, 790–792 (1978).
9. D. Lapenna, A. Mezzetti, S. de Gioia, S.D. Pierdomenico, F. Daniele and F. Cuccurullo, *Free Radical Biol. Med.* **19**, 849–852 (1995).
10. J.J. Strain, K.A. Thompson, M.E. Barker and P.G. McKenna, *Trace Elem. Med.* **7**, 25–27 (1990).
11. B. Lloyd, R.S. Lloyd and B.E. Clayton, *J. Epidemiol. Commun. Health* **37**, 213–217 (1983).
12. P.A. McAdam, D.K. Smith, E.B. Feldman and C. Hames, *Biol. Trace Elem. Res.* **6**, 3–9 (1984).
13. C.A. Swanson, M.P. Longnecker, C. Veillon, M. Howe, O.A. Levander, P.R. Taylor, P.A. McAdam, C.C. Brown, M.J. Stampfer and W.C. Willett, *Am. J. Clin. Nutr.* **52**, 858–862 (1990).
14. K.B. Schwarz, J. Cox, S. Sharma, F. Witter, L. Clement, S.S. Sehnert and T.H. Risby, *J. Nutr. Environ. Med.* **5**, 225–234 (1995).
15. P.M. Kuhnert, B.R. Kuhnert, P. Erhard, W.T. Brashear, S.L. Groh-Wargo, and S. Webster, *Am. J. Obstet. Gynecol.* **157**, 1241–1246 (1987).
16. F.M. Haste, O.G. Brooke, H.R. Anderson, J.M. Bland, A. Shaw, J. Griffin and J.L. Peacock, *Am. J. Clin. Nutr.* **51**, 29–36 (1990).
17. O. Faruque, M. R. Khan, M. Rahman and F. Ahmed, *Br. J. Nutr.* **73**, 625–632 (1995).
18. A. Brtkova, T. Magalova, K. Babinska, A. Bederova, *Biol. Trace Elem. Res.* **46**, 163–171 (1994).
19. E. Hoshino, R. Shariff, A. Van Gossum, J.P. Allard, C. Pichard, R. Kurian and K.N. Jeejeebhoy, *J. Parent. Ent. Nutr.* **14**, 300–305 (1990).
20. B. Frei, T.M. Forte, B.N. Ames, C.E. Cross, *Biochem. J.* **277**, 133–138 (1991).
21. P. Di Mascio, S. Kaiser and H. Sies, *Arch. Biochem. Biophys.* **274**, 532–538 (1989).
22. W. Stahl, H. Sies and A.R. Sundquist, in *Vitamin A in Health and disease,* R. Blomhoff, ed., Marcel Dekker, New-York, pp. 275–287 (1994).

69

THE EFFECT OF NUTRITIONAL SUPPLEMENTATION ON STROKE MORTALITY AND BLOOD PRESSURE

Results from the Linxian Nutritional Intervention Trials

Steven D. Mark, Wen Wang, Joseph F. Fraumeni, Jr., Jun-Yao Li,
Philip R. Taylor, Guo-Qing Wang, Wande Guo, Sanford M. Dawsey,
Bing Li, and William J. Blot

National Cancer Institute
6130 Executive Blvd.
Bethesda, Maryland 20892-7368

1. INTRODUCTION

Stroke disease is the second leading cause of death in Linxian, a county in north central China with one of the world's highest rates of esophageal/gastric cardia cancer (1). The determinants of stroke in Linxian and other parts of China are not well known, but dietary factors, including low vitamin and mineral intake, may be involved. In the mid-1980's two randomized trials were launched to test whether vitamin and mineral supplementation in this nutritionally deficient population would be effective in lowering cancer risks. Information on all deaths, including those from stroke, was ascertained. Measurements of blood pressure were taken prior to and at the end of the interventions. We have previously examined the effects of the interventions on stroke and hypertension (2, 3). In this paper we analyze the data from these trials in a uniform manner and evaluate the influence of the interventions and other pre-treatment factors such as sex and alcohol intake on stroke mortality and on hypertension.

2. MATERIALS AND METHODS

The design, methods of conduct, and primary endpoint analyses of the trials have been described in detail elsewhere (4–7). In the general population trial, 29584 adults aged 40–60 from four Linxian communes were randomly assigned to receive one of eight vitamin-mineral combinations. The groups were defined by the following combinations of four different

Therapeutic Uses of Trace Elements, edited by Nève et al.
Plenum Press, New York, 1996

Table 1. Nutrient combinations and doses used in the general population and dysplasia[1] trials

General population trial	Dose	Dysplasia trial	Dose
Factor A			
Vitamin A (palmitate; IU)	5000	Vitamin A (acetate; IU)	10000
Zinc (oxide; mg)	22.5	Zinc (sulfate; mg)	45
Factor B			
Riboflavin (mg)	3.2	Riboflavin (mg)	5.2
Niacin (mg)	40	Niacin (mg)	40
Factor C			
Ascorbic acid (mg)	120	Ascorbic acid (mg)	180
Molybdenum (yeast complex; μg)	30	Molybdenum (sodium molybdate; μg)	30
Factor D			
β-Carotene (mg)	15	β-Carotene (acetate; mg)	15
Selenium (selenium yeast; μg)	50	Selenium (sodium selenate; μg)	50
α-Tocopherol (mg)	30	α-Tocopherol (mg)	60

[1]See Table 2 for additional vitamins and minerals contained in the supplementation pill used for the dysplasia trial.

vitamin mineral combinations which we designate as factors A, B, C, D (Table 1): AB, AC, AD, BC, BD, CD, ABCD, or placebo. This choice of groups resulted in half the participants receiving each of the four nutrient combinations. The subjects that received and did not receive a given factor were balanced with respect to all the other nutrients. Supplementation began in March 1986 and continued through April 1991. In the dysplasia trial, 3318 adults aged 40–69 from three of these communes were cytologically diagnosed before the trial with esophageal dysplasia, and then randomly assigned to receive daily supplementation with 26 vitamins and minerals (Table 1, Table 2) or look-alike placebos during the period May 1985–April 1991. In both trials the supplements were delivered monthly, with compliance assessed

Table 2. Additional vitamins and minerals in the dysplasia trial[1] supplement

Nutrient	Doses used
Folic acid (μg)	800
Vitamin B-1 (thiamine mononitrate; mg)	5
Vitamin B-6 (pyridoxine HCl; mg)	6
Vitamin B-12 (cyanocobalamin; μg)	18
Vitamin D (IU)	800
Biotin (μg)	90
Pantothenic acid (calcium pantothenate; mg)	20
Calcium (dibasic calcium phosphate; mg)	324
Phosphorus (dibasic calcium phosphate; mg)	250
Iodine (potassium iodide; μg)	300
Iron (ferrous fumarate; mg)	54
Magnesium (magnesium oxide; mg)	200
Copper (cupric oxide; mg)	6
Manganese (manganese sulfate; mg)	15
Potassium (potassium chloride; mg)	15.4
Chloride (potassium chloride; mg)	14
Chromium (chromium chloride; μg)	30

[1]See Table 1 for additional vitamins and minerals contained in the supplementation pill used for the dysplasia trial.

by counting unused pills and by assaying nutrient levels in blood collected from random samples of participants every 3 months.

All deaths and all incident cancers in trial participants were ascertained through routine and special follow-up that ensured essentially complete reporting . The cause of non-cancer deaths, one category of which was stroke, was determined by local physicians and reviewed by experienced senior Chinese clinicians involved in this study. At baseline (prior to randomization) in 1984 (dysplasia trial) and 1985 (the general population trial) and again in the spring of 1991, questionnaires were administered and brief physical examinations were conducted. The baseline questionnaires included information on age, sex, cigarette smoking, and alcohol drinking. Physical examinations included standard mercury sphygmomanometer measurement of systolic and diastolic blood pressure, and measurement of weight and height. Approximately 80% of the trial participants still alive had end-of-trial blood pressure measurements. The ascertainment of blood pressure did not differ by treatment group. At the end of the trial participants were classified as being either "healthy" or in one of the following categories: 1) dead from stroke; 2) dead from other causes; 3) alive with cancer; 4) alive, cancer free, with systolic (systolic pressure $\geq$ 160) but not diastolic hypertension; 5) alive, cancer free, with diastolic (diastolic pressure $\geq$ 95) but not systolic hypertension; 6) alive, cancer free, with both systolic and diastolic hypertension. The cutoff values for systolic and diastolic pressure follow WHO criteria (8).

Relative risks (RR) and corresponding 95 percent confidence intervals (CI) for stroke mortality were calculated for the effect of treatment and for baseline risk factors using proportional hazards models (9). Since adjustment for baseline risk factors resulted in essentially no change in the estimates of the effect of the treatments on overall mortality or stroke mortality, the unadjusted estimates are presented. In the general population trial we present relative risk estimates both for the individual treatment groups (models contain indicator variables for each of the 7 groups that received vitamin/mineral supplementation) and for the four factors (models contain indicator variables for each factor). The analysis by factor imposes the assumption that the effects of the factors are additive on a log scale, whereas the group analysis makes no assumptions regarding factor interaction. The effect of the treatment group on 1991 systolic and diastolic blood pressure was estimated by linear regression. Polytomous logistic regression (10) was used to compare the odds in the treated versus the nontreated group for each of the six possible disease categories.

3. RESULTS

Characteristics of the participants at the start of the trial are shown in table 3. Both studies had a slight predominance of females. The median age of dysplasia trial participants (53 female, 55 male) was 2 years higher than the age of their general populations counterparts. About two-thirds of the men, but few women, in either trial smoked cigarettes (i.e., lifetime use six months or more). Around 10% of the females reported ever drinking in the past year, compared to slightly more than two-fifths of the men. Drinking was typically infrequent and in small amounts: in both studies over 90% of persons classified as drinkers reported drinking only a few time a year, with most of the remaining reporting drinking a few times per month. The distributions of systolic blood pressure, diastolic blood pressure, and body mass index (bmi=weight/height^2) were similar for participants in both the trials.

Dietary surveys conducted in this area have indicated that as many as 90 percent of adults have less than two-thirds the Chinese or U.S. recommended intake of vitamin A, riboflavin, and calcium (11). Plasma levels of vitamin A, many of the carotenoids, vitamin E, vi-

Table 3. Baseline characteristics by sex of participants in the dysplasia
and general population trials

	Dysplasia trial	General population trial
Female	56%	55%
Drink	8%	10%
Smoke	0.65%	0.20%
	median (interquartile range)	
Age	53 (47-49)	51 (44-58)
Sbp	130 (120-150)	130 (110-140)
Dbp	80 (70-90)	80 (70-90)
Bmi	20.2 (18.7-22.0)	21.9 (20.3-23.7)
Male	44%	45%
Drink	33%	40%
Smoke	65%	67%
	median (interquartile range)	
Age	55 (48-60)	53 (45-60)
Sbp	130 (110-140)	120 (110-140)
Dbp	80 (70-85)	80 (70-90)
Bmi	20.1 (19.0-21.3)	21.6 (20.9-22.9)

Sbp = systolic blood pressure; Dbp = diastolic blood pressure; Bmi = body mass index

tamin C, riboflavin, and zinc have also been found to be low compared to levels in for the
U.S. population (4, 12, 13). There were no appreciable differences in any of the serologic nutrient levels examine between subjects in the supplemented vs. placebo groups (5, 6).

As might be expected by virtue of their selection, the dysplasia trial participants had
higher cancer death rates and higher overall death rates than the general population trial participants (Table 4). Death rates for stroke, the second leading cause of death after cancer (5,
6) were similar in both trials (dysplasia trial; 3.0 per 1000 person years; general population
trial; 3.5 per 1000 person years).

In the dysplasia trial the group receiving supplements had an overall reduction in mortality of 7% (RR=0.93, 95% CI=0.75–1.16), with similar effects in males and females (Table
5). The intervention had a greater effect on stroke deaths (RR=0.63, 95% CI=0.75–1.16).
This was more pronounced on males (RR=.42, 95% CI=0.19–0.93) than on females (2)
(RR=0.93, 95% CI=0.44–1.98).

In the general population trial the effect on overall mortality (Table 6), was greatest for
factor D, with a risk reduction of 9% (RR=.91, 95% CI=0.84–0.99). There was no sex difference in treatment effect by factor (14) or group (data not shown). Table 7 shows the relative
risks of stroke by factor and group. Since estimates of factor effect, which make assumptions

Table 4. Cause-specific mortality rates in the dysplasia and general population trials

	Dysplasia trial (19,230 person-yrs)		General population trial (148,773 person yrs)	
	Number	Rate/1000 person-yrs	Number	Rate/1000 person-yrs
Total deaths	324	16.8	2127	14.2
Stroke deaths	57	3.0	523	3.5
Cancer deaths	176	9.2	792	5.3

Table 5. Dysplasia trial: relative risks of all deaths and stroke deaths

	Relative risk (95% CI)
All deaths	0.93 (0.75, 1.16)
Female	0.98 (0.69, 1.39)
Male	0.90 (0.68, 1.20)
Stroke deaths	0.63 (0.37, 1.07)
Female	0.93 (0.44, 1.98)
Male	0.42 (0.19, 0.93)

regarding the interactions of the factors, may not provide an adequate summary of the under-lying experience of the actual groups, we focus on the group effect estimates. In all 7 treatment groups there was a moderate benefit of the supplementation, with the largest reduction of 19% in stroke mortality occurring in the group taking factors A and D (RR=0.71, 95% CI=.50–1.00). The effect of sex on the variation by group or factor was small (data not shown).

In both the dysplasia trial (RR=1.5, 95% CI=0.87–2.4) and the general population trial (RR=1.7, 95% CI=1.32–2.12), males had increased stroke mortality compared to females (relative risks adjusted for smoking and drinking.) Table 8 lists the relative risks of stroke mortality according to other pre-treatment variables. The estimates in each trial were similar for all risk factors. The estimate of the relative risk associated with drinking remains largely unchanged when simultaneously adjusted for the other pre-treatment variables (dysplasia trial RR=0.56, 95% CI=0.24–1.27; general population trial RR=0.77, 95% CI=0.61–0.98). Adjustment increases the estimate of smoking effect in the general population trial (RR=1.09, 95% CI=0.85–1.40) and the estimate of BMI effect in the dysplasia trial (RR=1.05, 95% CI=0.59–1.85)

Among surviving participants in the dysplasia trial, males had an average 1991 diastolic pressure 2.02 mm HG lower in the treated than the control group (p=.009) and an average systolic pressure 1.33 mm HG lower (p=0.32) (2). For females, the decrease in blood pressure for the supplemented group was smaller with the average diastolic pressure 0.41 mm Hg lower and the average systolic pressure .74 lower (p=.53) (2). In the general population trial the average changes in blood pressure by group or factor were small and not statistically significant for either sex (3).

To simultaneously test and estimate the effect of the interventions on stroke mortality, other causes of mortality, hypertension, and non-fatal cancers, we created mutually exclusive and exhaustive outcome categories. With the trial's endpoint defined as the frequency of

Table 6. General population study: relative risks (95% CI) of all deaths according to factor and group

Factor			
A	B	C	D
1.00	0.98	1.09	0.91
(0.92, 1.09)	(0.90, 1.06)	(0.92, 1.10)	(0.84, 0.99)

Group						
CD	AC	BD	AB	BC	AD	ABCD
0.88	1.05	0.93	0.94	0.95	0.89	0.91
(0.74, 1.05)	(0.89, 1.24)	(0.79, 1.11)	(0.79, 1.11)	(0.81, 1.13)	(0.75, 1.05)	(0.76, 1.07)

Table 7. General population study: relative risks (95% CI) of stroke deaths according to factor and group

Factor

A	B	C	D
0.99	0.94	1.04	0.91
(0.84, 1.18)	(0.79, 1.11)	(0.88, 1.24)	(0.76, 1.07)

Group

CD	AC	BD	AB	BC	AD	ABCD
0.86	0.91	0.75	0.85	0.78	0.71	0.88
(0.62, 1.20)	(0.66, 1.27)	(0.53, 1.05)	(0.61, 1.18)	(0.55, 1.09)	(0.50, 1.00)	(0.64, 1.22)

counts in these categories, the dysplasia trial had, a statistically significant overall effect (p=0.05). This benefit was mainly among the males (2) who experienced half the risk of having both elevated systolic and diastolic pressures at the end of the study (RR=0.43, 95% CI=0.28–0.65) and a little more that one-third the risk of dying from stroke (RR=0.36, 95%=CI 0.16–0.83). In females the outcomes were not significantly different in the supplement and placebo groups (2). In the general population trial none of the treatment groups showed significant overall effects. Analyzing by factors, only factor D (p=.03) was significant (3). This was largely due to a reduction in non-stroke mortality (RR=0.90, 95% CI=0.81–1.00), stroke mortality (RR=0.90, 95% CI=0.74–1.08), and an increased prevalence of isolated diastolic hypertension (RR=1.23, 95% CI=1.06–1.43).

4. DISCUSSION

In both the general population and dysplasia trials in Linxian, every group that received any active vitamin and/or mineral supplementation had lower rates of stroke mortality than the corresponding placebo group. The mortality reduction was more marked in the dysplasia trial, especially in males, who experienced nearly a 60% reduction in stroke rates. The overall reductions were modest for the general population groups, with only the AD group achieving a level of reduction that approached statistical significance. There was no evidence that the treatment effects varied by sex. With respect to the end-of-trial blood pressure, males

Table 8. Relative risk of stroke associated with pre-treatment variables (Linxian, China 1985–1991)

	Dysplasia trial[1] RR (95% CI)	General population trial[2] RR (95% CI)
Systolic pressure per 5 mm Hg	1.21 (1.15, 1.24)	1.22 (1.20, 1.23)
Diastolic pressure per 5 mm Hg	1.25 (1.16, 1.32)	1.34 (1.31, 1.36)
Age per 5 years	1.90 (0.55, 2.34)	1.86 (1.75, 1.98)
Bmi[3] per 5 units	0.98 (0.54, 1.75)	1.20 (1.01, 1.42)
Drink	0.43 (0.16, 0.94)	0.56 (0.44, 0.71)
Smoke	1.00 (0.48, 2.12)	0.93 (0.73, 1.20)

[1] Estimates when each pre-treatment variable is entered separately into a Cox model stratified on sex and treatment group.
[2] Estimates when each pre-treatment variable is entered separately into a Cox model stratified on sex.
[3] Bmi = body mass index

in the supplemented group of the dysplasia trial had significantly lower blood pressure and significantly less hypertension than their untreated counterpart. No such beneficial effect was observed for the general population participants.

Though the differences between the outcomes of the two Linxian trials could be due to chance, the magnitude of the effect among males in the dysplasia trial, and the presence of an effect on both stroke and hypertension, warrants the consideration of other possibilities. With the exception of being slightly older and having a higher prevalence of cytological abnormalities predictive of esophageal and gastric cardia cancer, the dysplasia and general population trial participants were similar with respect to personal characteristics and underlying nutritional exposures. The interventions were different of course, with the treated group in the dysplasia trial receiving all the supplements taken by any of the general population participants plus additional mineral and vitamins. In the general population group that received all four factors, ABCD, the stroke mortality was not more favorable than those groups which received only 2 factors, so it is not simply a question of more being better in this nutritionally deprived population.

In addition to the supplements given in factors A, B, C, D, the actively treated group in the dysplasia trial received ten other mineral compounds, and seven other vitamins (six water soluble vitamins and vitamin D). A number of the cations, including potassium, calcium, and magnesium are reported to lower blood pressure and or risk of stroke (see 2 for a recent summary). Vitamins D and pyridoxine have been postulated to be of benefit for hypertension (15). The relationship of vitamins with stroke is less well studied (2, 16). There is, however, a growing body of evidence associating elevated serum (or plasma) levels of total homocysteine, and decreased serum levels and/or intake of folate, pyridoxine, or B12, with arterial occlusive disease (17–19). Folate, pyridoxine, and B12 are important cofactors in the metabolism of homocysteine, and supplementation with these vitamins, particularly folate, have been shown to lower homocysteine levels (17, 18). Homocysteine levels are approximately 20% higher in men than in women (17), which suggests a possible explanation for the greater treatment benefit experienced by males. With regard to risk of stroke in particular, prospective studies have ranged from showing a strong and positive association with serum level of homocysteine (20), to weak or non-existent associations (21, 22). Low serum folate itself has recently been reported to be a risk factor for ischemic stroke (24).

In both the dysplasia and general population trials the risks of stroke associated with baseline factors were similar. Of special interest is the protective effect associated with alcohol drinking. The associations of alcohol intake and stroke risk have been inconsistent. The available evidence suggests an increase in the risk of stroke with heavy drinking and an increased risk of hemorrhagic stroke will all levels of drinking (16). However, for ischemic and overall stroke there is contradictory evidence regarding whether low or moderate levels of drinking are associated with reduced stroke risk (24–27). In the Linxian studies we could not distinguish hemorrhagic from ischemic stroke, but population statistics suggest that approximately two-thirds of the strokes are ischemic (28). Those who consumed alcohol in Linxian were light or infrequent drinkers, and showed risk reductions of 30%. Whether this effect is due to alcohol consumption or some other unmeasured trait correlated with alcohol intake or abstention is unknowable.

In our continued follow-up of these study populations, and from the analyses of the biological specimens collected, we hope to further characterize the effects of the trial interventions. We are presently conducting a nested-case control study to explore the associations of serum vitamin and homocysteine levels with stroke, and to investigate whether these associations vary by the treatment received.

5. REFERENCES

1. W.J. Blot and J.Y. Li, *Natl. Cancer Inst. Monogr.* **69**, 29–34 (1985).
2. S.D. Mark, W. Wang, J.F. Fraumeni, J.Y. Li, P.R. Taylor, G.Q. Wang, W. Guo, S.M. Dawsey, B. Li and W.J. Blot, *Am. J. Epidemiol.* (In Press).
3. S.D. Mark, W. Wang, J.F. Fraumeni, J.Y. Li, P.R. Taylor, G.Q. Wang, W. Guo, S.M. Dawsey, B. Li and W.J. Blot, *Am. J. Epidemiol.* (Submitted).
4. B. Li, P.R. Taylor, J.Y. Li, S.M. Dawsey, W. Wang, J.A. Tangrea, B.Q. Liu, A.G. Ershow, S.F. Zheng, J.F. Fraumeni, Jr., et al, *Ann. Epidemiol.* **3**, 577–85 (1993).
5. W.J. Blot, J.Y. Li, P.R. Taylor, W. Guo, S. Dawsey, G.Q. Wang, C.S. Yang, S.F. Zheng, M. Gail, G.Y. Li, et al, *J. Natl. Cancer Inst.* **85**, 1483–92 (1993).
6. J.Y. Li, P.R. Taylor, B. Li, S. Dawsey, G.Q. Wang, A.G. Ershow, W. Guo, S.F. Liu, C.S. Yang, Q. Shen, et al, *J. Natl. Cancer Inst.* **85**, 1492–1498 (1993).
7. S.D. Mark, S.F. Liu, J.Y. Li, M.H. Gail, Q. Shen, S.M. Dawsey, F. Liu, P.R. Taylor, B. Li and W.J. Blot, *Int. J. Cancer* **57**, 162–166 (1994).
8. A. Zanchetti, J. Chalmers, K. Arakawa, et al, *Bull. World Health. Organ.* **71**, 503–517 (1993).
9. D.R. Cox, *J. R. Stat. Soc. B* **34**, 187–220 (1972).
10. A. Agresti, *Categorical Data Analysis*, John Wiley & Sons, New York (1990).
11. A.G. Ershow, S.F. Zheng, G. Li, J. Li, C.S. Yang and W.J. Blot, *J. Natl. Cancer Inst.* **73**, 1477–1481 (1984).
12. C.S. Yang, Y.H. Sun, Q.P. Yang, K.W. Miller, G.Y. Li, S.F. Zheng, J.Y. Li and W.J. Blot, *Natl. Cancer Inst. Monogr.* **69**, 23–27 (1985).
13. C.S. Yang, Y. Sun, Q.U. Yang, K.W. Miller, G.Y. Li, S.F. Zheng, A.G. Ershow, W.J. Blot and J.Y. Li, *J. Natl. Cancer Inst.* **73**, 1449–1453 (1984).
14. W.J. Blot, J.Y. Li, P.R. Taylor, W. Guo, S.M. Dawsey and Li, B., *Am. J. Clin. Nutr.* **62**, 1424s-1426s (1995).
15. K. Dakshinamurti and K.J. Lal, *World Rev. Nutr. Diet.* **69**, 40–73 (1992).
16. L.L. Bronner, D.S. Kanter and J.E. Manson, *N. Engl. J. Med.* **333**, 1392–1400 (1995).
17. M.R. Malinow, *J. Int. Med.* **236**, 603–617 (1994).
18. C.J. Boushey, S.A.A. Beresford, G.S. Omenn and A.G. Motulsky, *JAMA* **274**, 1049–1057 (1995).
19. J. Selhub, P.F. Jacques, A.G. Bostom, R.B. D'Agostino, P.W. Wilson, A.J. Belanger, D.H. O'Leary, P.A. Wolf, E.J. Schaefer and I.H. Rosenberg, *N. Engl. J. Med.* **332**, 286–291 (1995).
20. I.J. Perry, H. Refsum, R.W. Morris, S.B. Ebrahim, P.M. Ueland and A.G. Shaper, *Lancet* **346**, 1395–1398 (1995).
21. P. Verhoef, C.H. Hennekens, M.R. Malinow, F.J. Kok, W.C. Willett and M.J. Stampfer, *Stroke* **25**, 1924–1930 (1994).
22. G. Alfthan, J. Pekkanen, M. Jauhiainen, J. Pitkaniemi, M. Karvonen, J. Tuomilehto, J.T. Salonen and C. Ehnholm, *Atherosclerosis* **106**, 9–19 (1994).
23. W.H. Giles, S.J. Kittner, R.F. Anda, J.B. Croft and M.L. Casper, *Stroke* **26**, 1166–1170 (1995).
24. M.J. Stampfer, G.A. Colditz, W.C. Willett, F.E. Speizer and C.H. Hennekens, *N. Engl. J. Med.* **319**, 267–272 (1988).
25. H. Hansagi, A. Romelsjo, M. Gerhardsson de Verdier, S. Andreasson and A. Leifman, *Stroke* **26**, 1768–1773 (1995).
26. E. Beghi, G. Boglium, P. Cosso, G. Fiorelli, C. Lorini, M. Mandelli and A. Bellini, *Stroke* **26**, 1691–1696 (1995).
27. M. Gronbaek, A. Deis, T.I. Sorensen, U. Becker, P. Schnohr and G. Jensen, *Br. Med. J.* **310**, 1165–1169 (1995).
28. F.L. Shi, R.G. Hart, D.G. Sherman and C.H. Tegeler, *Stroke* **20**, 1581–1585 (1989).

70

EFFECTS OF ANTIOXIDANT VITAMIN AND TRACE ELEMENT SUPPLEMENTATION ON SELENIUM STATUS IN HEALTHY SUBJECTS

Results of a SU.VI.MAX Pre-Test

P. Preziosi,[1] J. Arnaud,[2] P. Galan,[1] A-M. Roussel,[2] M-J. Richard,[2] D. Malvy,[3] A. Paul-Dauphin,[4] S. Briancon,[4] A. Favier,[2] and S. Hercberg[1]

[1] Institut Scientifique et Technique de la Nutrition et l'Alimentation
Conservatoire National des Arts et Métiers
2 rue Conté, F-75003 Paris, France
[2] Laboratoire de Biochimie
CHRU de Grenoble, France
[3] Labo Santé Publique
CHRU de Tours, France
[4] Ecole de Santé Publique
CHRU de Nancy, France

1. INTRODUCTION

Antioxidant functions have been associated with decreased DNA damage, diminished lipid peroxidation and inhibited malignant transformation in vitro (1,2); further, they are associated epidemiologically, with a lower incidence of certain types of cancer and degenerative diseases such as ischemic heart disease and cataracts (3–5). Recently, nutritional surveys showed that a significant percentage of the affluent world population has a relatively low intake or borderline vitamin and/or trace element status. The SU.VI.MAX study was designed to quantify the preventive effect of a combination of antioxidant vitamins and trace elements (beta-carotene, vitamin C, vitamin E, selenium and zinc), in doses considered to be nutritional and non-pharmacological in terms of the incidence of cancer, heart disease, cataracts, infection and morbidity in a large adult population representative of an industrialized country (France). A randomized double-blind intervention trial involving these nutrients was undertaken over the course of 8 years. In preparing the SU.VI.MAX protocol, a pre-test was performed to assess the biological response to a 6 months supplementation with nutritional doses of antioxidant vitamins and trace elements.

Therapeutic Uses of Trace Elements, edited by Nève et al.
Plenum Press, New York, 1996

2. MATERIAL AND METHODS

The study was performed in 400 healthy subjects (166 males, 45–60 old and 235 females, 35–60 old). Volunteers were recruited via a short multi-media campaign, involving radio, newspapers and local journals. Volunteers were from four French urban areas (Paris, n = 149; Grenoble, n = 107; Nancy, n = 88 and Tours, n = 57). The study was a double-blind and placebo-controlled. Subjects were stratified by sex and age and randomly assigned to one of the two treatment groups. Group S received daily 20 mg zinc, 100 µg selenium, 6 mg beta-carotene, 30 mg vitamin E and 120 mg ascorbic acid for 6 months, while group P received a placebo. Each subject received one capsule per day for a period of 6 months. Supplements and placebo were identitical in appearance and were prepared especially for the study.

Biological markers of trace elements and vitamin status were measured initially, and then 3 and 6 months after the beginning of the supplementation. For each subject, 35 ml of whole blood were withdrawn by venipuncture between 7 and 8 a.m from fasting subjects. Serum zinc levels were determined (6) by flame atomic absorption spectrometry (Perkin Elmer 460, Norwalk, CT, USA). Serum selenium concentrations were determined on a Perkin Elmer 5100 (Norwalk, CT, USA) equipped with a HGA 600 furnace, an EDL lamp and Zeeman background correction (7). Blood cell selenium dependent glutathione peroxidase (GPX) was measured by a modified method of Gunzler et al (1974) using terbutyl hydroperoxide (Sigma Chemical Co, France) as substrate instead of hydrogen peroxide; results were expressed as µmol of NADPH (Boehringer-Mannhein, Germany) oxidised per minute per gram of hemoglobin. Reduced glutathione (GSSH) was assayed as previously described by Goldberg et al (1983). Malondyaldehyde (MDA), was measured using the Sobioda MDA kit (Grenoble, France) as described by Richard et al (8). Total glutathione (GSH) was determined by a modified method of Theodorus et al (9). Glutathione was determined using enzymatic cycling of GSH by means of NADPH and glutathione reductase coupled with DTNB. To assay oxidised glutathione (GSSG), GSH was masked by adding 10 µl of 2-vinyl-pyridine to 500 µl of deproteinised extract adjusted to pH 6 with triethanolamine. Superoxide dismutase (SOD) was measured using competition between the oxidative reaction of pyrogallol by superoxide radicals and dismutation of these radicals by SOD (10). Vitamin C status was evaluated by serum ascorbic acid determination using an automated method based on the continous flow principle, segmented with air bubbles (11). Serum retinol was measured by HPLC with normal phase (silicagel), isocratic elution with n-hexane/isopropanol (970:30) and detection by UV at 330 nm. Serum beta-carotene was measured by normal phase HPLC on Silicagel, isocratic elution with n-hexane/dioxane (990:10) and detection in visible light at 436 nm. Serum tocopherol was measured by normal phase HPLC on Silicagel, isocratic elution with n-hexane/ethyl acetate (930:70) and fluorescence detection with excitation at 298 nm and emission at 328 nm (12).

3. RESULTS AND DISCUSSION

Before supplementation, the two groups were similar with respect to the number of subjects, age, serum levels of vitamins and trace elements, and indicators of oxidative stress and antioxidant enzymes. At baseline, 14.4 % of men and 6 % of women had plasma ascorbic acid levels of less than 3.5 µg/L; 3.3 % of men and 14.8 % of women had plasma retinol concentrations of less than 401 µg/L (0.5 % lower than 287 µg/L); 23 % of men and 5 % of women had beta-carotene levels below 161 µg/L; 15.1 % of men and 23.8 % of women had

Table 1. Serum levels of vitamins and trace elements, and indicators of oxidative stress and antioxidant enzymes (mean ± SD)

	Males	Females
Plasma beta carotene (μg/L)	293 ± 184*	480 ± 343
Plasma retinol (μg/L)	649 ± 141*	521 ± 145
Plasma alpha-tocopherol (μg/L)	14.1 ± 3.8	12.8 ± 2.6
Serum zinc (μmol/L)	12.5 ± 1.8*	12.0 ± 1.7
Serum selenium (μmol/L)	1.12 ± 0.18	1.12 ± 0.18
Plasma GPx (U/L)	338 ± 45	335 ± 51
Erythrocyte SOD (U/mg Hb)	1.12 ± 0.10	1.12 ± 0.12
Erythrocyte GPx (U/g Hb)	42.5 ± 8.9*	45.5 ± 9.1
GSSG (μmol/L)	23 ± 15	20 ± 13
GSH (μmol/L)	875 ± 156	904 ± 146
GSSG/GSH + GSSG (%)	5.1 ± 3.5	4.3 ± 2.6
TBARS (μmol/L)	2.69 ± 0.38	2.71 ± 0.38

$* p < 0.05$

serum zinc of less than 10.7 μmol/L and 1.2 % of men and 2.1 % of women had serum selenium of less than 0.75 μmol/L. No subjects had plasma alpha-tocopherol under 4 mg/L. At T0, vitamin and trace element concentrations in blood were in agreement with a previous report. However, the beta-carotene concentrations in male subjects were lower than those previously reported in France. Selenium and alpha-tocopherol concentrations in blood were within reference ranges (Table 1).

After 3 months of supplementation, serum zinc and selenium significantly increased in group S (Table 2). Mean serum zinc and selenium concentrations were not significanly different at 3 and 6 months Compliance of subjects was very good (94 %) and tolerance of supplements was excellent. Finally, this pre-test confirmed that, after 3 months, supplementation by nutritional doses of selenium and zinc significantly increased the biochemical indicators of zinc and selenium status.

These data are consistent with other studies performed on different type of populations. Swanson et al showed that supplementation improved the concentration of Zn in serum in elderly healthy adults (13). In elderly subjects supplemented for one year, Bogden et al (14) reported an increase in plasma zinc levels in the group receiving 100 mg daily, but no change in the group receiving 15 mg daily. Some positive effects of selenium supplementation at moderate doses (100–200 μg/d during 2 months) have already been documented in healthy young Belgian subjects (15). In an elderly population, Peretz et al (16) demonstrated the efficiency of selenium-enriched yeast to increase plasma selenium concentrations.

Table 2. Effect of supplementation on serum selenium and zinc concentrations

	Groups	T0	Supplementation	
			3 months	6 months
Serum zinc (μmol/L)	S	12.2 ± 0.2	15.2 ± 5.2**	14.6 ± 3.1**
	P		12.6 ± 2.4	13.0 ± 2.1
Serum selenium (μmol/L)	S	1.12 ± 0.18	1.75 ± 0.23**	1.65 ± 0.19**
	P		1.11 ± 0.18	1.09 ± 0.20

$** p < 0.01$

4. REFERENCES

1. H. Sies, W. Stahl, A.R. Sundquist. *Ann. N. Y. Acad. Sci.* **669**, 7–20 (1992).
2. B. Ames. *Science.* **221**, 1256–1264 (1987).
3. G. Block. *Am. J. Clin. Nutr.* **53**, 270S-283S (1991).
4. L. Kohlmeier, S.B. Hastings. *Am. J. Clin. Nutr.* **62**, 1370S-1376S (1995).
5. A. Taylor, P.F. Jacques, E.M. Epstein. *Am. J. Clin. Nutr.* **62**, 1439S-1447S (1995).
6. J. Arnaud, J. Bellanger, F. Bienvenu, P. Chappuis, A. Favier. *Ann. Biol. Clin.* **44**, 77–87 (1986).
7. J. Nève, S. Chamart, L. Molle, in *Trace element analytical chemistry in medicine and biology.* Bratter P., Schamer P., eds., Walter de Gruyte, Berlin, pp. 1–10 (1987).
8. M-J. Richard, B. Portal, J. Meo, C. Coudray, A. Hodjian., A. Favier. *Clin. Chem.,* **38**, 704–10 (1992).
9. P. Theodorus, M. Akerboom, H. Sies. *Methods in Enzymology,* vol. 77, Academic press, Inc, New York (1981).
10. S. Marklund, G. Marklund. *Eur. J. Biochem.,* **47**, 469–474 (1974).
11. C.F. Bourgeois, R.R. Chartois, M.F. Counstans, P.R. George. *Analusis,* **9**, 519–525 (1989).
12. J.P. Vuilleumier, H.E. Keller, D. Gysel, F. Hunziker. *Internat. J. Vitam. Nutr. Res.,* **53**, 265–272 (1983).
13. C.A. Swanson, R. Mansourian, H. Dirren, C-H. Rapin. *Am. J. Clin. Nutr.,* **48**, 343–349 (1988).
14. J.D. Bogden, J.M. Oleske, M.A. Laverha, E.M. Munves, F.W. Kemp. *Am. J. Clin. Nutr.,* **48**, 655–663 (1988).
15. J. Nève, F. Vertongen, P. Capel. *Am.J. Clin. Nutr.,* **48**, 139–144 (1988).
16. A. Peretz, J. Nève, J. Desmedt, J. Duchateau, M. Dramaix, J.P. Famaey. *Am.J. Clin. Nutr.,* **53**, 1323–1328 (1991).

IRON STATUS OF A REPRESENTATIVE SAMPLE OF THE FRENCH ADULT POPULATION

Results from the SU.VI.MAX Study

P. Galan,[1] P. Preziosi,[1] M-J. M. Alferez,[2] A-M. Roussel,[3] D. Malvy,[4] A. Paul-Dauphin,[5] S. Briancon,[5] A. Favier,[3] and S. Hercberg[1]

[1] Institut Scientifique et Technique de la Nutrition et l'Alimentation
Conservatoire National des Arts et Métiers
2 rue Conté, F-75003 Paris, France
[2] Departamento de Fisiologia
Universidad de Granada, Spain
[3] Laboratoire de Biochimie
CHRU de Grenoble, France
[4] Labo Santé Publique
CHRU de Tours, France
[5] Ecole de Santé Publique
CHRU de Nancy, France

1. INTRODUCTION

In industrialized countries, there is growing interest in the association of iron metabolism disturbances with certain health problems (1–4). There exists two major disturbances of iron balance, iron deficiency and iron overload. Paradoxically, little information exists on the iron status of large representative samples of populations, particularly in European countries. Most information about iron nutritional status is based on studies of relatively small population groups (5,6). Few data are available on the iron status of the French population. However, in France, iron fortification of foods is not allowed (except in infant formulas and specialized dietetic products) and the use of iron supplements is therefore uncommon. In the present study, the iron status of a representative sample of the French adult population, issued from the SU.VI.MAX study, was assessed using biochemical indicators.

Therapeutic Uses of Trace Elements, edited by Nève et al.
Plenum Press, New York, 1996

2. MATERIAL AND METHODS

The SU.VI.MAX study is a large-scale 8-years long longitudinal study involving a double-blind supplementation protocol (antioxidant vitamins and trace elements at nutritional doses, versus placebo) given to a cohort of 15,000 adults selected by a screening process from a group of 80,000 volunteers througout France (women: 35–60 y; and men: 45–60 y). All volunteers are periodically followed up for nutritional intake, clinical events and biochemical status. At baseline, a venous blood sample was obtained to measure various biochemical indicators. Biochemical indicators of iron status were measured in 9,308 subjects: 6,241 women and 3,077 men. Hemoglobin concentration was evaluated using Réflotron (Boehringer Diagnostic); serum transferrin and serum ferritin levels were measured using a nephelometric assay (BNII Behring).

3. RESULTS AND DISCUSSION

Prevalences of iron depletion, anemia and iron-deficient anemia are presented in Figure 1. Iron depletion, recognized by a serum ferritin concentration of less than 15 µg/L, was found in 18% of adult women and in 1.9% of adult men. The prevalence was 22.7% in premenopausal women and 5.4% in post-menopausal women. Anemia (defined as a hemoglobin level of less than 130 g/L in men and less than 120 g/L in women) was observed in 6.9% of women and 2.8% of men. The prevalence of anemia was 8.1% in premenopausal women and in 4.1% in postmenopausal women. In women, more than half of the cases of anemias were associated with low serum ferritin concentrations; in men, this figure was 15%. Finally, we estimate that the prevalence of iron-deficient anemias was 3.9% in women and 0.4% in men.

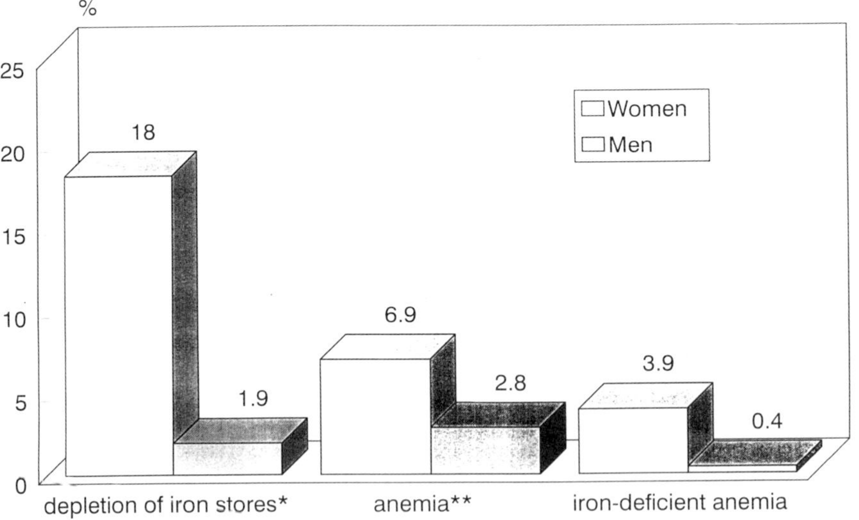

Figure 1. Prevalence of iron depletion, anemia and iron-deficient anemia in volunteers taking part in the SU.VI.MAX study.

While the consequences of anemia upon health are well known, the present study raises the question of the consequences of depletion of iron stores frequently observed in women. By means of longitudinal follow-up of volunteers in the SU.VI.MAX study (dietary intake, morbidity data), it will be possible to better understand the dietary determinants of iron status and to assess the potential deleterious effects of iron-depleted states (particularly upon immunity and risk of infection, physical capacity, asthenia, thermogenesis, etc.).

ACKNOWLEDGMENTS

This study was supported by the "club EPIFER" composed by Kellogg's, Besnier, Candia, Robapharm, Innothéra and Behring.

4. REFERENCES

1. S. Hercberg, P. Galan, H. Dupin. *Recent knowledge on iron and folate deficiency in the world.* Eds INSERM, Paris (1990).
2. Hallberg L, Rossander-Hulthen L. *Bibl. Nutr. Dieta*, **44**, 94–105 (1989).
3. P. Galan, S. Hercberg, Y. Touitou. *Comp. Biochem. Physiol.*, **77B**, 647–653 (1984).
4. S. Hercberg, P. Galan. *Acta Pediatr. Scand.*, **361**, 63–70 (1989).
5. P. Galan, S. Hercberg, Y. Soustre, M.C. Dop, H. Dupin. *Hum. Nutr. Clin. Nutr.*, **39C**, 279–287 (1985).
6. Y. Soustre , M.C. Dop, P. Galan, S. Hercberg. *Int. J. Vit. Nutr. Res.*, **56**, 281–286 (1986).
7. A. Dhur, S. Hercberg. *Bibl. Nutr. Diet.*, **44**, 106–113 (1989).
8. Hercberg S, Bichon., Galan P., Christides J.P., Carroget C., Potier de Courcy G. *Nutr. Rep. Int.* 915–930 (1987).
9. P. Preziosi, S. Hercberg, P. Galan, M. Devanlay, F. Cherouvrier, H. Dupin. *Ann. Nutr. Met.*, **38**, 192–202 (1994).

72

THE COPPER-TRANSPORTING ATPASES DEFECTIVE IN MENKES DISEASE AND WILSON DISEASE

Diane W. Cox

Research Institute, The Hospital for Sick Children
555 University Avenue, Toronto, Ontario, Canada M5G 1X8
Departments of Molecular and Medical Genetics, and Paediatrics
University of Toronto, Toronto, Canada
Department of Medical Genetics
University of Alberta, Edmonton, Canada

1. INTRODUCTION

The identification of genes defective in Menkes disease and in Wilson disease has provided a major breakthrough in our understanding of copper transport. A membrane transport protein is a key factor in the control of intracellular copper concentration. In addition to the high degree of identity between the genes for Menkes and Wilson diseases, a high degree of homology is shown with other metal transporting ATPases , first in bacteria (1) and more recently in yeast (2). The bacterial genes are generally present on plasmids, and replicate in the presence of high metal concentration in the surrounding environment, as provided by pollution, fungicides, or any other situation in which the metal content of the environment is increased. A particularly high degree of homology is seen with those bacteria which are resistant to either copper or cadmium, because of their efficiency in transporting these metals from the cell. A high level of amino acid identity is seen in the functionally important regions.

An examination of the simplified pathway of copper (Figure 1) helps to explain the abnormalities in both disorders. The transport, utilization and elimination of copper involves a series of processes, including absorption through the intestinal epithelial cells, transfer to albumin and copper histidine for blood transport, uptake by the liver, where incorporation into ceruloplasmin and excretion into the bile take place.

This report describes recent advances in two known disorders of copper transport, focusing particularly on Wilson disease and the studies in our laboratory.

Therapeutic Uses of Trace Elements, edited by Nève et al.
Plenum Press, New York, 1996

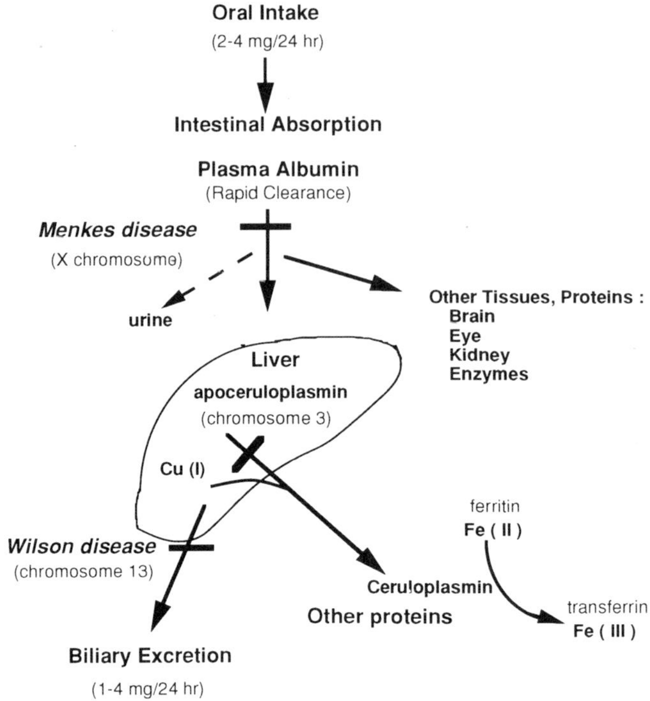

Figure 1. Simplified overview of the pathways for copper ion transport From Cox 1995, with permission from the American Journal of Human Genetics.

2. MENKES DISEASE

2.1. Features of the Disease

Menkes disease, an X-linked recessive disease, is a defect of transport across the intestinal membrane which prevents the exit of copper from intestinal cells (3,4). Excess copper can be stored in the low molecular weight, metal-inducible protein, metallothionein. Copper is then not available to bind to albumin and in trace amounts to amino acid such as copper histidine, for transport into other body tissues and essential enzymes. The result is a widespread copper deficiency, leading to insufficient levels of all of the copper containing enzymes, including lysyl oxidase involved in connective tissue and elastin cross linking, superoxide dismutase, cytochrome oxidase, tyrosinase and dopamine beta hydroxylase. The features of Menkes disease: abnormal connective tissue, twisted hair and arteries, neurological abnormalities, developmental delay and hypopigmentation can all be explained by the absence of these enzymes. Copper histidine shows some success in treatment of this condition, which, without treatment, results in death in early childhood (5).

The gene for Menkes disease (designated *ATP7A*) was initially cloned from a female affected with Menkes disease due to an interruption of the gene by a chromosome translocation breakpoint (6–8). The recognition of gene function was assisted by information on the copper transporting ATPase in bacteria (9), initially thought to be involved in potassium transport.

2.2. Mutations in the Gene

Approximately 20 per cent of the mutations in the Menkes disease gene are deletions several kb in length (6,7). These deletions are usually associated with forms of Menkes disease which are lethal within the first few years of life. Milder forms of the disease, including cutis laxa, are in some cases due to splice site defects, in which small amounts of normal product are formed (10,11). The mottled mouse mutants are due to defects in the orthologous gene in mice (11–13). There are many mutant alleles at this locus (blotchy, dappled, brindled), which result in a range of disease severity, depending upon the specific mutation.

3. WILSON DISEASE

3.1. Features of the Disease

Wilson disease (hepatolenticular degeneration) is an autosomal recessive disorder , with an incidence of about 1/30,000 in most populations (14). In Wilson disease, copper transport is normal to the point of entry into the liver, where the majority of copper is targeted. Metallothionein is induced and initially maintains the ingested copper in a relatively non toxic form. When this capacity is exceeded, further accumulation of hepatic copper results in damage to the hepatocytes, increased efflux of copper into the urine, and copper accumulation in a variety of body tissues. Copper is not incorporated into the serum protein ceruloplasmin, and excess is not excreted through the bile. As a result, copper accumulates in the liver, leading to hepatocyte damage and subsequent cirrhosis or liver failure. Copper also accumulates in brain and in the rim of the cornea, the latter resulting in characteristic brownish Kayser-Fleischer rings. Neurological damage occurs apparently from the toxic effects of copper in the brain, particularly in the lenticular nucleus. The varied signs and symptoms of Wilson disease, occurring between the ages of 3 and 50 years include cirrhosis of the liver, tremour, loss of speech, erythrocyte hemolysis, kidney abnormalities and occasionally psychiatric disorders. Treatment of this disease is generally accomplished by the use of chelating agents, such as penicillamine or triethylene tetramine, which remove the excess copper from plasma and in part from the liver, promoting its excretion via the urine. The chelating agents sometimes produce toxic side-effects, and excess copper depletion must be avoided. Alternatively, administration of zinc salts is also used to block copper absorption and prevent buildup of the copper to toxic levels (15). Our studies have focused on finding the basic defect in this disorder.

3.2. Cloning the Gene for Wilson Disease

The Wilson disease gene was localized to human chromosome 13 by co-segregation of the disease state with the red cell enzyme marker, esterase D, in several large Middle Eastern kindreds (16). Further linkage analysis using more polymorphic DNA markers in the 13q12–22 region, subsequently defined a region close to the retinoblastoma (RB) locus, flanked by the two random polymorphic markers D13S31 and D13S59, as the candidate region for the Wilson disease gene (17,18). We began a positional cloning strategy to identify the gene, first using genetic linkage studies in our series of 25 Canadian families with Wilson disease, to confirm the localization and, through recombinant individuals, to narrow the candidate region to about 1 to 2 megabases (19). We developed new markers in the region and derived a long-range restriction map using pulsed field gel electrophoresis (20). Yeast artificial chromosomes (YACs) were obtained, using four markers from our map, as probes to screen the library of YACs distributed by the Centre de Polymorphisme Humaine, Paris, France. DNA fragments from the ends of these YACs were then used as probes to obtain 19 overlapping YAC clones, all placed within the map generated by pulsed field gel electrophoresis.

Unlike the situation for a number of other genes identified by positional cloning, no patients were available with a helpful translocation to identify the location of the WND gene. When the Menkes disease gene was cloned and was shown to be expressed in many tissues with the notable exception of the liver, we formulated the hypothesis that Wilson disease might be due to a defect in a homologous gene. This was indeed the case and we identified a region of homology on several of the YACs, we had identified in the region of the gene, then on cosmids from a chromosome 14 library. Through cDNA selection, we assembled the cDNA for Wilson disease. The resulting gene (designated *ATP7B*) was 57% identical in amino acid sequence to the Menkes disease gene (21). In another study involving groups in Boston and New York, a cDNA from a brain library was found to be the WND gene, and represented a shorter alternative transcript expressed in low amounts in the brain (22).

The model we have proposed for the WND gene (23), based on its amino acid structure, is shown in Fig. 2. The major functional regions of the gene are conserved between Menkes disease and Wilson disease genes, and genes from bacteria and yeast. The transduction domain transfers the energy of ATP hydrolysis to cation transport. The aspartate residue within a conserved motif forms a phosphorylated intermediate during the cation transport cycle. A conserved hinge region lies at the C-terminal end of a large cytoplasmic ATP-binding domain. The cysteine-proline-cysteine motif, characteristic of the heavy metal transporting ATPases, is predicted to lie within the cation channel. There are six copper binding and eight transmembrane domains predicted. Several alternatively spliced shorter products can be produced(24,25).

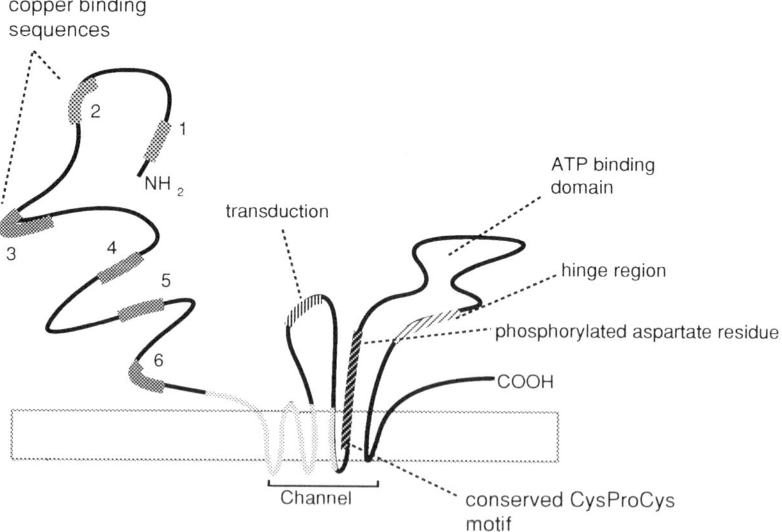

Figure 2. A model of the predicted product of the Wilson disease gene. The regions conserved in the Menkes disease gene and in bacterial genes are indicated. Shaded transmembrane domains are alternatively spliced. (Modified from Bull and Cox 1994 (23), with permission from Trends in Genetics).

3.3. Haplotype Studies with Microsatellite Markers

Highly polymorphic markers had been useful in initially identifying the region of the WND gene (26). The same markers are useful in helping to identify which mutation might be present in any given patient. Haplotypes are combinations of marked alleles that are located together on the same chromosome. A specific combination or haplotype tends to be associated with a specific mutation. Because no simple test is yet available to identify which mutation might be present, we currently haplotypes to guide our mutation testing. Care must be taken to ensure that the same allele designation is used for all patients.

The flanking markers are shown in Fig. 3. D13S31 and D13S59, restriction site length polymorphisms, are shown for reference. We use D13S316, D13S301 and D13S114 results in haplotyping (27).

We have found that some haplotypes of CA repeats are unique to Wilson disease chromosomes and are exceedingly rare or absent on normal chromosomes . The identification of these unusual haplotypes can lend support to the diagnosis of Wilson disease.

Testing of close markers has a major practical application. A small percentage of heterozygotes has been shown to have biochemical features indistinguishable from presymptomatic heterozygotes (28). Such heterozygotes apparently never develop symptoms of the disease and would therefore not need to be exposed to the potentially toxic effects of lifelong treatment on chelating agents. The pedigree in Fig. 4 indicates how markers can identify the status of sibs of a known patient. In the example shown, segregation of close markers indicates that each the two girls inherited different chromosomes from the mother, therefore the sister is not affected.

Completely reliable diagnosis is possible in most families. However this requires a DNA sample from at least one parent and an affected patient, or preferably both parents and the affected patient. This is currently the most reliable way to identify the genotype of sibs of an affected patient. However because of the very small chance of recombination between the gene and this closed markers, care must be taken that there is an informative marker flanking the gene on both sides.

3.4. The Mutations of the Wilson Disease Gene

Our method for mutation detection is single strand confirmation polymorphism (SSCP) analysis in which changes from normal in the mobility of 200 to 300 base pair fragments indicate the presence of a mutation. The appropriate fragment is then sequenced to identify the specific mutation present. The pattern of mutations is very different from that found in Menkes disease. We have not yet identified any large deletions in the gene. The majority of mutations identified in our patients are one or two base pair insertions or deletions. All types

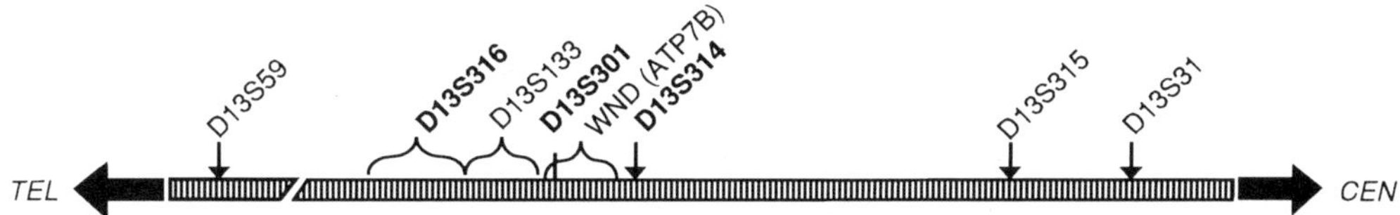

Figure 3. Approximate location of CA repeat markers located near the WND (*ATP7B*) gene. The markers shown in bold were isolated from cosmids containing the gene (21). Markers in bold are those used for haplotypes and lie within about 1 megabase. The flanking markers D13S31 and D13S59 are shown for reference.

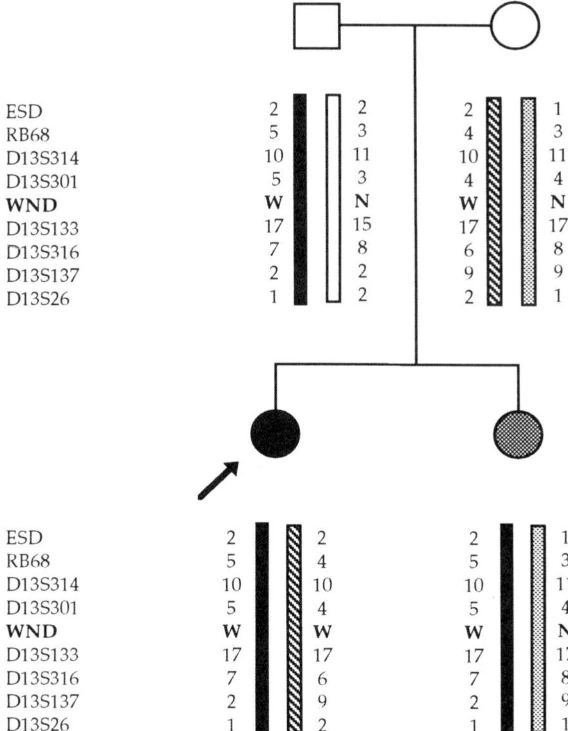

Figure 4. Diagnostic use of polymorphic DNA markers in a pedigree. DNA markers are listed in centromeric to telomeric order. Numbers represent alleles of each marker listed. The proband is shown as a filled circle (arrow) and the sib for diagnosis as a shaded circle.

of mutations are present: missense, nonsense, splice site in addition to small insertions or deletions (29). The mutations we have identified are shown in table 1. Additional mutations have been reported in Mediterranean populations (30).

Knowledge of all or most of the mutations present in various ethnic groups will dramatically improve diagnosis of Wilson disease, particularly in the early stages. As we learn more about the variety of clinical symptoms, we know that some patients can lack one or more of the common clinical features of the disease. Patients may lack Kayser-Fleischer rings, have normal serum ceruloplasmin concentration, and, in presymptomatic stages, have normal urinary copper excretion. Other diseases can produce high liver copper, compounding the difficulties of diagnosis.

3.5. Animal Models

In the Long-Evans Cinnamon (LEC) rat, a mutant identified in Japan from the Long-Evans (LE) strain, we have shown that the LEC rat carries a deletion of the last quarter of the gene, eliminating key functional regions, including the ATP binding fold (31). This rat is therefore an excellent model for the study of both pathology and treatment of Wilson disease.

The toxic milk mouse has also been shown to have a missense mutation in the Wilson disease gene (Theophilos et al., unpublished) Interestingly, in this mouse, the infant mouse is

Table 1. Mutations identified in our series of patients with Wilson disease

Mutation	Domain	Exon	Predicted effect	Ethnic group
Insertion/deletion				
1745insT	Cu6	5	frameshift	Indian
2158delA	Tm3	8	frameshift	Italian
2299insC	Tm4	8	frameshift	Italian, British
3085delAC	Phosph.	14	frameshift	British
3146delC [1]	ATP loop	14	frameshift	Indian
3400delC [1]	ATP loop	15	frameshift	Ukrainian
3626del4	ATP pocket	17	frameshift	Swedish
3648del6	ATP pocket	17	disrupts ATP binding	British
4091delTG	after Tm7	20	frameshift	British
Nonsense				
L936X	Tm5	12	truncates protein	Saudi, Greek
R1319X	ATP - Tm7	19	truncates protein	British
Splice site				
1708-1G-C	Cu6	5	skips exon 5	Indian
2575+1G-C	Td	10	includes intron 10	British
3556+1G-A	ATP pocket	16	includes intron 16	British
Missense				
M769V [2]	Tm4	8	disrupts Tm4	British
L777V [2]	Tm4	8	disrupts Tm4	British
R778L	Tm4	8	disrupts Tm4	Chinese
G943S	Tm5	12	disrupts Tm5	Bangladeshi
H1069Q [*]	loop motif	14	disrupts ATP binding	European
G1101R	ATP loop	15	disrupts ATP binding	Indian
I1102T	ATP loop	15	disrupts ATP binding	Indian
M1169V [2]	ATP loop	16	disrupts ATP hinge	British, Turkish
G1266K	ATP hinge	18	disrupts ATP hinge	French British
N1270S [1]	ATP Hinge	18	disrupts ATP hinge	Italian

Mutations are from (29) except those marked [1], described in (22).
[2] Conservative changed, described as a possible polymorphism (29), now believed to be a disease mutation.
Numerical designations have been altered from the original (29):three extra nucleotides were added in error at positions 1039,1044,1056 of the original sequence (21);32 amino acids (or 60 nucleotides) of exon 1 (24) have been added.

born copper-deficient and will die if suckled on an affected mother. Death is due to copper depletion, so apparently transport of copper to the fetus during pregnancy and from the mother's milk is impaired.

4. CONCLUSION

Discovery of the basic genetic defect in Wilson and Menkes diseases has greatly increased our understanding of copper transport. Many questions still remain. Furthermore, the role of the ATPase in incorporating copper into ceruloplasmin is still to be identified.

The connection between copper and iron transport is particularly interesting. When the Wilson or Menkes disease gene equivalent in yeast is defective, no incorporation into a ferroxidase similar to ceruloplasmin takes place (2). The result is toxicity due to lack of transport of iron. The same connection with iron transport has been shown in the condition aceruloplasminemia (32). No ceruloplasmin protein is produced, and as a result the major function of this protein as a ferroxidase in various tissues is lacking. The result is an overload of iron in liver, retina, pancreas and brain.

In addition to further studies on copper transport, a window has now been opened for the possibility of studying the transport of other heavy metals. One of the most exciting aspects of this story has been the discovery of similarity with bacterial metal transporting genes (23). Bacteria resistant to chromium, cobalt, nickel, zinc, arsenic, silver have been identified, but no equivalent genes have yet been identified in higher organisms. We are on the frontier of discoveries of metal transporting pathways.

REFERENCES

1. Odermatt, A., Suter, H., Krapf, R., and Solioz, M. *J. Biol. Chem.* **268**, 12775–12779 (1993)
2. Yuan, D. S., Stearman, R., Dancis, A., Dunn, T., Beeler, T., and Klausner, R. D. *Proc. Natl. Acad. Sci. USA* **92**, 2632–2636 (1995)
3. Danks, D. M. in *The Molecular and Metabolic Basis of Inherited Disease* (C. R. Scriver, A. L. Beaudet, W. S. Sly, and D. Valle, eds.) McGraw-Hill, New York (1995). 4125–4158.
4. Tumer, Z. and Horn, N. *Ann. Med.* **28**, (in press)
5. Sarkar, B., Lingertat-Walsh, K., and Clarke, J. T. R. *J. Pediat.* **123**, 828–830 (1993)
6. Vulpe, C., Levinson, B., Whitney, S., Packman, S., and Gitschier, J. *Nature Genet.* **3**, 7–13 (1993)
7. Chelly, J., Tumer, Z., Tonnesen, T., Petterson, A., Ishikawa-Brush, Y., Tommerup, N., Horn, N., and Monaco, A. P. *Nature Genet.* **3**, 14–19 (1993)
8. Mercer, J. F. B., Livingstone, J., Hall, B., Paynter, J. A., Begy, C., Chandrasekharappa, S., Lockhart, P., Grimes, A., Bhave, M., Siemieniak, D., and Glover, T. W. *Nature Genet.* **3**, 20–25 (1993)
9. Odermatt, A., Suter, H., Krapf, R., and Solioz, M. (1993) *Ann. N. Y. Acad. Sci.* 484–486
10. Levinson, B., Gitschier, J., Vulpe, C., Whitney, S., Yang, S., and Packman, S. *Nature Genet.* **3**, 6–11 (1993)
11. Das, S., Levinson, B., Whitney, S., Vulpe, C., Packman, S., and Gitschier, J. *Am J Hum Genet* **55**, 883–889 (1994)
12. Levinson, B., Vulpe, C., Elder, B., Martin, J., Verley, F., Packman, S., and Gitschier, J. *Nature Genet.* **6**, 369–378 (1991)
13. Martins da Costa, C., Baldwin, D., Portmann, B., Lolin, Y., Mowat, A. P., and Mieli-Vergani, G. (1992) *Hepatology* **15**, 609–615
14. Sarkar, B. in *Metal ions in biological systems* (Sigel, H., ed) Marcel Dekker, Inc., New York pp. 233–281 (1981),
15. Hoogenraad, T. U., Van Haltum, J., and Van der Hamer, C. j. A. *J Neurol Sci* **77**, 137–146 (1987)
16. Frydman, M., Bonné-Tamir, B., Farrer, L. A., Conneally, P. M., Magazanik, A., Ashbel, A., and Goldwitch, Z. *Proc. Natl. Acad. Sci. USA* **82**, 1819–1821
17. Bowcock, A. M., Farrer, L. A., Cavalli-Sforza, L. L., Hebert, J. M., Kidd, K. K., Frydman, M., and Bonné-Tamir, B. (1987) *Am J Hum Genet* **41**, 27–35 (1985)
18. Farrer, L. A., Bowcock, A. M., Hebert, J. M., Bonné-Tamir, B., Sternlieb, I., Giagheddu, M., St.George-Hyslop, P., Frydman, M., Lössner, J., Demelia, L., Carcassi, C., Lee, R., Beker, R., Bale, A. E., Donis-Keller, H., Scheinberg, I. H., and Cavalli-Sforza, L. L. *Neurology* **41**, 992–999 (1991)
19. Houwen, R. H. J., Thomas, G. R., Roberts, E. A., and Cox, D. W. *J Hepatol* **17**, 269–276 (1993)
20. Bull, P. C. and Cox, D. W. *Genomics* **16**, 593–598 (1993)
21. Bull, P. C., Thomas, G. R., Rommens, J. M., Forbes, J. R., and Cox, D. W. *Nature Genet.* **5**, 327–337 (1993)
22. Tanzi, R. E., Petrushkin, K. E., Chernov, I., Pellequer, J. L., Wasco, W., Ross, B., Romano, D. M., Parano, E., Pavone, L., Brzustowicz, L. M., Devoto, M., Peppercorn, J., Bush, A. I., Sternlieb, I., Piratsu, M., Gusella, J. F. Evgrafov, O., Penchaszadeh, G. K., Honig, B., Edelman, I. S., Soares, M. B., Scheinberg, I. H, and Gilliam, T. C. *Nature Genet* **5**, 344–350 (1995)
23. Bull, P. C. and Cox, D. W. *Trends Genet* **10**, 246–252 (1994)
24. Petrukhin, K. E., Lutsenko, S., Chernov, I., Ross, B. M., Kaplan, J. H., and Gilliam, T. C. *Hum Mol Genet* **3**, 1647–1656 (1994)
25. Thomas, G. R., Jensson, O., Gudmundsson, G., Thorsteinsson, L., and Cox, D. W. *Human Genet* **56**, 1140–1146 (1995)
26. Thomas, G. R., Roberts, E. A., Rosales, T. O., Moroz, S. P., Lambert, M. A., Wong, L. T. K., and Cox, D. W. *Hum Mol Genet* **2**, 1401–1405 (1993)
27. Thomas, G. R., Roberts, E. A., Walshe, J. M., and Cox, D. W. *Am J Hum Genet* **56**, 1315–1319 (1995)
28. Cox, D. W. and Billingsley, G. D. in *Genetics of Neuropsychiatric Diseases. Wenner-Gren International Symposium* (Wetterberg, L., ed), Macmillan, London pp. 167–177 (1989)

29. Thomas, G. R., Forbes, J. R., Roberts, E. A., Walshe, J. M., and Cox, D. W. *Nature Genet.* **9**, 210–217 (1994)
30. Figus, A., Angius, A., Loudianos, G., Bertini, C., Dessi, V., Loi, A., and Deiana, A. et al *Am J Hum Genet* **57**, 1318–1324 (1996)
31. Wu, J., Forbes, J. R., Shiene Chen, H., and Cox, D. W. *Nature Genet.* **7**, 541–545 (1994)
32. Harris, Z. L., Takahashi, Y., Miyajima, H., Serizawa, M., MacGillivray, R. T., and Gitlin, J. D., *Proc Natl Acad Sci (USA)* **92**, 2539–2543 (1989)

73

THE *ENTEROCOCCUS HIRAE* COPPER ATPASES

Structure, Function, and Regulation

P. Duda, D. Strausak, and M. Solioz

Department of Clinical Pharmacology
University of Berne
3010 Berne, Switzerland

1. INTRODUCTION

Copper is an essential element by functioning as a cofactor in many redox enzymes such as cytochrome c oxidase, lysyl oxidase, ascorbate oxidase, dopamine ß-hydroxylase, tyrosinases and superoxide dismutase. But copper is also very toxic to both eukaryotic and prokaryotic cells: it can initiate the formation of cell damaging radicals which oxidize proteins, DNA and biological membranes. Thus, homeostatic mechanisms have evolved to regulate intracellular copper concentration.

Several different copper regulating systems have been described so far, notably in *Escherichia coli* and *Pseudomonas syringae* (1,2), but the mode of copper transport remains unclear. Recently, primary, ATP-driven copper transport was described in the Gram-positive bacterium *Enterococcus hirae* (3). In this organism the intracellular copper concentration is regulated by two P-type copper ATPases, CopA and CopB. CopA serves in the uptake and CopB in the extrusion of copper. Subsequently, a number of other copper ATPases have been discovered. In humans, two genes that are defective in Menkes and Wilson disease, respectively, were found to encode putative copper ATPases (4–7). Based on sequence similarity, putative copper ATPases were also found in *Saccharomyces cerevisiae* (8,9), *Escherichia coli* (10), *Synechococcus* (11,12) and *Helicobacter pylori* (13).

All these ATPases differ significantly in their structure from the previously known P-type ATPases such as the Ca^{2+}-ATPases or the Na^+K^+-ATPases and form a distinct subclass, the CPX-type ATPases. Bacterial cadmium ATPases also appear to belong to the subclass of CPX-type ATPases (14–16). In this report we will discuss the structure, function, and regulation of the *E. hirae* copper ATPases and discuss other putative copper ATPases and the cadmium ATPases in the light of the current knowledge on copper homeostasis in *E. hirae*.

Therapeutic Uses of Trace Elements, edited by Nève et al.
Plenum Press, New York, 1996

2. RESULTS

2.1. Structure of the *cop*-Operon of *Enterococcus hirae*

We recently cloned and sequenced an operon involved in copper homeostasis in *Enterococcus hirae* (3). This *cop*-operon consists of the four genes *cop*YZAB (Fig. 1). *Cop*Y and *cop*Z code for two polar, regulatory proteins of 145 and 69 amino acids, and *cop*A and *cop*B encode P-type copper ATPases of 727 and 745 amino acids. Upstream of the first gene is a promotor region that contains inverted repeat sequences and could thus form a cruciform structure. The ATG initiation codon follows immediately this structure. The function of the promotor region and the four *cop*- genes was studied in some detail and will be described below.

2.2. Regulation of the *cop*-Operon

The expression of the cop-operon of E. hirae is regulated by the ambient copper level. Induction was observed by both high and low copper concentrations with minimal expression in 10 μM extracellular copper (17). The first two gene products of the *cop*-operon, CopY and CopZ, appeared to control this copper dependent expression. Wild-type cells that were disrupted in the *cop*Y-gene (Δ*cop*Y) showed a massive overexpression of both copper ATPases, whereas cells that were disrupted in the *cop*Z-gene (Δ*cop*Z) expressed CopA and CopB poorly (18). These data led to the conclusion that CopY acts as a repressor and CopZ as an inducer of the *cop*-operon.

2.3. Interaction of *cop*Y with the Promotor

The predicted gene product of the *cop*Y gene is a protein of 145 amino acids. It has limited sequence similarity to the repressor proteins MecI of *Staphylococcus epidermidis*,

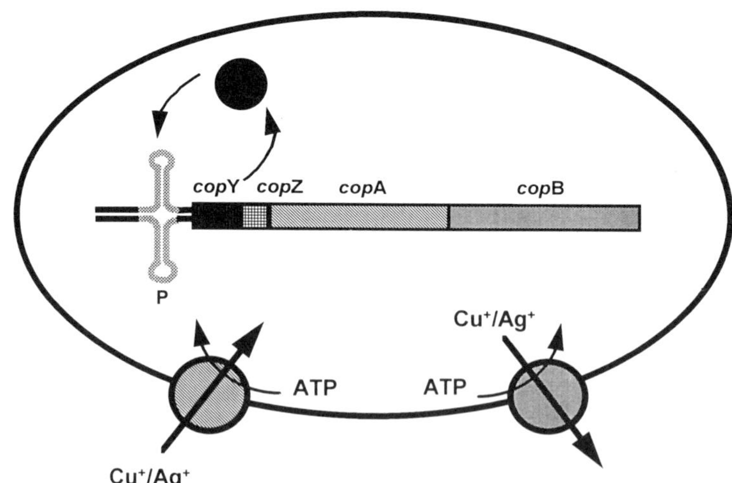

Figure 1. Schematic drawing of Enterococcus hirae with the cop-operon. The operon consists of the four genes, *cop*YZAB. *Cop*Y encodes a repressor protein and *cop*Z an activator protein of not clearly defined function. *Cop*A and *cop*B code for the two copper ATPases. P denotes the promotor that contains an inverted repeat and could thus form a cruciform structure. The ATG initaition codon would immediately follow this structure.

PenI of *Bacillus licheniformis* and BlaI of *Staphylococcus aureus* (19,20) that regulate β-lactamase expression in these organisms. That CopY is also a repressor could be shown by the direct interaction of CopY with the *cop*-promotor. CopY was overexpressed in *Escherichia coli* and purified to near homogeneity. The interaction of the purified repressor with the promotor region was shown in band shift assays as follows: DNA fragments of 530 basepairs encompassing the putative promotor region were incubated with purified repressor protein. The formation of DNA-protein complexes was visualized by the change in electrophoretic mobility of the radioactively labeled DNA band on polyacrylamide gels. Fig. 2 shows the result of such a band shift assay. Increasing concentrations of CopY lead to a shift of the DNA band. This shift occurs in two steps, suggesting that two monomers or two multimers of the repressor interact with the promotor sequence. CopY binds to the promotor region with an apparent K_m of 0.6 μM. Competition experiments with either cold promotor DNA or DNA carrying the promotor of the NaH-antiporter gene of *E. hirae* clearly demonstrate that CopY binds specifically to the *cop* promotor.

In DNaseI footprinting experiments we mapped the site of repressor-promotor interaction (Fig. 3). The area of DNA protected by CopY ranges from position -12 to -73 with respect to the ATG translational start codon. At position -42, there is a DNaseI-sensitive site, originating from exposed bases between the two bound repressor units. The -42 position corresponds to the transcriptional start site. The region protected by repressor binding thus encompasses the two inverted repeats and the transcriptional start site at position -42.

The promotor of the *cop*-operon contains an inverted repeat that has the potential to form a so called cruciform structure. Such cruciform structures are known to play a role in some DNA-protein interactions. The inverted repeats of the *cop*-promotor have the sequence TCGATTACA(G/T)TTGTAA. An ACA trinucleotide is invariant in most prokaryotic promotors (21). The crystal structure of the bacteriophage 434 repressor revealed that the side chains of the Gln-Gln-29 pair specifically interact with the ACA trinucleotide (22). A Gln-Gln pair is found in CopY at position 30/31 and it appears likely that this, too, is the site of interaction with ACA in the promotor region. Site-directed mutation will be used to demonstrate this interaction directly.

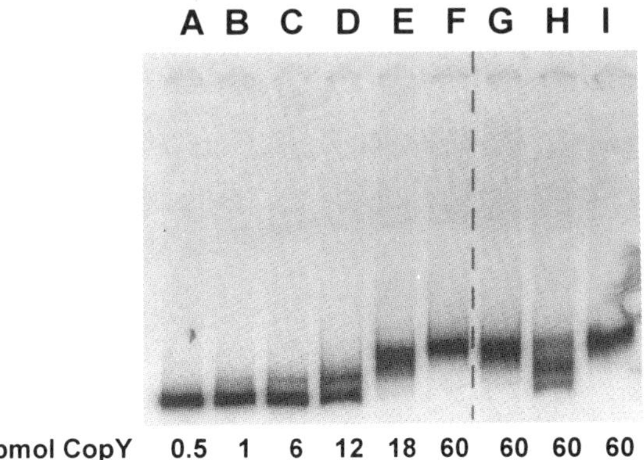

Figure 2. Interaction of CopY with the cop-promotor. 1.5 fmol of 530 bp DNA fragments carrying the *cop*-promotor were labeled with ^{32}P at the 3'-end of the sense strand and incubated with CopY under reducing conditions at RT for 30 min and subsequently resolved by gel electrophoresis on a low ionic strength 5% polyacrylamide gel. *Lanes A to F*, reaction with increasing amounts of CopY as indicated below the lanes; *lanes G and H*, competition with 30 and 100 fmol, respectively, of unlabeled *cop*-promotor DNA; *lane I*, competition with 100 fmol of unlabeled promotor DNA of the *E. hirae* NaH-antiporter gene

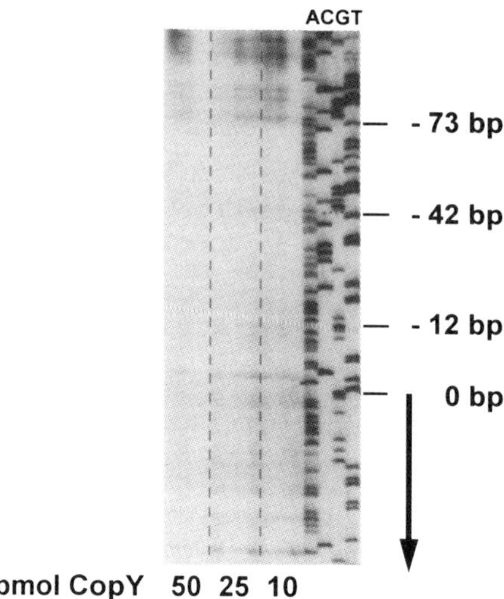

Figure 3. Localization of repressor-promotor interaction. 1.5 fmol of *cop*-promoter fragment, labeled at one end with ^{32}P, was incubated with the indicated amounts of CopY at room temperature for 30 min. 0.01U of DNaseI was added and incubated at 30°C for 2 min. The reactions were stopped by precipitation with ethanol and the reaction products resolved on 6% polyacrylamide sequencing gels. *Lanes A, C, G, T*, sequencing reaction according to Sanger.

2.4. Role of CopZ in Regulation

CopZ is a small, soluble protein of 69 amino acids. It contains the putative metal binding motif GMXCXXC, suggesting that CopZ is a copper binding protein. CopZ shares sequence identity with proteins that have similar metal binding motifs in the N-terminus, such as CopA of *E. hirae*, the human Menkes and Wilson copper ATPases, and mercuric reductases (18). The study of CopZ knock-out mutants had shown that this gene product is required for the expression of the *cop*-operon (18). However, purified CopZ protein did not in any observable way interact with the promotor-repressor complex in *in vitro* experiments (Strausak, unpublished results). Conceivably, CopZ acts as a metallothionein-like protein by binding excessive intracellular copper and thereby preventing cell damage. But an interaction of CopZ with the repressor and/or promotor or a complex thereof can not be excluded at present.

2.5. Model of the Regulation of the *cop*-Operon

We propose the model for the regulation of the *cop*-operon outlined in Fig. 4. When the copper concentration in the cell is limiting, CopY and CopZ have no copper bound and are in a soluble form, allowing transcription of the *cop*-operon at maximal rate. When the copper concentration reaches a physiological range, CopY binds copper and represses transcription by binding to the promotor. At toxic cytoplasmic copper concentrations, CopZ may sequester copper in a metallothionein like way, but could also interact in some way with the repressor and/or the promotor. Through which mechanism CopZ stimulates the induction of the *cop*-operon *in vivo* remains to be shown.

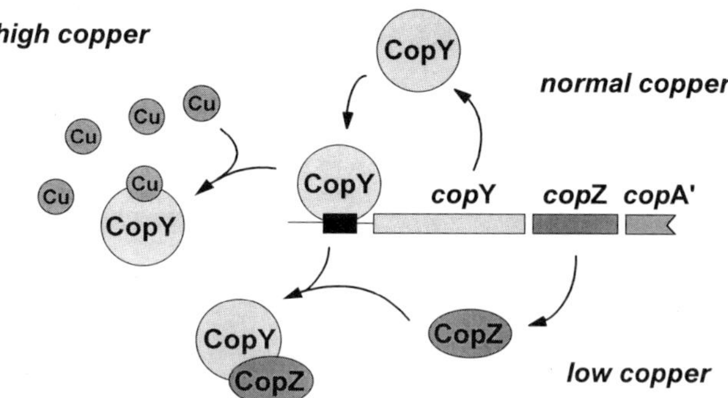

Figure 4. Model of the regulation of the *cop*-operon in *E. hirae*. Under physiological conditions, CopY binds to the repressor and shuts off transcription. When excessive copper is present, CopY is released. At limiting copper concentrations, CopZ may interact with CopY to release it from the DNA.

2.6. Function of the CopA and CopB ATPases

CopA and CopB were the first putative copper transporting ATPases to be described (3). Their sizes are predicted to be 727 and 745 amino acids, respectively, and they clearly belong to the class of P-type ATPases. Sequence similarity of the N-terminus of CopB to known copper binding proteins of *Pseudomonas syringae* had originally suggested an involvement in copper metabolism (23). Supportive evidence for this came from the study of mutant strains of *E. hirae*: null-mutation of the *cop*B gene (ΔcopB) increased sensitivity of the cells to high ambient copper while disruption of the *cop*A gene (ΔcopA) impaired survival under copper limiting conditions (17). From these observations, a model was derived, in which the CopA ATPase mediates copper import into the cell while the CopB ATPase is responsible for extrusion of excess copper from the cytoplasm.

Support for the role of CopB in export came from transport studies on whole cells. *E. hirae* loaded with ^{110m}Ag$^+$ extruded the radioactive silver by the action of the CopB ATPase (17). In these experiments, Ag$^+$ was a substitute for copper, which is the natural substrate of CopB.

CopB is the only copper ATPase for which copper transport has been shown biochemically (24). Native inside-out vesicles accumulated ^{64}Cu from the incubation medium. Accumulation by native inside-out membrane vesicles corresponds of course to extrusion in intact cells. Copper uptake was only observed when the vesicles were derived from *E. hirae* strains expressing the CopB ATPase. No copper accumulation was detected with ΔCopB or ΔCopAΔCopB membrane vesicles. Cu$^+$ rather than Cu^{2+} was transported, since accumulation only occurred under reducing conditions where copper is present as copper(I). ^{110m}Ag$^+$ was also a substrate for CopB. Transport occurred with an apparent K$_m$ for Cu$^+$ and Ag$^+$ of 1 µM and for ATP of 10 µM; v_{max} was 0.07 nmol/min/mg protein at pH 6.0 and was the same for Cu$^+$ and Ag$^+$ (24). Vanadate, a specific inhibitor of P-type ATPases, inhibited transport of copper and silver at 40 and 60 µM, respectively.

2.7. Purification of CopB

For further functional analysis, CopB was purified from membranes of a ΔCopY strain of *E. hirae*. This repressor-deficient strain overexpresses CopA and CopB about 50-fold (18).

Membranes from Y1 were prepared essentially as described (24) and proteins extracted with dodecyl-β-D-maltoside. The detergent extract was bound to a Ni-NTA agarose column and eluted with imidazole. This step removed the majority of all contaminating proteins. Final purification was achieved by anion-exchange chromatography on Mono Q Sepharose (Fig. 5). The purified CopB ATPase could be stored frozen for several month without significant loss of activity.

2.8. Properties of Purified CopB

The ATPase activity of purified, detergent-solubilized CopB ATPase, measured by detection of hydrolyzed P_i as described (25), was 30 to 50 nmol/min/mg. When the ATPase was reconstituted into asolectin proteoliposomes by the method of Apell et al. (26), a 10-fold increase in ATPase activity was observed. 50–80% of this activity could be inhibited by vanadate (Fig. 6). In most, but not all experiments, vanadate concentrations below 10 μM stimulated the ATPase activity. At any rate, the purified, reconstituted CopB ATPase appears suitable to study the transport functions of this interesting enzyme in a defined system.

2.9. Evolution of P-Type ATPases

While copper transport by CopB of *E. hirae* has unequivocally been demonstrated, the function of other putative copper ATPases remains to be established. Much attention has recently been attracted to the field of copper transport by the discovery of two human homologues of the *E. hirae* copper ATPases: the Menkes and the Wilson ATPases (4–7,27). These proteins appear to be responsible for cellular copper homeostasis and mutations in them cause either excessive copper accumulation in liver and brain as in Wilson's disease or copper deficiency in certain brain cells as in Menkes' disease.

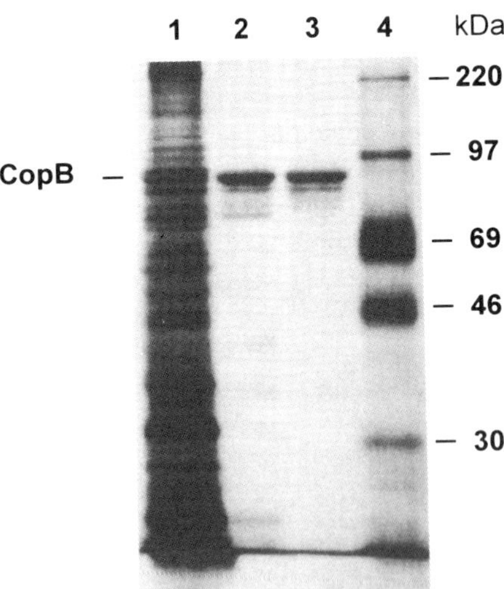

Figure 5. Purification of the CopB ATPase. Samples from the different purification steps were analyzed by gel electrophoresis on a 10% SDS-polyacrylamide gel and the proteins visualized by silver staining. Lane 1, 50 μg extracted membrane protein; lane 2, 5 μg of the CopB-containing fraction eluted from the Ni-NTA-column; lane 3, 5 μg of CopB eluted from the Mono Q column; lane 4, marker proteins of the molecular weights indicated on the Figure in kDa. Protein concentrations were assessed by the method of Bradford (29).

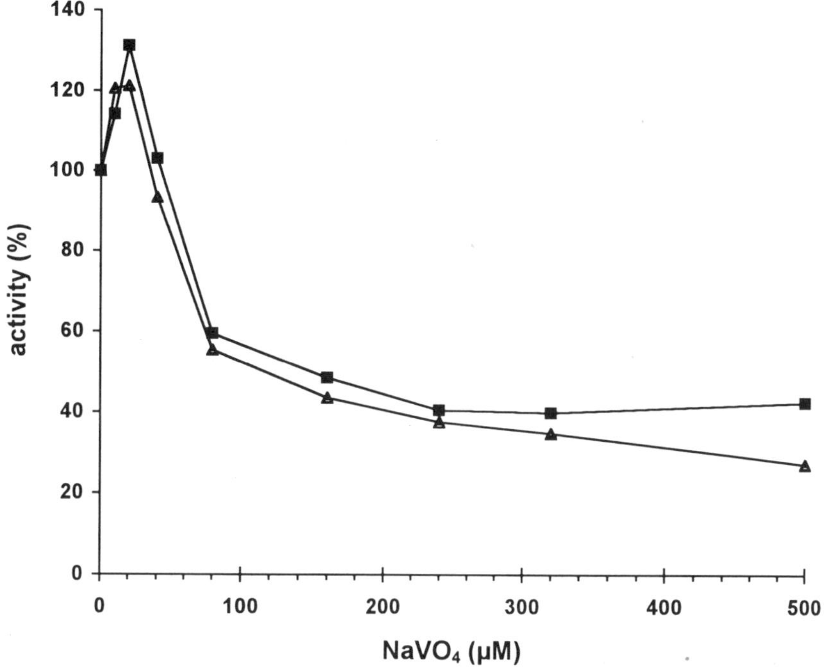

Figure 6. Inhibition of CopB ATPase activity by Na_2VO_4. ATP hydrolysis by CopB was measured during 60 min in the absence or presence of the indicated vanadate concentrations in buffer containing either 2 µM $CuSO_4$ (Δ) or 2 µM $AgNO_3$ ($\square$). Activity is given as percent of controls without vanadate.

Currently, the sequence of over a dozen putative copper ATPases from species as diverse as bacteria, yeast, and humans are known. Clearly, all the copper ATPases belong to the large family of P-type ATPases. The copper ATPases share with the P-type ATPases the highly conserved motifs involved in the interaction with ATP and the formation of an aspartyl-phosphate intermediate during the reaction cycle (28). Without a three-dimensional structure of a P-type ATPase available, one has to contend with model building and predictions

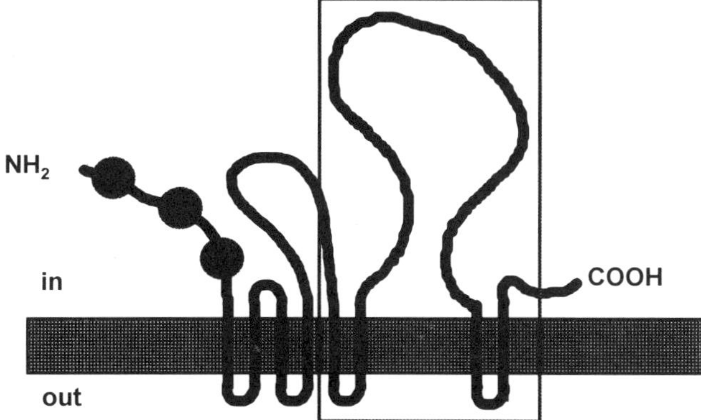

Figure 7. Predicted structure of CPX-type ATPases. The P-type-specific core domain is boxed. The number of heavy metal binding sites (filled circles) in the hydrophilic N-terminus varies from one to six in the different copper ATPases and from one to two in the cadmium ATPases.

based on the primary structure. Fig. 7 shows a crude model for the membrane topology of a copper ATPase as accepted by most workers in the field.

The putative copper ATPases as well as the closely related bacterial cadmium ATPases (14) differ from the non-heavy metal ATPases by possessing a number of features not present in non-heavy metal ATPases, namely: (i) cysteine- or histidine-rich heavy metal binding motifs in the hydrophilic N-terminus, (ii) a conserved CPX-motif (Cys-Pro-Cys/His/Ser) in the putative ion channel domain, (iii) a conserved His-Pro 34 to 43 amino acids downstream of the CPX-motif, (iv) two additional predicted transmembrane helices at the N-terminus, and (v) only two transmembrane helices at the C-terminus instead of the four to six helices predicted for non-heavy metal P-type ATPases. Not surprisingly, then, phylogenetic analysis of all the P-type ATPases reveals the existence of a distinct subclass of the P-type ATPases, formed by the copper and cadmium ATPases (Fig. 8). Based on the most striking common feature of all heavy metal transporting ATPases, the CPX-motif in the ion channel domain, we propose this distinct subclass of P-type ATPases to be called CPX-type ATPases.

ACKNOWLEDGMENT

This work was supported by Grant 32–37527.92 from the Swiss Nationl Foundation and by a grant from the Roche Research Foundation.

REFERENCES

1. C. Cervantes and F. Gutierrez-Corona, *FEMS Microbiol. Rev.* **14**, 121–138 (1994).
2. D. A. Cooksey, *FEMS Microbiol. Rev.* **14**, 381–386 (1994).
3. A. Odermatt, H. Suter, R. Krapf and M. Solioz, *Ann. N. Y. Acad. Sci.* **671**, 484–486 (1992).
4. C. Vulpe, B. Levinson, S. Whitney, S. Packman and J. Gitschier, *Nature Genet.* **3**, 7–13 (1993).
5. J. F. B. Mercer, J. Livingston, B. Hall, J. A. Paynter, C. Begy, S. Chandrasekharappa, P. Lockhart, A. Grimes, M. Bhave, D. Siemieniak and T. W. Glover, *Nature Genet.* **3**, 20–25 (1993).
6. R. E. Tanzi, K. Petrukhin, I. Chernov, J. L. Pellequer, W. Wasco, B. Ross, D. M. Romano, E. Parano, L. Pavone and L. M. Brzustowicz, *Nature Genet.* **5**, 344–350 (1993).
7. P. C. Bull, G. R. Thomas, J. M. Rommens, J. R. Forbes and D. W. Cox, *Nature Genet.* **5**, 327–337 (1993).
8. M. R. Rad, L. Kirchrath and C. P. Hollenberg, *Yeast* **10**, 1217–1225 (1994).
9. D. Fu, T. J. Beeler and T. M. Dunn, *Yeast* **11**, 283–292 (1995).
10. C. Trenor, III, W. Lin and N. C. Andrews, *Biochem. Biophys. Res. Commun.* **205**, 1644–1650 (1994).
11. L. T. Phung, G. Ajlani and R. Haselkorn, *Proc. Natl. Acad. Sci. U. S. A.* **91**, 9651–9654 (1994).
12. K. Kanamaru, S. Kashiwagi and T. Mizuno, *Mol. Microbiol.* **13**, 369–377 (1994).
13. Z. Ge, K. Hiratsuka and D. E. Taylor, *Molec. Microbiol.* **15**, 97–106 (1995).
14. G. Nucifora, L. Chu, T. K. Misra and S. Silver, *Proc. Natl. Acad. Sci. U. S. A.* **86**, 3544–3548 (1989).
15. M. Lebrun, A. Audurier and P. Cossart, *J. Bacteriol.* **176**, 3049–3061 (1994).
16. M. Lebrun, A. Audurier and P. Cossart, *J. Bacteriol.* **176**, 3040–3048 (1994).
17. A. Odermatt, R. Krapf and M. Solioz, *Biochem. Biophys. Res. Commun.* **202**, 44–48 (1994).
18. A. Odermatt and M. Solioz, *J. Biol. Chem.* **270**, 4349–4354 (1995).
19. T. Himeno, T. Imanaka and S. Aiba, *J. Bacteriol.* **168**, 1128–1132 (1986).
20. V. Wittman and H. C. Wong, *J. Bacteriol.* **170**, 3206–3212 (1988).
21. G. B. Koudelka, S. C. Harrison and M. Ptashne, *Nature* **326**, 886–888 (1987).
22. J. E. Anderson, M. Ptashne and S. C. Harrison, *Nature* **326**, 846–852 (1987).
23. A. Odermatt, H. Suter, R. Krapf and M. Solioz, *J. Biol. Chem.* **268**, 12775–12779 (1993).
24. M. Solioz and A. Odermatt, *J. Biol. Chem.* **270**, 9217–9221 (1995).
25. P. Lanzetta, A., L. Alvarez, J., P. Reinach, S. and O. Candia, A. *Anal. Biochem.* **100**, 95–97 (1979).
26. H.-J. Apell and M. Solioz, *Biochim. Biophys. Acta*, **1017**, 221–228 (1990).
27. J. Chelly, Z. Tumer, T. Tonnesen, A. Petterson, Y. Ishikawa Brush, N. Tommerup, N. Horn and A. P. Monaco, *Nature Genet.* **3**, 14–19 (1993).
28. P. L. Pedersen and E. Carafoli, *Trends Biochem. Sci.* **12**, 186–189 (1987).
29. M. M. Bradford, *Anal. Biochem.* **72**, 248–254 (1976).

74

MOLECULAR GENETICS OF WILSON DISEASE

Study of 12 Families

M. Bost,[1,2] M. Accominotti,[3] A. Lachaux,[4] F. Régnier,[4] G. Chazot,[2,5] and
A. Vandenberghe[1,6]

[1] Laboratoire de Neurogénétique, Hôpital de l'Antiquaille
Lyon, France
[2] Trace Element - Institute for UNESCO
Lyon, France
[3] Laboratoire de Biochimie - Analyse des Traces
[4] Service de Gastroentérologie Infantile, Hôpital Edouard-Herriot
Lyon, France
[5] Hôpital Neurologique, Lyon, France
[6] Institut des Sciences Pharmaceutiques et Biologiques
Lyon, France

1. INTRODUCTION

Wilson disease (WD) or "hepatolenticular degeneration" is an autosomal recessive disorder of copper (Cu) transport. Early diagnosis and treatment with copper-chelating agents (penicillamine) or by blocking intestinal Cu absorption (zinc salts) are essential to prevent Cu accumulation and irreversible damage to tissues such as liver and brain. Diagnosis is usually based on low ceruloplasmin and plasma copper, increased copper excretion in urine, and copper deposits in the cornea (Kayser-Fleisher ring) and liver. However, wide variations may occur in the biochemical tests and sometimes diagnosis is not entirely reliable.

The genetic locus for the WD gene has been assigned to 13q14–21 by linkage analysis (1). The gene product has been identified as Cu transporting P-type ATPase (2) and about 40 disease specific mutations (3,4) have been described. Carrier detection by analysis of mutations is difficult because of allelic heterogeneity. On the contrary, it is now greatly facilitated in families, by the development of several highly polymorphic microsatellite markers closely linked and in linkage disequilibrium with the WD locus (5), if one affected sibling and his parents are available for genotyping.

In this paper, we report a study of 12 WD pedigrees (16 WD patients), well described clinically and biologically in which one child is affected; genetic analysis was performed using microsatellite markers and by detecting the most frequent WD-specific mutations.

2. MATERIALS AND METHODS

2.1. Presentation of WD Patients Studied

Peripheral blood was collected from 7 French, 1 French and Moroccan, 1 Italian, 1 Portugese, 1 Moroccan families living in France and from 1 Spanish family living in Spain. The WD patients presented different clinical manifestations: 12 with a hepatic form, 2 with a neurological form, 1 with a hepatic and neurological form, 1 with an ophthalmological presentation. The age of onset varied from 5 to 32 y. The range of Cu biochemical parameters (plasma copper, ceruloplasmin concentrations, urinary copper excretion and liver copper levels) presented heterogeneous values: ceruloplasmin: 60–180 mg/l; serum copper: 2.06–21.8 µmol/l; urinary copper after penicillamine: 0.74–1163 µmol/24h; hepatic copper: 8.23–40.7 µmol/g dry matter.

2.2. DNA Analysis

DNA was extracted from whole blood (20 ml) collected in EDTA by a phenol/chloroform method. WD gene flanking markers used in this study are D13S227, D13S228, D13S314, D13S315 and D13S316. The PCR amplification of these markers was carried out using pairs of specific primers (5, 6) in 50 µl total volume containing 50 mM KCl; 10 mM Tris pH 8.0; 10 mg/ml BSA; 1.5 mM $MgCl_2$; 200 µmol each of dCTP, dGTP, dTTP, dATP; 2.5 U *Taq* polymerase (ATGC), 100 ng of genomic DNA, 20 pmol of each primer.

Amplification was performed in a Perkin Elmer programmable thermal controller: 10 min denaturation at 94 °C, then 30 cycles of 30 s denaturation at 94 °C, 30 s annealing at the appropriate temperature (D13S227 = 58 °C; D13S228 = 58°C; D13S314 = 62 °C; D13S315 = 58 °C; D13S316 = 62 °C) and 30 s extension at 72 °C then 10 min final extension at 72 °C. The PCR samples were then submitted to an electrophoresis through 7% denaturing polyacrylamide gels which were silver-stained.

2.3. Detection of Mutation

Each individual exon (7, 9 ,14 and 18) was amplified using primers (3) complementary to the DNA sequences flanking the exon-intron boundaries, under conditions identical to those used for microsatellite marker analysis, and labelled with 1 pmol of ^{32}P-dCTP.

For SSCP ("Single Strand Conformation Polymorphism") analysis, 5 µl of the PCR product were mixed with 2 µl of a solution containing 95% formamide, 20 mM EDTA, 0.05% bromophenol blue, 0.05% xylene cyanol. The samples were denatured at 110 °C, cooled on ice and then applied on a 6 % polyacrylamide gel containing or not 4% glycerol, 90 mM Tris borate and 10 mM EDTA. Electrophoresis was performed with a sequencing apparatus at a constant temperature of 20 °C and a constant power of 8 W for 17 h. After electrophoresis, the gels were dried and exposed to film.

3. RESULTS

3.1. DNA Analysis with Microsatellite Markers

We carried out presymptomatic testing in a Portugese family in which an adolescent was affected (hepatic transplantation) and the carrier status of her brother was unknown due to uncertain biochemical parameters of Cu accumulation (ceruloplasmin: 120 mg/l; serum Cu: 7.25 µmol/l; basal urinary Cu: 2.26 µmol/24 h; urinary Cu after penicillamine: 19.7 µmol/24 h; liver Cu: 2.84 µmol/g). We used microsatellite markers (Figure 1) to establish an indirect molecular diagnosis of WD for this child.

3.2. Mutation Detection

Mutation screening for WD patients in the 12 families detected one mutation on exon 14 (Figure 2).

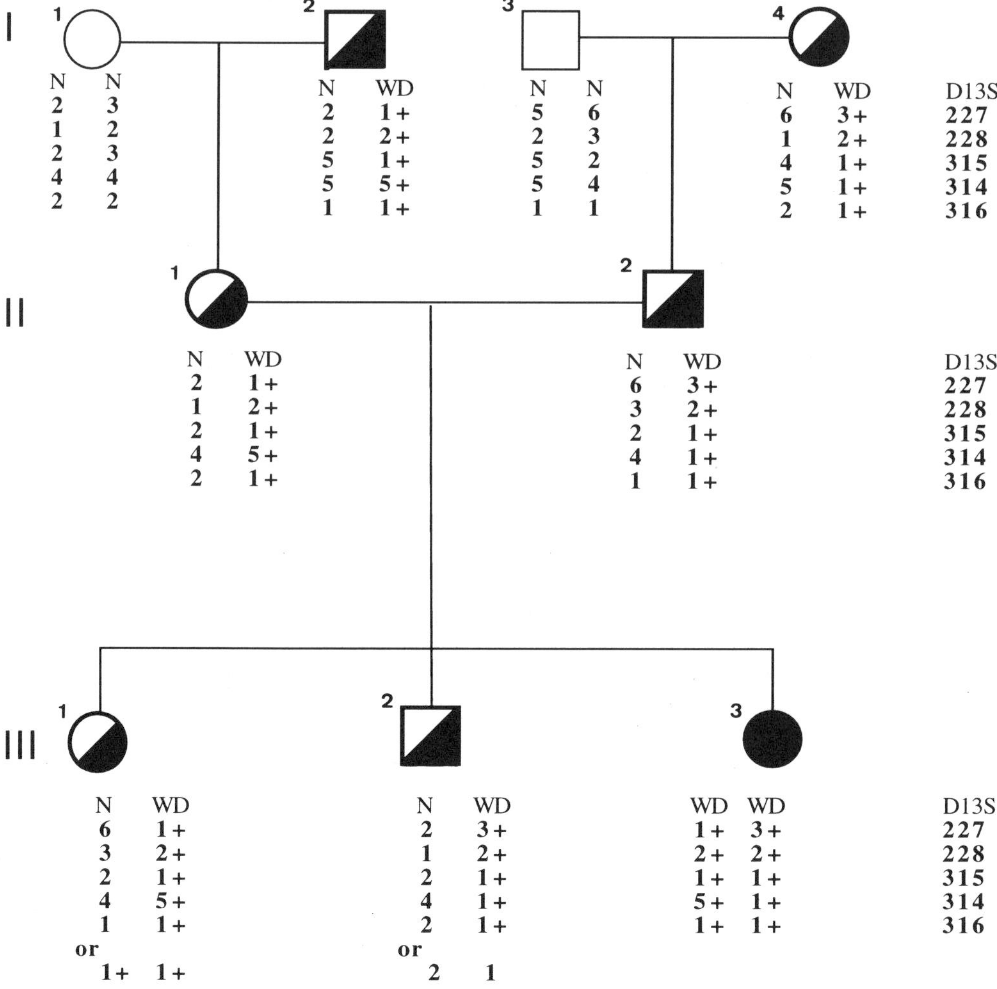

Figure 1. Pedigree with allelic distribution of all the microsatellite markers used in the study. Patients are shown as shaded symbols and heterozygotes as half-shaded symbols ; N = normal chromosome, WD = Wilson disease chromosome.

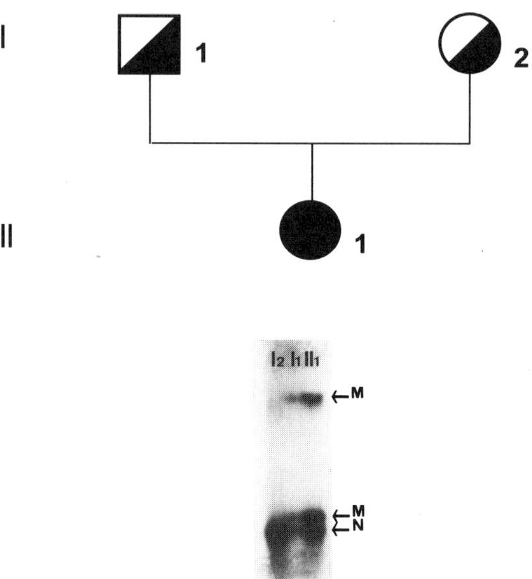

Figure 2. Detection of WD mutation in exon 14 by SSCP analysis. (N = normal allele, M = mutant allele.)

4. DISCUSSION

The Portugese WD family studied (Figure 1) was not informative for D13S316 because the father (II$_2$) was homozygous for this marker. However, the microsatelllite markers D13S227, D13S228, D13S314 and D13S315 enabled us to conclude that the patient's brother (III$_2$), who had uncertain biochemical parameters of Cu metabolism, is heterozygous for WD. In the other 11 families, the use of microsatellite markers allowed the detection of 9 carriers and 4 asymptomatic WD homozygotes. This DNA test has various advantages over other diagnostic methods. Firstly, it has been argued that WD cannot be accurately diagnosed on the basis of biochemical Cu determinations in patients under the age of 1 to 2 years because these values remain unchanged in young WD children as Cu overload in tissues is not yet excessive. Secondly, invasive liver biopsy which is often necessary to accurately confirm diagnosis, can be avoided. Further, the identification of presymptomatic WD homozygotes would enable therapeutic action with available Cu chelating agents (penicillamine) or zinc to prevent symptoms before irreparable liver or neurological damage. SSCP analysis detected a mutation in exon 14 (which will be sequenced later) in 1 WD family (a 7 year-old; hepatic form). One mutation detected on exon 14 for 12 unrelated patients represents nearly 8%; this does not confirm published results for the His 1070 → Gln mutation (28%) if this is the mutation in question. The difference could be due either to heterogeneous clinical symptomatology or the varied geographical origins of our patients; however, the number of WD families involved is insufficient for statistical analysis. In another family, we suspect a mutation but cannot conclude: 3 WD homozygous children (13, 5 and 2 y; hepatic form) presented the same SSCP pattern for exon 14 and this was unexpected in relation to the mother's pattern; the mother's mutation could not be detected on this pattern and must therefore be located on another exon. Unfortunately, the father refused to be examined; only sequencing analysis will provide the answer, and allow us to determine the functional significance of mutations detected by SSCP analysis. The main advantages of SSCP analysis are its simplicity and relative sensitivity (90%); no additional steps are required after PCR and the gel system is reliable. This method provides mutation screening as the first step in direct WD diagnosis. To detect both WD mutations (from the mother and father), exons need to be screened one by one. In our laboratory, the amplification products are rendered radioactive as we have a wide experience of these methods. Direct mutation analysis would provide an important diagnostic tool for potential patients with no family history of WD.

The availability of many close, flanking, polymorphic microsatellite markers in linkage disequilibrium makes haplotype analysis possible and in turn indicates mutation screening for WD patients. This genetic study (haplotype and mutation) will help elucidate the clinical and biological heterogeneity of the disease.

5. REFERENCES

1. M. Frydman, B. Bonne-Tamir, L.A. Farrer, P.M. Conneally, A. Magazanik, A. Ashbel, Z. Goldwitch, *Proc. Natl. Acad. Sci.* USA **82**, 1819–1921 (1985).
2. K. Petrukhin, S. Lutsenko, I. Chernov, B.M. Ross, J.H. Kaplan, T.C. Gilliam, *Hum. Mol. Genet.* **3**, 1647–1656 (1994).
3. G.R. Thomas, J.R. Forbes, E.A. Roberts, J.M. Walshe, D.W. Cox, *Nature Genet.* **9**, 210–217 (1995).
4. A. Figus, A. Angius, G. Loudianos *et al*, *Am. J. Hum. Genet.* **57**, 1318–1324 (1995).
5. G.R. Thomas, P.C. Bull, E.A. Roberts, J.M. Walshe, D.W. Cox, *Am. J. Hum. Genet.* **54**, 71–78 (1994).
6. E.A. Stewart, A. White, J. Tomfohrde *et al*, *Am. J. Hum. Genet.* **53**, 864–873 (1993).

75

ABNORMAL FEATURES OF THE METABOLISM AND CELLULAR BIOLOGY OF COPPER IN MENKES DISEASE

Their Use in the Post and Antenatal Diagnosis

P. Guiraud, M. J. Richard, and A. Favier

Laboratoire de Biochimie CCHU A. Michallon, BP 217X
38043 Grenoble Cedex 9, France

1. INTRODUCTION

Menkes disease was described for the first time in 1962 by J.H. Menkes (1). This is an X-linked recessive disorder affecting copper metabolism. This disease appears as a convulsive encephalopathy, the main clinical manifestations being a neurological degeneration, a mental retardation, a convulsive state and "kinky" depigmented hairs. A fatal issue commonly occurs before the age of five. Some biological parameters are significantly modified: serum copper and ceruloplasmin levels are dramatically decreased. Cellular uptake and retention of copper are abnormally increased in enterocytes, lymphoblasts and fibroblasts. The exon structure of the Menkes gene has been recently determined. This gene encodes a predicted copper binding P-type ATPase (2–4).

Only few European laboratories are performing the antenatal diagnosis of Menkes disease. Since 1981, our laboratory is studying copper metabolism and we have successively developed in France the post- and antenatal diagnosis of Menkes disease by measuring the ^{64}Cu incorporation and release in cultured skin fibroblasts, amniocytes and trophoblasts of patients. In these cells, total intracellular copper is significantly increased as well as the percentage of copper retention (80 to 100%). The progress in the genetic characterization of the disease will perhaps lead to detect heterozygotes, which is not possible with the methodology presently used for the diagnosis. A reliable molecular biology-based diagnosis of the Menkes disease will probably be developed in a near future.

This presentation summarizes our studies and the recent knowledge concerning the cellular and molecular biology of copper and their implications in the improvement of the Menkes disease diagnosis.

Therapeutic Uses of Trace Elements, edited by Nève et al.
Plenum Press, New York, 1996

2. COPPER METABOLISM AND MENKES DISEASE

As all metals included in the group of trace elements, copper (Cu) is a metal present in all tissues of living organisms at a relatively constant low level and submitted to a fine homeostatic regulation. A perturbation in the metabolism of Cu may lead to structural and functional abnormalities of biomolecules resulting in severe physiological disturbances. Damages resulting from a nutritional deficiency are corrected by a supplementation.

The recommended daily oral ingestion ranges from 0.4 to 3 mg, but the common ingested quantity is usually 2 to 5 mg. About 30% to 50% of ingested Cu are absorbed at stomach, duodenum and jejunum levels. Several inhibitors are known such as phytates, plant proteins, ascorbic acid, and metals (Ag, Cd, Fe, Hg, Mo, Zn). Absorption of Cu is facilitated by citrate, histidine and cystein. Cu is mainly eliminated in the bile and feces while urine, sweat, saliva and hairs are secondary pathways (Figure 1). Cu is found in all tissues, the richest organs being brain, heart and liver, followed by bone, eyes, kidneys and skin. Fine mechanisms concerning Cu uptake, distribution inside cells and release out of organelles or cells remain unclear. However, Cu seems to be absorbed by enterocytes as Cu-L aminoacid complexes, mostly Cu-Histidine. The transport pathway across the membrane as well as the role of metallothioneins (MT) in the Cu absorption by enterocytes are not exactly known. In the enterocytes, MT play an important role as inactive storage form of Cu and in its transport across intestinal mucosa. The uptake of Cu by hepatocytes involves an active transport system including a membrane receptor for Cu-albumin complexes. In liver, Cu is combined to ceruloplasmin. Thus, in plasma 96% of Cu is present in ceruloplasmin, Cu-albumin, Cu-pep-

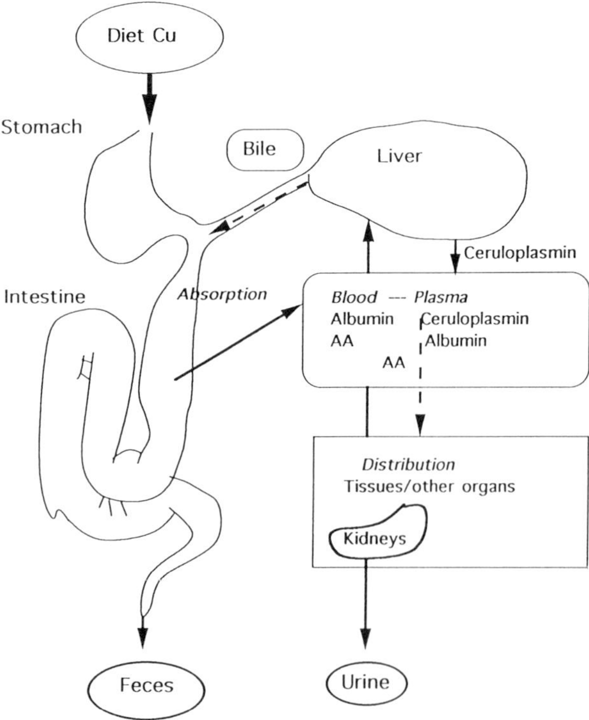

Figure 1. General copper metabolism.

tides and Cu-aminoacids representing only 4%. The main physiological markers for the Cu status are cupremia (0.75–1.35 mg/l) and ceruloplasminemia (0.16–0.60 g/l).

Nutritional Cu deficiency is rare. Chronic or acute intoxications by Cu have been reported but have mainly a professional origin. Two human genetic disorders of Cu transport reflect the conflicting requirements of Cu homeostasis: Menkes disease (deficiency syndrome), Wilson disease (chronic intoxication syndrome).

The diverse catalytic properties of Cu ions, usually involving oxidation-reduction systems and a Cu^+/Cu^{2+} oxidation-reduction cycle, have made this metal an essential element in human nutrition (5). The unique properties of the essential trace element Cu require a delicate cellular balance between necessity and toxicity. Over thirty known proteins, as fundamental as the cytochrome oxidases and as specialized as dopamine β-hydroxylase, use the oxidative capacity of Cu. But that same oxidative potential can cause extensive cellular damages (Table 1) through free radical generation and direct oxidation of lipids, proteins and nucleic acids leading to cell death via membrane damage and impairment of cell functions (6, 7).

Cellular response to Cu must, therefore, be carefully regulated to minimize the side-effects of this metabolite whilst maintaining an efficient balance to meet cellular requirements. The ability of cells and organs to maintain such a homeostasis is a prerequisite for normal growth which is not achieved for example in the case of patients with Menkes disease.

The Menkes steely or kinky hair syndrome is a lethal X-linked neurodegenerative disorder with tissue specific Cu sequestration and reduced serum Cu and ceruloplasmin-Cu oxidase leading to death in early childhood (1, 8). The disease is an inherited syndrome of Cu metabolism: patients show specific Cu deficiency symptoms, the main phenotypic features being a lack of keratinization and pigmentation of hair, degenerative changes of the elastic tissue in the aorta and blood vessels, general connective tissue manifestations such as cutis laxa and scorbutic changes (9, 10). In addition, progressive psychomotor retardation with seizures and temperature instability are seen and the affected males rarely survive for more than 3 years. The disease shows clinical heterogeneity, the mildest form being probably the occipital horn syndrome (11). Most of the clinical features can be explained by the dysfunction of important Cu-requiring enzymes (Table 2), including dopamine β hydroxylase, cytochrome c oxidase, superoxide dismutase, and lysyl oxidase (12). In X-linked Menkes disease, the reduced activity of numerous Cu-containing proteins may be responsible for the diverse clinical findings of progressive neurologic degeneration, poor temperature regulation, connective tissue defects, pallor, distinctive steely or kinky hair and death in early childhood (1,9). At the cellular level, defective Cu export from cells such as enterocytes or fibroblasts results in the characteristic phenotype of Cu accumulation observed in cultured cells (8,13–15), the most important exception concerning hepatocytes. In patients, Cu deficiency results from the trapping of Cu in intestinal mucosa, kidneys and connective tissue, accompanied by failure of its distribution to other tissues.

Table 1. Copper: an antioxidant or prooxidant trace element

Copper antioxidant proteins	Copper as a prooxidant metal
Ceruloplasmin	$Cu^+ + O_2 \longrightarrow Cu^{2+} + O_2^{\cdot-}$
Metallothioneins (Cd, Cu, Zn)	$Cu^+ + H_2O_2 \longrightarrow Cu^{2+} + OH^{\cdot} + OH^-$
Cu/Zn superoxide dismutase (SOD)	$Ascorbate + 2Cu^{2+} \longrightarrow$
Amino acid-Cu complexes ("SOD-like")	$dehydroascorbate + 2Cu^+ + 2H^+$

Table 2. Biological roles of copper and related symptoms in Menkes disease

Biological function	Protein	Related symptoms
Erythropoiesis (iron mobilization)	Ceruloplasmin	Anemia, neutropenia
Cellular respiration	Cytochrome C oxidase	Decrease in cell energetic pool (ATP), hypothermia
Antioxidant	Ceruloplasmin Metallothioneins Superoxide dismutase	Increase of oxidative damages, lesion of cellular membranes and biomolecules
Extracellular matrix (connective tissue...)	Lysyl oxidase	Spontaneous fractures, arterial wall fragility, cardiac arythmy, kinky hairs (pili torti...), cutis laxa...
Nervous system	Dopamine β hydroxylase	Myelin synthesis decreased Neurologic disorders
Pigmentation	Tyrosinase	Melanin synthesis decreased Depigmentation (hair, skin...)

(other troubles : hypercholesterolemia, mental and growth retardation...)

In contrast, the clinical presentation of the autosomal recessive Wilson disease (location on the chromosome 13) results from Cu toxicity. Decreased Cu export from the liver results in Cu-induced chronic liver disease and contributes to pathologic changes in other tissues, especially in the brain, kidneys and eyes (12).

Thus, defects in Cu export underlie both the Cu deficiency of Menkes disease and the toxicity of Cu excess in Wilson disease. The underlying molecular defects, however, have remained unknown till 1993.

In january 1993, three independant research teams proposed a candidate gene for Menkes disease (Xq13.3) with evidences that it encodes a Cu- transporting ATPase (2–4). At the end of the same year, the Wilson disease gene (13q14.3) was reported as a putative Cu transporting P-type ATPase similar to the Menkes gene (16–18).

3. DIAGNOSIS OF MENKES DISEASE

The Cu disturbances associated with Menkes disease include low serum Cu and ceruloplasmin levels, and abnormal tissue binding of Cu *in vivo* as well as *in vitro*. Increased Cu binding has been reported in several cell types among which cultured fibroblasts and amniotic fluid cells, and chorionic villi (8, 14, 19). These features provide markers that have been the basis of both postnatal and antenatal diagnosis methods of the disease.

In fibroblasts, the determination of total intracellular Cu level when grown in the usual culture medium (RPMI 1640/15% fetal calf serum) is significantly increased in patients, while heterozygotes show great variations (Figure 2). No increase in total intracellular Cu level is observed in cultured amniotic cells preventing this parameter to be used in antenatal diagnosis.

Diagnosis of Menkes disease has been developed only in few laboratories in Europe. The method used consists in growing either skin fibroblasts from patients (for postnatal confirmation of the diagnosis or carrier study) or amniotic fluid cells (antenatal diagnosis). Cells are then incubated in culture medium containing the radioactive isotope ^{64}Cu. After 10 and 20 hours of incubation in a water saturated, 5% CO_2 incubator at 37 °C, the ^{64}Cu incorporated is estimated by counting the radioactivity in the harvested cells . Then the retention of ^{64}Cu by the cells is assessed by recording the radioactivity remaining in the cells after 8 hours of incubation in fresh ^{64}Cu-devoided culture medium (Figure 3). In pathologic cases,

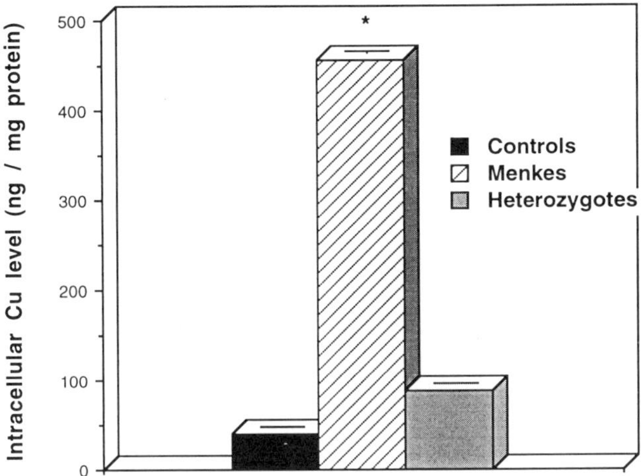

Figure 2. Intracellular copper level in skin fibroblasts. Values are means ± SEM; *: results significantly different from control, p < 0.001 (Student t test); controls n = 68, Menkes n = 18, heterozygotes n = 9.

both the level of ^{64}Cu incorporated and the level of ^{64}Cu retained by the cells are very significantly increased (Figure 4).

The main limitations for this method are the ^{64}Cu short half-life (13 hours), unavailability for the detection of heterozygotes, the requirement of an amniocentesis before 17th week for patient and controls, early diagnosis (11th week) possible by measuring copper in fresh chorionic villi but with a high risk of contamination by copper.

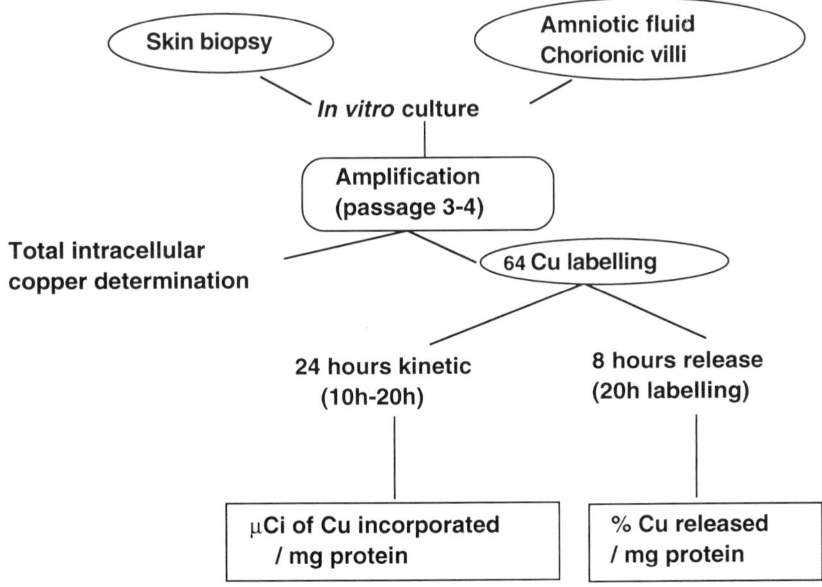

Figure 3. Diagnosis of Menkes disease.

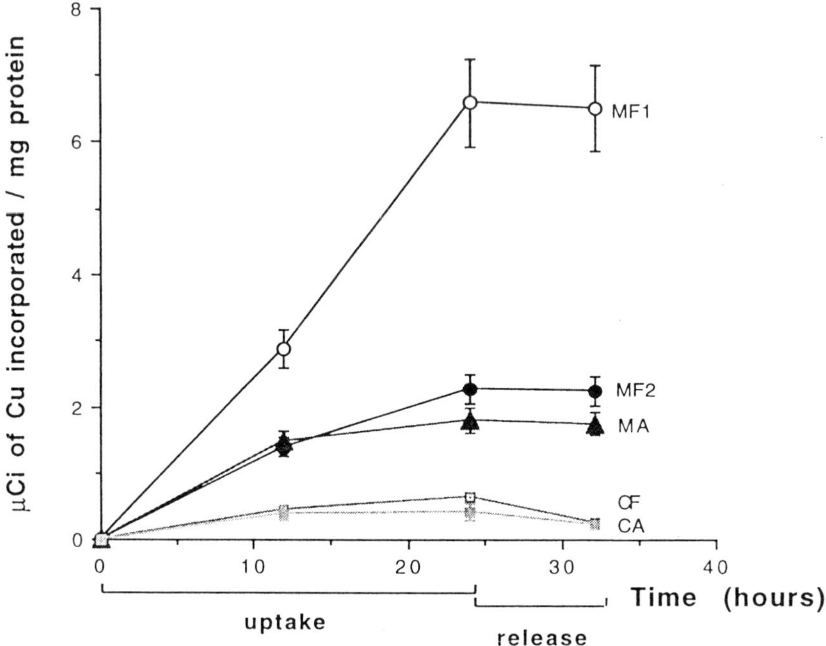

Figure 4. Time-course study of radioactive copper incorporation by skin fibroblasts and amniotic fluid cells. MF1 and MF2: Menkes fibroblasts, divided into two groups, giving a high response (MF1) or a low response (MF2); MA: Menkes amniocytes; CF: control fibroblasts; CA: control amniocytes. Values are means ± SD.

Menkes disease is due to a recessive gene, which is located on the X-chromosome and will be expressed only in the absence of the normal allele. Accordingly, all males who carry the mutant gene will be affected and can be identified by means of the genetic marker. As a result of Lyonization, female carriers will, however, rarely show obvious manifestations of the disease. Identification of female carriers is therefore an important preventive measure (20).

4. PERSPECTIVES

The proposition of a candidate gene for Menkes disease is probabely of considerable importance for the diagnosis and prevention of the disease. Isolation of the Menkes disease gene has enabled the analysis of genetic defects in affected individuals. However results obtained for the moment are not conclusive as deletion have been found only in 13% of the patients for a region analyzed corresponding to 2/3 of the 4.5 kb translated sequence of the 8.5 kb Menkes transcript, results obtained for putative carriers were also inconsistent (21). A survey of 230 unrelated patients has revealed a deletion or rearrangement detectable by Southern blot analysis in 45 individuals (2, 22). The recent characterization of the complete exon-intron structure of the Menkes disease gene will perhaps allow to provide means of an easier and reliable DNA-based methodology for diagnosis of Menkes disease and especially for carrier detection. Determination of the intron sequences flanking each exon will enable identification of the genetic defect directly from genomic DNA using PCR-based methods (22).

5. REFERENCES

1. J.H. Menkes, M. Alter, G. Steigleder, D.R. Weakley and J.H. Sun, *Pediatrics* **29**, 764–779 (1962).
2. J. Chelly, Z. Tümer, T. Tonnesen, A. Petterson, Y. Ishikawa-Brush, N. Tommerup, N. Horn and A.P. Monaco, *Nature Genet.* **3**, 14–19 (1993).
3. J.F.B. Mercer, J. Livingstone, B. Hall et al., *Nature Genet.* **3**, 20–25 (1993).
4. C. Vulpe, B. Levinson, S. Whitney, S. Packman and J. Gitschier, *Nature Genet.* **3**, 7–13 (1993).
5. E.J. Underwood, in *Trace Elements in Human and Animal Nutrition*, 4th edn., New York: Academic Press, pp. 56–108 (1977).
6. D.M. Miller, G.R. Buettner and S.D. Aust, *Free Rad. Biol. Med.* **8**, 95–108 (1990).
7. E.R. Stadtman, *Free Rad. Biol. Med.* **9**, 315–325 (1990).
8. D.M. Danks, E. Cartwright, E.J. Stevens and R.R.W. Townley, *Science* **179**, 1140–1142 (1973).
9. D.M. Danks, P.E. Campbell, B.J. Stevens, V. Mayne and E. Cartwright, *Pediatrics* **50**, 188–201 (1972).
10. N. Horn, T. Tonnesen and Z. Tümer, *Brain Pathol.* **2**, 351–362 (1992).
11. B. Levinson, J. Gitschier, C. Vulpe, S. Whitney, S. Yang and S. Packman, *Nature Genet.* **3**, 6 (1993).
12. D.M. Danks, in *The metabolic basis of inherited disease*, J.R. Scriver, A.L. Beaudet, W.S. Sly and D. Valle, eds., McGraw-Hill, New York, pp. 1411–1431 (1989).
13. K. Heydorn, E. Damsgaard, N. Horn, M. Mikkelsen, I. Tygstrup, S. Vestermark and J. Weber, *Hum. Genet.* **29**, 171–175 (1975).
14. N. Horn, K. Heydorn, E. Damsgaard, I. Tygstrup and S. Vestermark, *Clin. Genet.* **14**, 186–187 (1978).
15. N. Horn and O.A. Jensen, *Ultrastructural pathol.* **1**, 237–242 (1980).
16. P.C. Bull, G.R. Thomas, J.M. Rommens, J.R. Forbes and D.W. Cox, *Nature Genet.* **5**, 327–337 (1993).
17. K. Petrukhin, S.G. Fischer, M. Pirastu et al., *Nature Genet.* **5**, 338–343 (1993).
18. R.E. Tanzi, K. Petrukhin, I. Chernov et al., *Nature Genet.* **5**, 344–350 (1993).
19. T. Tonnesen, A.M. Gerdes, E. Damsgaard, P. Miny, W. Holzgreve, F. Sondergaard and N. Horn, *Prenatal Diag.* **9**, 159–165 (1989).
20. N. Horn, *J. Inher. Metab. Dis.* **6**, 59–62 (1983).
21. Z. Tümer, T. Tonnesen and N. Horn, *J. Inher. Metab. Dis.* **17**, 267–270 (1994).
22. Z. Tümer, B. Vural, T. Tonnesen, J. Chelly, A.P. Monaco and N. Horn, *Genomics* **26**, 437–442 (1995).

76

F-SSCP SCREENING FOR TWO COMMON MUTATIONS HIS1070GLN AND GLY1267LYS IN FRENCH WILSON PATIENTS, AND REPORT OF TWO NOVEL MUTATIONS

G. Liu,[1] B. Aral,[1] I. Ceballos-Picot,[1] C. Franvel,[2] P. Lecoz,[3] and P. Chappuis[2]

[1] CNRS URA1335, Hopital Necker-EM
149 Rue de Sèvres 75015 Paris France
[2] Hopital Lariboisière
Laboratoire Central de Biochimie
[3] Service de Neurologie
2 Rue Ambroise-Paré, 75475 Paris Cedex 10, France

1. INTRODUCTION

Wilson disease (WD) is an autosomal recessive disorder of copper metabolism with an incidence of between 1:35 000 and 1:100 000 (1). The disorder manifests itself as chronic liver disease and/or later on, with neurological symptoms, due to gradual accumulation of copper (2,3). The gene responsible for Wilson disease has recently been mapped to chromosome 13 and cloned (4,5); it encodes a P-type ATPase involved in copper transport across membranes. To date more than 25 mutations have been identified in the ATP7B gene responsible for WD and among them two very common mutations have been reported. The first, His1070Gln, is a C to A transversion in exon 14 of the gene, with an allele frequency of 28 % in the Northern European population. The other, Gly1267Lys, is a G to A transversion in exon 18 of the gene, with an allele frequency of 10% in the Northern European population (6). We investigated the prevalance of these 2 mutations in 40 French WD patients from 22 unrelated French families using fluorescence-based Single-Strand Conformational Polymorphisms (F-SSCP) analysis with subsequent direct sequencing of the abnormally shifted exons.

2. MATERIALS AND METHODS

A total of 22 unrelated French families with 40 symptomatic WD patients were investigated. Most of these patients had a neurological and/or psychiatric symptomatology; some of

Therapeutic Uses of Trace Elements, edited by Nève et al.
Plenum Press, New York, 1996

them also presented with a hepatic form of the disease. The ultimate diagnosis of WD in each patient was subsequently proven by standard biochemical methods such as a lowered serum ceruloplasmin concentration as well as either increased 24 hr urinary copper excretion or a raised liver copper concentration. The presence of a Kayser-Fleischer ring was also helpful in the diagnostic criteria of the disease. High molecular weight DNA was isolated from the peripheral blood leucocytes of patients by salt extraction. The probands in each family were

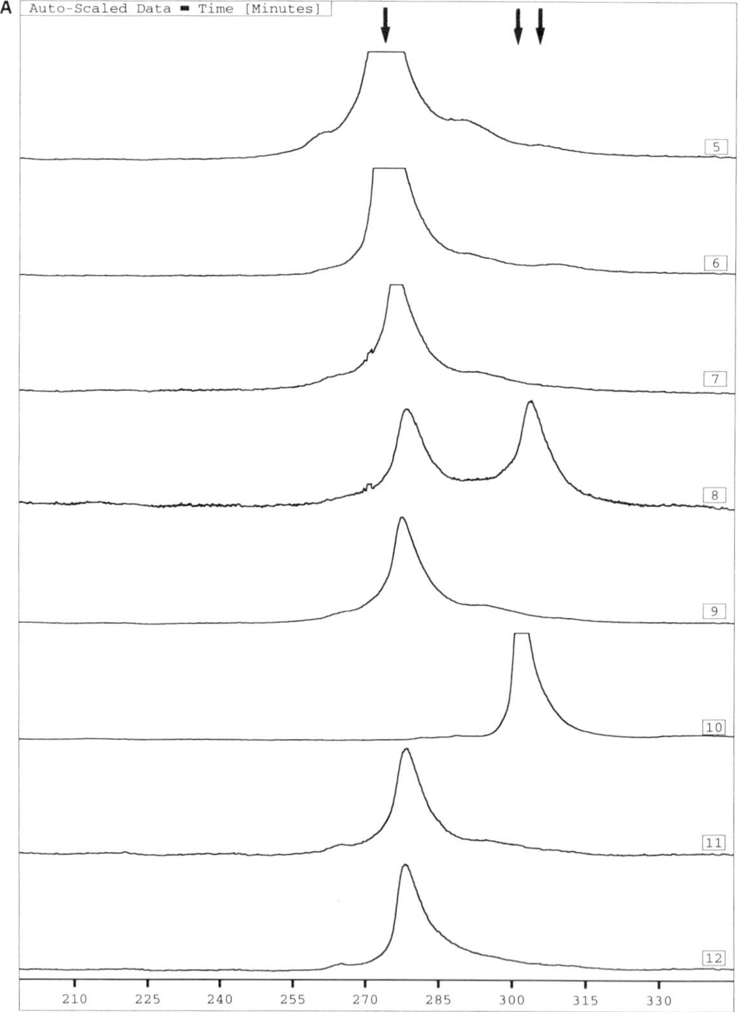

Figure 1. (A) Exon 14 F-SSCP analysis of of the ATP7B gene from eight WD disease patients (numbered from 5 to 12). The unique peak (marked by a unique arrow) detected in each trace corresponds to the end-labelled strand of the normal alleles; the double arrows indicate the position of the same strand from a mutated diseased allele. The unique peak in the trace #10 corresponds to a homozygous His1070Gln mutation. (B) Automated sequence analysis of PCR products from exons 14 and 18 of the ATP7B gene. Sequences of the corresponding exons with their accompanying amino acid codes (below) are depicted. (B1) In exon 14, the A-to-C transversion mutation results in a E1065A missense mutation. The arrow shows the heterozygous position. (B2) In exon 18, the G-to-A transition mutation replacing Gly1243 with Glu (G1243E). The arrow shows the unambiguous mutated sequence.

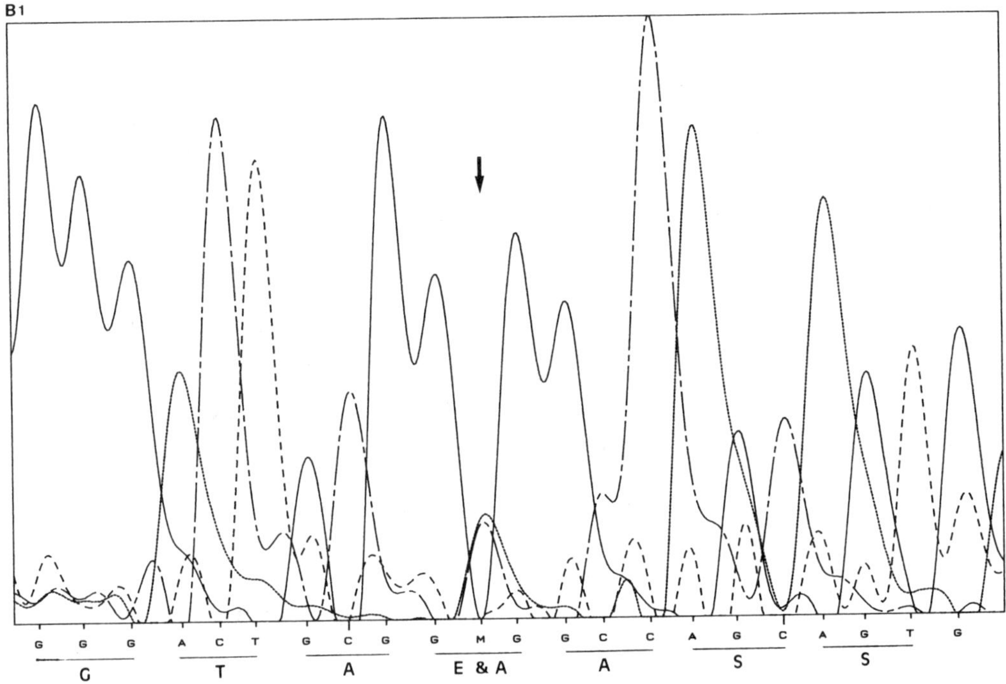

B2

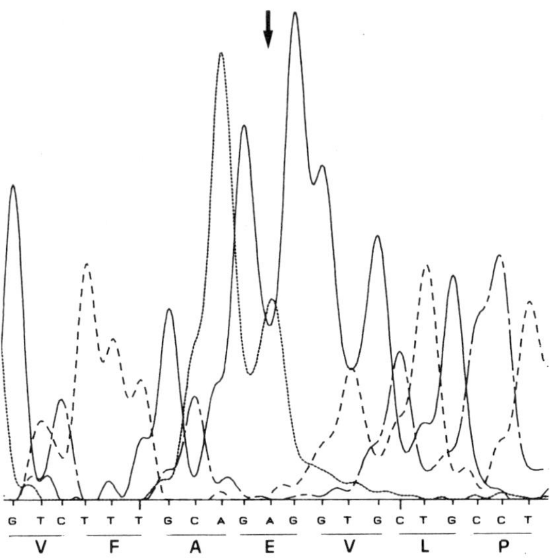

Figure 1. (cont.)

specifically tested for the exon 14 and exon 18 mutations of the Wilson disease gene (6). Both exons were amplified under conditions identical to those described by Thomas *et al.* (7). We performed a fluorescence-based Single-Strand Conformational Polymorphisms (F-SSCP) (8) analysis in order to screen for mutated exons. Patient samples exhibiting peak shifts in exon 14 and/or exon18 relative to normal samples on F-SSCP were subjected to direct cycle-sequencing using the SequiTherm^{TM} Long-Read kit (Epicentre Technologies) to identify the mutation.

3. RESULTS

A total of 40 symptomatic WD patients belonging to 22 unrelated French families were tested by semi-automated DNA-based diagnostic techniques for His1070Gln and Gly1267Lys mutations in the ATP7B gene. Exons 14 and 18 were amplified by PCR and mutated exons screened by fluorescence-based Single-Strand Conformational Polymorphisms (F-SSCP) analysis. (Fig. 1A). Our results confirm those of previous reports based on F-SSCP analysis giving similar frequencies for both exons. Regarding exon 14 analysis, our patients showed either a normal pattern (n = 30), or homozygosity for a peak shift (n = 4), or an heterozygosity for the two patterns (n = 6). Consequently, in our population, the overall percentage of patients presenting a mutation in exon 14 was 25 %. Direct sequence analysis of shifted exon 14 showed that 10 patients had the previously described His1070Gln mutation (6), and one patient had a Glu1065Ala missense mutation (Fig. B1). Exon 18 analysis revealed 4 heterozygosities for a peak shift (10%) and the rest of the patients had a normal pattern (n = 36). To our surprise, the direct sequence analysis of shifted exons 18 indicated that these 4 patients carried the same Gly1243Glu missense mutation (Fig. B2) instead of the previously described common mutation Gly1267Lys (6).

4. DISCUSSION

Our initial screening program for mutations in both exons 14 and 18 from 22 unrelated WD families has identified 3 mutations, one of which has been previously described (6). These mutations are solely missense mutations. Four previously described mutations (6), 3088delAC and 3149delC in exon 14 and Gly1267Lys (in fact, this mutation is probably Gly1266Arg, personnal communication) and Asn1271Ser in exon 18, were not detected. Instead, 25 % of our French patients (10 out of 40) carry the previously described change His1070Gln (6), the most common mutation in our families, including four homozygous individuals; one patient carries a novel, previously undescribed mutation which is a Glu1065Ala missense mutation in exon 14; the glutamic acid residue could be identified as a potential "hotspot" point because another mutation affecting the same codon has been de-

Table 1. Exons 14 and 18 mutations of the ATP7B gene in 40 French WD patients presenting with a neurological form of the disease

Mutation	Exon	Heterozygote	Homozygote	Frequency (%)
His1070Gln	14	6	4	25
Glu1065Ala	14	1	0	3
Gly1243Glu	18	4	0	10

scribed by Figus et al. (9) as a Glu1065Lys mutation in a turkish WD patient. This mutation is expected to alter the ATP loop of the protein which forms a specific secondary structure and consequently affect the ATP binding (10). Among the 4 Gly1243Glu missense mutation detected in exon 18, three patients were unrelated. This mutation is predicted to disrupt the ATP hinge of the ATP7B protein. All patients carrying the most common mutation, His1070Gln, presented a neurologic and/or psychiatric form of WD, with late age of onset of the symptoms (> 20 years old); three homozygous patients for this mutation also presented with early abdominal manifestations but their ceruloplasmin levels were not very low and their response to the chelation treatment was rapid and effective. The patient with the Glu1065Ala mutation and the 4 patients with Gly1243Glu missense mutations in exon 18, also had a neurological form of the disease. The results related to mutations are summarized in Table 1.

Given the high frequency of occurrence of the His1070Gln mutation in our WD population presenting the neurological form of the disease, it can be speculated that this mutation, in a compound heterozygous state, is associated with late, neurological presentation of WD.

In the future, we will concentrate our efforts on the investigation of other exons of the WD gene using the F- SSCP method with subsequent direct sequencing.

5. REFERENCES

1. D.M. Danks, in *Metabolic Basis of Inherited Disease*, A.L. Beaudet, W.S. Sly, D. Valle, eds., McGraw-Hill, New York, pp. 1411–1431 (1989).
2. R.H.J. Houwen, J. Van Hattum, T.U. Hoogenraad, *Neth. J. Med.* **43**, 26–37 (1993).
3. G.J. Brewer, V. Yuzbasiyan-Gurkan, *Medicine* **71**, 139–164 (1992).
4. P.C. Bull, G.R. Thomas, J.M. Rommens, J.R. Forbes, D.W. Cox, *Nature Genet.* **5**, 327–337 (1993).
5. R.E. Tanzi, K. Petrukhin, I. Chernov, et al., *Nature Genet.* **5**, 344–350 (1993).
6. G.R. Thomas, J.R. Forbes, E.A. Roberts, J.M. Walshe, D.W. Cox, *Nature Genet.* **9**, 210–217.
7. G.R. Thomas, P.C. Bull, E.A. Roberts, J.M. Walshe, D.W. Cox, *Am. J. Hum. Genet.* **54**, 71–78 (1994).
8. B. Aral, M. Coudé, J. Aupetit, I. Ceballos-Picot, P. Kamoun, B. Chadefaux-Vekemans, *Meths. Mol. Cell Biol.* **5**, 237–241 (1995).
9. A. Figus et al., *Am. J. Hum. Genet.* **57**, 1318–1324 (1995).
10. D.H. MacLennan, D.M. Clarke, T.W. Loo, I.S. Skerjanc, *Acta Physiol. Scand.* **146**, 141–150 (1992).

INDEX